The AVIATION DICTIONARY

For Pilots and Aviation Maintenance Technicians

© Jeppesen Sanderson Inc., 2003
All Rights Reserved
55 Inverness Drive East, Englewood, CO 80112-5498
ISBN 0-88487-318-8

A

A-stage — In composites, the initial stage of mixing the two parts of a thermosetting resin system together. The material is soluble in some liquids and fusible.

A chamber — One of the chambers of a pressure-type carburetor regulator unit. A chamber regulates air-inlet pressure from the air intake.

A check — Primarily associated with large commercial aircraft, the lowest level of inspection in a Continuous Airworthiness Maintenance Program. The entire program consists of A, C, and D checks.

A&B system — An emergency brake design used in some larger multiple-disc brake systems. These brakes have a number of cylinders, with alternate cylinders connected either to the aircraft's "A" system or to the aircraft's "B" system. Normal brake actuation takes over if one system fails.

A&P technician — An aircraft technician certified by the Federal Aviation Administration as having met the experience and knowledge requirements for certification. An A&P technician is qualified to return licensed United States airplanes to service after performing certain inspections and maintenance procedures. Increasingly referred to as AMTs (Aviation Maintenance Technicians).

A, B, C, D, E chambers — The five chambers that make up a diaphragm-controlled unit of a pressure type carburetor. Chamber A is regulated air-inlet pressure. Chamber B is boost venturi pressure. Chamber C contains metered fuel pressure. Chamber D contains unmetered fuel pressure. Chamber E is fuel pump pressure.

A-battery — A dry cell battery with a voltage of 1.5 to 6.0 volts and capable of supplying a reasonable amount of current.

abbreviated briefing — In meteorology, a shortened briefing to supplement mass disseminated data.

abbreviated IFR flight plans — An authorization by ATC requiring pilots to submit only that information needed for the purpose of ATC. It includes only a small portion of the usual IFR flight plan information. In certain instances, this may be only aircraft identification, location, and pilot request. Other information may be requested if needed by ATC for separation/control purposes. It is frequently used by aircraft which are airborne and desire an instrument approach or by aircraft which are on the ground and desire a climb to VFR-on-top.

abeam — An aircraft is "abeam" a fix, point or object when that fix, point, or object is approximately 90 degrees to the right or left of the aircraft track. Abeam indicates a general position rather than a precise point.

abeam fix — A fix positioned approximately 90 degrees from a navaid along a route of flight.

abnormal procedures training (APT) — The use of flight training devices to teach pilots how to handle procedures such as engine failures, inflight fires, and other abnormal situations. This training is completed on the ground and is the safest means of allowing a pilot to experience abnormal procedures.

abort — To terminate a preplanned aircraft maneuver; e.g., an aborted takeoff.

aborted start — In gas turbine engines, termination of the engine starting cycle when combustion (light-off) does not occur within a prescribed time limit.

abradable seal — A general description of knife-edge seals which can wear away slightly and still function. Such seals in a gas turbine engine abrade slightly to produce a close fit.

abradable shroud — Generally a honeycomb-type turbine shroud ring set into the outer turbine case. Can abrade without engine degradation if the turbine blades creep and contact the shroud.

abrade — To scrape or wear away a surface or a part by mechanical or chemical action.

abrasion — 1. An area of roughened scratches or marks usually caused by foreign matter between moving parts or surfaces. 2. The wearing or rubbing away of a surface by a substance used for grinding, grating, polishing, etc.

abrasive — A substance used to wear away surfaces by the use of friction. In grinding wheels, the abrasives most commonly used are silicon carbide or aluminum oxide.

abrasive blasting — The removal of carbon and other deposits from machine parts using a high velocity blast of air that contains fine particles of abrasive sand, glass bead, or walnut shell.

abscissa — The horizontal reference line of a graph or curve by which a point is located with reference to a system of coordinates.

absolute altimeter — A radar altimeter used to indicate the exact height of an aircraft over the terrain. See also radio altimeter.

absolute altitude — The actual height above the surface of the earth, either land or water.

absolute ceiling — The maximum height above sea level at which an aircraft can maintain level flight under standard atmospheric conditions.

absolute humidity — The actual water vapor present in a given volume of air.

absolute instability — In meteorology, the state of an atmospheric layer when the actual temperature lapse rate exceeds the dry adiabatic lapse rate. An air parcel receiving an initial upward displacement in an absolute unstable layer will accelerate away from its original position.

absolute pressure — Pressure above zero pressure as read on a barometer type instrument, i.e. Standard Day, 14.7 psia.

absolute pressure controller (APC) — An instrument that regulates the maximum turbocharger compressor discharge pressure in a reciprocating engine turbocharger system. (34 + or -.5 in. Hg to critical altitude, approximately 16,000 ft)

absolute temperature — Temperature referenced from absolute zero. (-273.18°C or -459.6°F) There are two absolute scales in use, the Rankine scale using Fahrenheit degrees and the Kelvin scale using Celsius degrees.

absolute value — The value of a number without considering its sign (whether it is plus or minus).

absolute zero — The temperature (-273.18°C) at which molecular motion ceases.

abstraction — General rather than specific. For example, aircraft is an abstraction; airplane is less abstract; jet is more specific; and jet airliner is even more specific.

AC 43.13-1B — An Advisory Circular in book form issued by the Federal Aviation Administration (and reprinted by others) that covers acceptable methods, techniques, and practices for aircraft inspection and repair. The procedures described in this advisory circular are considered by the Federal Aviation Administration to be acceptable data for aircraft inspection, maintenance, and alteration.

AC fittings — Air Corps fittings. Replaced by the AN (Army/Navy) standard and MS (Military Standard) fittings. AN fittings have a slight shoulder between the cone and the first thread. AC fittings do not have this shoulder. Other differences include sleeve design and the pitch of the threads.

AC plate resistance — The internal resistance of a vacuum tube to the flow of alternating current. AC plate resistance is measured in ohms.

AC/DC — Electrical components that can operate equally well on alternating current or on direct current electricity.

accelerate — To increase the speed of an object.

accelerate stop distance available (ASDA) — The length of the take-off run available plus the length of the stopway, if provided.

accelerated life test — An operational test used to predict the service life a system or component will have under normal operating conditions.

accelerate-go distance — In multi engine aircraft, the distance required to accelerate to liftoff speed or V1 (depending on the pilot's operating handbook (POH)), experience an engine failure, and complete the takeoff and climb to clear a 50-foot obstacle.

accelerate-stop distance — In multi engine aircraft, the sum of the distances necessary to (1) accelerate the airplane from a standing start to V_{EF} with all engines operating; (2) accelerate the airplane from V_{EF} to V_1, assuming the critical engine fails at VEF; and (3) come to a full stop from the point at which V_1 is reached. The light twin pilot's operating handbook (POH) may base the distance on an engine failure occurring at a specified takeoff decision speed (V_1) or liftoff speed.

accelerate-stop distance available — The runway plus stopway length declared available and suitable for the acceleration and deceleration of an airplane aborting a takeoff.

accelerating agent — A substance used to hasten a chemical action or change.

accelerating pump — A small pump in a carburetor used to supply a momentarily rich mixture to the engine when the throttle is suddenly opened. This prevents hesitation during the time of transition between operating on the idle system and the main metering system.

accelerating system — An accelerating system used in float carburetors that supplies extra fuel during increases in engine power. This is usually accomplished by a small fuel pump called an accelerating pump.

accelerating well — A secondary tank built into the main oil tank. The well, or hopper, retains only that portion of fluid being circulated through the engine. Thus, oil warm-up is hastened during engine warm-up. The well also makes oil dilution practical. Also referred to as temperature accelerating well.

acceleration — The increase in velocity of an object. Acceleration is usually expressed in terms such as feet per second per second.

acceleration check — A maintenance check calculating the time an engine takes to spool-up from idle to rated power without hesitation or evidence of backfire.

acceleration due to gravity — The acceleration of a freely falling body due to the attraction of gravity, expressed as the rate of increase of velocity per unit of time. In a vacuum the rate is 32.2 feet per second near sea level.

acceleration error — An error inherent in magnetic compasses, caused by the force of acceleration acting on the dip compensating weight when the aircraft accelerates or decelerates on an easterly or westerly heading. In compasses compensated for flight in the Northern Hemisphere, when the aircraft accelerates on an easterly or westerly heading, the compass gives the indication that the aircraft is turning to the north. When the aircraft decelerates on either of these headings, the compass gives the indication that the aircraft is turning to the south.

acceleration of gravity — The rate of increase in speed of a freely falling body due to the attraction of gravity. Expressed as the rate of increase of velocity per unit of time (32.17 feet per second per second at sea level and 45° latitude). The acceleration decreases with an increase in altitude until it becomes zero upon leaving the earth's gravitational field.

acceleration thermostat — A bimetallic probe positioned in the exhaust stream of an auxiliary power unit. When overheated, it expands to dump a Pb fuel control signal and reduce fuel flow.

acceleration well — An enlarged space around the discharge nozzle of some float-type carburetors. When the throttle is opened suddenly, this fuel is rapidly discharged from the main discharge nozzle.

accelerator — A substance added to a catalyzed resin to shorten its curing time.

accelerator system — A system in an aircraft carburetor that supplies additional fuel to the engine when the throttle is opened suddenly.

accelerator winding — A series winding used in vibrating-type voltage regulators, which, when the points open, decreases the magnetic field immediately, allowing the points to close more rapidly.

accelerometer — A sensitive instrument calibrated in G-units that measures the amount of force exerted by acceleration on a body. One G-unit force is equal to the weight of the object.

acceptance test — A test performed on an airplane or piece of equipment to ensure it is in the condition specified in the purchase contract. Large and expensive aircraft are given extensive acceptance tests before a customer accepts them.

acceptor atom — An impurity atom in a semiconductor material that will receive or accept electrons. Germanium with an acceptor impurity is called P-type germanium because it has a positive nature.

access cover — See access panel.

access door — A door that provides entry into or exit from an aircraft. It also provides access to servicing points and manually operated drains.

access panel — A panel on an aircraft that can be removed easily to facilitate inspection and maintenance.

accessories — Components that are used with an engine, but are not a part of the engine itself. Units such as magnetos, carburetors, generators, and fuel pumps are commonly installed engine accessories.

accessory drive gearbox — Provides mounting space for engine accessories. Also referred to as main gearbox.

accessory gear trains — Drive system containing both spur and bevel-type gears. Used in different types of engines for driving engine components and accessories

accessory section — The part of an engine that provides the necessary mounting pads for accessory units such as magnetos, fuel pumps, oil pumps, and generators.

accident — An unintended event or circumstance.

accommodation — The time required to focus on a distant object after looking at the instrument panel.

accretion — The production of a precipitation particle when a supercooled water droplet freezes as it collides with a snowflake or a smaller ice particle. Such particles may become the nucleus of a hailstone.

accumulated error — The sum of all of the errors that occur in the operation of a system or in the manufacturing of a part.

accumulator — A hydraulic component consisting of two chambers separated by a piston, diaphragm, or bladder. Compressed air in one chamber holds pressure on hydraulic fluid in the other chamber, allowing the fluid to be stored under pressure. An accumulator is used to assist in bringing the propeller out of the feathered position by providing a burst of oil pressure to the hub when the control lever is moved out of the feather position.

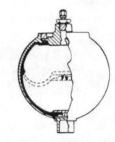

accumulator precharge — Compressed air stored in the air chamber of a hydraulic accumulator prior to introducing hydraulic pressure into the system.

accuracy — The state of being exact and free from mistakes. Conformance to a standard without error.

accurate — Free from error.

ace — A term that originated in World War I to acknowledge pilots who shot down five or more enemy aircraft.

acetone — Liquid ketone (C_3H_6O). A flammable, fast-evaporating solvent used in many types of aircraft finishes. Also used as a cleaning fluid.

acetylene cylinder — A seamless steel shell with welded ends, approximately 12 inches in diameter and 36 inches long. A fully charged acetylene cylinder of this size contains approximately 225 cu. ft. of gas at pressures up to 250 PSI.

acetylene gas — A flammable, colorless non-toxic gas that has a distinctive, disagreeable odor. Calcium carbide is made to react with water to produce acetylene. Mixed with oxygen in the proper proportions and ignited, acetylene gas will produce temperatures that range from 5,700°F to 6,300°F for welding purposes.

acid — A chemical substance that contains hydrogen, has a characteristically sour taste, and is prone to react with a base or an alkali to form a salt and to accept electrons from the alkali.

acid diluent — A constituent of a wash primer used to mildly etch the surface of the metal being primed. Provides a good bond between the finishing system and the metal.

acid-resistant paint — A paint that resists the etching effects of acid. Used on battery boxes and surrounding areas.

acknowledge — During communications, this indicates: Let me know that you have received my message.

acknowledge [ICAO] — During communications, this indicates: Let me know that you have received and understood this message.

acorn nut — A dome-shaped nut with a threaded hole that does not go completely through, producing a finished and smooth appearance. Acorn nuts and cap nuts are terms used interchangeably.

acrobatic category airplane — An aircraft certificated for flight without restrictions, except those found necessary as a result of flight tests.

acrobatic flight — An intentional maneuver involving an abrupt change in an aircraft's attitude, an abnormal attitude, or abnormal acceleration not necessary for normal flight.

acrobatics — Flight maneuvers such as loops and rolls that are not considered to be necessary for normal utility category flight.

acrylic — A glossy, transparent thermoplastic material used for cast or molded parts such as aircraft windshields and side windows.

acrylic lacquer — An aircraft finish that consists of an acrylic resin base and certain volatile solvents.

acrylic nitrocellulose lacquer — A common topcoat for aircraft, available either as a matte or glossy finish.

acrylic resin — A clear thermoplastic produced by polymerizing acrylic acid. Used for cast and molded aircraft windshields, windows, and parts, or as a coating and adhesive.

acrylic vitro lacquer finish — An aircraft finish applied in a specified sequence. It includes a wash primer coat, modified zinc chromate primer coat, and an acrylic nitrocellulose lacquer topcoat.

activated charcoal — Used as a filter for liquids and as a medium to absorb gases.

active current — Current in an AC circuit that is in phase with the voltage. Active current produces true power.

active detection systems — A detection system that transmits a signal such as radar as opposed to a passive detection system that only receives a signal.

active electrical component — An electrical part that controls current or voltage for switching or amplification.

active runway — See runway in use, active runway, and/or duty runway.

actual calculated landing time (ACLT) — A flight's frozen calculated landing time. An actual time determined at freeze calculated landing time (FCLT) or meter list display interval (MLDI) for the adapted vertex for each arrival aircraft based upon runway configuration, airport acceptance rate, airport arrival delay period, and other metered arrival aircraft. This time is either the vertex time of arrival (VTA) of the aircraft or the tentative calculated landing time (TCLT)/ACLT of the previous aircraft plus the arrival aircraft interval (AAI), whichever is later. This time will not be updated in response to the aircraft's progress.

actual navigation performance (ANP) — A measure of the current estimated navigational performance. Also referred to as Estimated Position Error (EPE). See also required navigational performance.

actuating cylinder — A cylinder and piston arrangement used to convert hydraulic or pneumatic pressure into work by the fluid under pressure moving the piston. The force applied is equal to the piston area times the pressure on the fluid. Actuating cylinders can be double-action or single-action actuating types.

actuating horns — The levers to which control cables are attached to move the control surfaces.

actuator — A mechanical device for moving or controlling something. The action may be linear, rotary, or oscillating. Many actuators are actuated by either hydraulic or pneumatic pressure.

actuator piston — The movable part of a hydraulic or pneumatic linear actuator.

acute angle — An angle of less than 90°. Also referred to as a closed angle.

adapter — A device that fits one component to another.

Adcock radio antenna — A directional radio transmitting antenna made up of two vertical conductors from which electromagnetic energy radiates. The conductors are connected so that they radiate signals in opposite phases.

additional services — Advisory information provided by ATC which includes but is not limited to the following:
a. Traffic advisories.
b. Vectors, when requested by the pilot, to assist aircraft receiving traffic advisories to avoid observed traffic.
c. Altitude deviation information of 300 feet or more from an assigned altitude as observed on a verified (reading correctly) automatic altitude readout (Mode C).
d. Advisories that traffic is no longer a factor.
e. Weather and chaff information.
f. Weather assistance.
g. Bird activity information.
h. Holding pattern surveillance.

Additional services are provided to the extent possible contingent only upon the controller's capability to fit them into the performance of higher priority duties and on the basis of limitations of the radar, volume of traffic, frequency congestion, and controller workload. The controller has complete discretion for determining if he/she is able to provide or continue to provide a service in a particular case. The controller's reason not to provide or continue to provide a service in a particular case is not subject to question by the pilot and need not be made known to him/her.

additives — Materials that are mixed into a two-part resin system to improve the properties of the system.

address (computer) — A location within a computer's memory where data is located.

adequate vis ref (adequate visual reference) — Runway markings or runway lighting that provides the pilot with adequate visual reference to continuously identify the take-off surface and maintain directional control throughout the take-off run.

adhesion — The forming of a tight bond, usually in reference to surface coatings and adhesives.

adhesive — A substance applied to two mating surfaces to bond them together by surface attachment.

adhesive film — In composites, premixed adhesives cast onto a thin plastic film. Requires refrigerated storage.

adiabatic — Air compression occurring without loss or gain of heat.

adiabatic change — A physical change in state or condition that occurs within a material with no loss or gain of heat.

adiabatic cooling — A process of cooling the air through expansion. For example, as air moves up a slope it expands with the reduction of atmospheric pressure and cools as it expands.

adiabatic heating — A process of heating dry air through compression. For example, as air moves down a slope it is compressed, as it is compressed it warms.

adiabatic lapse rate — The decrease in temperature that occurs with changes in altitude when no heat is added to or taken from the air. It is nominally 5.4°F per 1,000 ft.

adiabatic process — In meteorology, a closed system where temperature, pressure, and density vary without any gain or loss of heat.

adjacent — In mathematics, the two sides of a triangle that have a common angle.

adjust — To change something in order to make it more satisfactory or to make it operate better.

adjustable stabilizer — A stabilizer that can be adjusted in flight to trim the airplane, thereby allowing the airplane to fly hands-off at any given airspeed.

adjustable-pitch propeller — A propeller with blades whose pitch can be adjusted on the ground with the engine not running, but which cannot be adjusted in flight. Also referred to as a ground adjustable propeller.

adjustable-split die — A tool used for cutting external threads on round stock. The die is split on one side and an adjusting screw is used to spread the die to adjust the fit of the threads.

adjusting idle mixture — Adjusting carburetor idle mixture tailored for the best performance of the particular engine and installation.

ADM — Aeronautical decision making. A systematic mental process used by aircraft pilots to consistently determine the best course of action in response to a given set of circumstances.

Administrator — The Federal Aviation Administrator or any person to whom he/she has delegated his/her authority in the matter concerned.

admittance — A measure of the ease with which alternating current can flow in an electrical circuit. Admittance is the current divided by the voltage, and is the reciprocal of impedance. Measured in siemens.

advance — To move forward.

advanced composites — A fibrous material embedded in a resin matrix. The term "advanced" applies to those materials, which have superior strength and stiffness and the process in which they are manufactured. Advanced composites are generally the ones used structurally on an aircraft.

advanced firing — See advanced timing.

advanced ground instructor — A person certificated by the FAA who is authorized to provide: ground training in the aeronautical knowledge areas that are required for issuance of any certificate or rating; ground training required for any flight review; and a recommendation for a knowledge test required for the issuance of any certificate.

advanced timing — Ignition when takes place before the piston reaches top dead center.

advancing blade — The blade moving in the same direction as the helicopter or gyroplane. In rotorcraft that have counterclockwise main rotor blade rotation as viewed from above, the advancing blade is in the right half of the rotor disc area during forward movement.

advection — The horizontal transport of air or atmospheric properties. In meteorology, advection is sometimes referred to as the horizontal component of convection.

advection current — An air current that move horizontally over a surface.

advection fog — Fog that forms when moist air is moved horizontally across a surface that is cold enough to cool the air to a temperature that is below its dew point.

adverse yaw — A condition of flight in which the nose of an airplane starts to move in the direction opposite of the intended turn. It is caused by the downward deflected aileron producing induced drag. Often called aileron drag.

advise intentions — During communications, this indicates: Tell me what you plan to do.

advisory — Advice and information provided to assist pilots in the safe conduct of flight and aircraft movement.

advisory frequency — The appropriate frequency to be used for Airport Advisory Service

advisory route (ADR) — A designated route along which air traffic advisory service is available. NOTE: Air traffic control service provides a much more complete service than air traffic advisory service; advisory areas and routes are therefore not established within controlled airspace, but air traffic advisory service may be provided below and above control areas.

advisory service — Advice and information provided by a facility to assist pilots in the safe conduct of flight and aircraft movement. .

aerated — **1.** Mixed with air. When lubricating oil is used in an engine, it mixes with air, and it is said to be aerated. **2.** Supplied with air or exposed to the circulation of air.

aeration — The process of mixing air into a liquid.

aerial — Of or relating to an aircraft in flight. It is used in terms such as aerial photography.

aerial photograph — Any photograph made from an aircraft in flight.

aerial refueling — A procedure used by the military to transfer fuel from one aircraft to another during flight.

aerodrome — A defined area on land or water (including any buildings, installations and equipment) intended to be used either wholly or in part for the arrival, departure, and movement of aircraft.

aerodrome beacon [ICAO] — Aeronautical beacon used to indicate the location of an aerodrome from the air.

aerodrome control service [ICAO] — Air traffic control service for aerodrome traffic.

aerodrome control tower [ICAO] — A unit established to provide air traffic control service to aerodrome traffic.

aerodrome elevation [ICAO] — The elevation of the highest point of the landing area.

aerodrome flight information service (AFIS) — A directed traffic information and operational information service provided within an aerodrome flight information zone, to all radio equipped aircraft, to assist in the safe and efficient conduct of flight.

aerodrome traffic circuit [ICAO] — The specified path to be flown by aircraft operating in the vicinity of an aerodrome.

aerodrome traffic frequency (ATF) — A frequency designated at an uncontrolled airport. An ATF is used to ensure all radio equipped aircraft operating within the area, normally within a 5 NM radius of the airport, are listening on a common frequency. The ATF is normally the ground station frequency. Where a ground station does not exist, a common frequency is designated. Radio call sign is that of the ground station, or where no ground station exists, a broadcast is made with the call sign "Traffic Advisory." Jeppesen charts list the frequency and the area of use when other than the standard 5 NM.

aerodrome traffic zone (ATZ) — An airspace of detailed dimensions established around an aerodrome for the protection of aerodrome traffic.

aerodynamic balance — The portion of a control surface on an airplane that extends ahead of the hinge line. This utilizes the airflow about the aircraft to aid in moving the surface.

aerodynamic blockage thrust reverser — A configuration of thrust reverser used in turbojet engines in which thin airfoils or obstructions are placed in the engine's exhaust stream to duct the high-velocity exhaust gases forward. This decreases the airplane's landing roll.

aerodynamic braking — The generation of aerodynamic drag used to reduce the roll after landing or to allow the aircraft to descend at a steep angle without building up excessive airspeed. Examples would include speed brakes and spoilers to steepen glide paths and reduce landing roll. The reverse pitch on propellers and reverse thrust on turbine engines are also used in reducing landing roll.

aerodynamic center — The point within the airfoil section located at a point approximately one-fourth of the way back from the leading edge. It is the point at which the (pitching) moment coefficient is relatively constant for all angles of attack.

aerodynamic center of horizontal tail — The point at which the flow of air over the horizontal stabilizer creates a force which pushes the tail up or down.

aerodynamic coefficients — Non-dimensional coefficients for aerodynamic forces and moments.

aerodynamic contrail — As an aircraft moves through moist air the forces created by dynamic flow over the lifting surfaces cause the surrounding atmosphere to reach saturation, to form a cloud like trail. Usually this is generated by high-performance aircraft.

aerodynamic design point — In turbine engines, the most efficient compression ratio that occurs at altitude.

aerodynamic drag — Drag caused by turbulent airflow on an airfoil such as a wing, propeller, or compressor blade.

aerodynamic factors — **1.** Those factors that affect the amount of lift or drag produced by an airfoil. **2.** The forces acting on a propeller while rotating through the air as it transforms the rotary power of the engine into thrust.

aerodynamic heating — The temperature rise caused by high-speed air flowing over an aerodynamic surface.

aerodynamic lift — The upward force caused by high-speed air flowing over an airfoil.

aerodynamic shape — The shape of an object with reference to the airflow over it. Certain shapes cause air pressure differentials which produce lift; others are designed for minimum airflow resistance.

aerodynamic twisting force (ATF) — One of the five forces acting on a rotating propeller. The aerodynamic twisting force tends to twist the blade angle toward the feather position.

aerodynamic twisting moment — A rotational force applied to an object due to aerodynamic loads on the object. Usually a concern in the design of propellers, but also to a lesser extent for wing design. It occurs when the center of lift is ahead of the center of rotation.

aerodynamics — The science of the action of air on an object, and with the motion of air on other gases. Aerodynamics deals with the production of lift by the aircraft, the relative wind, and the atmosphere.

Aerofiche — Registered trade name for a form of microfiche used in the aircraft industry. Two hundred eighty-eight frames of information may be placed on a single 4" x 8" card of film.

aeronaut — A person who operates or travels in airships or balloons.

aeronautical beacon — A visual NAVAID displaying flashes of white and/or colored light to indicate the location of an airport, a heliport, a landmark, a certain point of a Federal airway in mountainous terrain, or an obstruction.

aeronautical chart — A map used in air navigation containing all or part of the following: topographic features, hazards and obstructions, navigation aids, navigation routes, designated airspace, and airports. Common aeronautical charts include:
a. Sectional Aeronautical Charts (1:500,000) — Designed for visual navigation of slow or medium speed aircraft. Topographic information on these charts features the portrayal of relief, and a judicious selection of visual check points for VFR flight. Aeronautical information includes visual and radio aids to navigation, airports, controlled airspace, restricted areas, obstructions and related data.
b. VFR Terminal Area Charts (1:250,000) — Depict Class B airspace which provides for the control or segregation of all the aircraft within the Class B airspace. The chart depicts topographic information and aeronautical information which includes visual and radio aids to navigation, airports, controlled airspace, restricted areas, obstructions, and related data.
c. Jeppesen Class B Airspace Charts — Provide aeronautical information for orientation purposes by depicting airways and navaids used to assist in determining the aircraft's position relative to the vertical and lateral limits of the Class B airspace. Also include flight procedures and VFR approach control frequencies for each Class B airspace. Charts are identified by the principal city and state using an index number of 10-1A.
d. World Aeronautical Charts (WAC) (1:1,000,000) — Provide a standard series of aeronautical charts

covering land areas of the world at a size and scale convenient for navigation by moderate speed aircraft. Topographic information includes cities and towns, principal roads, railroads, distinctive landmarks, drainage, and relief. Aeronautical information includes visual and radio aids to navigation, airports, airways, restricted areas, obstructions and other pertinent data.

e. Enroute Low Altitude Charts — Provide aeronautical information for enroute instrument navigation (IFR) in the low altitude stratum. Information includes the portrayal of airways, limits of controlled airspace, position identification and frequencies of radio aids, selected airports, minimum enroute and minimum obstruction clearance altitudes, airway distances, reporting points, restricted areas, and related data. Area charts, which are a part of this series, furnish terminal data at a larger scale in congested areas.

f. Enroute High Altitude Charts — Provide aeronautical information for enroute instrument navigation (IFR) in the high altitude stratum. Information includes the portrayal of jet routes, identification and frequencies of radio aids, selected airports, distances, time zones, special use airspace, and related information.

g. Jeppesen Area Navigation Enroute Charts — Provide aeronautical information for flight planning and flying, IFR or VFR, random area navigation routes in the U.S. Depict VORTACs (and VORDMEs) with four cardinal radials marked with 10-mile ticks to establish rho/ theta (distance and bearing) values for describing and plotting area navigation waypoints. Charts also include VORs, station declination VORTAC antenna elevation, FSS and ARTCC communications, controlled airspace, minimum off-route altitudes, airports, special use airspace, ARTCC boundaries, times zones and state boundaries. Airport and Facility Listings include primary airport coordinates, elevation, identifier and bearing/distance from nearby VORTAC/VORDME facilities including facility frequency, identifier, and antenna elevation.

h. Instrument Approach Procedures (IAP) Charts — Portray the aeronautical data which is required to execute an instrument approach to an airport. These charts depict the procedures, including all related data, and the airport diagram. Each procedure is designated for use with a specific type of electronic navigation system including NDB, TACAN, VOR, ILS/ MLS, and RNAV. These charts are identified by the type of navigational aid(s) which provide final approach guidance.

i. Instrument Departure Procedure (DP) Charts — Designed to expedite clearance delivery and to facilitate transition between takeoff and enroute operations. Each DP is presented as a separate chart and may serve a single airport or more than one airport in a given geographical location.

j. Standard Terminal Arrival (STAR) Charts — Designed to expedite air traffic control arrival procedures and to facilitate transition between enroute and instrument approach operations. Each STAR procedure is presented as a separate chart and may serve a single airport or more than one airport in a given geographical location.

k. Airport Taxi Charts — Designed to expedite the efficient and safe flow of ground traffic at an airport. These charts are identified by the official airport name: i.e. Ronald Reagan Washington National Airport.

aeronautical chart [ICAO] — A representation of a portion of the earth, its culture and relief, specifically designated to meet the requirements of air navigation.

aeronautical decision making (ADM) — A systematic approach to the mental process used by aircraft pilots to consistently determine the best course of action in response to a given set of circumstances.

Aeronautical Information Manual (AIM) — A primary FAA publication whose purpose is to instruct airmen about operating in the National Airspace System of the U.S. It provides basic flight information, ATC Procedures and general instructional information concerning health, medical facts, factors affecting flight safety, accident and hazard reporting, and types of aeronautical charts and their use. Previously referred to as the Airman's Information Manual.

aeronautical information publication (AIP) [ICAO] — A publication issued by or with the authority of a State and containing aeronautical information of a lasting character essential to air navigation.

Aeronautical Radio, Incorporated (ARINC) — A provider of transportation communications and systems engineering solutions for aviation, airports, defense, government, and transportation. In aviation, ARINC provides standards for communications compatibility.

aeronautics — The science of making and flying airplanes. A term that applies to anything that is in any way associated with the design, construction, or operation of an aircraft. Aerodynamics and aerostatics are both branches of aeronautics.

aerosol — A liquid that is broken up into tiny drops divided into extremely fine particles and dispersed or sprayed into the air by the use of a propellant such as carbon dioxide, nitrogen, or Freon.

aerospace — Space from the Earth's surface extending outward beyond the earth into space.

aerospace industry — That portion of our economy associated with such devices as aircraft, space ships, missiles, and their associated parts.

aerospace vehicle — Any controllable device capable of flight in the aerospace.

aerostat — A device such as a balloon or dirigible that is supported in the air by displacing more than its weight.

aerostatics — The branch of science that deals with the generation of lift by the displacement of air by a body lighter than the air it displaces. Balloons and dirigibles that are filled with hot air or gas fall under the science of aerostatics.

affective domain — A grouping of learning levels associated with a person's attitudes, personal beliefs, and values, which range from receiving through responding, valuing, and organization to characterization.

affirmative — In communications, this indicates: Yes.

aft — To the rear, back, dorsal, or tail of the aircraft.

aft flap — The rear section of a triple-slotted, segmented wing flap.

after bottom center (ABC) — In a reciprocating engine, the amount of crankshaft rotation after the piston has passed the bottom of its stroke. Expressed in degrees.

after top center (ATC) — In a reciprocating engine, the amount of crankshaft rotation after the piston has passed the top of its stroke. Expressed in degrees.

afterburner — A portion of a jet engine in which additional fuel is sprayed into the hot, oxygen-rich exhaust, where it burns and produces additional thrust. Afterburners provide a great amount of additional thrust with a minimum of additional weight.

after-firing — A condition often resulting from either too rich fuel/air mixture or unburned fuel being pumped into the exhaust system of a reciprocating engine and ignited when it comes in contact with some hot component. Sometimes referred to as afterburning or torching.

afterglow — The glow that remains on the phosphorescent screen of a cathode ray tube after the electron beam passes.

aft-fan engine — A turbofan engine that has a fan constructed as an extension of the turbine blades.

age hardening — The process of increasing the hardness of a metal after the heat-treat process. Age hardening occurs at room temperature and continues for a period of several days until the metal reaches its fully hardened state.

aggression — Because of safety concerns or social structures, students may display the defense mechanism of aggression. They may ask irrelevant questions, refuse to participate in class activities, or disrupt the group.

aging — See age hardening.

agitate — To stir or shake something in order to mix its ingredients.

AGL altitude — The vertical elevation above ground. Expressed in feet.

agonic line — An irregular imaginary line across the surface of the earth along which the magnetic and geographic poles are in line, and where there is no variation error.

agricultural aircraft — Aircraft that are specifically designed and built for use in the application of chemicals to crops for insect and weed control.

aileron — A primary control surface located near the wing tip that makes up part of the total wing area. Ailerons are operated by the lateral motion of the controls and cause rotation of the aircraft about the longitudinal axis.

aileron angle — The angle of displacement of an aileron from its neutral, or trailing, position.

aileron spar — A spar that extends only part of the span of the wing and provides a hinge attachment point for the aileron. Also referred to as "false spar."

aileron station — Distances measured outboard from the root end of an aileron, parallel to the aileron spar.

air — A mixture of gases that comprises the earth's atmosphere. Pure dry air contains approximately 78% nitrogen and 21% oxygen. The remaining 1% consists of argon, carbon dioxide, hydrogen, and traces of neon and helium. Dry air weighs 0.07651 pounds per cubic foot at sea level with a temperature of 59°F and has a barometric pressure of 14.69 PSI at 40° latitude.

air adapters — A component in a centrifugal compressor gas turbine engine. Its purpose is to deliver air from the diffuser to the individual can-type combustion chambers at the proper angle.

air, ambient — The atmospheric air surrounding all sides of the aircraft or engine. Expressed in units of lbs./sq. inch or in. Hg.

air bleed — 1. A small hole in the fuel passage between the float bowl and the discharge nozzle of a float carburetor. The hole introduces air into the liquid fuel and serves as an aid to atomization. 2. Used in gas turbine engines for a variety of purposes and is taken from the engine's compressor section.

air brake — A plate or series of plates that can be projected into the airplane's slipstream to provide turbulence and drag to slow the airplane during descent, glide, or when maneuvering. Air brakes differ from flaps in that they produce no useful lift. Also referred to as speed brakes.

air capacitor — A capacitor that uses air as the dielectric.

air carrier — A person who undertakes directly by lease, or other arrangement, to engage in air transportation.

Air Carrier District Office (ACDO) — An FAA field office serving an assigned geographical area, staffed with Flight Standards personnel serving the aviation industry and the general public on matters related to the certification and operation of scheduled air carriers and other large aircraft operations.

air commerce — Transportation by aircraft of persons or property for hire or compensation.

air commerce — Interstate, overseas, or foreign air commerce or the transportation of mail by aircraft or any operation or navigation of aircraft within the limits of any Federal airway or any operation or navigation of aircraft which directly affects, or which may endanger safety in, interstate, overseas, or foreign air commerce.

air conditioning — The process of treating air to control simultaneously its temperature, humidity, cleanliness, and distribution to meet the requirements of a conditioned space. In combination with pressurization, complete environmental control is possible.

air conditioning system — A system consisting of cabin air conditioning and pressurization that supplies conditioned air for heating and cooling the cockpit and cabin spaces. This air also provides pressurization to maintain a safe, comfortable cabin environment.

air cycle cooling system — One of several cooling systems consisting of an expansion turbine, an air-to-air heat exchanger, and various valves that control airflow through the system. Used to provide a comfortable atmosphere within the aircraft cabin.

air cycle machine — An air conditioning system That uses compressor bleed air to condition air and pressurize the cabin. Primarily used in jet turbine powered aircraft.

air defense emergency — A military emergency condition declared by a designated authority. This condition exists when an attack upon the continental U.S., Alaska, Canada, or U.S. installations in Greenland by hostile aircraft or missiles is considered probable, is imminent, or is taking place.

air defense identification zone (ADIZ) — The area of airspace over land or water, extending upward from the surface, within which the ready identification, the location, and the control of aircraft are required in the interest of national security.
a. Domestic Air Defense Identification Zone — An ADIZ within the United States along an international boundary of the United States.
b. Coastal Air Defense Identification Zone — An ADIZ over the coastal waters of the United States.
c. Distant Early Warning Identification Zone (DEWIZ) — An ADIZ over the coastal waters of the State of Alaska.

ADIZ locations and operating and flight plan requirements for civil aircraft operations are specified in FAR 99.

air density — The density of the air in terms of mass per unit volume. Dense air has more molecules per unit volume than less dense air. The density of air decreases with altitude above the surface of the earth and with increasing temperature.

air filter — A filtering device that prevents dust and dirt from entering the intake or induction system.

air filter system — An air filter system normally consists of a filter and a door that can either allow the air to be filtered through it or bypass the filter. When the filter system is operating, air is drawn through a louvered access panel that does not face directly into the airstream. With this entrance location, considerable dust is removed as the air is forced to turn and enter the duct.

air gap — The space between the reluctor and the speed sensor on some shaft rotational speed detection systems. The reluctor is a toothed gear or cam that causes variations in the magnetic field surrounding the pickup device. This varying field can be transmitted into RPM readings. The spacing (air gap) between these parts is critical, as the magnetic fields involved are very small.

air impingement — A fault that resembles haze in an enamel or lacquer paint finish. It is caused by microscopic-size bubbles that form when paint is applied with too high an atomizing air pressure.

air impingement starter — A starter used on small gas turbine engines in which a stream of high-pressure compressed air is directed onto the blades of the compressor or the turbine in order to rotate the engine for starting.

air inlet — A portion of a turbine engine designed to conduct incoming air to the compressor section with a minimum energy loss resulting from drag or ram pressure loss.

air lock — A pocket of trapped air that blocks the flow of fluid.

air mass — A widespread mass of air having similar characteristics (e.g., temperature) which usually helps to identify the source region of the air. Fronts are distinct boundaries between air masses.

air metering force — The force used in Bendix pressure carburetors and fuel injection systems in which venturi and ram air pressures control the amount of fuel metered.

air navigation facility — Any facility used in, available for use in, or designed for use in, aid of air navigation, including landing areas, lights, any apparatus or equipment for disseminating weather information, for

signaling, for radio-directional finding, or for radio or other electrical communication, and any other structure or mechanism having a similar purpose for guiding or controlling flight in the air or the landing and take-off of aircraft.

air refueling — A method of refueling aircraft while in flight in order to extend the airplane's range. The military uses flying tankers to allow aircraft to fly extended missions.

air route surveillance radar (ARSR) — Air route traffic control center (ARTCC) radar used primarily to detect and display an aircraft's position while enroute between terminal areas. The ARSR enables controllers to provide radar air traffic control service when aircraft are within the ARSR coverage. In some instances, ARSR may enable an ARTCC to provide terminal radar services similar to but usually more limited than those provided by a radar approach control.

air route traffic control center (ARTCC) — A facility established to provide air traffic control service to aircraft operating on IFR flight plans within controlled airspace and principally during the enroute phase of flight. When equipment capabilities and controller workload permit, certain advisory/assistance services may be provided to VFR aircraft.

Air Route Traffic Control Center (ARTCC) — Provides enroute air traffic control guidance primarily for aircraft operating on IFR flight plans.

air scoop — 1. A hooded opening to an engine carburetor or other device used to receive the ram air during flight, which in turn, increases the amount of air taken into the structure. 2. A specially designed scoop or duct that guides air to the carburetor and intake manifold of a reciprocating engine induction system.

air seal — A seal used to keep air from passing out of a housing unit. Usually air seals are thin rotating or stationary rims designed to act as air dams to reduce airflow leakage between the gas path and the internal engine or over blade tips.

air start — The process of starting an aircraft engine in flight. In an air start, aerodynamic forces cause the propeller or the compressor to turn the engine. A starter is not generally used during an air start.

air starter — See air-turbine starter.

air strip — See airfield.

air taxi — Used to describe a helicopter/VTOL aircraft movement conducted above the surface but normally not above 100 feet AGL. The aircraft may proceed either via hover taxi or flight at speeds more than 20 knots. The pilot is solely responsible for selecting a safe airspeed/altitude for the operation being conducted.

air temperature control — An air control door or valve, near the entrance of the carburetor, which admits alternate heated air to the carburetor to prevent carburetor ice.

air temperature gauge — A gauge that indicates the temperature of the air before it enters the carburetor. The temperature reading is sensed by a bulb located in the air intake passage to the engine.

air traffic — Aircraft operating in the air or on an airport surface, exclusive of loading ramps and parking areas.

air traffic [ICAO] — All aircraft in flight or operating on the maneuvering area of an aerodrome.

air traffic clearance — An authorization by air traffic control, for the purpose of preventing collision between known aircraft, for an aircraft to proceed under specified traffic conditions within controlled airspace. The pilot-in-command of an aircraft may not deviate from the provisions of a visual flight rules (VFR) or instrument flight rules (IFR) air traffic clearance except in an emergency or unless an amended clearance has been obtained. Additionally, the pilot may request a different clearance from that which has been issued by air traffic control (ATC) if information available to the pilot makes another course of action more practicable or if aircraft equipment limitations or company procedures forbid compliance with the clearance issued. Pilots may also request clarification or amendment, as appropriate, any time a clearance is not fully understood, or considered unacceptable because of safety of flight. Controllers should, in such instances and to the extent of operational practicality and safety, honor the pilot's request. FAR Part 91.3(a) states: "The pilot-in-command of an aircraft is directly responsible for, and is the final authority as to, the operation of that aircraft." The pilot is responsible to request an amended clearance if ATC issues a clearance that would cause a pilot to deviate from a rule or regulation, or in the pilot's opinion, would place the aircraft in jeopardy.

air traffic control — A service operated by an authoritative body to promote the safe, orderly, and expeditious flow of air traffic.

air traffic control (ATC) — A service provided by the FAA to promote the safe, orderly, and expeditious flow of air traffic.

air traffic control assigned airspace (ATCAA) — Airspace of defined vertical/lateral limits, assigned by ATC, for the purpose of providing air traffic segregation between the specified activities being conducted within the assigned airspace and other IFR air traffic.

air traffic control clearance — An authorization by air traffic control, for the purpose of preventing collision between known aircraft, for an aircraft to proceed under specified traffic conditions within controlled airspace.

air traffic control clearance [ICAO] — Authorization for an aircraft to proceed under conditions specified by an air traffic control unit.

Note 1: For convenience, the term air traffic control clearance is frequently abbreviated to clearance when used in appropriate contexts.

Note 2: The abbreviated term clearance may be prefixed by the words taxi, takeoff, departure, enroute, approach or landing to indicate the particular portion of flight to which the air traffic control clearance relates.

air traffic control service — See air traffic control.

air traffic control service [ICAO] — A service provided for the purpose of: **a.** Preventing collisions: **1)** Between aircraft, and **2)** On the maneuvering area between aircraft and obstructions; and **b.** Expediting and maintaining an orderly flow of air traffic.

air traffic control specialist — A person authorized to provide air traffic control service.

air traffic control system command center — An Air Traffic Tactical Operations facility consisting of four operational units.

a. Central Flow Control Function (CFCF). Responsible for coordination and approval of all major inter-center flow control restrictions on a system basis in order to obtain maximum utilization of the airspace.

b. Central Altitude Reservation Function (CARF). Responsible for coordinating, planning, and approving special user requirements under the Altitude Reservation (ALTRV) concept.

c. Airport Reservation Office (ARO). Responsible for approving IFR flights at designated high density traffic airports (John F. Kennedy, LaGuardia, O'Hare, and Washington National) during specified hours.

d. ATC Contingency Command Post. A facility which enables the FAA to manage the ATC system when significant portions of the systems's capabilities have been lost or are threatened.

air traffic service — A generic term meaning:
a. Flight Information Service
b. Alerting Service
c. Air Traffic Advisory Service
d. Air Traffic Control Service
1) Area Control Service,
2) Approach Control Service, or
3) Airport Control Service.

air transportation — Interstate, overseas, or foreign air transportation or the transportation of mail by aircraft.

airborne — The condition of an airplane, glider, or balloon when it is off the ground.

airborne delay — Amount of delay to be encountered in airborne holding.

airborne intercept radar — Radar contained in fighter-type aircraft to locate and track airborne targets.

airborne moving target indicator — A feature of airborne radar that electronically filters out targets that are either stationary or moving at less than a preset speed.

airborne navigation equipment — A phrase embracing many systems and instruments. These systems include VHF omnirange (VOR), instrument landing systems, distance-measuring equipment, automatic direction finders, Doppler systems, inertial navigation systems, Global Positioning Systems (GPS), and others.

airborne weather radar — An electronic device used to see objects in darkness, fog, or storms, as well as in clear weather. The range and relative position are indicated on the radar scope.

air-breathing engine — An engine that requires an intake of air to supply the oxygen needed to operate. Reciprocating and turbine engines are both air-breathing engines.

air-cool — To remove excess heat from an object by transferring it directly into the airstream.

air-cooled blades and vanes — Hollow airfoils in the hot section that receive air from the cold section so they can operate in a much higher temperature environment. Holes, sometimes referred to as gill holes, film holes, or tip holes, direct air back to the gas path.

air-cooled oil cooler — A heat exchanger in the lubrication system of an aircraft engine that removes heat from the oil and transfers it into the air that flows through the cooler.

air-cooled turbine blades — Hollow turbine wheel blades of certain high-powered gas turbine engines that are cooled by passing compressor bleed air through them.

air-core electrical transformer — A transformer made up of two or more coils wound on a core of non-magnetic material. Air-core transformers are normally used for radio-frequency alternating current.

aircraft — Any weight-carrying device designed to be supported by the air or intended to be used for flight in the air.

aircraft accident — Any damage or injury that occurs when an aircraft is moving with the intention of flight.

Aircraft Addressing and Reporting System (ACARS) — A digital data network using aircraft VHF and/or HF radio to transmit and receive data to and from airline operations. The system is used by the airlines to maintain contact with their aircraft around the world and by pilots to obtain data such as weather.

aircraft alteration — The modification of an aircraft, its structure, or its components that changes the physical or flight characteristics of the aircraft. Alterations are classified as major or minor, in accordance with Federal Aviation Regulations, Part 43.

aircraft approach category — A grouping of aircraft based on a speed of 1.3 times the stall speed in the landing configuration at maximum gross landing weight. An aircraft shall fit in only one category. If it is necessary to maneuver at speeds in excess of the upper limit of a speed range for a category, the minimums for the next higher category should be used. For example, an aircraft which falls in Category A, but is circling to land at a speed in excess of 91 knots, should use the approach Category B minimums when circling to land. The categories are as follows:
a. Category A — Speed less than 91 knots.
b. Category B — Speed 91 knots or more but less than 121 knots.
c. Category C — Speed 121 knots or more but less than 141 knots.
d. Category D — Speed 141 knots or more but less than 166 knots.
e. Category E — Speed 166 knots or more. Category E includes only certain Military Aircraft and is not included on Jeppesen Approach Charts.

aircraft basic operating weight — The established basic weight of an aircraft available for flight without its fuel and payload.

aircraft battery — A source of electrical energy for an aircraft that can be used for starting. The battery also serves as an auxiliary source of power when the engine generator is inoperable.

aircraft cable — Strands of wire formed into a helical or spiral shape. Cable designations are based on the number of strands and the number of wires in each strand. The most common aircraft cables are 7x7 and 7x19.

aircraft checkouts — An instructional program designed to familiarize and qualify a pilot to act as pilot in command of a particular aircraft type.

aircraft classes — For the purposes of Wake Turbulence Separation Minima, ATC classifies aircraft as Heavy, Large, and Small as follows:
a. Heavy — Aircraft capable of takeoff weights of more than 255,000 pounds whether or not they are operating at this weight during any particular phase of flight.
b. Large — Aircraft of more than 41,000 pounds, maximum certificated takeoff weight, up to 255,000 pounds.
c. Small — Aircraft of 41,000 pounds or less maximum certificated takeoff weight.

aircraft conflict — Predicted conflict, within URET CCLD (User Request Evaluation Tool Core Capability Limited Deployment), of two aircraft, or between aircraft and airspace. A Red alert is used for conflicts when the predicted minimum separation is 5 nautical miles or less. A Yellow alert is used when the predicted minimum separation is between 5 and approximately 12 nautical miles. A Blue alert is used for conflicts between an aircraft and predefined airspace.

aircraft damage liability coverage — A policy that covers any damage to the aircraft, and works much like collision insurance on a car.

aircraft dope — A colloidal solution of cellulose acetate or nitrate, combined with sufficient plasticizers to produce a smooth, flexible, homogeneous film. The dope increases a fabric's tensile strength, air tightness, weatherproofing, and tautness.

aircraft engine — An engine that is used or intended to be used for propelling aircraft. It includes turbosuperchargers, appurtenances, and accessories necessary for its functioning, but does not include propellers.

aircraft inspection — A systematic check of an aircraft and its components. The purpose of an aircraft inspection is to detect any defects or malfunctions before they become serious. Annual inspections, 100-hour inspections, progressive inspections, and preflight inspections are common types of aircraft inspections.

aircraft lighting system — A system that provides illumination for both exterior and interior use. The system includes lighting of instruments, cockpits, cabins and other sections occupied by crewmembers and passengers as well as exterior lights for landing and ground taxiing.

aircraft list (ACL) — A view available with URET CCLD (User Request Evaluation Tool Core Capability Limited Deployment) that lists aircraft currently in or predicted to be in a particular sector's airspace. The view contains textual flight data information in line format and may be sorted into various orders based on the specific needs of the sector team.

aircraft listings — Information sheets published by the FAA that contain essential information on particular models of aircraft.

aircraft log — A record containing the operational or maintenance history of the aircraft.

aircraft operating weight — The basic weight of an aircraft plus the weight of the crewmembers, equipment, fuel, oil, and passengers.

aircraft pitch — the movement about an aircraft's lateral or pitch axis. Movement of the cyclic forward or aft causes the nose of the helicopter or gyroplane to pitch up or down.

aircraft plumbing — The hoses, tubing, fittings, and connections used to transfer fluids through an aircraft.

aircraft quality — Indicates that aircraft equipment or materials are to be produced under closely controlled, special, and restricted methods of manufacture and inspection.

aircraft records — Documentation of the maintenance performed and the flight time on an aircraft, its engines, or its components.

aircraft repair — Restoration of an aircraft and/or its components to a condition of airworthiness after a failure, damage, or wear has occurred.

aircraft rigging — The final adjustment and alignment of the various components of an aircraft to give it the proper aerodynamic characteristics.

aircraft roll — The movement of the aircraft about its longitudinal axis. Movement of the cyclic right or left causes the helicopter or gyroplane to tilt in that direction.

aircraft steel structure — A truss-type fuselage frame usually constructed of steel tubing welded together in such a manner that all members of the truss can carry both tension and compression loads.

aircraft surge launch and recovery (ASLAR) — Procedures used at USAF bases to provide increased launch and recovery rates in instrument flight rules conditions. ASLAR is based on:
a. Reduced separation between aircraft which is based on time or distance. Standard arrival separation applies between participants including multiple flights until the DRAG point. The DRAG point is a published location on an ASLAR approach where an aircraft landing second in a formation slows to a predetermined airspeed. The DRAG point is the reference point at which MARSA applies as expanding elements effect separation within a flight or between subsequent participating flights.
b. ASLAR procedures shall be covered in a Letter of Agreement between the responsible USAF military ATC facility and the concerned FAA facility. Initial Approach Fix spacing requirements are normally addressed as a minimum.

aircraft tires — A rubber cushion filled with compressed air that fits around a wheel. Tires help absorb the shock and roughness of landings and takeoffs; they also support the weight of the aircraft while on the ground and provide the necessary traction for braking and stopping aircraft upon landing.

aircraft welding — The process of joining metal by fusing the materials together while they are in a plastic or molten state. There are three general types of welding: gas, electric arc, and electric resistance.

aircraft wooden structures — An aircraft structure in which wood is used as the structural material.

air-dry — The process of removing moisture from a material by exposing it to the air.

airfield — Any area in which aircraft may land, take off, and park. An airfield may also be called an airstrip. The term airfield includes the buildings, equipment, and maintenance facilities used to store or service aircraft.

airflow over wing section — Air flowing over the top surface of the wing must reach the trailing edge of the wing in the same amount of time as the air flowing under the wing. The greater velocity of air traveling a larger distance over the top surface produces lift.

airfoil — Any surface designed to obtain a desirable reaction from the air through which it moves. The airfoil converts movement of air over its surfaces into a force useful for flight. Wings, control surfaces, propeller blades, and helicopter rotors are examples of airfoils.

airfoil profile — The outline of an airfoil section such as a wing.

airfoil section — The cross-sectional shape of an airfoil, viewed as if it were sliced vertically in a fore-and-aft plane.

airframe — The fuselage, booms, nacelles, cowlings, fairings, airfoil surfaces (including rotors but excluding propellers and rotating airfoils of engines), and landing gear of an aircraft and their accessories and controls.

airframe technician — Any person certified by the FAA to perform maintenance or inspections on the airframe of certificated aircraft.

airline — A company or organization that operates aircraft for the transportation of persons or cargo.

airliner — A large, transport-type aircraft used in air commerce for the transportation of passengers or cargo.

Airloc fastener — A patented form of cowling fastener in which the actual locking is done by turning a steel cross-pin in a spring steel receptacle.

airman — A person involved in flying, maintaining, or operating aircraft.

airman certificate — A certificate issued by the FAA authorizing a person to perform certain aviation-related duties. Certificates are issued to pilots, technicians, and parachute riggers.

Airman's Information Manual — Provides flight information and air traffic control procedures for the National Airspace System. Now referred to as the Aeronautical Information Manual.

airman's meteorological information — See AIRMET.

airmass — An extensive body of air having fairly uniform properties of temperature and moisture within a horizontal plane.

airmass thunderstorm — A "nonsevere" or "ordinary" thunderstorm produced by local airmass instability. May produce small hail, wind gusts less than 50 knots. See also severe thunderstorm.

airmass wind shear — Wind shear that develops near the ground at night under fair weather conditions in the absence of strong fronts and/or strong surface pressure gradients.

AIRMET — An advisory pertinent to aircraft with limited capabilities, containing information on:
1. moderate icing,
2. moderate turbulence,
3. sustained surface winds of 30 knots or more,
4. ceilings less than 1,000 feet and/or visibility less than 3 miles affecting 50 percent of the area at one time, and
5. extensive mountain obscuration.

Issued only to amend the area forecast concerning weather phenomena which are of operational interest to all aircraft and potentially hazardous to aircraft having limited capability because of lack of equipment, instrumentation, or pilot qualifications. AIRMETs concern weather of less severity than that covered by SIGMETs or Convective SIGMETs.

air-oil separator — A device in the vent portion of the lubrication system of a gas turbine engine that separates any oil from the air before the air is vented overboard.

airplane — An engine-driven, heavier-than-air, fixed-wing aircraft that is supported in flight by the dynamic reaction of the air against its wings.

airplane checkouts — An instructional program designed to familiarize and qualify a pilot to fly an aircraft not previously familiar.

airport — An area of land or water, including any associated buildings and facilities, used or intended to be used for the landing and takeoff of aircraft.

airport advisory area — The area within 10 statute miles of an airport where a flight service station is located, but where there is no control tower in operation.

airport arrival rate (AAR) — A dynamic input parameter specifying the number of arriving aircraft which an airport or airspace can accept from the ARTCC per hour. The AAR is used to calculate the desired interval between successive arrival aircraft.

airport departure rate (ADR) — A dynamic parameter specifying the number of aircraft which can depart an airport and the airspace can accept per hour.

airport elevation — The highest point of an airport's usable runways measured in feet from mean sea level.

airport information aid — See airport information desk.

airport information desk — An airport unmanned facility designed for pilot self-service briefing, flight planning, and filing of flight plans.

airport lighting — Various lighting aids that can be installed at an airport. Types of airport lighting include:
a. Approach Light System (ALS) — An airport lighting facility which provides visual guidance to landing aircraft by radiating light beams in a directional pattern by which the pilot aligns the aircraft with the extended centerline of the runway on his final approach for landing. Condenser Discharge Sequential Flashing Lights/Sequenced Flashing Lights may be installed in conjunction with the ALS at some airports.

Types of Approach Light Systems are:
1) ALSF-1. — Approach Light System with Sequenced Flashing Lights in ILS Cat-I configuration.
2) ALSF-2. — Approach Light System with Sequenced Flashing Lights in ILS Cat-II configuration. The ALSF-2 may operate as an SSALR when weather conditions permit.
3) SSALF — Simplified Short Approach Light System with Sequenced Flashing Lights.
4) SSALR — Simplified Short Approach Light System with Runway Alignment Indicator Lights.
5) ALSF — Medium Intensity Approach Light System with Sequenced Flashing Lights.
6) MALSR — Medium Intensity Approach Light System with Runway Alignment Indicator Lights.
7) LDIN — Lead-in-light system: Consists of one or more series of flashing lights installed at or near ground level that provides positive visual guidance along an approach path, either curving or straight, where special problems exist with hazardous terrain, obstructions, or noise abatement procedures.
8) RAIL — Runway Alignment Indicator Lights (Sequenced Flashing Lights which are installed only in combination with other light systems).
9) ODALS — Omni-directional Approach Lighting System consists of seven omni-directional flashing lights located in the approach area of a nonprecision runway. Five lights are located on the runway centerline extended with the first light located 300 feet from the threshold and extending at equal intervals up to 1,500 feet from the threshold. The other two lights are located, one on each side of the runway threshold, at a lateral distance of 40 feet from the runway edge, or 75 feet from the runway edge when installed on a runway equipped with a VASI.

b. Runway Lights/Runway Edge Lights — Lights having a prescribed angle of emission used to define the lateral limits of a runway. Runway lights are uniformly spaced at intervals of approximately 200 feet, and the intensity may be controlled or preset.

c. Touchdown Zone Lighting — Two rows of transverse light bars located symmetrically about the runway centerline normally at 100 foot intervals. The basic system extends 3,000 feet along the runway.
d. Runway Centerline Lighting — Flush centerline lights spaced at 50-foot intervals beginning 75 feet from the landing threshold and extending to within 75 feet of the opposite end of the runway.
e. Threshold Lights — Fixed green lights arranged symmetrically left and right of the runway centerline, identifying the runway threshold.
f. Runway End Identifier Lights (REIL) — Two synchronized flashing lights, one on each side of the runway threshold, which provide rapid and positive identification of the approach end of a particular runway.
g. Visual Approach Slope Indicator (VASI) — An airport lighting facility providing vertical visual approach slope guidance to aircraft during approach to landing by radiating a directional pattern of high intensity red and white focused light beams which indicate to the pilot that he is "on path" if he sees red/white, "above path" if white/white, and "below path" if red/red. Some airports serving large aircraft have three-bar VASIs which provide two visual glide paths to the same runway.
h. Boundary Lights — Lights defining the perimeter of an airport or landing area.

airport marking aids — Markings used on runway and taxiway surfaces to identify a specific runway, a runway threshold, a centerline, a hold line, etc. A runway should be marked in accordance with its present usage such as:
a. Visual.
b. Nonprecision instrument.
c. Precision instrument.

airport movement area safety system (AMASS) — A software enhancement to ASDE radar which provides logic predicting the path of aircraft landing and/or departing, and aircraft and/or vehicular movements on runways. Visual and aural alarms are activated when logic projects a potential collision.

airport reference point (ARP) — A point on the airport designated as the official airport location. Usually the approximate geometric center of all usable runway surfaces.

airport reservation office — Office responsible for monitoring the operation of the high density rule. Receives and processes requests for IFR operations at high density traffic airports.

airport rotating beacon — A visual NAVAID operated at many airports. At civil airports, alternating white and green flashes indicate the location of the airport. At military airports, the beacons flash alternately white and green, but are differentiated from civil beacons by dual-peaked (two quick) white flashes between the green flashes.

airport surface detection equipment — Radar equipment specifically designed to detect all principal features on the surface of an airport, including aircraft and vehicular traffic, and to present the entire image on a radar indicator console in the control tower. Used to augment visual observation by tower personnel of aircraft and/ or vehicular movements on runways and taxiways.

airport surveillance radar (ASR) — Approach control radar used to detect and display an aircraft's position in the terminal area. ASR provides range and azimuth information but does not provide elevation data. Coverage of the ASR can extend up to 60 miles.

airport taxi charts — Designed to expedite the efficient and safe flow of ground traffic at an airport. These charts are identified by the official airport name: e.g., Ronald Reagan Washington National Airport.

airport traffic area — No longer a designator of airspace. Generally superseded by Class B, C, and D designations according to the type of airport. See Class B, Class C, and Class D.

airport traffic control service — A service provided by a control tower for aircraft operating on the movement area and in the vicinity of an airport.

airport traffic control tower — A terminal facility that uses air/ground communications, visual signaling, and other devices to provide ATC services to aircraft operating in the vicinity of an airport or on the movement area. Authorizes aircraft to land or takeoff at the airport controlled by the tower or to transit the Class D airspace area regardless of flight plan or weather conditions (IFR or VFR). A tower may also provide approach control services (radar or nonradar).

airport/facility directory — A publication designed primarily as a pilot's operational manual containing all airports, seaplane bases, and heliports open to the public including communications data, navigational facilities, and certain special notices and procedures. This publication is issued in seven volumes according to geographical area.

airscrew — A British term for aircraft propeller.

airship — An engine-driven, lighter-than-air aircraft that can be steered.

airspace — The space lying above a certain geographical area.

airspace conflict — Predicted conflict of an aircraft and active Special Activity Airspace (SAA).

airspace hierarchy — Within the airspace classes, there is a hierarchy and, in the event of an overlap of airspace: Class A preempts Class B, Class B preempts

Class C, Class C preempts Class D, Class D preempts Class E, and Class E preempts Class G.

airspeed — The speed of an aircraft relative to its surrounding air mass. The unqualified term "airspeed" means one of the following: **1.** Indicated Airspeed — The speed shown on the aircraft airspeed indicator. This is the speed used in pilot/controller communications under the general term "airspeed." **2.** True Airspeed — The airspeed of an aircraft relative to undisturbed air. Used primarily in flight planning and enroute portion of flight. When used in pilot/controller communications, it is referred to as "true airspeed" and not shortened to "airspeed."

airspeed indicator — A differential air pressure gauge that measures the difference between ram, or impact air pressure, and the static pressure of the air to indicate the speed of the aircraft through the air.

airstart — The starting of an aircraft engine while the aircraft is airborne, preceded by engine shutdown during training flights or by actual engine failure.

airstream direction detection — A unit of an angle-of-attack indicating system. The airstream direction detector contains the sensing element that measures local airflow direction relative to the angle of attack by determining the angular difference between local airflow and the fuselage reference plane.

air-to-air missile — A missile launched from an aircraft toward an airborne target.

air-to-surface missile — A missile launched from an aircraft toward a target on the ground.

air-turbine starter — A large volume of compressed air from an auxiliary power unit or bleed air from an operating engine that is directed into the air-turbine starter. This air spins the turbine inside the starter, and the starter, which is geared to the main engine compressor, spins the engine fast enough for it to start.

airway — A Class E airspace area established in the form of a corridor, the centerline of which is defined by radio navigational aids.

airway [ICAO] — A control area or portion thereof established in the form of corridor equipped with radio navigational aids.

airway beacon — A light signal used to mark airway segments in remote mountain areas. The light flashes Morse Code to identify the beacon site.

Airworthiness Certificate — A certificate issued by the FAA to all aircraft that have been proven to meet the minimum standards set down by the Federal Aviation Regulations.

Airworthiness Directive — A regulatory notice sent out by the FAA to the registered owner of an aircraft informing the owner of a condition that prevents the aircraft from continuing to meet its conditions for airworthiness. Airworthiness Directives (AD notes) must be complied with within the required time limit, and the fact of compliance, the date of compliance, and the method of compliance must be recorded in the aircraft's maintenance records.

airworthy — To be airworthy, an aircraft or one of its component parts must meet two criteria: **a.** Conform to its TC (Type Certificate). Conformity to type design is considered attained when the aircraft configuration and the components installed are consistent with the drawings, specifications, and other data that are part of the TC, and would include any STC's (Supplemental Type Certificates) and field approved alterations (337's). **b.** Must be in a condition for safe operation. This refers to the condition of the aircraft relative to wear and deterioration, (been maintained, annual/100 hour inspections, etc.).

albedo — The reflectivity of the Earth and its atmosphere.

Alclad — A clad structural aluminum alloy. Alclad is a corrosion protection coating of pure aluminum that is rolled onto the alloy sheet in the rolling mill. It makes up approximately 5% of the thickness on each side.

alcohol — A colorless, volatile, flammable liquid produced by the fermentation of certain types of grain, fruit, or wood pulp. Alcohol is used as a cleaning fluid, as a solvent in many aircraft finishes, and as a fuel for certain types of specialized engines. Isopropyl alcohol is used in some anti-icing systems for propellers, windshields, and carburetors.

alcohol deicing — The act of preventing or controlling ice formation by spraying alcohol onto a surface, in the case of windshields, or into the inlet airstream of a carburetor.

ALERFA (alert phase) [ICAO] — A situation wherein apprehension exists as to the safety of an aircraft and its occupants.

alert — A notification to a position that there is an aircraft-to-aircraft or aircraft-to airspace conflict, as detected by Automated Problem Detection (APD).

alert area — Special use airspace which may contain a high volume of pilot training activities or an unusual type of aerial activity.

alert notice — A request originated by a flight service station (FSS) or an air route traffic control center (ARTCC) for an extensive communication search for overdue, unreported, or missing aircraft.

alerting service — A service provided to notify appropriate organizations regarding aircraft in need of search and rescue aid and assist such organizations as required.

algebra — A branch of mathematics that uses letters or symbols to represent numbers in formulas and equations.

algebraic expression — A quantity made up of letters, numbers, and symbols. The parts of the expression that are separated by a plus or a minus sign are called the terms of the expression. An algebraic expression that has only one term is called a monomial; an algebraic expression that has two or more terms is called a polynomial.

algorithm — A system or procedure used in solving a problem.

alignment — The arrangement or position of parts in the correct relationship to each other.

alignment pin — Installed in a helicopter rotor blade to serve as an index when aligning the blades of a semi-rigid rotor system.

alignment tool — A nonmetallic adjustment tool used to align (adjust) electronic circuits that would be adversely affected by a metallic device.

alkali — A chemical substance, usually the hydroxide of a metal. An alkali has a characteristically bitter taste and is prone to react with an acid to form salt, supplying electrons to the acid.

alkaline — Having the property of reacting with an acid to form a salt and of giving up electrons to the acidic material.

alkaline cell — An electrochemical cell that uses powdered zinc as the anode, powdered graphite and manganese dioxide as the cathode, and potassium hydroxide as the electrolyte. An alkaline cell has an open-circuit voltage of 1.5 volts, and it has from 50% to 100% more capacity than a carbon-zinc cell of comparable size.

alkyd resin — A synthetic resin used as the base for certain enamels and primers.

Allen head bolt — A bolt or screw with a hexagonal receptacle in its head to accommodate an Allen wrench for turning. Also referred to as an internal hex fastener.

Allen wrench — A hexagonal-shaped tool used to turn an Allen screw. Shaft is usually "L" shaped to provide leverage.

alligator clip — A spring-loaded clip with long, narrow jaws and meshing teeth. It is used on the end of an electrical wire to make temporary connections in an electrical circuit.

allowable — Permissible.

allowance — The permissible dimensional difference between the mating parts of a machine.

alloy — 1. In metallurgy, a substance of two more combined elements, one or more which is metal. 2. In composites, a blend of polymers or copolymers with other polymers or elastomers. Also referred to as a polymer blend.

alloy steel — Steel into which certain chemical elements have been mixed. Alloy steel has different characteristics from those of simple carbon steel.

alloying agent — A chemical element used to change the characteristics of a base metal to form an alloy.

all-weather spark plug — A shielded spark plug for use in an aircraft reciprocating engine. The ceramic insulator is recessed into the shell so that a resilient collar (sometimes referred to as a cigarette) on the ignition harness lead can provide a watertight seal. All-weather spark plugs are identified by their 3/4", 20-thread-per-inch shielding.

alnico — An alloy of iron, aluminum, nickel, and cobalt. Alnico has an extremely high permeability and excellent retentivity for use in magnets.

Alodine — A registered trademark for a conversion coating chemical that forms a hard, unbroken aluminum oxide film, chemically deposited on a piece of aluminum alloy. Alodining serves the same function as anodizing, but does not require an electrolytic bath. It conforms to specification MIL-C-5541 B.

along track distance (LTD) — The distance measured from a point-in-space by systems using area navigation reference capabilities that are not subject to slant range errors.

alpha — The first letter of the Greek alphabet. Often used to denote something that is first.

alpha cutoff frequency — Rated cutoff frequency of a transistor at which gain (alpha) decreases to .707 of the gain measured at low-frequency.

alpha hinge — The hinge at the root of a helicopter rotor blade that allows the tip of the blade to move back and forth in its plane of rotation. The axis of the alpha hinge is perpendicular to the plane of rotor rotation. Also referred to as a lead-lag hinge.

alpha mode — Propeller range when operating in a forward thrust condition. See also beta mode.

alpha mode of operation — The operation of a turboprop engine that includes all of the flight operations, from takeoff to landing. Alpha operation is typically 95% to 100% of the engine operating speed. See also alpha range.

alpha particle — A positively charged nuclear particle that has the same mass as the nucleus of a helium atom. Alpha particles consist of two protons and two neutrons.

alpha range — The pitch of a turbopropeller system that maintains a constant RPM of the engine in flight idle conditions.

alpha transistor operation — A measure of emitter-to-collector current gain in a transistor that is connected in a common-base amplifier circuit. The alpha of a junction transistor is never greater than one; its output is always less than its input.

alpha wave detector — A device used to measure and display alpha brain waves.

alpha waves — Waves produced by the human brain when it is relaxed.

alphanumeric — Consisting of numbers and letters.

alphanumeric display — Letters and numerals used to show identification, altitude, beacon code, and other information concerning a target on a radar display.

alternate aerodrome [ICAO] — An aerodrome to which an aircraft may proceed when it becomes either impossible or inadvisable to proceed to or to land at the aerodrome of intended landing.
Note: The aerodrome from which a flight departs may also be an enroute or a destination alternate aerodrome for the flight.

alternate airport — An airport at which an aircraft may land if a landing at the intended airport becomes inadvisable.

altimeter — An instrument that indicates flight altitude by sensing pressure changes and displaying altitude in feet or meters.

altimeter setting — The barometric pressure reading used to adjust a pressure altimeter for variations in existing atmospheric pressure or to the standard altimeter setting (29.92 inches of mercury, 1013.2 hectopascals or 1013.2 millibars).

altimeter setting indicator — A precision aneroid barometer used to determine the local current altimeter setting.

altitude — The height of a level, point, or object measured in feet Above Ground Level (AGL) or from Mean Sea Level (MSL).
a. MSL Altitude — Altitude expressed in feet measured from mean sea level.
b. AGL Altitude — Altitude expressed in feet measured above ground level.
c. Indicated Altitude — The altitude as shown by an altimeter. On a pressure or barometric altimeter it is altitude as shown uncorrected for instrument error and uncompensated for variation from standard atmospheric conditions.

altitude engine — A reciprocating aircraft engine having a rated takeoff power that is producible from sea level to an established higher altitude.

altitude readout — An aircraft's altitude, transmitted via the Mode C transponder feature, that is visually displayed in 100-foot increments on a radar scope having readout capability.

altitude reservation — Airspace utilization under prescribed conditions normally employed for the mass movement of aircraft or other special user requirements which cannot otherwise be accomplished. ALTRVs are approved by the appropriate FAA facility.

altitude restriction — An altitude or altitudes, stated in the order flown, which are to be maintained until reaching a specific point or time. Altitude restrictions may be issued by ATC due to traffic, terrain or other airspace considerations.

altitude restrictions are cancelled — Adherence to previously imposed altitude restrictions is no longer required during a climb or descent.

altocumulus (Ac) — Mid-level principle cloud type occurring in layers or patches, the elements of which appear as small fleecy, rounded clouds. Altocumulus can contain supercooled water droplets or ice crystals.

altocumulus castellanus — Altocumulous clouds that show vertical development resembling the turrets of a castle. Clouds with vertical development indicate instability at the altitude of the clouds.

alumina — An oxide of aluminum (Al_2O_3). Alumina occurs in nature in the form of corundum, emery, sapphires, or bauxite, a type of clay.

aluminium — A British term for aluminum.

aluminizing — 1. A form of corrosion protection for steel parts. 2. A metal coating process that bonds either a corrosion-resistant or a wear-resistant surface to a base metal. Older aircraft used aluminum coating for hot section parts.

aluminum — A metallic chemical element with a symbol of Al and an atomic number of 13. Aluminum is a bluish, silvery-white metal that is lightweight, malleable, and ductile. It is the chief metal used in aircraft construction. It is produced from the clay bauxite, which is a form of aluminum oxide. In its natural form, aluminum is soft and weak, but it can be alloyed with copper, magnesium, manganese, and zinc to give it strength. Aluminum is a good conductor of both electricity and heat and is a good reflector of heat and light. Pure aluminum is highly resistant to corrosion.

aluminum alloy — Pure aluminum to which one or more alloying elements has been added to increase its hardness, toughness, durability, and resistance to fatigue.

aluminum electrolytic capacitor — An electrolytic (liquid dielectric) capacitor that has aluminum plates. The "plates" can be windings of aluminum sheet crimped together at the ends of the spirals in order to lessen the inductance of the capacitor. Electrolytic capacitors must be used on direct current only. If subjected to alternating current they will heat and possibly explode.

aluminum oxide — A compound of aluminum and oxygen (Al_2O_3). It is extremely hard and is used as an abrasive.

aluminum paste — Extremely small flakes of aluminum metal suspended in a substance to make a paste. Aluminum paste is mixed with clear dope to make aluminum-pigmented dope. Aluminum dope is applied over clear dope on aircraft fabric to prevent ultraviolet rays of the sun from damaging the clear dope and fabric underneath.

aluminum welding — The welding of aluminum and aluminum alloys used in aircraft construction using equipment and techniques acceptable to the Federal Aviation Administration.

aluminum wool — Shavings of aluminum metal that are formed into a pad. Aluminum wool can be used to remove corrosion products from aluminum alloy parts and also to smooth out minor scratches from the surface of aluminum sheets or tubing.

aluminum-pigmented dope — Clear aircraft dope in which extremely tiny flakes of aluminum metal are suspended. When sprayed on aircraft fabric, the flakes leaf out to form an opaque covering, protecting the fabric and clear dope from the harmful effects of the sun's ultraviolet rays.

amalgam — A mixture of different elements. Often used to indicate a mixture of mercury with some other metal.

amalgamate — To combine, join, or mix ingredients.

amber — A hard, yellowish, translucent, fossilized tree resin. Often used in jewelry.

ambient — The condition of the atmosphere as it exists at the time of observation.

ambient air — See ambient.

ambient pressure — The pressure of the air that surrounds an object.

ambient temperature — The temperature of the air that surrounds an object.

ambiguity — Something that does not have a clear meaning.

American Society of Testing Materials (ASTM) — A not-for-profit organization that provides a global forum for the development and publication of voluntary consensus standards for materials, products, systems, and services. Many standards in the aircraft industry are ASTM standards.

American Standards — Dimensional standards for fasteners that are issued by the American Standard Association.

American Wire Gauge (AWG) — The standard used for measuring the diameter of round wires and the thickness of non-ferrous metal sheets. Also referred to as the Brown and Sharpe gauge.

AMM (aircraft maintenance manual) A manual developed by the aircraft manufacturer that includes information prepared for the AMT or technician who performs work on units, components, and systems while they are installed on the airplane. It is normally supplied by the manufacturer and approved by the FAA as part of the original process of certification. It will contain the required instructions for continued airworthiness that must accompany each aircraft when it leaves the factory. An Aircraft Maintenance Manual can also be a manual developed by a Part 125 operator as part of their specific operating manual. As such the FAA does not specifically approve the manual. Also referred to as an MMM or Manufacturers Maintenance Manual.

ammeter — An electrical measuring instrument used to measure electron flow in amperes. Ammeters that measure very small rates of flow are called milliammeters (thousandths of an ampere) or microammeters (millionths of an ampere).

ammeter shunt — A low-resistance resistor installed in parallel with an ammeter to allow the meter to read a flow of current that exceeds the current limit of the instrument. The ammeter, acting as a millivoltmeter, measures the voltage drop across the shunt, and indicates, on a scale, the amount of current flowing through the circuit.

ammonia — An invisible gas made up of one atom of nitrogen and three atoms of hydrogen (NH_3). Ammonia becomes a liquid at -28°F, and it freezes at -107°F. Ammonia is used to case-harden steel by a process called nitriding.

amorphous — Without shape.

ampere (A) — A measure of electron flow. One ampere is equal to a flow of one coulomb (6.28 billion billion electrons) past a point in one second. One ampere is the amount of current that can be forced through one ohm of resistance by a pressure of one volt.

ampere turn — A measure of magnetomotive force (mmf) of an electromagnet. It is the force produced when one ampere of current flows through one turn of wire in a coil. One ampere turn is equal to 1.26 gilberts.

ampere-hour capacity — A measure or rating of a battery that indicates the capacity of electrical energy the battery can supply. One ampere-hour is the product of the current flow in amperes, multiplied by the length of time, in hours, that the battery can supply this current.

ampere-hour meter — An electrical measurement instrument that measures the rate of current flow per unit of time.

amphibian — An aircraft designed to land on and take off from either land or water.

amphibious floats — Floats that can be attached to an aircraft to allow it to operate from either land or water. Retractable wheels are mounted in the floats and can be extended for operation on the land.

amplification — The increase in either voltage or current that takes place in a device or in an electrical circuit.

amplification factor — The ratio of the output amplitude of an electrical or electronic circuit to its the input amplitude.

amplifier — An electronic device that increases the amplitude of a signal relative to the amplitude of its input.

amplitude — The magnitude or amount a value changes from its at-rest condition, or its normal condition, to its maximum condition. In wave motion, one half the distance between the wave crest and the wave trough.

amplitude modulation — A system of varying the amplitude in a radio-frequency carrier wave so that it can carry information.

AMS specifications — Specifications for aircraft components that conform to established engineering and metallurgical practices in the aircraft industries. AMS specifications are developed by the SAE Aeronautics Committee.

AN aeronautical standard drawings — Dimensional standards for aircraft fasteners developed by the Aeronautical Standards Group. AN is the part number prefix for all fasteners that are described in these drawings.

AN fittings — A series of fittings for flared tubing, using a 37° flare angle and having a small shoulder between the ends of the threads and the beginning of the flare cone. Included in the listing of parts whose design and material have been approved by both the United States Air Force and Navy and are acceptable for use in civilian and military airplanes.

AN hardware — Standard hardware items such as bolts, nuts, washers, etc., whose design and material has been approved by both the United States Air Force and Navy and are acceptable for use in civilian and military airplanes.

A-N radio range — An early, obsolete navigational aid. Two antennas radiate signals that are heard on a low-frequency receiver as an "N" (dah-dit) on one side of the desired track and an "A" (dit-dah) on the other. When on track, the signal received sounds as a steady tone.

anaerobic resin — A single-component polyester resin that hardens when all air is restricted from it.

analog — A physical variable that keeps a fixed relationship with another variable as it changes. For example, the position of the hands of a clock keep a fixed relationship with time. It is because of this relationship that we can tell the time of day by knowing the positions of the hands of the clock. The position of the clock hands is an analog of time.

analog computer — An electronic computer that operates by converting different levels of voltage or current into numerical values.

analog data — Data represented by a continuously varying voltage or current.

analog-to-digital conversion — An alteration that changes analog information into a digital form.

analyzer, engine — A portable or permanently installed instrument, whose function is to detect, locate, and identify engine operating abnormalities such as those caused by a faulty ignition system, detonation, sticking valves, poor fuel injection, etc.

anchor light — A white light displayed on boats or seaplanes indicating that they are anchored.

anchor nut — A nut riveted or welded to a structure in such away that a screw or bolt can be screwed into it. An anchor nut does not have to be held with a wrench to keep it from turning.

AND gate — A logic device whose output is high only if all inputs are high..

anemometer — An instrument that measures the velocity of moving air. One type of anemometer uses a series of hemispherical metal cups mounted on arms on a shaft. The air blows the cups and rotates the shaft. A counter measures and converts this into wind speed that may be displayed in feet per second, meters per second, kilometers per hour, miles per hour, or knots. Other anemometers convert wind speed to a reading by use of impellers or by measuring the effect of moving air on a hot wire.

anemometer — An instrument for measuring wind speed.

aneroid — 1. A sealed flexible container that expands or contracts in relation to the surrounding air pressure. It is used in an altimeter or a barometer to measure the pressure of the air. 2. A thin disc-shaped box or capsule, usually metallic, which is partially evacuated of air and sealed. It expands or contracts with changes of the surrounding air or gas.

aneroid barometer — An instrument for measuring atmospheric pressure, its key component is a partially

21

evacuated cell which changes dimensions in proportion to the change in atmospheric pressure.

angle — A geometric figure formed by two lines or two plane surfaces extending from the same point.

angle drill — A drilling tool in which the twist drill is held at an angle to the spindle of the drill motor.

angle of attack — **1.** The acute angle formed between the relative wind striking an airfoil and the zero lift line of the airfoil. The chord line of the airfoil is often substituted for the zero-lift line. **2.** (Absolute) The angle of attack of an airfoil, measured from the attitude of zero lift. **3.** (Critical) The angle of attack at which the flow about an airfoil changes abruptly as shown by corresponding abrupt changes in the lift and drag. **4.** (For infinite aspect ratio) The angle of attack at which an airfoil produces a given lift coefficient in a two-dimensional flow. Also referred to as "effective angle of attack." **5.** (Turbine compressor) The acute angle formed between the chord line of the compressor blades and the direction of the air that strikes the blades.

angle of attack indicator system — Detects the local angle of attack of the aircraft from a point on the side of the fuselage and furnishes reference information to an angle-of-attack indicator.

angle of azimuth — An angle measured radially and horizontally clockwise from north (0°).

angle of departure — In communications, the angle between a transmitter's signal propagation and a horizontal plane.

angle of head — In countersunk heads, the included angles of the conical under portion or bearing surface, usually 82° or 100°.

angle of incidence — 1. The acute angle that the wing chord makes with the longitudinal axis of the aircraft. **2.** The angle at which blades are set into the compressor disk. A fixed angle in all cases except the variable pitch fan. Angles set for optimum airflow at altitude cruise and RPM.

angle of refraction — The angle between a refracted beam as it passes through the refracting material (e.g., water) and a line perpendicular to the surface of the refracting material.

angle of roll — The angle through which an aircraft must be rotated about its longitudinal axis in order to bring its lateral axis into the horizontal plane. Also referred to as the angle of bank.

angle of stabilizer setting — The acute angle between the longitudinal axis of an airplane and the chord of the stabilizer. The angle is positive when the leading edge is higher than the trailing edge.

angle of wing setting — The acute angle between the plane of the wing chord and the longitudinal axis of the airplane. The angle is positive when the leading edge is higher than the trailing edge. Also referred to as angle of incidence.

angle of yaw — The acute angle between the direction of the relative wind and the plane of symmetry of an aircraft.

angled gearbox — In helicopters, the same function as a transfer gearbox. Receives its name because the driveshaft is angled, usually 90° up toward the main rotor.

angular acceleration — The rate at which a rotating object increases its rotational speed.

angular measurement — The measured rotational displacement between two lines that project from the same point.

angular momentum — The product of an object's mass directed along a rotating axis.

angular type piston pump — A pump with an angular housing that causes a corresponding angle to exist between the cylinder block and the drive shaft plate to which the pistons are attached. It is this angular configuration of the pump that causes the pistons to stroke as the shaft is turned.

angular velocity — **1.** The rate of change of an angle as a shaft rotates. Expressed in revolutions per minute or radians per second. **2.** The velocity of an object located at a given distance of, and rotating about, a center point. Expressed in radians per second.

anhydrous — A material that does not contain water.

anion — A negative ion that moves toward an anode in an electrolysis process.

anisotropic — In composites, fibers are placed in different directions to respond to the stresses applied in different directions.

anneal — To soften by means of heat treatment.

annealed wire — Wire softened by heat treatment. Necessary because the process of drawing the wire through dies causes it to be work hardened.

annealing — A method of heat treatment in which a metal is softened, losing some of its hardness. See also annealing process.

annealing process — Heating of an alloy to a temperature called solid solution temperature. This is followed by allowing it to cool slowly at a controlled rate through its critical range for the purpose of inducing softness. This results in the removal of former heat-treatment strain hardening and internal stresses.

annual inspection — A complete inspection of an aircraft and engine, required by the Federal Aviation

Regulations, to be accomplished every twelve calendar months on all certificated aircraft. Only an A&P technician holding an Inspection Authorization can conduct an annual inspection.

annual rings — The rings that appear in the end of a tree log. The more rings there are, and the closer they are together, the stronger the wood.

annular combustor — 1. Annular refers to ring shaped. Therefore, a ring-shaped or cylindrical one-piece combustion liner inside a combustion outer case. 2. A cylindrical one-piece combustion chamber, sometimes referred to as a single basket-type combustor.

annular, basket type — Combustion chamber. One of several basic types used in turbine engines. It consists of a housing and a liner. The liner is a one-piece shroud (combustion chamber) extending all the way around the outside of the turbine shaft housing. Fuel is sprayed from nozzles mounted around a full manifold into the inner liner of the combustor. Here it is mixed with air from the compressor and burned.

annular, can type — Combustion chamber. One of several basic types used in turbine engines. Each of the can-type combustion chambers consist of an outer case or housing, within which there is a perforated stainless steel combustion chamber liner or inner liner. Interconnected tubes join each can for flame propagation that spreads combustion during the initial starting operation.

annulus — Opening between two concentric rings, e.g. the space between the compressor disk and outer case could be referred to as the compressor annulus.

annunciator panel — A set of warning lights in direct view of a pilot in a cockpit. The lights are identified by the name of the system they represent, and they are usually covered with a colored lens to show the meaning of the condition they announce. Red lights are used to indicate a dangerous condition, amber lights show that some system is armed or active, and green lights represent a safe condition. Sometimes referred to as a master warning system.

annunciator system — See annunciator panel.

anode — 1. The positive plate of an electrochemical combination, such as a battery or electroplating tank. Electrons leaving the anode cause it to be less negative, or positively charged. When electrons leave an anodic material, the chemical composition of the anode changes from a metal to a salt caused by the reaction with the electrolyte. In the process, the anode is corroded or eaten away. 2. The electrode in a vacuum tube or a semiconductor diode to which the electrons travel after they leave the cathode. The anode in a vacuum tube (an electron tube) is called the plate. The anode of a semiconductor diode is the end that is made of P-type material, and it is not marked (the cathode has the mark). In the diode symbol, the anode is shown by the arrowhead.

anode current — The current measured at the anode of an electronic device.

anode of a chemical cell — Within a battery (chemical cell), electrons flow from the anode to the cathode. This creates an area of less negative (more positive) charge. (Externally, the electrons flow from the cathode to the anode.)

anodic — The positive component of an electrolytic cell.

anodizing — The formation of a hard, unbroken film of aluminum oxide on the surface of an aluminum alloy. This film is electrolytically deposited by using the aluminum as the anode and chromic acid as the electrolyte.

anoxia — A severe case of hypoxia (a lack of oxygen), which can cause permanent damage to the brain.

antenna — An electrical circuit designed to radiate and receive electromagnetic energy. Antennas vary in shape and design depending upon the specific purpose of the antenna and the frequency to be transmitted or received.

antenna coupler — A transformer used between a radio receiver or transmitter and the antenna to optimize the amount of power passed between them.

antenna current — A measurement of radio-frequency current in an antenna.

antenna duplexer — A device that allows two transmitters to simultaneously use a single antenna.

antenna lens — A device used to focus microwaves onto a microwave antenna.

antenna matching — The process of matching the impedance of a radio antenna with the impedance of the transmission line that carries the signal from the radio transmitter to the antenna.

antenna wire — A wire with a low electrical resistance and a high tensile strength. Copperweld, which is a form of wire that has copper plated over a core of strong steel wire, is often used as antenna wire.

antiblush thinner — A slow-drying thinner that is used in conditions of high humidity to prevent blushing of the aircraft dope.

anticollision light — A flashing light on the exterior of the aircraft used to increase the visibility of the aircraft.

anticyclone — An area of high atmospheric pressure that has a closed circulation and when viewed from above, the circulation is clockwise in the Northern Hemisphere and counterclockwise in the Southern Hemisphere.

anticyclone — An area of high atmospheric pressure which has a closed circulation that is anticyclonic, i.e., as viewed from above, the circulation is clockwise in the Northern Hemisphere, counterclockwise in the Southern Hemisphere.

anticyclonic flow — In the Northern Hemisphere the clockwise flow of air around an area of high pressure and a counterclockwise flow in the Southern Hemisphere.

antidetonant fluid — A fluid such as a water/alcohol mix that enables more power to be obtained from the engine when injected into the fuel/air stream. The fluid itself does not increase the engine power; it merely replaces formerly excess fuel, allowing for a cooler operating engine. The water/alcohol dissipates heat more rapidly than fuel.

antidotes — A viable alternative to hazardous attitudes in aeronautical decision-making. Each of the five hazardous attitudes has an associated antidote, which should be memorized and employed to minimize their effects.

antidrag wire — A diagonal, load-carrying member of a Pratt truss wing. It runs from the rear spar inboard to the front spar outboard, and it opposes tensile loads tending to pull forward on the wing.

antifreeze — A chemical added to a liquid for the purpose of lowering its freezing point.

antifriction bearings — Ball or roller bearings that have a special low drag quality.

antiglare paint — A black or dark paint that dries to a dull or matte finish. It is applied to a surface to prevent glare from impairing the aircraft crew's vision.

anti-icing — The prevention of the formation of ice on a surface. Ice may be prevented by using heat or by covering the surface with a chemical that prevents water from reaching the surface. Anti-icing should not be confused with deicing, which is the removal of ice after it has formed on the surface.

anti-icing equipment — Aircraft equipment used to prevent structural icing.

anti-icing fluid — A fluid comprised of alcohol and glycerin, used to prevent the formation of ice on the leading edge of propellers, in the throat of a carburetor, or on the windshield.

anti-icing system — 1. Any system or method used to provide heat or supply anti-icing fluid to critical external surfaces in order to prevent ice formation. 2. A system in a gas turbine engine in which some of the hot compressor bleed air is routed through the engine air inlet system to warm it and prevent ice from forming.

antiknock rating — The rating of fuel that refers to the ability of the fuel to resist detonation.

antileak check valve — A check valve used to prevent oil tank seepage to lower portions of the lube system during periods on engine inactivity. Anti-leak check valves hold the oil in the tank against the pull of gravity. When the oil pump puts a low pressure on the check valve, it opens and allows oil to flow from the tank into the engine.

antimissile missile — A missile used to destroy another missile.

antimony — A silvery, metallic element with a symbol of Sb and an atomic number of 51. Often used as an alloying agent with lead for use in lead-acid batteries.

antipropeller end — The end of an engine away from the propeller.

anti-servo tab — An adjustable tab attached to the trailing edge of a stabilator that moves in the same direction as the primary control. It is used to make the stabilator less sensitive.

antiskid system — A system of controls for aircraft brakes that releases the hydraulic pressure to the brake in the event the wheel begins to lock up or skid.

antitear strips — Strips of aircraft fabric that are laid over the wing rib under the reinforcing tape before the fabric is stitched.

antitorque pedals — In rotorcraft, the pedals used to control the pitch of the tail rotor or air diffuser in a NOTAR® system. The pilot controls the pitch of the anti-torque rotor or the diffuser to position the helicopter about its vertical axis.

antitorque rotor — In rotorcraft, a rotor turning in a plane perpendicular to that of the main rotor and parallel to the longitudinal axis of the fuselage. It is used to control the torque of the main rotor and to provide movement about the yaw axis of the helicopter.

antiwindmilling brake — A friction brake fitted as an accessory to the main gearbox of some older engines. Seldom seen today. Note that the turboprops include a brake in their propeller mechanisms, but it is not the friction brake described here.

anvil — A hard-faced block on which a material is hammered or shaped.

anvil cloud — The top portion of a cumulonimbus (Cb) consisting primarily of cirrus cloud and a spread-out area that, together, resemble an anvil.

anxiety — Mental discomfort that arises from the fear of anything, real or imagined. May have a potent effect on actions and the ability to learn from perceptions.

aperiodic damping — Damping that prevents an object from over swinging, or moving past its at-rest position. Also referred to as dead-beat damping.

aperiodic-type compass — A magnetic compass in which the floating magnet assembly is fitted with damping vanes to increase the period of its oscillations.

aperture — A device used to control the bandwidth of an antenna.

API scale — A scale that has been developed by the American Petroleum Institute to measure the specific gravity of a liquid.

apogee — The point at which an orbiting vehicle is the greatest distance from the center of the object it is circling.

apparent power — In an AC circuit, the product of RMS (Root Mean Square) current and RMS voltage, expressed in volt-amperes

apparent weight — The weight of the object plus or minus any outside influence. The apparent weight of an object immersed in water would be the weight of the object minus any buoyancy. The apparent weight of an object in an elevator would be the weight of the object plus or minus the effects of acceleration due to the elevator moving upward or downward.

appliance — Any instrument, equipment, mechanism, part, apparatus, or accessory, including communications equipment, used or intended to be used in operating or controlling an aircraft in flight. An appliance is installed or attached to the aircraft, but is not part of the engine, airframe, or propeller.

applicability — Something that applies to and/or affects another.

application — A basic level of learning where the student puts something to use that has been learned and understood.

application step — The third step of the teaching process, where the student performs the procedure or demonstrates the knowledge required in the lesson. In the telling-and-doing technique of flight instruction, this step consists of the student doing the procedure while explaining it.

approach — The flight of an airplane just preceding the landing.

approach clearance — Authorization by ATC for a pilot to conduct an instrument approach. The type of instrument approach for which a clearance and other pertinent information is provided in the approach clearance when required.

approach control — A terminal air traffic control facility providing approach control service.

approach control facility — A terminal ATC facility that provides approach control service in a terminal area.

approach control service — Air traffic control service provided by an approach control facility for arriving and departing VFR/IFR aircraft and, on occasion, enroute aircraft. At some airports not served by an approach control facility, the ARTCC provides limited approach control service.

approach fix — From a database coding standpoint, an approach fix is considered to be an identifiable point in space from the intermediate fix (IF) inbound. A fix located from the initial approach fix (IAF) to the intermediate fix is considered to be associated with the approach transition or feeder route.

approach gate — An imaginary point used within ATC as a basis for vectoring aircraft to the final approach course. The gate will be established along the final approach course 1 mile from the final approach fix on the side away from the airport and will be no closer than 5 miles from the landing threshold.

approach light system (ALS) — An airport lighting facility which provides visual guidance to landing aircraft by radiating light beams in a directional pattern by which the pilot aligns the aircraft with the extended centerline of the runway on his final approach for landing. Condenser Discharge Sequential Flashing Lights/Sequenced Flashing Lights may be installed in conjunction with the ALS at some airports.

approach lights — High intensity lights along the approach end of an instrument runway that aid the pilot in the transition from instruments to visual flight at the end of an instrument approach.

approach sequence — The order in which aircraft are positioned while on approach or awaiting approach clearance.

approach speed — The recommended speed contained in aircraft manuals used by pilots when making an approach to landing. This speed will vary for different segments of an approach as well as for aircraft weight and configuration.

approach with vertical guidance (APV) — An instrument approach based on a navigation system that is not required to meet the precision approach standards of ICAO Annex 10 but provides course and glidepath deviation information. Raro-VNAV, LDA with glidepath, LNAV/VNAV and LPV are APV approaches.

approach/departure control service — An air traffic control service provided by an approach control facility for arriving and departing VFR/IFR aircraft and, on occasion, enroute aircraft.

appropriate ATS authority [ICAO] — The relevant authority designated by the State responsible for providing air traffic services in the airspace concerned. In the United States, the "appropriate ATS authority" is the Program Director for Air Traffic Planning and Procedures, ATP-1.

appropriate authority — **1.** Regarding flight over the high seas: the relevant authority is the State of Registry. **2.** Regarding flight over other than the high seas: the relevant authority is the State having sovereignty over the territory being overflown.

appropriate obstacle clearance minimum altitude — Any of the following: Minimum IFR Altitudes MIA, Minimum Enroute IFR Altitude MEA, Minimum Obstruction Clearance Altitude MOCA, Minimum Vectoring Altitude MVA.

appropriate terrain clearance minimum altitude — Any of the following: Minimum IFR Altitudes MIA, Minimum Enroute IFR Altitude MEA, Minimum Obstruction Clearance Altitude MOCA, Minimum Vectoring Altitude MVA.

approved — Unless used with reference to another person, means approved by the Administrator.

approved data — Data which may be used as an authorization for the techniques or procedures for making a repair or an alteration to a certificated aircraft. Approved data may consist of such documents as Advisory Circular 43.13-1B and -2A, Manufacturer's Service Bulletins, a manufacturer's kit, instructions, Airworthiness Directives, or specific details of a repair issued by the engineering department of the manufacturer.

approved inspection system — A maintenance program consisting of the inspection and maintenance necessary to maintain an aircraft in airworthy condition in accordance with approved Federal Aviation Administration practices.

approved parachute — A parachute manufactured under a type certificate or a technical standard order.

approved pilot school — Pilot schools that: are approved by the FAA, must conduct flight and ground training under specific guidelines in FAR Part 141, and meet rigid operational requirements. Graduates of these schools are permitted certification with less total flight experience than that specified in Part 61.

approved repair station — A facility approved by the Federal Aviation Administration for certain types of repair to certificated aircraft.

approved type certificate — An approval issued by the Federal Aviation Administration for the design of an airplane, engine, or propeller. This certifies that the product has met at least the minimum design standards.

approximate — Close to, but not exactly correct. Located close together.

apron — A defined area on an airport or heliport intended to accommodate aircraft for purposes of loading or unloading passengers or cargo, refueling, parking, or maintenance. With regard to seaplanes, a ramp is used for access to the apron from the water.

Aqua-dag — A non-petroleum lubricant used for components in an oxygen system. Oil or other petroleum products cannot be used with oxygen system components because of petroleum's propensity to ignite when in contact with high concentrations of oxygen.

Arabic numerals — The numbers 1, 2, 3, 4, 5, 6, 7, 8, 9, 0, and any combination thereof.

aramid — A fiber that is Aromatic Polyamide. Kevlar® is the brand name which is manufactured by Dupont. There are other companies which also weave aramid fabrics for aircraft use.

arbor for balancing a prop — A low-friction spindle or axle used to support the center of a propeller during maintenance procedures to ensure the blades of a propeller all weigh the same.

arbor press — A press with either a mechanically or hydraulically operated ram. Often used to press bearings into their race.

arc — 1. A portion of the circumference of a circle. 2. A sustained luminous discharge of electricity across a gap. 3. The track over the ground of an aircraft flying at a constant distance from a navigational aid by reference to distance measuring equipment (DME).

arc cosine — Inverse cosine function, that is, the angle for which the cosine is calculated. Also written as arc cos or $\cos^{-1}$

arc lamp — A source of light produced by an electric arc. The arc is produced when electrons flow through ionized gases between two electrodes.

arc sine — Inverse sine function, that is, the angle for which the sine is calculated. Also written as arc sin or $\sin^{-1}$

arc tangent — Inverse tangent function, that is, the angle for which the tangent is calculated. Also written as arc tan or $\tan^{-1}$

arc welding — Welding in which the heat required to melt the metal is produced by an electric arc.

arch — A normally curved structure that spans an opening. In architecture, it is self supporting with lower members supporting higher members.

Archimedes' principle — The principle of buoyancy, which states that a body immersed in a fluid is buoyed up with a force equal to the weight of the fluid it displaces.

arctic air — Air with its origins in an arctic region. Normally much colder than the region it subsequently passes over.

arctic airmass — An airmass with characteristics developed mostly in winter over Arctic surfaces of ice and snow. Surface temperatures are basically, but not always, lower than those of polar air.

area — The number of square units in a surface.

area control center [ICAO] — An air traffic control facility primarily responsible for ATC services being provided IFR aircraft during the enroute phase of flight. The U.S. equivalent facility is an air route traffic control center (ARTCC).

area navigation (RNAV) — A system that provides enhanced navigational capability to the pilot. RNAV equipment can compute the airplane position, actual track and ground speed and then provide meaningful information relative to a route of flight selected by the pilot. Typical equipment will provide the pilot with distance, time, bearing and crosstrack error relative to the selected "TO" or "active" waypoint and the selected route. Several distinctly different navigational systems with different navigational performance characteristics are capable of providing area navigational functions. Present day RNAV includes INS, LORAN, VOR/DME, and GPS systems. Modern multi-sensor systems can integrate one or more of the above systems to provide a more accurate and reliable navigational system. Due to the different levels of performance, area navigational capabilities can satisfy different levels of required navigational performance (RNP). The major types of equipment are:

a. VORTAC referenced or Course Line Computer (CLC) systems, which account for the greatest number of RNAV units in use. To function, the CLC must be within the service range of a VORTAC.

b. OMEGA/VLF, although two separate systems, can be considered as one operationally. A long-range navigation system based upon Very Low Frequency radio signals transmitted from a total of 17 stations worldwide.

c. Inertial (INS) systems, which are totally self-contained and require no information from external references. They provide aircraft position and navigation information in response to signals resulting from inertial effects on components within the system.

d. MLS Area Navigation (MLS/RNAV), which provides area navigation with reference to an MLS ground facility.

e. LORAN-C is a long-range radio navigation system that uses ground waves transmitted at low frequency to provide user position information at ranges of up to 600 to 1,200 nautical miles at both enroute and approach altitudes. The usable signal coverage areas are determined by the signal-to-noise ratio, the envelope-to-cycle difference, and the geometric relationship between the positions of the user and the transmitting stations.

f. GPS is a space-based radio positioning, navigation, and time-transfer system. The system provides highly accurate position and velocity information, and precise time, on a continuous global basis, to an unlimited number of properly equipped users. The system is unaffected by weather, and provides a worldwide common grid reference system.

area navigation (RNAV) approach configuration — **a. Standard T** — An RNAV approach whose design allows direct flight to any one of three initial approach fixes (IAF) and eliminates the need for procedure turns. The standard design is to align the procedure on the extended centerline with the missed approach point (MAP) at the runway threshold, the final approach fix (FAF), and the initial approach/ intermediate fix (IAF/IF). The other two IAFs will be established perpendicular to the IF. **b.** Modified T — An RNAV approach design for single or multiple runways where terrain or operational constraints do not allow for the standard T. The "T" may be modified by increasing or decreasing the angle from the corner IAF(s) to the IF or by eliminating one or both corner IAF's. **c.** Standard I — An RNAV approach design for a single runway with both corner IAFs eliminated. Course reversal or radar vectoring may be required at busy terminals with multiple runways. **d.** Terminal Arrival Area (TAA) — The TAA is controlled airspace established in conjunction with the Standard or Modified T and I RNAV approach configurations. In the standard TAA, there are three areas: straight-in, left base, and right base. The arc boundaries of the three areas of the TAA are published portions of the approach and allow aircraft to transition from the en route structure direct to the nearest IAF. TAAs will also eliminate or reduce feeder routes, departure extensions, and procedure turns or course reversal. **1)** Straight-In Area — A 30NM arc centered on the IF bounded by a straight line extending through the IF perpendicular to the intermediate course. **2)** Left Base Area — A 30NM arc centered on the right corner IAF. The area shares a boundary with the straight-in area except that it extends out for 30NM from the IAF and is bounded on the other side by a line extending from the IF through the FAF to the arc. **3)** Right Base Area — A 30NM arc centered on the left corner IAF. The area shares a boundary with the straight-in area except that it extends out for 30NM from the IAF and is bounded on the other side by a line extending from the IF through the FAF to the arc.

area navigation [ICAO] — A method of navigation which permits aircraft operation on any desired flight path within the coverage of station-referenced navigation aids or within the limits of the capability of self-contained aids, or a combination of these.

area navigation high route — An area navigation route within the airspace extending upward from, and including, 18,000 feet MSL to flight level 450.

area navigation low route — An area navigation route within the airspace extending upward from 1,200 feet above the surface of the earth to, but not including, 18,000 feet MSL.

area navigation/RNAV — A method of navigation that permits aircraft operations on any desired course within the coverage of station referenced navigation signals or within the limits of self contained system capability.

area of decision — The most critical time for an engine failure to occur in a multi-engine airplane, which is just after liftoff while the airplane is accelerating and climbing over immediate obstacles. The area of decision exists between the point where obstacle clearance speed is reached and landing gear retracted, and the point where the single-engine best angle-of-climb speed (Vxse) is reached. An engine failure in this area requires an immediate decision to abort takeoff or to continue climbing. To make an intelligent decision in case of a failure within this area of decision, one must consider aircraft performance with the following conditions: runway length, obstruction height, field elevation, density altitude, air temperature, headwind, takeoff weight, and pilot proficiency.

area weight — In composites, the weight of fiber reinforcement per unit area (width × length) of tape or fabric.

areas of operation — Phases of the practical test arranged in a logical sequence within the PTS. Each area of operation has several task listings to be evaluated during the flight. Areas of operation are based on the corresponding flight proficiency requirements in the FARs.

argon — A gaseous element with a symbol of Ar and an atomic number of 18. It is an inert gas often used in welding, to shield the molten metal from oxidation.

ARINC — An acronym for Aeronautical Radio, Inc., a corporation largely owned by a group of airlines. ARINC is licensed by the FCC as an aeronautical station and contracted by the FAA to provide communications support for air traffic control and meteorological services in portions of international airspace.

arithmetic — A mathematical system dealing with real numbers. Consists of addition, subtraction, multiplication, and division.

arithmetic sum — The result of adding the absolute values (without regard to sign) of real numbers.

arm — The horizontal distance in inches between the reference datum line and the center of gravity of an object. If the object is behind the datum, the arm is positive, and if in front, the arm is negative.

armature — The rotating or moving component of a magnetic circuit. Motors, generators, and alternators have armatures.

armature core — In an electrical apparatus (motor, generator, servo, etc.), a laminated soft iron about which the armature coils are wound.

armature gap — In a motor or generator, the space between the armature and the field poles.

armature reaction — The distortion of the generator field flux by the current flowing in the windings of the armature. Armature reaction causes the brushes to pick up current from the armature at a point on the commutator where there is a potential difference. This causes the brushes to spark.

armed — A condition in which a device is made ready for actuation.

Armed Forces — The Army, Navy, Air Force, Marine Corps, and Coast Guard, including their regular and reserve components and members serving without component status.

Army Aviation Flight Information Bulletin — A bulletin that provides air operation data covering Army, National Guard, and Army Reserve aviation activities.

aromatic gasoline — Gasoline that has had its anti-detonation characteristics improved by blending in aromatic additives such as benzene, toluene, or xylene.

aromatics — Chemical compounds such as toluene, xylene, and benzene. These aromatics may be blended with gasoline to improve its anti-detonation qualities. Aromatics are also known to soften rubber hoses, diaphragms, and seals used in fuel metering system components.

arresting system — A safety device consisting of two major components, namely, engaging or catching devices, and energy absorption devices for the purpose of arresting both tail hook and/or nontail hook-equipped aircraft. It is used to prevent aircraft from overrunning runways when the aircraft cannot be stopped after landing or during aborted takeoff. Arresting systems have various names; e.g., arresting gear, hook device, wire barrier cable.

arrival aircraft interval — An internally generated program in hundredths of minutes based upon the AAR. AAI is the desired optimum interval between successive arrival aircraft over the vertex.

arrival center — The ARTCC having jurisdiction for the impacted airport.

arrival delay — A parameter which specifies a period of time in which no aircraft will be metered for arrival at the specified airport.

arrival routes (ICAO) — Routes on an instrument approach procedure by which aircraft may proceed from the enroute phase of flight to the initial approach fix.

arrival sector — An operational control sector containing one or more meter fixes.

arrival sector advisory list — An ordered list of data on arrivals displayed at the PVD/MDM of the sector which controls the meter fix.

arrival sequencing program — The automated program designed to assist in sequencing aircraft destined for the same airport.

arrival stream filter (ASF) — An on/off filter that allows the conflict notification function to be inhibited for arrival streams into single or multiple airports to prevent nuisance alerts.

arrival time — The time an aircraft touches down on arrival.

arsenic — A chemical element with the symbol of As and an atomic number of 33. An extremely small amount of arsenic is alloyed with silicon or germanium to make N-type semiconductor material.

articulated connecting rod — A link rod that connects the pistons in a radial engine to the master rod. There is one less articulated rod than there are cylinders in each row of cylinders in a radial engine since one piston is attached to the master rod.

articulated rod assembly — See articulated connecting rod.

articulated rotor — In rotorcraft, a rotor system in which each of the blades is connected to the rotor hub in such a way that it is free to change its pitch angle, and move up and down and fore and aft in its plane of rotation.

artificial aging — A process of increasing the strength of a heat-treated aluminum alloy by holding it at an elevated temperature after it has been solution heat treated. Artificial aging is Also referred to as precipitation heat treatment. See also precipitation heat treatment.

artificial feel — A force feedback or cushioning effect that is used in the automatic flight control systems of some aircraft. Artificial feel produces an opposition to the movement of the controls proportional to the aerodynamic load that is acting on the control surfaces.

artificial horizon — A flight instrument in which a bar or a display is held in a constant relationship with the earth's horizon. An artificial horizon provides the pilot with a visual reference when the natural horizon is not visible.

artificial radio antenna — A device attached to the output of a radio transmitter when adjusting the transmitter. The artificial antenna has the same impedance as the antenna, but the radio signal put into the artificial antenna is not radiated.

asbestos — A fiber of magnesium silicate that has a high resistance to fire. It has good insulating qualities.

A-scan — A display on which the indications appear similar to the capital letter "A" along a horizontal time line. Found in the displays of ultrasonic test equipment and sometimes radar.

ASCII (American Standard Code for Information Interchange) — A format of data used as a standard to transmit information between computers. Consists of 128 characters that are each made up of seven binary bits. Pronounced "as-kee."

ash — The solid residue remaining after a substance has burned.

ashless dispersant oil (AD) — A popular mineral oil used as a lubricant for aircraft reciprocating engines. Ashless dispersant oil does not contain any metallic ash-forming ingredients. It has additives in it that keep the contaminants that form in the oil dispersed throughout the oil so they will not join together and clog the oil filters.

aspect ratio — 1. The ratio of the span of the wing to its chord. The aspect ratio of a tapered wing is found by dividing the square of the wing span by its area. 2. In composites, the ratio of length to diameter of a reinforcing fiber.

asphalt — A heavy, brownish-black mineral that is found in crude oil. Asphalt is used as a base for some acid-resistant paints.

aspheric — In optics, an element such as a lens that is not perfectly spherical.

aspirator — A device for producing suction or for moving or collecting materials by suction.

assembly — The fitting together of parts to form a complete structure or unit.

assembly break — A joint in the structure of an aircraft that is formed when two sub-assemblies are removed from their jigs and joined together to form a single unit.

assembly drawing — An aircraft drawing that shows a group of parts laid out in the relationship they will have when they are assembled.

assembly line — An arrangement of work stations in an aircraft factory that allows certain functions to be performed on the aircraft being manufactured.

assigned radio frequency — The frequency of a carrier signal produced by a radio transmitter. The frequency is assigned by the Federal Communications

Commission for a particular transmitter or particular type of transmission.

associated conditions — The conditions found on performance charts in the pilot's operating handbook (POH). The information on the chart is based on the given associated conditions and any other conditions do not provide the same performance.

astable — Not stable. An astable electronic device has two conditions of temporary stability, but no condition of permanent stability.

astable multivibrator — A free-running multivibrator. They contain two devices such as transistors to control the flow of electrons. When one transistor is conducting, the other is not conducting. The two transistors alternate between a condition of conducting and not conducting.

A-stage — In composites, the intermediate stage in the reaction of the two parts of the resin system after being mixed. The resin system reacts to heat by softening. The resin in a prepreg material is usually in the B-stage before the curing process.

astern — A location or direction to the rear or rear part of an aircraft.

ASTM specifications — Standards developed by the American Society for Testing Materials. See also American Society of Testing Materials (ASTM).

astrocompass — An optical device used to determine direction by reference to the stars (celestial navigation).

astrodome — A hemispherical window, usually in the ceiling of an aircraft whereby a navigator can use an astrocompass to determine heading or a sextant to determine position.

astronaut — A person who travels beyond the earth's atmosphere.

astronautics — The field of science that deals with space flight and includes the design, construction, and operation of space vehicles and their related support activities.

astronomical twilight — The period of time before sunrise or after sunset when the sun is not more than 18° below the horizon.

astronomical unit — One astronomical unit is the average distance between the earth and the sun. Approximately 93 million miles.

astrophysics — The field of science that deals with the physical characteristics of heavenly bodies. Characteristics include their mass, density, temperature, size, luminosity, and origin.

asymmetrical — A condition in which the shape of a body is NOT the same on both sides of its center line.

asymmetrical airfoil — An airfoil that has a different curve above the chord line than it has on the underside.

asymmetrical lift — A condition of unequal lift produced by the rotor disc of a helicopter in motion. The advancing blade travels at its peripheral speed plus the forward speed while the airspeed of the retreating blade is the peripheral speed minus the forward speed.

asymmetrical thrust — In single-engine airplanes, left turning tendencies are caused by both asymmetrical propeller loading (P-factor) and torque. Multi-engine airplanes have an even greater tendency to turn during climbs or other high angle-of-attack maneuvers due to the additional engine and propeller. The position of these engines in relation to the airplane's centerline causes asymmetrical propeller loading to exert a more forceful turning moment.

asynchronous — In electronics, signals generated as required by the equipment instead of by a timing device.

ATA system — Airline Transport Association's standardized format for maintenance manuals.

ATC advises — Used to prefix a message of noncontrol information when it is relayed to an aircraft by other than an air traffic controller.

ATC assigned airspace — Airspace of defined vertical/lateral limits, assigned by ATC, for the purpose of providing air traffic segregation between the specified activities being conducted within the assigned airspace and other IFR air traffic.

ATC clearance — An authorization by air traffic control, for the purpose of preventing collision between known aircraft, for an aircraft to proceed under specified traffic conditions within controlled airspace. The pilot-in-command of an aircraft may not deviate from the provisions of a visual flight rules (VFR) or instrument flight rules (IFR) air traffic clearance except in an emergency or unless an amended clearance has been obtained. Additionally, the pilot may request a different clearance from that which has been issued by air traffic control (ATC) if information available to the pilot makes another course of action more practicable or if aircraft equipment limitations or company procedures forbid compliance with the clearance issued. Pilots may also request clarification or amendment, as appropriate, any time a clearance is not fully understood, or considered unacceptable because of safety of flight. Controllers should, in such instances and to the extent of operational practicality and safety, honor the pilot's request. FAR Part 91.3(a) states: "The pilot-in-command of an aircraft is directly responsible for, and is the final authority as to, the operation of that aircraft." The pilot is responsible to request an amended clearance if ATC issues a clearance that would cause a pilot to deviate from a rule or regulation, or in the pilot's opinion, would place the aircraft in jeopardy.

ATC clears — Used to prefix an ATC clearance when it is relayed to an aircraft by other than an air traffic controller.

ATC instructions — Directives issued by air traffic control for the purpose of requiring a pilot to take specific actions such as "Turn left heading two five zero," "Go-around," or "Clear the runway."

ATC preferred route notification — URET CCLD (User Request Evaluation Tool Core Capability Limited Deployment) notification to the appropriate controller of the need to determine if an ATC preferred route needs to be applied, based on destination airport.

ATC preferred routes — Preferred routes that are not automatically applied by Host.

ATC requests — Used to prefix an ATC request when it is relayed to an aircraft by other than an air traffic controller.

ATCSCC delay factor — The amount of delay calculated to be assigned prior to departure.

athodyd — Aero-thermodynamic duct. An open tube shaped to produce thrust when fuel is ignited inside. Fuel is added to incoming air as the athodyd moves through the air at a high speed. This burning causes air expansion that speeds up the air and produces thrust.

athwartships — At right angles to the longitudinal axis of an aircraft.

atmosphere — The envelope of gases that surrounds the Earth.

atmospheric electricity — Any electrical phenomena in the Earth's atmosphere, usually occurring between clouds or between clouds and ground.

atmospheric moisture — The presence of H_2O in any one or all of the states: water vapor, water, or ice.

atmospheric noise — A noise (commonly referred to as "static") on the radio. It is caused by the discharge of atmospheric electricity between clouds or between clouds and the ground.

atmospheric pressure — The pressure that is exerted on the surface of the earth by the air that surrounds the earth. Within standard conditions, at sea level, the atmospheric pressure is 14.69 PSI.

atmospheric sounding — A measure of atmospheric variables aloft, usually pressure, temperature, humidity and wind.

atmospheric stability — Describes a state in which an air parcel will resist vertical displacement, or once displaced (for instance by flow over a hill) will tend to return to its original level.

atom — The smallest particle of an element, consisting of a positively charged nucleus orbited by one or more negatively charged electrons.

atomic fission — The splitting of the nucleus of a heavy atom such as uranium. This creates nuclei of two or more lighter atoms releasing a tremendous amount of energy in the form of heat and light.

atomic fusion — The combining of two nuclei of light atoms to create a single nucleus of a heavy atom. This releases a tremendous amount of energy in the form of heat and light.

atomic number — The number of protons in the nucleus of an atom. The atomic number of an element determines the position of the element in the periodic table of elements.

atomic weight — The average weight of a single atom in a chemical element.

atomize — To reduce a liquid to a fine spray.

atomizing nozzle — A device that converts a liquid to a fine spray.

ATR racking system — A widely accepted size and mounting standard for airborne electronic equipment.

ATS route — A specified route designated for channeling the flow of traffic as necessary for the provision of air traffic services. NOTE: The term "ATS Route" is used to mean variously, airway, advisory route, controlled or uncontrolled route, arrival or departure route, etc.

attention — An element that helps your students gain interest in the lesson.

attenuate — To diminish the intensity or strength of a signal, for instance of a radar beam. In electronics, the volume control changes (attenuates) the sound level without changing any other characteristics of the sound.

attenuation — **1.** A reduction in the strength of a signal, the flow of current, flux, or other energy in an electronic system. **2.** In radar meteorology, any process which reduces intensity of radar signals.

attenuator — A device in electronics that changes the level of a signal's intensity without changing any other characteristics of the signal.

attitude — **1.** The position of an aircraft as determined by the relationship of its axes and a reference, usually the earth's horizon. **2.** A personal motivational predisposition to respond to persons, situations, or events in a given manner that can, nevertheless, be changed or modified through training as a sort of mental shortcut to decision making.

attitude gyro — A gyro-actuated flight instrument that displays the attitude of the aircraft relative to the earth's horizon.

attitude indicator — An instrument that gives a pilot an artificial reference of the airplane's attitude in pitch and roll with respect to the earth's surface.

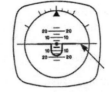

attitude management — The ability to recognize one's own hazardous attitudes and the willingness to modify them as necessary through the application of an appropriate antidote thought.

attraction — A force acting mutually between particles of matter, tending to draw them together.

audio frequency (af) — Frequency in a range that can normally be heard by the human ear, ranging from about 16 Hz to 16,000 Hz.

audio sweeps — Tones in the audio band that increase and decrease in frequency (sweep). A siren is an audio sweep as is the signal transmitted by an Emergency Locator Transmitter (ELT).

audio-frequency amplifier — An electronic amplifier that is capable of amplifying alternating current with a frequency in the range of human hearing.

audio-frequency oscillator — An electronic device used to create an alternating current in the audio range, generally from 16-16,000 Hz.

audio-frequency vibrations — Vibrations in a range heard by the human ear, generally from 16-16,000 Hz.

audiovisual system — A communication system utilizing both sound and pictures. Film and television are examples of audiovisual systems.

auditory learners — Students who acquire knowledge best by listening.

augmentation — Any designed method of increasing the basic thrust of an engine for a short period such as for takeoff or combat; usually accomplished by coolant injection or afterburning.

augmenter — A device used to draw cooling air through an engine cowling by the use of a low pressure created by the rapid moving exhaust gases as they leave the exhaust pipes.

augmenter tube — A specially shaped tube mounted around the exhaust tail pipe of an aircraft reciprocating engine. When exhaust gases flow through the augmenter tube, they produce a low pressure in the engine compartment that draws cooling air into the compartment.

aural radio range — A navigational system in which the pilot positions the airplane in accordance with audio signals over a radio receiver. The A-N range is an example of an aural radio range. As a note, such a range is now obsolete.

aural warning system — A bell or horn type warning sounding system that alerts the pilot during an abnormal takeoff condition, landing condition, pressurization condition, mach-speed condition, an engine or wheel well fire, calls from the crew call system, and calls from the SELCAL system.

aurora — An eerie nighttime illumination that consists of streamers or arches of light appearing in the upper atmosphere of a planet's magnetic polar regions. Auroras are caused by the emission of light from atoms excited by electrons as they accelerate along the planet's magnetic field lines. Two such auroras are the aurora borealis, which occurs in the Northern Hemisphere, and the aurora australis, which occurs in the Southern Hemisphere.

austenite — A supersaturated solution of carbon that exists in iron at high temperatures.

authorized — Approved by the Federal Aviation Administration to act or perform certain functions.

authorized instructor — According to FAR Part 61.3, a person who holds a valid ground instructor certificate issued under Part 61 or Part 143 when conducting ground training in accordance with the privileges and limitations of his or her ground instructor certificate; or a person who holds a current flight instructor certificate issued under Part 1 when conducting ground training or flight training in accordance with the privileges and limitations of his or her flight instructor certificate; or a person authorized by the Administrator to provide ground training or flight training under SFAR No. 58, or Part 61, 121,135, or 142 when conducting ground training or flight training in accordance with that authority.

auto lean — A lean fuel-air mixture whose ratio is kept constant by an automatic mixture control in the carburetor.

auto rich — A rich fuel-air mixture whose ratio is kept constant by an automatic mixture control in the carburetor.

autoclave — A pressure vessel in which the air inside can be heated to a high temperature and the pressure raised to a high value in order to decrease the amount of time needed to cure plastic resins. Improves the quality of the curing process.

autoclave molding — In composites, a manufacturing method that uses an autoclave. The composite assembly is placed into an autoclave at 50 to 100 psi to consolidate the laminate by removing entrapped air and excess resin.

autofeather — A portion of a propeller control system that causes the propeller to feather automatically if the engine on which it is mounted is shut down during flight.

autogyro — Gyroplane. Autogiro. A heavier-than-air rotorcraft whose rotor is spun by aerodynamic forces that act on the blades. No engine power is used to drive the rotor in flight.

autoland approach — An autoland approach is a precision instrument approach to touchdown and, in some cases, through the landing rollout. An autoland approach is performed by the aircraft autopilot which is receiving position information and/or steering commands from onboard navigation equipment.

Note: Autoland and coupled approaches are flown in VFR and IFR. It is common for carriers to require their crews to fly coupled approaches and autoland approaches (if certified) when the weather conditions are less than approximately 4,000 RVR.

automated information transfer — A precoordinated process, specifically defined in facility directives, during which a transfer of altitude control and/or radar identification is accomplished without verbal coordination between controllers using information communicated in a full data block.

automated mutual-assistance vessel rescue system — A facility which can deliver, in a matter of minutes, a surface picture (SURPIC) of vessels in the area of a potential or actual search and rescue incident, including their predicted positions and their characteristics.

automated problem detection (APD) — An Automation Processing capability that compares trajectories in order to predict conflicts.

automated problem detection boundary (APB) — The adapted distance beyond a facilities boundary defining the airspace within which URET CCLD (User Request Evaluation Tool Core Capability Limited Deployment) performs conflict detection.

automated problem detection inhibited area (APDIA) — Airspace surrounding a terminal area within which APD is inhibited for all flights within that airspace.

automated radar terminal systems — The generic term for the ultimate in functional capability afforded by several automation systems. Each differs in functional capabilities and equipment. ARTS plus a suffix Roman numeral denotes a specific system. A following letter indicates a major modification to that system. In general, an ARTS displays for the terminal controller aircraft identification, flight plan data, other flight associated information; e.g., altitude, speed, and aircraft position symbols in conjunction with his radar presentation. Normal radar co-exists with the alphanumeric display. In addition to enhancing visualization of the air traffic situation, ARTS facilitate intra/inter-facility transfer and coordination of flight information. These capabilities are enabled by specially designed computers and subsystems tailored to the radar and communications equipment and operational requirements of each automated facility. Modular design permits adoption of improvements in computer software and electronic technologies as they become available while retaining the characteristics unique to each system.

a. ARTS II — A programmable nontracking, computer aided display subsystem capable of modular expansion. ARTS II systems provide a level of automated air traffic control capability at terminals having low to medium activity. Flight identification and altitude may be associated with the display of secondary radar targets. The system has the capability of communicating with ARTCCs and other ARTS II, IIA, III, and IIIA facilities.

b. ARTS IIA — A programmable radar-tracking computer subsystem capable of modular expansion. The ARTS IIA detects, tracks, and predicts secondary radar targets. The targets are displayed by means of computer-generated symbols, ground speed, and flight plan data. Although it does not track primary radar targets, they are displayed coincident with the secondary radar as well as the symbols and alphanumerics. The system has the capability of communicating with ARTCCs and other ARTS II, IIA, III, and IIIA facilities.

c. ARTS III — The Beacon Tracking Level (BTL) of the modular programmable automated radar terminal system in use at medium to high activity terminals. ARTS III detects, tracks and predicts secondary radar-derived aircraft targets. These are displayed by means of computer-generated symbols and alphanumeric characters depicting flight identification, aircraft altitude, ground speed and flight plan data. Although it does not track primary targets, they are displayed coincident with the secondary radar as well as the symbols and alphanumerics. The system has the capability of communicating with ARTCC's and other ARTS III facilities.

d. ARTS IIIA — The Radar Tracking and Beacon Tracking Level (RT&BTL) of the modular, programmable automated radar terminal system. ARTS IIIA detects, tracks and predicts primary as well as secondary radar-derived aircraft targets. This more sophisticated computer-driven system upgrades the existing ARTS III system by providing improved tracking, continuous data recording and fail-soft capabilities.

automated surface observation system (ASOS) — A surface weather observing system implemented by the National Weather Service, the Federal Aviation Administration and the Department of Defense. It is designed to support aviation operations and weather forecast activities. The ASOS provides continuous minute-by-minute observations and performs the basic observing functions necessary to generate an aviation routine weather report (METAR) and other aviation weather information. ASOS information may be transmitted over a discrete VHF radio frequency or the voice portion of a local NAVAID.

automated terminal tracking system (ATTS) — ATTS is used to identify the numerous tracking systems including ARTS IIA, ARTS IIE, ARTS IIIA, ARTS IIIE, STARS, and M-EARTS.

automated unicom — Provides completely automated weather, radio check capability and airport advisory information on an Automated UNICOM system. These systems offer a variety of features, typically selectable by microphone clicks, on the UNICOM frequency. Availability will be published in the Airport/Facility Directory and approach charts.

automated weather observing system (AWOS) — An automated weather reporting system which transmits local real-time weather data directly to the pilot.
a. AWOS-A only reports altimeter setting.
b. AWOS-1 Usually reports altimeter setting, wind data, temperature, dewpoint and density altitude.
c. AWOS-2 Reports same as AWOS-1 plus visibility.
d. AWOS-3 Reports the same as AWOS-2 plus cloud/ceiling data.

automatic — An operation that occurs by itself. An automatic operation has all the necessary signals built into it so that it will perform its function without any external decision or manipulation.

automatic adjusters — A portion of the return spring system of disk brakes that maintain a constant clearance between the disk and the brake linings as the lining wears. The automatic adjusters allow the brake piston to move back a specific amount each time the brake is released.

automatic altitude report — An aircraft's altitude, transmitted via the Mode C transponder feature, that is visually displayed in 100-foot increments on a radar scope having readout capability.

automatic altitude reporting — That function of a transponder which responds to Mode C interrogations by transmitting the aircraft's altitude in 100-foot increments.

automatic carrier landing system — U.S. Navy final approach equipment consisting of precision tracking radar coupled to a computer data link to provide continuous information to the aircraft, monitoring capability to the pilot, and a backup approach system.

automatic direction finder (ADF) — An aircraft radio navigation system which senses and indicates the direction to a L/MF nondirectional radio beacon (NDB) ground transmitter. Direction is indicated to the pilot as a magnetic bearing or as a relative bearing to the longitudinal axis of the aircraft depending on the type of indicator installed in the aircraft. In certain applications, such as military, ADF operations may be based on airborne and ground transmitters in the VHF/UHF frequency spectrum.

automatic flight control system (AFCS) — A complete instrument system including an automatic pilot that is coupled with radio navigation and approach equipment. An aircraft with an AFCS can be flown in a completely automatic mode.

automatic frequency control (AFC) — A circuit in a radio receiver that keeps the receiver tuned to a desired frequency within specific limits.

automatic gain control (AGC) — An electronic circuit within a radio receiver that keeps the output volume relatively constant.

automatic mixture control (AMC) — A device in a fuel metering system (carburetor or a fuel injection system) that keeps the fuel-air mixture ratio constant as the density of the air changes with altitude.

automatic pilot — An automatic flight control system that keeps an aircraft in level flight or on a set course. Automatic pilots can be given course guidance by the human pilot, or they may be coupled to a radio navigation signal.

Automatic Terminal Information Service (ATIS) — The continuous broadcast of recorded noncontrol information in selected terminal areas. Its purpose is to improve controller effectiveness and to relieve frequency congestion by automating the repetitive transmission of essential but routine information, e.g., "Los Angeles information Alfa. One three zero zero Coordinated Universal Time. Weather, measured ceiling two thousand overcast, visibility three, haze, smoke, temperature seven one, dew point five seven, wind two five zero at five, altimeter two niner niner six. I-L-S Runway Two Five Left approach in use, Runway Two Five Right closed, advise you have Alfa."

automatic volume control (AVC) — A circuit that regulates a volume relative to the strength of the input signal.

automatic-reset circuit breaker — An electrical circuit protection device that opens a circuit during a current overload and then automatically resets itself and restores the circuit when the overload is no longer present. Automatic reset circuit breakers are installed in some electric motors, but they are not approved for use in aircraft.

automaton — A machine that operates in a preset and relatively unsupervised manner. A robot is an example of an automaton.

autopilot — Those units and components that furnish a means of automatically controlling the aircraft.

Autopilot Flight Director System (AFDS) — An integrated system of autopilot, flight instrumentation, and feedback to flight controls. Individual components of the system are controlled from the autopilot system.

autorotation — A rotorcraft flight condition in which the lifting rotor is driven entirely by action of the air when the rotorcraft is in motion.
 a. Autorotative Landing/Touchdown Autorotation — Used by a pilot to indicate that the landing will be made without applying power to the rotor.
 b. Low Level Autorotation — Commences at an altitude well below the traffic pattern, usually below 100 feet AGL and is used primarily for tactical military training.
 c. 180 degrees Autorotation — Initiated from a downwind heading and is commenced well inside the normal traffic pattern. "Go-around" may not be possible during the latter part of this maneuver.

autorotation region — The portion of the rotor disk of a helicopter that produces an autorotative force.

autorotative force — An aerodynamic force that causes an autogyro or helicopter rotor to turn when no power is supplied to it.

Autosyn — A remote-indicating instrument or system based upon the synchronous-motor principle, in which the angular position of the rotor of one motor at the measuring source is duplicated by the rotor of the indicator motor. Used in fuel-quantity or fuel-flow measuring systems, position-indicating systems, etc.

autotransformer — A single winding transformer having a carbon brush that can tap off any number of turns for the secondary. It produces variable voltage AC output.

auxiliary — A supplement, or an addition, to a main unit.

auxiliary flight surfaces — Lift-modifying devices on an airfoil, such as flaps, slots, or slats.

auxiliary fuel pump — An electrically operated fuel pump used to supply fuel to the engine for starting, takeoff, or in the case of engine-driven fuel pump failure. Auxiliary fuel pumps are also used to pressurize the fuel in the line to the engine-driven fuel pump in order to prevent vapor lock at altitude when the fuel is warm.

auxiliary hydraulic pump — A hydraulic pump that is used as an alternate source of hydraulic pressure for emergencies, or to produce hydraulic pressure when the aircraft engines are not in operation.

auxiliary ignition units — An auxiliary ignition system that facilitates engine starting. The auxiliary device is incorporated in the ignition system to provide a high ignition voltage. Reciprocating engine starting systems normally include one of the following types of auxiliary starting systems: booster coil, induction vibrator, impulse coupling, or other specialized retard-breaker and vibrator-starting systems.

auxiliary power unit (APU) — A gas turbine, usually located in the aircraft fuselage, whose purpose is to provide either electrical power, air pressure for starting main engines, or both. Similar in design to ground power units.

auxiliary pump — Any pump used as an alternate source or in an emergency situation. An auxiliary pump provides assistance or support to the main pump.

auxiliary rotor — A rotor that serves either to counteract the effect of the main rotor torque on a rotorcraft or to maneuver the rotorcraft about one or more of its three principal axes.

auxiliary view — A view used in an aircraft drawing that is made at some angle to one of the three views of an orthographic drawing. It is used to show details that would not otherwise be visible.

available landing distance (ALD) — The amount of runway remaining when operating at a controlled airport where land and hold short operations (LAHSO) are in effect. Pilots may accept such a clearance provided that the pilot-in-command determines that the aircraft can safely land and stop within the ALD.

avalanche diode — Zener diode. Semiconductor diode in which reverse breakdown voltage current causes the diode to develop a constant voltage. Used for voltage regulation.

avalanche voltage — The reverse voltage required to cause a zener diode to break down.

average value — Sine wave alternating current. 0.637 times the peak value of alternating current or voltage, measured from the zero reference line.

aviation — The branch of science, business, or technology that deals with any part of the operation of machines that fly through the air.

aviation maintenance technician (AMT) — An aviation maintenance technician is a person certificated by the FAA to work on aircraft structures and engines. Also referred to as an A&P (airframe and powerplant) mechanic.

aviation medical examiner (AME) — A person to whom the FAA delegates authority to examine applicants for, and holders of, airman medical certificates to determine whether or not they meet the medical standards for its issuance. AMEs also issue or deny airman medical certificates based upon whether or not they meet the applicable medical standards. The medical standards are found in Title 14 of the Code of Federal Regulations Part 67.

aviation medicine — A special field of medicine that establishes standards of physical fitness for pilots.

aviation physiology — The study of the biological processes and functions of life and living matter, and the physical and chemical aspects involved in those processes as it relates to aviation.

aviation safety counselors — Volunteers within the aviation community who share their technical expertise and professional knowledge as a part of the FAA Aviation Safety Program.

aviation safety reporting system (ASRS) — The ASRS investigates the causes underlying a reported event, and incorporates each report into a database, which provides information for research regarding aviation safety and human factors. Each report is held in strict confidence and the FAA cannot use ASRS information in enforcement actions against those who submit reports.

aviation shears — Compound-action hand shears used for cutting sheet metal. They normally come in sets of three: one that cuts to the left, one that cuts to the right, and one that cuts straight. Also referred to as aviation snips.

aviation turbulence — Bumpiness in flight.

aviation weather — Specific characteristics of weather that pertain to flight or to the operation of aircraft.

Aviation Weather Service — A service provided by the National Weather Service (NWS) and FAA which collects and disseminates pertinent weather information for pilots, aircraft operators, and ATC. Available aviation weather reports and forecasts are displayed at each NWS office and FAA FSS.

avionics — AVIation electrONICS. Airborne electronic equipment.

Avogadro's principle — A principle of physics, that states that under equal pressure and temperature, equal volumes of all gases will contain equal numbers of molecules.

avoirdupois weight — The system of weight for measuring the weight of most substances. In avoirdupois weight, one pound is equal to 453.6 grams.

awl — A sharp-pointed tool that is used to make holes in soft materials such as leather, plastic, or wood.

axes of an aircraft — Three mutually perpendicular imaginary lines about which an aircraft is free to rotate. The longitudinal axis passes through the aircraft's center of gravity from front to rear. The lateral axis passes through the aircraft's center of gravity from wingtip to wingtip, and the vertical axis passes through the aircraft's center of gravity from top to bottom.

axial — Motion along a real or imaginary straight line on which an object supposedly or actually rotates. Turbine engine centerline.

axial flow — The straight-through flow of a fluid. In an axial-flow compressor, the air flows through the compressor parallel to the engine and the stages of compression do not essentially change the direction of the flow.

axial flow compressor — In gas turbine engines, a compressor with the airflow parallel to the axis of the engine. The numerous compressor stages raise pressure of air but essentially make no change in direction of airflow.

axial flow turbine engine — A turbine engine in which the air is compressed by a series of rotating airfoils. The airflow through the engine is essentially in a straight line.

axial load — A load on a bearing that acts parallel to the shaft supported in the bearings. The thrust load produced by a propeller is an axial load. Axial loads are usually carried by ball bearings or by tapered roller bearings and are carried into the engine crankcase through the thrust bearing.

axial loading — An aerodynamic force that tries to move the compressor forward. Axial loading is supported in a gas turbine engine by ball bearings.

axial velocity — The speed of a gas traveling in a circular path. Usually applied to the speed combustion gases as they rotate through the combustor section of a turbine engine.

axial winding — In composites, a manufacturing method using filament-winding equipment. In axial winding the filaments are parallel to the axis.

axial-centrifugal compressor — A combination axial and centrifugal compressor usually fitted together with the axial portion as the front stages and the centrifugal portion as the rear stage. This is a popular design for smaller turbine engines.

axial-lead resistor — A disparate electronic component that provides a given amount of resistance to a circuit. The wire leads of an axial-lead resistor extend from the ends in a direction that is parallel to the axis of the resistor.

axis — A straight line about which a body can rotate.

axis of symmetry — An imaginary center line about which a body or object is symmetrical in either weight or area.

axis-of-rotation — In rotorcraft, the imaginary line about which the rotor rotates. It is represented by a line drawn through the center of, and perpendicular to, the tip-path plane.

axle — The shaft on which one or more wheels are mounted and about which the wheels are free to rotate.

axonometric projection — A projection used in mechanical drawing that shows a solid rectangular object inclined in a way that three of its faces are visible. An isometric projection is a form of axonometric projection.

azimuth — The angular measurement in a horizontal plane and in a clockwise direction.

B

babbitt metal — An alloy of tin, lead, copper, and antimony used for lining engine bearings. It is used because of its exceptional low-friction qualities.

back — The curved portion of a propeller blade that corresponds to the curved upper surface of an airfoil.

back current — The current that flows in a semiconductor (such as a transistor or diode) when the device is reverse biased.

back plate — A floating plate on which the wheel cylinder and brake shoes attach on an energizing-type brake.

back pressure — **1.** In a reciprocating engine, the pressure caused by the exhaust system that opposes the evacuation of the burned gases from the cylinders. **2.** Pressure in an aft direction when applied to the control wheel or stick of an airplane.

back voltage — Counter-electromotive force generated in a conductor by the action of changing lines of flux cutting across the conductor. As the magnetic field produced by the changing current builds up and decays, it cuts across the conductor and induces a voltage in it. The polarity of this induced voltage is opposite to that of the voltage that caused the original current to flow.

backfire — **1.** In welding, a momentary backward flow of the gases at the torch tip that causes the momentary burning back of a flame into the tip, followed by a snap or pop, then either a relight or the flame extinguishing. A backfire may be caused by touching the tip against the work, by overheating the tip, or by operating the torch at other than recommended pressures. **2.** A burning or explosion within a reciprocating engine when the fuel-air mixture in the induction system is ignited by gases that are still burning inside the cylinder when the intake valve opens.

backfiring — See backfire.

background noise — An unwanted signal found in an electronic device.

backhand welding — The technique of pointing the torch flame toward the finished weld and moving away in the direction of the unwelded area, welding the edges of the joint as it is moved. The welding rod is added to the puddle between the flame and the finished weld.

backing — In meteorology, change of wind direction in a counterclockwise sense in the northern hemisphere (for example, northwest to west) with respect to either space or time; opposite of veering.

backing plate — A reinforcing plate used when making a sheet metal repair. It is often a doubler.

backlash — The clearance measured between the meshing teeth of two gears. Usually units such as selector valves, pumps, etc. are required to have a predetermined amount of backlash. However, too much movement due to backlash can be undesirable.

backlash check — Measuring of the amount of movement of one gear when the gear meshed with it is held rigid. It is an indication of the amount of space between meshed gears.

backsaw — A fine-toothed wood cutting saw that has a stiff lip along its upper edge. Backsaws are used to make cuts across a board.

backscatter — The signal, reflected back to the transmitter/receiver, when a radar beam hits an object. Also referred to as an echo.

back-taxi — A term used by air traffic controllers to taxi an aircraft on the runway opposite to the traffic flow. The aircraft may be instructed to back-taxi to the beginning of the runway or at some point before reaching the runway end for the purpose of departure or to exit the runway.

backup ring — An anti-extrusion ring used on the side of an O-ring packing away from the pressure. This stiff ring, usually made of some material such as Teflon®, prevents the resilient O-ring from being extruded by hydraulic pressure into the space between the cylinder wall and the piston.

bacteria — Microscopic plant life that lives in the water entrapped in fuel tanks. The growth of bacteria in jet aircraft fuel tanks causes a film, which holds water against the aluminum alloy surface. This can result in corrosion.

bactericide — An agent that is used to destroy bacteria.

baffle — **1.** A series of partitions inside an aircraft fuel tank. These baffles have holes in them that allow the fuel to feed to the tank outlet, but they keep the fuel from surging enough to uncover the fuel outlet. **2.** A partition separating the upper portion of the tank that prevents oil pump cavitation as the oil tends to rush to the top of the tank during periods of deceleration. The baffle is fitted with a weighted swivel outlet control valve, which is free to swing below the baffle. The valve is normally open, but closes when the oil in the bottom of the tank rushes to the top of the tank during deceleration. **3.** Sheet metal shields used to direct the flow of air between and around the cylinders of an air-cooled reciprocating engine. Also referred to as cylinder deflectors. **4.** A structure used to impede, regulate, or alter the flow direction of a gas, fluid, or sound.

bag molding — A method of applying pressure to a piece of laminated plastic material so that all of the layers are held in tight contact with each other. The reinforcing material is injected with plastic resin, which is laid up over a rigid mold in as many layers as are needed. A sheet of flexible, air-tight plastic material is placed over the mold and the edges are sealed to form a bag over the part. The entire assembly is then placed in an autoclave for curing. In the absence of an autoclave, a vacuum pump can be attached to the bag, and the air pumped out. The atmospheric pressure pressing on the outside of the bag supplies the needed force.

bag side — In composites, the side of a part that is cured against the vacuum bag.

bagging — In composites, applying an impermeable layer of film over an uncured part and sealing the edges so that a vacuum can be drawn.

bailout bottle — Small oxygen cylinder connected to the oxygen mask supplying several minutes of oxygen. It can be used in case of primary oxygen system failure or if an emergency bailout at high altitude became necessary.

Bakelite — A phenol resin, often used as electrical insulation, made by the Bakelite Corporation.

baking soda — The common term for bicarbonate of soda ($NaHCO_3$).

balance — A state of equilibrium.

balance bridge — A method of measuring resistance. It consists of three known resistances and an unknown resistance. When the known resistances are adjusted so the bridge is balanced, the output is zero. The unknown resistance can then be calculated using the relationship R1:R2=R3:R4. See also Wheatstone bridge.

balance chamber — An internal air chamber in a turbine engine used to absorb some of the compressor axial loading "thrust."

balance checks — A check or series of inspections performed on rotating components after overhaul to statically and dynamically check for correct balancing.

balance point — 1. A point within an object at which the sum of all its components have zero rotation. 2. The point about which an object will balance.

balance pressure torch — A welding torch where the oxygen and acetylene are both fed to the torch at the same pressure. The openings to the mixing chamber for each gas are equal in size, and the delivery of each gas is independently controlled.

balance tab — An auxiliary control mounted on a primary control surface, which automatically moves in the direction opposite the primary control to provide an aerodynamic assist in the movement of the control. Sometimes referred to as a servo tab.

balance, static — 1. A condition of balance which does not involve any dynamic forces. 2. When a body will stand in any position as the result of counterbalancing and/or reducing the heavy portions, it is said to be in standing or static balance.

balanced actuator — A hydraulic or pneumatic actuator having the same area on each side of the actuator piston. Fluid power into the actuator produces the same amount of force in either direction of piston movement.

balanced amplifier — An electronic amplifier that has two output circuits that are equal, but opposite in phase. Also referred to as a push-pull amplifier.

balanced control surface — A primary control surface with an overhang ahead of the hinge line that provides an aerodynamic assist to reduce control pressures.

balanced design — In filament-winding of composites, a winding pattern so designed that the stresses in all filaments are equal.

balanced laminate — In composites, each layer except the 0/90° is placed in plus and minus pairs around the centerline. These plies do not have to be adjacent to each other.

balanced transmission line — A radio frequency transmission line matched in impedance between each conductor and ground, and between conductors and the electronic equipment to which the conductors are connected. Neither conductor is grounded.

balancing — The act of performing a balance procedure using prescribed methods.

ball bearing — An anti-friction bearing consisting of grooved inner and outer races and one or more sets of steel balls held in a sheet metal retainer. Ball bearings can be designed to support thrust loads as well as radial loads.

ball bearing assembly — Consists of grooved inner and outer races, one or more sets of balls, and, in bearings designed for disassembly, a bearing retainer. They are used for supercharger impeller shaft bearings and rocker arm bearings in some engines. Special deep-groove ball bearings are used in aircraft engines to transmit propeller thrust to the engine nose section.

ball check valve — A check valve in a fluid power system that uses a spring-loaded steel ball and a seat to allow flow in one 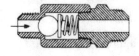 direction only. The ball is forced tightly against its seat by fluid flowing into the valve from the end that contains the spring, thereby stopping the fluid flow through the valve. Fluid flowing into the valve from the ball end forces the ball off its seat, allowing flow through the valve.

ball joint — A flexible expansion joint used in an aircraft engine exhaust system to allow relative movement of the parts as a result of their expansion and contraction.

ball peen hammer — A hammer with one side of its head shaped like a ball.

ballast — 1. Permanently installed weight in an aircraft used to bring the center of gravity into the allowable range. 2. An electrical circuit component designed to stabilize current flow. 3. In gliding, used to describe any system that adds weight to the glider. Performance ballast employed in some gliders increases wing loading using releasable water in the wings (via integral tanks or water bags). This allows faster average cross-country speeds. Trim ballast is used to adjust the flying CG, often necessary for light-weight pilots. Some gliders also have a small water ballast tank in the tail for optimizing flying CG.

ballast lamp — A resistance-type lamp connected into a circuit in a series in order to maintain a constant current. As current increases, the filament gets hotter and creates a higher resistance, lowering the current until a balance (and constant current) is achieved.

ballistic missile — A self-propelled long-range missile, which is guided by preset mechanisms as it goes upward, but is free falling as it comes down.

balloon — A lighter-than-air, non-steerable aircraft that is not engine-driven. Its rising capability comes from gases or hot air that fill the gas envelope.

balsa wood — The light, strong wood of the balsa, a tropical tree. It is sliced across its grain and sandwiched between two face sheets of thin metal or fiberglass to make rigid, lightweight panels.

balun — A transformer used to match a balanced antenna to an unbalanced transmission line.

bamboo — A species of tree consisting of light weight, strong, tubular sections. Used in some early aircraft construction.

banana plug — A mechanism used to make a temporary connection to an electrical circuit. The contacts of a banana plug are springs that have the general shape of a banana. These springs press out against the walls of the banana jack in order to make a low-resistance contact.

band — A range of electro-magnetic frequencies.

band saw — A controllable-speed power saw used to cut wood, plastics, or metal. The band saw blade is in the form of a narrow strip of steel with teeth along one edge. The ends of the blade are welded together to form a continuous loop, and the loop passes over two large wheels, one above the saw table and the other below the table.

band-pass filter — An electronic filter that passes a band of frequencies while rejecting all frequencies above or below the band.

band-reject filter — An electronic filter that rejects a specific band of frequencies while passing those above and below the band.

bandwidth — The difference between the maximum and minimum frequencies in a band.

bank — To incline or tilt an airplane about its longitudinal axis.

bank indicator — A flight instrument consisting of a curved glass tube filled with a liquid similar to kerosene and enclosing a round ball. When the aircraft is horizontal, the ball is located in the lowest part of the tube; as the aircraft banks, gravity holds the ball at the lowest point as the tube rotates from side to side. The tube can be calibrated to show the angle of banking. It can also indicate the relationship between the force of gravity and centrifugal force in a turn. If the bank angle is correct for the rate of turn, the ball will stay in the center of the tube. However, if the angle of bank is too steep for the rate of turn, the ball will roll to the inside of the turn. If the angle of bank is not steep enough for the rate of turn, the ball will roll to the outside of the turn. A bank indicator is built into the face of a turn and slip indicator.

bar — A metric unit of pressure equal to 1,000,000 dynes per square centimeter. Pressure is often measured in meteorological services in millibars, which is $1/_{1,000}$ of a bar. The standard absolute pressure of the atmosphere at sea level is 1013.2 millibars.

bar folder — A forming machine used for making bends or folds along edges of metal sheets. It is best suited for folding small hems, flanges, seams, and edges. Most bar folders have a capacity for metal up to 22 gauge thickness.

bar graph — A graph used to show relationships between different values. Each value in a bar graph is represented by a bar of an appropriate length.

bare conductor — An electrical conductor that is not protected with any type of insulating material.

barnstormers — Early aviation pioneers who traveled from town to town piloting their planes in sight-seeing flights with passengers or in exhibition stunts. Some were clowns and characters, but others were more serious and showed the promises for the future of aircraft transportation. Barnstorming included acrobatics and stunts such as picking up handkerchiefs off the ground with hooks attached to the wing tips etc.

barograph — An instrument used to measure absolute pressure. Barographs are often sealed and carried in an aircraft in order to make a permanent record of the altitude reached by the aircraft.

barometer — An instrument used to measure atmospheric pressure. Used in forecasting weather or measuring the height above sea level. The two principle types are mercury and aneroid.

barometric altimeter — An instrument that measures altitude above mean sea level using an internal aneroid barometer to compare pressure of the ambient air in relation to known lapse rates of pressure as altitude varies with height. Also referred to as a pressure altimeter.

barometric pressure — Pressure existing above zero pressure. Measured in millibars or inches of mercury.

barometric tendency — The change in barometric pressure over a set period of time. For example, "The barometric tendency in the past three hours has been a 1 mb rise" Also referred to as pressure tendency.

barrel — The part of a reciprocating engine cylinder made of a steel alloy forging. The inner surface of the barrel is hardened to resist wear of the piston and the piston rings which bear against it during operation. In some instances the barrel will have threads on the outside surface at one end so that it can be screwed into the cylinder head.

barrel roll — An airplane flight maneuver in which the airplane rolls around a distant visual point with a constant angular displacement from the point throughout the roll.

barriers to effective communication — Things which impede communication, such as lack of common experience, or confusion between the symbol and the symbolized object. Other examples include overuse of abstractions and interference.

base — 1. The center electrode of a transistor. The signal is normally applied to the base. 2. The electrode between the emitter and the collector in a bipolar transistor. 3. A bipolar transistor electrode that normally receives the signal.

base leg — A flight path at right angles to the landing runway off its approach end. The base leg normally extends from the downwind leg to the intersection of the extended runway centerline.

base line — A line used as a basis for measuring.

base metal — The metal to which alloying agents are added.

baseball stitch — A hand stitch similar to that used to sew the cover on a baseball. It is used for hand sewing of aircraft fabric.

BASIC (computer) — A programming language used by digital computers. BASIC is an acronym for Beginners All-purpose Symbolic Instruction Code.

basic empty weight (GAMA) — Standard empty weight (airframe, engines, and all items of operating equipment that have fixed locations and are permanently installed, including fixed ballast, hydraulic fluid, unusable fuel, and full engine oil, in addition to optional and special equipment that has been installed. (Except aircraft certified under CAR Part 3, the standard empty weight does not include oil.)

basic empty weight — The weight of the standard aircraft, operational equipment, unusable fuel, and full operating fluids, including full engine oil.

basic fuel system — The tanks, booster pumps, lines, selector valves, strainers, engine-driven pumps, and pressure gauges that make up an aircraft's basic fuel system.

basic ground instructor — A person authorized by the FAA to provide: ground training in the aeronautical knowledge areas required for issuance of a recreational pilot certificate, private pilot certificate, or associated ratings under Part 61; ground training required for a recreational pilot and private pilot flight review; and recommendation for a knowledge test required for issuance of a recreational pilot certificate or private pilot certificate under Part 61.

basic load — The load on a structural member or part in any condition of static equilibrium of an airplane. When a specific basic load is expressed, the particular condition of equilibrium must be indicated in the context.

basic magneto — A high-voltage generating device in a reciprocating engine. It is adjusted to give maximum voltage at the time the points break and ignition occurs. It must also be synchronized accurately to the firing position of the engine.

basic maneuvers — Straight-and-level flight, climbs, descents, and turns. These four maneuvers form the foundation of the development of all piloting skills. Each flight maneuver, regardless of its complexity, is composed of combinations of the basic maneuvers.

basic need — A perception factor that describes a person's ability to maintain and enhance the organized self.

basic radar service — A radar service for VFR aircraft that includes safety alerts, traffic advisories and limited radar vectoring, as well as aircraft sequencing at some terminal locations.

basic size — The size from which the limits of size are derived by the application of allowances and tolerances.

basic weight — The weight of an aircraft, its power plant, and all of the fixed equipment. It includes unusable fuel and undrainable oil for aircraft not certified under FAR Part 23 (aircraft certified under FAR Part 23 include full oil as part of empty weight).

basket weave — In composites, a woven reinforcement where two or more warp threads go over and under two or more filling threads in a repeat pattern. The basket weave is less stable than the plain weave but produces a flatter and stronger fabric. It is also a more pliable fabric than the plain weave and maintains a certain degree of porosity.

bastard file — A double-cut, metal-working file that has course cutting teeth. There are five grades of cuts from the coarsest to the finest: coarse cut, bastard cut, second cut, smooth cut, and dead-smooth cut.

batch — In composites, material that was made with the same process at the same time having identical characteristics throughout. Also referred to as a lot.

bathtub capacitor — A bathtub-shaped capacitor that is sealed in a metal container.

battery — A device made up of a number of individual cells used to store electricity by converting it into chemical energy. Electrons are caused to flow from one pole, the anode, to another pole, the cathode, by a chemically produced potential difference.

battery analyzer — A transformer rectifier unit used to charge nickel-cadmium batteries. The analyzer has a built-in load bank as well as timers, indicators, and controls for deep-cycling and recharging these batteries.

battery bus — The electrical tie point in an airplane where power from the battery is distributed to the various loads.

battery charger — A power supply that converts alternating current into direct current for charging batteries.

battery charger/analyzer — A power supply that converts alternating current into direct current for charging batteries. The charger/analyzer is a special device with a timer, load bank, and monitoring equipment for complete servicing of aircraft batteries.

battery ignition system — An ignition system that uses a battery as its source of energy, rather than a magneto. This system is similar to that used in an automobile. A cam driven by the engine opens a set of points to interrupt current in a primary circuit. The resulting collapsed magnetic field induces a high voltage in the secondary circuit, which is directed by a distributor to the proper cylinder.

baud — In computers, a measure of data transmission rate. One baud is equal to one bit per second.

bauxite — A clay-like substance that is the source of aluminum. To extract the aluminum, the bauxite is changed into alumina (aluminum oxide). Then the alumina is reduced to metallic aluminum by an electrolytic process.

bay — Any specific compartment in the body of an aircraft. It may also refer to a portion of a truss, or fuselage, the area between adjacent bulkheads, struts, or frame positions.

bayonet — Something that is detachable and can be easily put on or removed.

bayonet exhaust pipe — The elongated and flattened end of the exhaust pipe of a reciprocating engine. It is designed to minimize exhaust noise prevent exhaust valve warpage by maintaining a relatively constant temperature at the exhaust ports of the cylinders.

bayonet exhaust stack — See bayonet exhaust pipe.

bayonet gauge — A dipstick-type gauge used to measure the quantity of a liquid such as oil or hydraulic fluid.

bayonet thermocouple — A thermocouple used to indicate engine temperature. The bayonet probe fits into an adapter that is screwed into the cylinder.

bayonet thermocouple probe — A pickup for cylinder head temperature that presses into an adapter screwed into the side of a cylinder head. Used for measuring cylinder head temperature on an air-cooled aircraft engine.

B-battery or B-power supply — In electronic vacuum tubes, positive plate voltage is called B+ voltage. The source of voltage for these plates is referred to as B-battery or B-power supply.

beacon — See radar, nondirectional beacon, marker beacon, airport rotating beacon, aeronautical beacon, and/or airway beacon.

bead — 1. A trough-like impression formed in a sheet metal member for the purpose of stiffening the member. 2. A raised rounded ridge formed near the end of a piece of rigid tubing. A hose is slipped over the end of the tube, then the hose clamp is installed between the end of the hose and the bead. The bead keeps the tube from being pulled from the hose.

bead heel — On tires, the outer bead edge that fits against the wheel flange.

bead seat area — The highly stressed portion of a wheel where the bead of the tire seats against the wheel.

bead thermistor — A component in a fire detection system that signals the presence of a fire or an overheat condition. The beads in the detector are wetted with a eutectic salt, which possesses the characteristic of suddenly lowering its electrical resistance as the sensing element reaches its alarm temperature. The lowered resistance starts the fire-warning procedure by turning on the fire-warning light and sounding the fire-warning bell.

bead toe — On tires, the inner bead edge closest to the aircraft tire center line.

bead welds — The part of a weld that joins edges of metal parts that have been heated and melted together to form one solid piece when solidified. The bead sticks up above the surface of metal that has been welded. Usually some additional metal is added to the weld, in the form of a wire or rod, to build up the weld seam to a greater thickness than the base metal. The characteristics of a bead on a good weld are: height uniformity, a smooth and uniform ripple on its surface, and an even blend into the base metal.

beads — On tires, steel wires embedded in rubber and wrapped in fabric. The beads anchor the carcass plies and provide firm mounting surfaces on the wheel.

beam — **1.** A supporting structural member in any construction designed to withstand loads in both shear and bending. **2.** A radio signal sent continuously in one direction.

beam power tube — An electron tube that utilizes directed electron beams to add to its power-handling capability. Beam power tubes are power amplifier tubes, rather than a voltage amplifier.

beam radio antenna — A directional radio transmitting antenna that radiates its energy in a narrow beam.

bearing — **1.** An angular measurement of direction from an airplane in flight and a known point. **2.** A surface that supports and reduces friction between moving parts.

bearing area — The cross-section area of the bearing load member on a sample.

bearing burnishing — An aircraft engine run-in process that creates a highly polished surface on new bearings and bushings installed during overhaul. The burnishing is usually accomplished during the first periods of the engine run-in at comparatively slow engine speeds.

bearing cage — A thin sheet-metal separator that holds the bearing rollers equally spaced around the race. The cage should not contact either of the races.

bearing cone — The assembly, which consists of a tapered, hardened steel, cone-shaped bearing race that fits over the axle, the rollers, and the cage that holds the rollers in position.

bearing cup — The steel race of a roller bearing that is shrink-fitted into the bearing cavity of the wheel.

bearing degausser — A device that removes magnetism from bearings and other engine components.

bearing failure — The failure of a riveted joint caused by the sheets tearing at the rivet holes rather than the rivets shearing.

bearing field detector — A device that detects magnetism in rotating engine components.

bearing friction — Friction caused by a bearing.

bearing heater tank — An oil bath heater used to expand a bearing so that it can be hand fitted over its shaft journal.

bearing navigation — The horizontal direction of one object in relation to another object.

bearing pressurizing — The process of increasing air pressure in the bearing pockets by admitting compressor air.

bearing race — The hardened steel surface upon which anti-friction bearings ride.

bearing rollers — Hardened steel rollers that support the wheel. They roll between a hardened steel, cone-shaped race on the axle and a hardened steel race, the cup, inside the wheel.

bearing scratch detector — A hand-held ball bearing tipped tool, which is passed over bearing surfaces. If the ball finds a depression, the bearing is usually rejected.

bearing seal — A device used in turbine engines to seal the lubricating oil in the bearing cavity. Usually located at main bearings.

bearing stack — A group of thrust-type bearings placed one on top of the other to form a stack. Primarily used to allow propeller blades to rotate in the hub under high centrifugal loads.

bearing strain — In composites, the ratio of the deformation of the bearing hole, in the direction of the applied force, to the pin diameter.

bearing strength — The force required to pull a rivet through the edge of the sheet or to elongate the hole. The bearing strength of a material is affected by both its thickness and by the size of the rivet.

bearing stress — In composites, the applied load in pounds divided by the bearing area.

bearing sump — The compartment housing the engine main bearings, formed by bearing seals on either side of the bearing on the shaft. Seals are used to control the inward leakage of gas path air.

bearing support — The inner hub of a major engine case, which is supported by struts and houses a main bearing.

bearing surface — A surface that supports and reduces friction between moving parts or a moving load. Bearing surfaces are ordinarily treated in several ways to decrease the friction between the surface and the moving load.

bearing wheel — An airplane wheel that has tapered, roller-type bearings that consist of a bearing cone, rollers with a retaining cage, and a bearing cup, or outer race. Each wheel has the bearing cup, or race, pressed into place and is often supplied with a hubcap

beat — A low frequency vibration produced when two sources of vibration act on the same object at the same time. In a multi-engine airplane, if two engines have slightly different RPMs, the airframe vibrations caused by these engines will produce a very noticeable beat. This beat is caused by the difference in the frequency of the two vibrations.

beat-frequency oscillator (BFO) — A variable frequency electronic oscillator designed to produce a signal frequency that is mixed with another frequency in order to develop an intermediate frequency or an audio frequency that can be heard.

Beaufort scale — In meteorology, a scale used to describe wind force, ranging from 0 to 12, 0 (zero) represents less than 1 MPH, and 12 represents speeds of more than 72 MPH, or hurricane force.

beef up — To strengthen or reinforce.

beehive spring — A hard steel retaining spring used to hold a rivet set in a pneumatic rivet gun. Prevents the set from being driven out of the gun. It derives its name, beehive, from its shape.

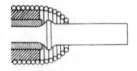

beep button — A switch on the collective control used to trim a helicopter turbine engine by increasing or decreasing the steady state RPM. Also referred to as a beeper button.

beeswax — A substance secreted by bees. It is used to coat rib lacing cord to protect it from moisture and prevent slippage.

before bottom center (BBC) — The degrees of crankshaft travel before the piston stops at the bottom of its stroke.

before top center (BTC) — The degrees of crankshaft travel before the piston reaches the top of its stroke.

behaviorism — Theory of learning that learning is dependent on particular behaviors being positively reinforced by someone other than the student.

bel — A unit used to express the ratio of two values of power. The number of bels is the logarithm to the base 10 of the power ratio.

bell gear — The large stationary gear used in a spur gear-type planetary reduction gearing system.

bellcrank — A double lever in an aircraft control system used to change the direction of motion. Bellcranks are commonly used in an aileron system to change spanwise movement into chordwise movement to move the control surface.

bellmouth — A turbine engine air inlet duct having a flared or convergent shape used to direct air into a gas turbine engine. The shape of the bellmouth increases the efficiency of the incoming air to the engine.

bellows — Circular, pleated, or corrugated capsules or compartments used to measure pressure. They may be either evacuated or filled with a specific pressure of inert gas and exposed to the pressure to be measured. Their dimensional change is measured as the pressure surrounding them varies.

Bellville washer — A lock washer made in the form of a cupped steel washer.

below minimums — Weather conditions below the minimums prescribed by regulation for the particular action involved; e.g., landing minimums, takeoff minimums.

belt frame — A circumferential fuselage frame usually having a channel or hat section.

bench check — A functional check performed on a part that has been removed from an aircraft to determine its condition of serviceability. The equipment is set up on a test bench and operated to find out whether or not it is functioning as it should.

bench plate — A flat, cast iron plate built into a bench used for working sheet metal. Holes in the bench plate support stakes that are used to form the sheet metal.

bench timing — A functional procedure to time a magneto by setting the breaker points, and for checking the rotor for the E-gap position. This process is done in the shop before installation in the engine.

bend allowance — The amount of material actually used in the bend of sheet metal. This amount of metal must be added to the overall length of the layout pattern to assure adequate metal for the bend. Bend allowance depends on four factors: the degree of the bend, the radius of the bend, the thickness of the metal, and the type of metal used. The amount of material in the bend is usually found by using a bend allowance chart.

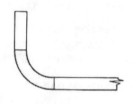

bend allowance chart — A chart used to save time in calculating the bend allowance of sheet metal. Formulas and charts for various angles, radii of bends, material thicknesses, and other factors have been established to make up the chart.

bend radius — The radius of the bend as measured on the inside of the curve.

bend tangent line (BL) — A line made on a sheet metal layout to indicate the beginning of the bend, and the line at which the metal stops curving. All the space between the bend tangent lines is the bend allowance.

bending — The stresses in an object caused by a load being applied to one end while the other is restrained. This results in a tensile load on one side of the material and a compressive load on the other.

bending strength — The resistance of a material to curving under load and bending stresses.

Bendix fuel injection system — A continuous-flow fuel metering system that measures engine air consumption and uses airflow forces to control fuel flow to the engine.

benzene — A colorless, volatile, flammable, aromatic hydrocarbon liquid (C_6H_6). Used as a solvent in aircraft finishing materials.

Bernoulli's principle — In physics, the interrelation between pressure, velocity, and gravitational effects in moving fluids. It states that for the steady flow of a frictionless and incompressible fluid, the total energy (consisting of the sum of the kinetic energy due to the velocity, the potential energy due to elevation in a gravitational field, and the pressure energy given by the pressure divided by the density) is a constant along the flow path. An increase in velocity at constant elevation must therefore be matched by a decrease in pressure. This principle is used to explain the lift of an airfoil, the theory of carburetors, etc.

Bernoulli's theorem — The principle that states static pressure and velocity (RAM) pressure of a gas or fluid passing through a duct (at constant subsonic flow rate) are inversely proportional, i.e. total pressure does not change.

beryllium — A hard metallic chemical element with a symbol of Be and an atomic number of 4. Found in combination with other alloys.

beryllium bronze — An alloy of copper which is combined with approximately 3% beryllium.

best angle-of-climb airspeed — The best angle-of-climb airspeed (V_X) will produce the greatest gain in altitude for horizontal distance traveled.

best economy mixture — The fuel-air mixture used in reciprocating engines to achieve the greatest range of flight. It can only be used with reduced power, as it does not have the additional fuel needed for cooling.

best glide speed (best L/D speed) — The airspeed that results in the least amount of altitude loss over a given distance. This speed is determined from the performance polar. The manufacturer publishes the best glide (L/D) airspeed for specified weights and the resulting glide ratio. For example, a glide ratio of 36:1 means that the glider will lose one (1) foot of altitude for every 36 feet of forward movement in still air at this airspeed.

best power mixture — The fuel-air mixture ratio used to allow the engine to produce its maximum power. The best power fuel-air mixture ratio is richer than the ratio that is used for the best economy. It uses an excess of fuel to provide for cooling.

best rate-of-climb airspeed — The best rate-of-climb airspeed (V_Y) produces the maximum gain in altitude per unit of time.

beta — The current gain of a transistor when it is connected as a grounded emitter amplifier. Beta is the ratio of the change in collector current to the change in base current when the collector voltage is held constant.

beta mode — The reverse pitch area of propeller. Used for slowing the aircraft and maneuvering on the ground. See also alpha mode.

beta particle — A particle emitted from the nucleus of an atom during radioactive decay.

beta range — The pitch range of a turbopropeller system for ground handling and reversing whereby the propeller can be operated to provide either zero or negative thrust.

beta tube — An oil passage and control valve that connects crankcase oil to the propeller hub. Enables variable propeller pitch including beta mode.

bevel — An angle other than a right angle.

bevel gears — A pair of toothed wheels having angled surfaces and whose shafts are not parallel. Bevel gears permit a shaft to drive another shaft that is not parallel to it.

bezel — The rim that holds the glass in an instrument case.

B-H curve — A curve that shows the association between the flux density (B) in a piece of magnetized material and the magnetizing force (H), which is needed to produce the flux density.

biannual — Occurring twice each year.

bias — **1.** A cut, fold, or seam made diagonally to the warp or fill threads of a fabric. **2.** An electrical reference used to establish the operating condition of a semiconductor device or an electron tube. **3.** In composites, a 45° angle to the warp threads. Fabric can be formed into contoured shapes by using the bias.

bias current — The current that flows in the emitter-base circuit of a transistor.

bias voltage — The DC voltage placed on the grid of an electron tube.

bias, forward — The polarity relationship between a power supply and a semiconductor that allows conduction.

bias, reverse — The polarity relationship between a power supply and a semiconductor that does not allow conduction.

bias-cut surface tape — Aircraft covering surface tape that is cut at a 45° angle to the length of the tape from selvage edge to selvage edge.

bicarbonate of soda — Common baking soda ($NaHCO_3$). It is used as a neutralizing agent for spilled battery acid.

bicycle gear — A landing gear that supports the main weight of the aircraft on wheels in line with each other along the length of the fuselage. The wings are supported by smaller outriggers near the wing tips.

bidirectional antenna — A radio antenna that produces maximum signal strength in two directions, 180° apart.

bi-directional cloth — In composites, a cloth in which the fibers run in various directions. Usually woven together in two directions.

bi-directional laminate — In composites, a laminate with the fibers oriented in more than one direction.

bifilar — A system developed by Sikorsky for dampening rotor vibration.

bifilar resistor — A resistor wound of wire, which is doubled back on itself in order to decrease the amount of inductance in the resistor.

bifilar transformer — An electrical transformer in which the primary and the secondary are wound side by side to increase the coefficient of coupling between the windings.

bifurcated duct — A split exhaust duct used on turbofan or lift fan engines.

bilge — The lowest part of an aircraft structure where water, dirt, and other debris accumulate.

bill of material — A list of the materials and parts necessary for the fabrication or assembly of a component or a system.

billet — A bar of semi-finished iron or steel nearly square in section.

billow cloud — A cloud layer having a "herring bone" appearance, these nearly parallel lines of clouds are oriented at right angles to the wind shear.

bimetallic circuit breaker — An electrical circuit protective device that consists of a sandwich of two metals having dissimilar expansion characteristics. When current exceeds the breaker rating, one of the metals expands more than the other causing the sandwich to warp and break the circuit. Upon cooling, the circuit can be reconnected, sometimes automatically.

bimetallic element — A device using two different metals joined together to produce either a mechanical bending or an electrical voltage as the temperature varies.

bimetallic strip — Two dissimilar metals, such as chromel and constantan that are in close proximity to each other, used in fire detection systems.

binary coded decimal — In computing, a system of encoding numbers in which each number is represented by a series of four binary digits.

binary number — A number in the binary number system that consists of only two digits: zero and one.

binding post — A subassembly used for clamping or holding electrical conductors in a rigid position. It commonly consists of a screw having a collar head or body with one or more clamping screws.

binoculars — A form of hand-held optical instrument used to look at objects that are far away. Binoculars have a set of magnifying lenses for each eye. Prisms are used to get a high degree of magnification in a short physical length. Also referred to as field glasses.

binomial — An algebraic expression that contains two terms connected by a plus or minus sign.

bioastronautics — The science that deals with the medical and biological aspects of astronautics.

biochemistry — Chemistry that deals with the chemical compounds and processes involved with living organisms.

biocidal action — The function of certain fuel additives that kill microbes and bacteria living in water in aircraft fuel tanks. This prevents scum that would promote corrosion in these tanks. Biocidal additives are also put in aircraft dope that is used on cotton or linen fabric to kill the bacteria that can destroy organic fabrics.

biocidal agent — A chemical combination destructive to certain types of living organisms.

biodegradable — A condition of a material that allows it to be broken down into simple products by the action of certain types of microorganisms.

biophysics — The interdisciplinary study of biological phenomena and problems using the principles and techniques of physics.

biplane — An airplane having two wings, one placed above the other.

bipolar transistor — The term used to describe either an NPN or a PNP transistor.

birch — One of several high-grade woods used in the manufacturing of fixed-pitch wooden propellers.

bisector of a line — A position on a line that divides it into two segments of equal length.

bisector of an angle — A line that divides an angle into two equal angles.

bismaleimide (BMI) — In composites, a polyimide resin that cures at a very high temperature, and has a very high operating temperature range in the 550 – 600° F range, and some around the 700°F range. These are more difficult to cure because of the moisture emissions during the cure may cause voids or delaminations.

bismuth — A hard, brittle, grayish-white, trivalent, metallic chemical element that has the symbol Bi and atomic number 83. Used as an alloying agent for changing the characteristic of certain metals. It is also used to dope silicon or germanium to make a P-type semiconductor material.

bistable — A condition that exists in a circuit in which either of two conditions may exist as a steady state.

bistable circuit — A circuit that has two stable conditions. The circuit will operate in the condition selected until it is intentionally changed.

bistable multivibrator — An oscillator circuit that uses two transistors of which only one transistor conducts at a time. When the first transistor stops conducting, the second transistor automatically starts to conduct.

bistatic radar — A radar system that uses separate transmitter and receiver antennas.

bit — One unit of a binary number.

bitumen — An asphaltic residue that remains after the fractional distillation of crude oil. Asphalt and tar are two commonly used bitumens.

bituminous paint — A heavy, thick, tar-based paint used as an acid-resistant paint to reduce the corrosive action of fumes and spilled electrolytes in battery compartments.

black box — A piece of electronic equipment that may be removed and replaced as a single unit.

black ice — Transparent ice that forms on black pavement, making it difficult to see. It may be caused by the refreezing of melted water or from freezing rain. Also a thin sheet of transparent ice that forms on the surface of water.

black light — Ultraviolet light with rays that are in the lower end of the visible spectrum. While more or less invisible to the human eye, black lights excite, or make visible, certain materials such as fluorescent dyes.

bladder-type fuel cell — A neoprene impregnated fabric bag installed in a portion of the aircraft structure to form a cell and used to hold fuel.

blade — In gas turbine engines, a rotating airfoil utilized in a compressor as a means of compressing air or in a turbine for extracting energy from the flowing gases.

blade alignment — In rotorcraft, an adjustment procedure, used on semi-rigid rotor systems, to place the blades in proper positions on the lead-lag axis of the rotor system. Blade alignment is sometimes referred to as chordwise balance.

blade angle — The angle between the plane of propeller rotation and the face of the propeller blade.

blade angle check and adjusting — A method used to check the blade-angle setting at a predetermined blade station. The blade angle is checked using a device called a Universal Protractor.

blade antenna — A wide-band, quarter-wavelength antenna used on aircraft for communications or navigation in the ultra-high or very-high frequency bands.

blade back — The cambered side of a propeller blade that corresponds to the curved upper surface of an airfoil, similar to that of an aircraft wing. The opposite side of the blade face.

blade base — The portion of the blade where the contoured section meets the root area. Also referred to as the blade platform.

blade beam — A paddle-shaped lever having a slot shaped to fit the cross section of a propeller blade. Used for manually turning propeller blades. Blade beams are also referred to as blade wrenches.

blade blending — A process used to remove small shallow scratches or dents of turbine blades using mild abrasive materials and sanding techniques. Blending requires maintaining the original contour and shape of the blade within prescribed limits.

blade butt — The root end of a propeller blade that fits into the hub of a propeller assembly.

blade chamber — The top or convex side of a rotating airfoil such as a compressor blade.

blade chord line — An imaginary line drawn through the blade from the leading edge to the trailing edge.

blade climbing — In rotorcraft, a condition when one or more blades are not operating in the same plane of rotation during flight. Might not exist during ground operation.

blade coning — In rotorcraft, an upward sweep of rotor blades as a result of lift and centrifugal force.

blade cross over — See blade climbing.

blade cuff — A metal, wood, or plastic fairing installed around the shank of a propeller blade to carry the airfoil shape of the blade all of the way to the propeller hub. The airfoil shape of the cuff pulls cooling air into the engine nacelle.

blade dampener — A shock absorbing mechanism installed between a helicopter rotor blade and the hub to diminish or dampen blade movement on the lead-lag axis.

blade droop — The angle of the spanwise axis of the helicopter rotor at rest with only the forces of gravity acting on it.

blade face — The flat portion of a propeller blade, resembling the bottom portion of an airfoil.

blade feather/feathering — In rotorcraft, the rotation of the blade around the spanwise (pitch change) axis.

blade fillet — The portion of the blade closest to the base or platform. Usually an area where the least damage is allowed.

blade flap — In rotorcraft, the ability of the rotor blade to move in a vertical direction. Blades may flap independently or in unison.

blade flapping — The movement of helicopter rotor blades, about a horizontal hinge, in which the blades tend to rise and descend as they rotate. Blade flapping tends to minimize asymmetrical lift by increasing the angle of attack of the retreating blade while decreasing the angle of attack of the advancing blade.

blade grips — The part of a helicopter rotor hub into which the blades are attached by a lead-lag hinge pin. Blade grips are sometimes referred to as blade forks.

blade inspection method (BIM) — A system using an indicator and inert gas to detect rotor blade cracking. Used by Sikorsky Helicopter.

blade inspection system (BIS) — A method used by Bell Helicopter to determine if rotor blades have cracked.

blade lead or lag — In rotorcraft, the fore and aft movement of the blade in the plane of rotation. It is sometimes called hunting or dragging.

blade loading — In rotorcraft, the load imposed on rotor blades, determined by dividing the total weight of the helicopter by the combined area of all the rotor blades.

blade root — The portion of a propeller blade that fits into the propeller hub. The blade root is also called the blade butt.

blade section — A cross section of a propeller blade made at any point by a plane parallel to the axis of rotation of the propeller and tangent at the center of the section to an arc drawn with the axis of rotation as its center.

blade shank — The thick, rounded portion of a propeller blade near the hub.

blade span — In rotorcraft, the length of a blade from its tip to its root.

blade stall — A condition of the rotor blade in flight when it is operating at an angle of attack greater than the maximum angle of lift. This occurs at high forward speed to the retreating blade and to all blades during "settling with power."

blade station — Reference points on the blade measured in inches from the center of the propeller hub. Blade station measurements are used to identify locations along the blade of a propeller.

blade sweeping — An adjustment of the dynamic chordwise balance in which one or both blades are moved aft of the alignment point.

blade tabs — Fixed tabs mounted on the trailing edge of helicopter rotor blades for track adjustment.

blade tip — In rotorcraft, the part of the blade the furthest from the hub of the rotor.

blade track — In rotorcraft, the relationship of the blade tips in the plane of rotation. Blades that are in track will move through the same plane of rotation.

blade tracking — 1. The process of determining the position of the tips of the propeller blades relative to each other. 2. The mechanical procedure used to bring the blades of the rotor in satisfactory relationship with each other under dynamic conditions so that all blades rotate in a common plane.

blade twist — The variation in the angle of incidence of a blade between the root and the tip. The amount of thrust that is produced by a propeller blade is determined by the pitch angle of the blade at each blade station, and by the speed at which the blade is moving through the air. In order to maintain a constant amount of thrust along the blade, the blade angle must be twisted; however, some degree of twist can also be caused by aerodynamic forces.

blade wrench — A paddle-shaped lever having a slot shaped to fit the cross section of a propeller blade. Used for manually turning propeller blades. Blade beams are also referred to as blade wrenches.

blade-disk — A forged, one-piece compressor or turbine blade and disk as opposed to separate blades fitted into a disk.

blank — To cut out the surplus material from a part, prior to finishing.

blank blade — The identification of one blade of a helicopter during electronic balancing. It is the blade with the single interrupter. The target blade will have the double interrupter.

blanket — 1. A sheet of insulation material in the cabin and passenger compartments used to aid in suppressing noise. 2. A shroud covering for airplane heat ducts.

blanket method of recovering — The method of applying fabric to an aircraft structure so that the fabric is wrapped around the structure and attached by sewing or cementing it in place to either the trailing edge of a wing or to the longerons of a fuselage.

blanking — The process of forcing stock material through a cutout in a die plate with a die that is slightly smaller, and the same shape, as the cutout in the die plate. Creates a shaped piece (blank) of the same shape as the punch.

blast fence — A sturdy structure used to prevent jet blast damage to equipment and personnel located behind operating aircraft.

bleed — In composites, an escape passage at the parting line of a mold (like a vent, but deeper), which allows material to escape, or bleed out.

bleed air — Compressed air tapped from the compressor stages of a turbine engine by use of ducts and tubing. Bleed air can be used for deice, anti-ice, cabin pressurization, heating, and cooling systems.

bleed orifice — A calibrated orifice used to bleed down or adjust the pressure in a system.

bleed valve — In a turbine engine, a flapper valve, a popoff valve, or a bleed band designed to bleed off a portion of the compressor air to the atmosphere. Used to maintain blade angle of attack and provide stall-free engine acceleration and deceleration.

bleeder — In composites, a layer of material used during the manufacture or repair of a part to allow entrapped air and resin to escape. It is removed after curing. It also serves as a vacuum valve contact with the part.

bleeder current — A current drawn from a source through a bleeder resistor to ground. Used to stabilize the output voltage.

bleeder resistance — A permanently connected resistor connected across the output of a power supply and designed to "bleed off" a small portion of the current.

bleeder resistor — The resistor of a voltage divider through which the smallest amount of current flows. This resistor is generally selected so that the current through it is about 10% of the total circuit current. Also used in power supplies to stabilize the output voltage or to bleed the voltage from capacitors after a component is turned off.

bleeding — **1.** The act of removing air from a system. **2.** A maintenance procedure for purging the fuel system of air locks and to aid in flushing any traces of preservative oil from a pressure carburetor. **3.** A maintenance procedure in which air is removed from the hydraulic fluid in the brake system of an aircraft. If there is any air in the fluid in hydraulic brakes, the air will compress when the brakes are applied, the brakes will feel "spongy," and their effectiveness will be reduced.

bleedout — In composites, excess resin that flows out during the curing process, usually into a bleeder cloth. Sometimes appears during the filament winding process if the fiber has been through a resin bath.

blemish — A defect or injury mark that damages an object or diminishes its value.

blending — A metal filing and stoning procedure used to recontour damaged compressor and turbine blades to an aerodynamic shape. A fine stone is used to blend (smooth) the reworked area into the original surface of the blade.

blimp — A non-rigid airship.

blind flight — An early aviation term for instrument flying.

blind rivet — A rivet designed to be used in sheet metal structure where it is not possible to use a bucking bar for riveting.

blind speed — The rate of departure or closing of a target relative to the radar antenna at which cancellation of the primary radar target by moving target indicator (MTI) circuits in the radar equipment causes a reduction or complete loss of signal. See also blind velocity.

blind spot — An area that is not visible.

blind spot — **1.** An area from which radio transmissions and/or radar echoes cannot be received. **2.** Portion of the airport not visible from the control tower.

blind transmission — A transmission from one station to other stations in circumstances where two-way communication cannot be established, but where it is believed that the called stations may be able to receive the transmission.

blind velocity [ICAO] — The radial velocity of a moving target such that the target is not seen on primary radars fitted with certain forms of fixed echo suppression.

blind zone — See blind spot.

blink Zyglo — A method of Zyglo inspection wherein a part is cleaned and soaked with a fluorescent penetrant for an appropriate length of time. The part is rinsed and all of the penetrant is cleaned from its surface. The part is vibrated while it is being examined under a black light. If the vibration opens up a crack that has accepted some of the penetrant, the crack will show up as a blinking light.

blinker — Oxygen-flow indicator acting as a moveable shutter, opening and closing with each breath.

blister — **1.** An enclosed raised spot on the surface of a finish on a metal. It may be filled with vapor or with products of corrosion. **2.** In composites, an undesirable rounded elevation of the surface of a plastic, and somewhat resembling in shape a blister on the human skin.

blizzard — A severe weather condition characterized by low temperatures and strong winds bearing a great amount of snow, either falling or picked up from the ground.

block — To secure from, or release an airplane for flight. It includes actions by ground crew personnel who aid the pilot in parking, mooring, or releasing by handling the wheel chocks, gear pins, etc.

block diagram — A functional diagram of a system in which the units are represented by squares that describe the functions of the unit and show its relationship to the other units of the system. Arrows between the blocks show the direction of the flow of energy or information within the system. Block diagrams do not show any of the actual components.

block heater — An electrical heater embedded in die that is used for hot dimpling sheet metal.

block plane — A small, hand-held carpenter's tool used for smoothing the surface of wood.

block test — An operational test of an aircraft engine when the engine is installed in a test cell to determine its condition.

blocked — Phraseology used to indicate that a radio transmission has been distorted or interrupted due to multiple simultaneous radio transmissions.

blocking capacitor — A capacitor that has high impedance to DC and low frequency AC. However, it has low impedance to the AC signal being passed through the circuit.

block-to-block time — The lapsed time between an airplane leaving the ramp for the purpose of flight and its returning after landing.

bloom — A bar of iron or steel hammered or rolled from an ingot.

blow molding — A plastic molding process in which a hollow tube of thermoplastic material is heated inside a mold. Air pressure is applied to the inside of the tube, and the soft plastic material is forced out against the walls of the mold. The outside of the part takes the form of the inside of the mold.

blowback — In rotorcraft, the tendency of the rotor disc to tilt aft in forward flight as a result of flapping.

blowdown turbine — A power recovery device used on the Wright R-3350 engine that is driven by the exhaust gases from the engine, and coupled through a clutch to the engine crankshaft. Also referred to as Power-recovery turbine (PRT).

blower — **1.** A mechanical device such as a fan that is used to move a column of air. **2.** An internal gear-driven supercharger in an aircraft reciprocating engine. Blowers are used to increase the pressure of the fuel/air mixture after it has passed through the carburetor, and to improve the distribution of the fuel-air mixture to all of the cylinders.

blower clutch — A unit in a two-speed supercharger system of a reciprocating engine that can be driven at two different speeds by means of a clutch.

blower section — The blower section of an aircraft reciprocating engine crankcase that houses the internal, gear-driven supercharger.

blow-in doors — In gas turbine engines, spring-loaded doors located ahead of the first stage of the compressor. These doors are spring-loaded to hold them closed, but under conditions of low airspeed and high engine power, they open automatically to allow more air to enter the compressor. Blow-in doors help prevent compressor stall. Also referred to as auxiliary air-intake doors.

blowing dust — Dust particles picked up locally from the surface and blown about in clouds or sheets.

blowing sand — Sand picked up locally from the surface and blown about in clouds or sheets.

blowing snow — Snow picked up from the surface by the wind and carried to a height of 6 feet or more.

blowing spray — Water particles picked up by the wind from the surface of a large body of water.

blown boundary layer control — A system used to decrease aerodynamic drag on the surface of a wing. Blown boundary layer control uses high-velocity air blown through ducts or jets to energize the boundary layer.

blow-out plug — A safety plug or disc on the outside skin of an aircraft fuselage near the installation of high pressure oxygen and CO_2, or other fire extinguisher agents. It is designed to rupture and discharge its contents overboard if, for any reason, the pressure of the gas in the cylinders rises to a dangerous value. Colored disks in the blow-out plugs identify the system that has been relieved in this manner.

blowtorch — A small burner having a device to intensify combustion by means of a blast of air or oxygen. Usually includes a fuel tank pressurized by a hand pump. Used for soldering, welding, and glass blowing.

blue arc — An instrument marking that indicates an operating range. For example, the blue arc might indicate the manifold pressure gauge range in which an engine can be operated with the carburetor control set at automatic lean.

blueprint — An engineering drawing used to convey the construction or assembly of objects with the help of lines, notes, abbreviations, and symbols. Blueprints are made by placing a tracing of the drawing over a sheet of chemically treated paper and exposing it to a strong light for a short period of time. When the exposed paper is developed, it turns blue. The inked lines of the tracing now show as white lines on a blue background, thus its name blueprint. Blueprints have been replaced in many engineering departments by prints.

blush — The white or grayish cast that forms on a lacquer or dope film, which has been applied under conditions of too high humidity. It is actually nitrocellulose, which has precipitated from the finish.

BMEP Indicator — An engine instrument that measures output shaft torque and converts it to brake mean effective pressure (BMEP).

B-nut — A nut used to connect a piece of flared tubing to a threaded fitting. B-nuts are used with a sleeve that is slipped over the tubing before the tubing is flared. The B-nut forces the sleeve tight against the flare, sealing against the flare cone of the male fitting. The B-nut derives its nickname from its predecessor (no longer in use), which was called an A-nut.

board-foot — A commercial unit of measurement used to measure lumber. One board-foot is the amount of lumber in a piece of 1-inch thick wood that measures 1' x 1.'

bob weight — A mechanical weight in the elevator control system of some airplanes. In some aircraft it is used to apply a nose-down force on the elevator control system. This force is counteracted by an aerodynamic force that is caused by the elevator trim tab. If the aircraft slows down enough that the aerodynamic force on the trim tab is lost, the bob weight forces the nose down, and the airplane picks up speed. In some other aircraft, a bob weight is used to counteract rapid control inputs to prevent overstressing the aircraft.

bogie landing gear — The landing gear of an aircraft that uses tandem wheels connected by a central strut. Aircraft having bogie landing gear are sometimes supported by outrigger wheels mounted far out on the wing when the aircraft is parked.

bogus parts — Parts that are not approved for use in aircraft. Bogus parts are often illegally marked so as to appear to be authorized parts. The use of bogus parts compromises the safety of an aircraft and makes it unairworthy.

boiling — The process whereby water changes state to vapor throughout a fluid. Occurs when saturation vapor pressure equals the total air pressure.

boiling point — Temperature at which a liquid changes to vapor. The boiling point of water, under standard conditions, is 212°F, or 100°C.

bolt — An externally threaded fastener with an enlarged head on one end and threads on the other end.

bolt bosses — The enlarged portion of a casting or forging where the bolts pass through.

bomb tester — A spark plug tester in which the plug is exposed to approximately 200 PSI of air pressure. High voltage is applied to the center electrode cavity of the spark plug, and the electrodes are observed to see the type and amount of spark being produced. Plugs that can spark in this atmosphere are considered to be acceptable for use in the aircraft engine.

bond — An attachment of one material to another or of a finish to the metal or fabric.

bond ply — In composites, the ply or fabric patch which comes in contact with the honeycomb core.

bond strength — In composites, the stress required to pull apart two plies or from the ply to the core. The amount of strength of the adhesion.

bonded structure — A structure whose parts are joined together by chemical methods rather than mechanical fasteners. Honeycomb material, laminated fiberglass, and composite materials are examples of aircraft bonded structure.

Bonderizing — The registered trade name for a patented process of covering steel parts with a phosphate coating to protect the parts from corrosion.

bonding — 1. A procedure used in joining parts by using adhesives rather than any form of mechanical fastener. 2. A method of electrically connecting all the components of an aircraft structure together so that static electricity cannot build up on one part of the structure to create a voltage that is high enough to allow it to jump to another part causing radio interference.

bonding agent — An adhesive used to bond structure parts together.

bonding braid — A soft annealed, tinned copper flat braid used for battery grounding or bonding strap for electrical equipment.

bonding jumper — A low-resistance wire or electrical connection used to electrically ground a component or structure to an airframe. Bonding jumpers carry the return current from an electrical component back to the battery.

bonnet assembly — The operating head of a fire extinguisher, which contains an electrically ignited powder charge used to rupture a disc and release the extinguishing agent.

bookmark — A means of saving addresses on the World Wide Web for easy future access. Usually done by selecting a button on the web browser screen, it saves the current web address so it does not have to be input again into the computer (web addresses are often a lengthy series of characters).

boom — A spar or outrigger connecting the tail surfaces or auxiliary equipment to the main supporting structure of an aircraft.

boost — **1.** To assist. **2.** An older term synonymous with manifold pressure.

boost charge — A constant-voltage charge applied to a discharged battery installed in an airplane to restore a charge sufficient to start the engine.

boost pump — An electrically driven fuel pump, usually of the centrifugal type, located in one of the fuel tanks. It is used to provide fuel to the engine for starting and providing fuel pressure in the event of failure of the engine driven pump. It also pressurizes the fuel lines to prevent vapor lock.

boost system — A hydraulically actuated system that aids the pilot in operation of the flight controls.

boost venturi — A small venturi whose discharge end is at the throat of the main venturi, and which surrounds the main discharge nozzle of a float-type carburetor. It increases the pressure drop for a given airflow.

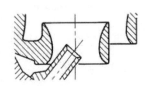

boosted brake — A form of brake power source using a master cylinder in which the hydraulic pressure from the aircraft hydraulic power system is used to aid the pilot in applying force to the master cylinder. This boost, or assistance, is automatically applied when the pressure required at the brake is greater than the pilot can produce with foot pressure alone.

boosted control system — A hydraulically actuated system that aids the pilot in operation of the flight controls.

booster coil — A transformer coil used with a vibrator to produce a high voltage at the spark plugs during starting.

booster magneto — A small, high-voltage magneto used to produce a hot spark for starting reciprocating engines. The output for the booster magneto is fed into a trailing finger on the distributor. This spark follows the normal ignition spark.

boot — A telescoping rubber seal. Sometimes placed over a cable that passes through a bulkhead separating a pressurized and non-pressurized section, or around the portion of a strut where it telescopes into another section.

bootstrapping — **1.** Technique with which something is brought into the desired state through its own action. Derived from the term "picking oneself up by the bootstraps." **2.** A condition in a turbocharged engine when a turbocharger system senses small changes in temperature or RPM and continually changes the turbocharger output in an attempt to establish equilibrium. Typically occurs during part-throttle operation and is characterized by a continual drift or transient increase in manifold pressure.

bore — **1.** Diameter of an engine cylinder. **2.** The internal diameter of a pipe, cylinder, or hole.

borescope — An optical tool with which a visual inspection can be made inside an area that is otherwise impossible to see. It consists of a light, mirrors, and lenses.

boric acid — A white crystal that can be dissolved in water to make a weak acid solution. Boric acid is used to neutralize spilled electrolyte from nickel-cadmium batteries.

boring — A process of increasing the size of a hole in a piece of material by cutting it with a rotary cutting tool.

boron — A non-metallic chemical element with a symbol of B and an atomic number of 5. When it is used to dope silicon or germanium, it produces a P-type material. Boron fibers are also used to add stiffness and strength to some of the composite structural materials use in modern aircraft.

boron filament — In composites, a strong, lightweight fiber used as a reinforcement. A tungsten-filament core with boron gas deposited on it. Has a high strength to weight ratio.

boss — An enlarged or thickened part of a forging or casting to provide additional material for strength at its attaching point.

bottle bar — A special bucking bar recessed to hold a rivet set. It is used in reverse riveting.

bottled gas — Any of the gases kept under pressure (acetylene, propane, oxygen, and nitrogen) in heavy steel containers.

bottom dead center (BDC) — The crankshaft position when the piston is at the bottom extreme of its stroke, and the crank pin is below and directly in line with the wrist pin and the center of the crankshaft.

bottoming reamer — A reamer used to smooth and enlarge holes to exact size, but having no starting taper. A bottoming reamer completes the reaming of blind holes by finishing the hole to size nearly to the bottom of the hole.

bottoming tap — A tap used to cut full threads at the bottom of a blind hole. The bottoming tap is not tapered. It is used after the hole has been partially tapped with a

tapered tap.

bounce — A condition where the breaker points of a magneto tend to bounce after they close. This is caused by a weak breaker spring.

boundary layer — **1.** The layer of air between the surface of an object, such as an airfoil, and the freestream air. At the surface of an object air particles are slowed to a relative velocity of near zero due to the viscosity of the air. Surrounding this area, the air gradually increases in speed until it reaches the velocity of the freestream air at a given distance from the surface. **2.** The layer of the earth's atmosphere from surface to approximately 2,000 feet AGL, where surface friction influences are large.

boundary layer control — A method of removing random flowing air from the immediate surface of an airfoil caused by the turbulent flow of the boundary layer. Boundary layer control is obtained by either adding energy to it through vortex generators (added energy delays the airflow from separating from the airfoil) or by sucking or vacuuming it off through tiny holes on the surface of the airfoil.

boundary lights — Lights defining the perimeter of an airport or landing area.

Bourdon tube — The mechanism in a pressure gauge consisting of a flat or elliptical cross-sectioned tube bent into a curve or spiral. When pressure is applied, the tube attempts to straighten. The amount the tube straightens is proportional to the amount of pressure inside the tube, and as it straightens, it moves a pointer across the instrument dial.

bow or camber — The amount that a side of a surface deviates from being straight.

bow wave — **1.** The v-shaped disturbance through a fluid such as water or air as the result of being displaced by the foremost point of an object moving through it. **2.** A shock wave that forms immediately ahead of an aircraft that is flying at a speed faster than the speed of sound.

Bowden cable — A control system that uses a spring steel wire, enclosed inside a helically wound wire casing, used to transmit both pushing and pulling motion to the device being actuated. Often used as throttle cables or mixture control cables.

bowline knot — A knot used to tie down an aircraft and to start the rib-stitching when attaching fabric to an aircraft structure. A properly tied bowline knot will not slip yet is easy to untie.

box brake — A metal-forming machine similar to a leaf (cornice) brake. It is used to form all four sides of a box by allowing the sides that have been bent up to fit between the fingers of the clamp while the last bends are being made. A box brake is also called a finger brake.

box spar — A design for wood spars in the shape of a box or a square. The top and bottom of the box are called the caps of the spar, and the sides of the box are called the webs of the spar.

box wrench — A wrench with an enclosed end that has six, eight, or twelve points. It can be used to tighten or loosen nuts and bolts, and can easily fit into close spaces and can be used to apply a greater amount of force than can be applied with an open-end wrench.

boxbeam wing — A wing construction made in the form of a box, which uses two main longitudinal members with connecting bulkheads to furnish additional contour and strength.

boxing of paint — A mixing procedure in which the paint is poured back-and-forth between two containers until the pigment and the base paint are completely mixed.

Boyle's law — A gas law, which states that at a constant temperature the volume of the gas will vary inversely as its pressure changes.

brace — A device that holds parts together, or in place. Something that gives support.

braced wing — A wing that requires external bracing and is not self-supporting. See cantilever for contrasting definition.

bracketing — A technique for navigating by VOR in a crosswind, where the course is maintained by a series of turns into the crosswind to regain and maintain the desired course.

brad — A thin wire nail or spike that has a small-diameter and a small barrel-shaped head.

braid — The rubber-coated, woven metal fabric reinforcing cord used to encase hydraulic flex hoses.

braided shield — A covering of braided metal over one or more insulated conductors to form shielded cable. This braid intercepts the magnetic field produced by the alternating current and keeps the field from causing radio interference.

braiding — Weaving of fibers into a tubular shape instead of a flat fabric

brake — **1.** A device inside an aircraft wheel used to apply friction to the wheel to slow or stop its rotation. Wheel brakes slow the aircraft down during taxiing and landing. Types of brakes used on aircraft are in four general categories: shoe, expander tube, single disc, and multiple disc. **2.** A metal-working shop tool that is used to make straight bends across sheets of metal. Brakes can be adjusted to make bends with the proper bend radius and correct number of degrees.

brake back plate — A retainer plate to which the wheel cylinder and the brake shoes attach.

brake caliper — The clamp in a disc brake system that grips the brake disc. When pressure is applied to the brake, the calipers apply pressure to the disc to produce the braking action.

brake horsepower — The actual horsepower delivered to the propeller shaft of an aircraft engine. Prior to electronic bench testing, horsepower calculated by measuring the amount of resistance against a flywheel brake. This method is no longer used but the term remains an industry standard.

brake horsepower — The power delivered at the propeller shaft (main drive or main output) of an aircraft engine.

brake line — The mark on a flat sheet of metal that is set even with the nose of the radius bar of a cornice brake and serves as a guide in bending. The brake line can be located by measuring out one radius from the bend tangent line closest to the end that is to be inserted under the nose of the brake. Also called a sight line.

brake lining — A material with a high coefficient of friction and the ability to withstand large amounts of heat. It acts as the wearing surface in aircraft brakes.

brake mean-effective pressure (BMEP) — A computed value of the average pressure that exists in the cylinder of an engine during the power stroke. BMEP is measured in pounds per square inch and is used to compute the amount of power the engine is developing.

brake specific fuel consumption (BSFC) — The number of pounds of fuel burned per hour to produce one horsepower in a reciprocating engine.

braking action — A report of conditions on the airport movement area providing a pilot with a degree/quality of braking that might be expected. Braking action is reported in terms of good, fair, poor, or nil.

braking action advisories — An air traffic control advisory issued to arriving and departing aircraft to inform them about degraded braking action.

branching — A programming technique, which allows users of interactive video, multimedia courseware, or online training to choose from several courses of action in moving from one sequence to another.

brass — A metal alloy consisting of copper and zinc.

Brayton cycle — The name given to the thermodynamic cycle of a gas turbine engine to produce thrust. This is a varying volume, constant pressure, cycle of events and is commonly called the constant-pressure cycle. A more recent term is continuous combustion cycle because of the four continuous and constant events: intake, compression, expansion (including power), and exhaust.

brazier-head rivet — A form of aircraft rivet with a large thin head. Its specification is AN 455. These rivets have been superseded in use by AN 470 (MS 20470) universal head rivets.

brazing — Refers to a group of metal-joining processes in which the bonding material is a nonferrous metal or alloy with a melting point higher than 800°F. It is a method of joining two pieces of metal by wetting their surface with a molten alloy of copper, zinc, and tin. The brazed joint has more strength than a soldered joint but less than a welded joint.

breadboard (electronics) — A structure that allows electronic circuits to be constructed with components that are not permanently connected. Allows circuits to be tested and modified before being put into a permanent configuration.

break line — A line used in drawings to indicate that a portion of the object is not shown.

break-away point — Refers to an intentionally weakened point on a shaft such as in a dual element fuel pump. The shear point is designed to break away if one element becomes jammed, leaving the other element still functioning.

break-before-make switch — A double-throw switch that breaks one circuit before it makes contact with the other circuit.

breakdown voltage — The voltage at which the dielectric is ruptured, or the voltage level in a gas tube at which the gas becomes ionized and starts to conduct.

breaker assembly — A mechanism used in high-tension magneto ignition systems to automatically open and close the primary circuit at the proper time in relation to piston position in the cylinder to which an ignition spark is being furnished. The interruption of the primary current flow is accomplished through a pair of breaker contact points.

breaker contact — A pair of electrical contacts that are opened and closed by a cam in the magneto for the purpose of timing the ignition of a reciprocating engine.

breaker point bounce — A condition caused by a weak breaker point spring. It is a fault in which the breaker points in an aircraft magneto bounce open rather than remaining closed when the cam follower moves off the cam lobe.

breaker points — Interrupter contacts in the primary circuit of a magneto or battery ignition system. They are opened by a cam the instant the highest current flows in the primary circuit, thus producing the maximum rate of collapse of the primary field.

breakers — On tires, extra layers of reinforcing nylon chord fabric are placed under the tread rubber to protect casing plies and strengthen tread area. Breakers are considered an integral part of the carcass construction.

breakout — **1.** In composites, when drilling or cutting the edges of a composite part, the fibers may separate or break. **2.** A technique to direct aircraft out of the approach stream. In the context of close parallel operations, a breakout is used to direct threatened aircraft away from a deviating aircraft.

breast drill — A drill designed to hold a larger size twist drill than the hand drill, and is used to drill relatively large holes in wood. A breast plate affixed at the upper end of the drill permits the use of body weight to increase the pressure on the drill.

breather — **1.** A vent in an engine oil system that keeps pressure within the tank the same as the atmospheric pressure. **2.** In composites, a loosely woven fabric that does not come in contact with the resin and used to provide venting and pressure uniformly under a vacuum cure. Breather material is used under the vacuum valve to allow the air to be evacuated inside the vacuum bagged part. Removed after curing.

breather pressure system — The breather pressurizing system ensures a proper oil spray pattern from the main bearing oil jets and furnishes a pressure head to the scavenge system.

breather pressurizing valve — An aneroid-operated valve and a spring loaded blowoff valve. Pressurization is provided by compressor air that leaks by the seals and enters the oil system. At sea level pressure the breather pressurizing valve is open. It closes gradually with increasing altitude and maintains an oil system pressure sufficient to assure oil flows similar to those at sea level.

breech chamber — The component of a self-contained cartridge/pneumatic starter for gas turbine engines. A solid propellant cartridge is placed in the breech chamber. When electrically ignited, it produces a low pressure, high-volume pneumatic source to turn the turbine blades fast enough to enable starting the engine.

bridge circuit — An electrical circuit that contains four impedances connected in such a way that their schematic diagram forms a square. One pair of diagonally opposite corners is connected to an input device, and the other two corners are connected to the output device.

bridge rectifier — An electrical rectifier circuit using four diodes arranged in a bridge circuit to change AC to DC.

bridging — In composites, this term can refer to plies of fabric over a curved edge that don't come in full contact with the core material. It is also used to describe excess resin that has formed on edges during the curing process.

briefing — An oral presentation where the speaker presents a concise array of facts without inclusion of extensive supporting material.

brine — A solution of salt (sodium chloride) and water used as a quenching agent in the heat treatment of metal. Greater hardness is obtained from quenching in salt brine, and less in oil, than is obtained by the use of water.

Brinell hardness test — A test used to determine the hardness of a metal by forcing a hardened steel sphere into the surface with a given force. The diameter of the indention, measured with a special microscope, is directly related to the hardness of the material.

brinelling — Indentations in bearing races usually caused by high static loads or application of force during installation or removal. They are usually rounded or spherical due to the impressions left by contacting balls or the rollers of the bearing.

bristle brush — A brush with non-metallic bristles constructed of short, stiff, coarse hair or filament. Bristle brushes are used to clean metal parts to prevent metal particles from becoming embedded in the structural metal, which could cause dissimilar metal corrosion.

British Thermal Unit (BTU) — A unit of heat. One BTU equals the heat energy required to raise one pound of water one degree Fahrenheit (e.g. one pound of jet fuel contains approximately 18,600 BTU).

brittleness — The propensity of a material to break when bent, deformed, or hammered. Brittleness is the resistance to change in the relative position of the molecules within the material.

broaching — The process of removing metal by pushing or pulling a cutting tool, called a broach, along the surface.

broad-band antenna — An antenna constructed in such a way as to receive or transmit a wide range of frequencies.

broadcast — The transmission of radio waves without intent of receiving information. One way communication.

broken-line — A graph using sharp, abrupt changes in the information line.

broken-line graph — A graph that represents the way in which values change. The horizontal axis of the graph represents one value, and the vertical axis represents another value. Straight lines are used to connect points that show true values at each plotted point.

bromochlorodifluoromethane (Halon 1211) — Chemical formula ($CBrClF_2$). A liquefied gas with a UL toxicity rating of 5, used as a fire extinguishing agent. It is colorless, non-corrosive and evaporates rapidly without leaving residue. It does not freeze or cause cold burns and will not harm fabrics, metals, or other materials it contacts. Halon 1211 acts rapidly on fires by producing a heavy blanketing mist that eliminates air from the fire and interferes chemically with the combustion process.

bromotrifluoromethane (Halon 1301) — Chemical formula (CF_3Br). A liquefied gas with a UL toxicity rating of 6, used as a fire extinguishing agent. It has all the characteristics of Halon 1211. The significant difference between the two is that Halon 1211 throws a spray similar to CO_2 while Halon 1301 has a vapor spray that is more difficult to direct.

bronze — An alloy of copper and tin. Used for bearing surfaces.

brush — **1.** A component device in an electric generator or motor designed to provide an electrical contact between a stationary conductor and a rotating element. Brushes are made of a carbon compound that contacts each segment of a rotating commutator. The brush conducts voltage between the rotating armature and an external circuit. **2.** A device composed of bristles or hairs fastened to a wood or plastic handle. Brushes are used to apply paint or other substances to a surface.

brush guard — A protective device used to guard the tail rotor blades of a helicopter from damage during ground operations.

brush run-in — The procedure of running a motor or generator for a time after installing new brushes to allow the brushes to shape themselves to the commutator. Also referred to as brush break-in.

brushing — A motion that barely touches an object in passing.

BTU — See British thermal unit.

bubble — A small space filled with a volume of air or gas entrapped in a liquid.

bubble memory (computer) — Non-volatile memory. Consists of a thin layer of material normally magnetized in one direction. When an oppositely polarized magnetic field is applied to a circular area of this substance, the area is reduced to a smaller circle, or bubble. The bit of information is thus stored until changed by application of a new magnetic field.

bubble octant — A celestial navigation instrument, like the sextant, that uses a bubble level in the octant to provide an artificial horizon that allows a navigator to find the angle between a line tangent to the earth's surface (the horizon) and a line to the stars that are used for navigation.

bucker — The person holding the bucking bar used to upset a rivet.

bucket — Accepted jargon for turbine blade.

bucket root — A method of turbine disk blade retention in which the blade root has a stop made on one end of the root so that the blade can be inserted and removed in one direction only, while on the opposite end is a tang. This tang is bent to secure the blade in the disk.

bucket wheel — Slang for turbine wheel in turbine engines.

bucking — The coordinated process between the bucking bar holder and the pneumatic rivet gun operator, whereby a shop head is formed on a solid rivet. A special hardened steel bar is held against the rivet shank, which has been inserted into a hole drilled in metal to be joined, while the pneumatic hammer is held at the rivet head during the hammering.

bucking bar — A tool made of alloy steel stock that is held against the shank end of a rivet while the shop head is being formed.

buckle — A bend or kink in or on a surface of a metal structure. Caused by the failure of the part under a compressive load or excessive localized heating.

buckle line — In composites, on a honeycomb core, it is a line of collapsed cells with undistorted cells on either side. It usually is found on the inside of the radius on a formed core.

buckled areas — Localized areas in a turbine engine combustion chamber liner in which small areas have been heated to an extent to cause the area to buckle.

buckling — In composites, a failure of the fabric in which it deflects up or down rather than breaking.

buffer — **1.** Any device used to absorb shock. **2.** Used to isolate an input or to strengthen a signal in a digital electronic circuit to match inputs and outputs having differing values.

buffer amplifier — An amplifier in a transmitter circuit designed to isolate the oscillator section from the power section thus preventing a frequency shift or otherwise operate improperly.

buffet — A series of waves or blows caused by imbalance such as those that can occur with flight controls. The aircraft feels as though it were being hit with a series of blows, shocks, or waves.

buffeting — Erratic movement of aircraft controls caused by the turbulent flow of air over the surfaces.

bug — An unexpected malfunction.

bug light — A tool made up of an electrical wire, flashlight battery, and bulb used to check the continuity of an electrical circuit.

building block concept — Concept of learning that new skills and knowledge are best based on a solid foundation of previous experience or knowledge. As knowledge and skill increase, the base expands, supporting further learning. Also referred to as blocks of learning.

build-up and vent valve — A manually operated valve on a liquid oxygen converter. In the build-up position, pressure is allowed to reach a preset value and excess pressure is vented into the atmosphere. In the vent position, gas is vented into the atmosphere without pressure buildup.

bulb angle — An extruded angle of metal with a rounded edge resembling a bulb on one of the legs.

bulb root — A means by which turbine engine rotor blades are attached to the rotor hub. The base of the blade is cylindrical and larger than the rest of the blade. This fits into a mated hole in the rotor hub.

bulb temperature — A unit of a carburetor induction system that monitors the air inlet temperature to be sure the inlet temperature does not exceed the maximum value specified by the engine manufacturer.

bulbed Cherrylock® rivet — A special form of blind rivet manufactured by the Cherry division of Textron Inc., in which the stem is locked into the hollow shank by a special locking collar that swages into a groove in the stem.

bulb-fit — A design of compressor blade attachment to the disk shaped in a bulb fashion.

bulkhead — A structural partition in a fuselage or wing. Bulkheads usually divide the fuselage or wing into bays and provide additional strength to the structure.

bumping — The shaping or forming of sheet metal by hammering or pounding. During this process, the metal is supported by a dolly, sandbag, or die.

bundled cable — Any number of individually insulated electrical wires tied together with lacing cord or with special plastic wire-wrapping straps.

bungee — A shock-absorber cord made from natural rubber strands encased in a braided cover of woven cotton cords and then treated to resist oxidation and wear.

bungee cord — An elastic cord encased in a braided cloth cover that holds and protects the rubber, yet allows the rubber to stretch. Bungee cords are used in some of the simpler aircraft landing gears to assist in retracting the landing gear and to absorb shock.

buntover — The tendency of a gyroplane to pitch forward when rotor force is removed.

buoyancy — The property of an object that allows it to float on the surface of a liquid, or ascend through and remain freely suspended in a compressible fluid such as the atmosphere.

burble — A breakdown of the laminar airflow over an airfoil caused by too high an angle of attack. The result is an increase in drag and a loss of lift.

burble point — The angle of attack at which burbling first occurs on an airfoil.

burn down coat — A coat of dope with some of its thinner replaced with retarder. The coat is sprayed on a blushed area to soften and reflow the surface in order to remove the blush.

burned areas — Localized areas of a turbine engine combustion chamber liner that have been heated to an extent to cause visible damage.

burned surface — A condition resulting from high surface temperatures with relatively low pressures and accompanied by heat discoloration. This condition may or may not mark the surface. The cause is usually improper clearance or insufficient lubrication. Areas affected are bearings or journals.

burner — The section of a turbine engine into which fuel is injected and burned. The point within the engine where combustion occurs.

burner cans — Any number of individual combustion chambers in the combustion section of a gas turbine engine.

burner compartment — A section of the cowling behind which the burner section of a turbine engine is located.

burner pressure (Pb) — Static pressure signal used as a measure of mass airflow through the engine and sent to the fuel control unit for fuel scheduling purposes.

burning — **1.** Surface damage due to excessive heat. It is usually caused by improper fit, defective lubrication, or over temperature operation. **2.** The combustion process that occurs when fuel is mixed with air and ignited.

burning in — An electronic components manufacturing process in which the equipment is operated for a specified period of time in order to stabilize the operating characteristics of the components.

burning point — The lowest temperature at which a petroleum product in an open container will ignite and continue to burn when an open flame is held near its surface.

burnish — To polish a metal surface by rubbing it with a smooth, extremely hard tool, called a burnishing tool. A lubricant is usually required. Small scratches can be smoothed by burnishing.

burnishing — The process of polishing a surface by sliding contact with a smooth, harder surface. Displacement or removal of metal does usually not occur during burnishing.

burr — A sharp or roughened projection of metal usually resulting from machine processing.

burring — The forming of a sharp or roughened projection usually as a result of a machining operation.

burst RPM — The RPM at which the blades of a turbine motor will separate from the rotor due to excessive centrifugal loads.

bus — A main electrical power circuit to which a number of component circuits connect.

bus bar — An electrical power distribution point to which several circuits may be connected. It is often a solid metal strip having a number of terminals installed on it.

bushing — A removable cylindrical lining for an opening used to minimize resistance and serve as a guide.

butt fusion — A method of joining two pieces of thermoplastic material. Butt fusion is done by heating the ends of the two pieces until they are in a molten state and forcing them together before they cool and harden.

butt joint — A welded joint made by placing two pieces of material edge to edge, so that there is no overlapping, and then welding them to form one piece. The types of butt joints are: flanged, plain, single bevel, and double bevel.

butt rib — The last rib at the inboard end of an airfoil. The rib on a wing which is closest to the wing attachment fittings.

butterfly tail — A design that combines the vertical and horizontal surfaces of the empennage. The shape is that of a "V."

butterfly valve — A damper or valve consisting of a disk turning about one of its diameters to control the flow of fluid in a round tube.

buttock line — A measurement of width, left or right of, and parallel to, the longitudinal axis. Also referred to as BL or butt line.

butyl — The trade name of a synthetic rubber product made by the polymerization of isobutylene. It withstands such potent chemicals as Skydrol hydraulic fluid.

butyrate dope — A finish for aircraft fabric consisting of a film base of cellulose fibers dissolved in acetic and butyric acids, with the necessary plasticizers, solvents, and thinners.

Buys-Ballot's law — A law of meteorology which states that if a person stands in the Northern Hemisphere with the wind striking at their back, the center of the low-pressure area, around which the wind is blowing, is ahead and to the left.

buzz — In turbine engines, an airflow instability that occurs when a shock wave is alternately swallowed and regurgitated by the inlet. At its worst, the condition can cause violent fluctuations in pressure through the inlet, which may result in damage to the inlet structure or, possibly, to the engine itself.

bypass capacitor — A capacitor that provides a low-impedance path for alternating current to bypass a circuit component, when the component is being used to produce a DC voltage drop.

bypass duct — A cold airstream duct. Also referred to as a fan exhaust duct on a turbofan engine.

bypass jacket — An annular bypass around an oil cooler through which oil flows when it does not need cooling.

bypass jet — A form of turbojet engine in which a portion of the compressor air is bypassed around the combustion chamber and into the tailpipe.

bypass ratio — **1.** The ratio of the mass airflow in pounds per second through the fan section of a turbofan engine to the mass airflow that passes through the gas generator portion of the engine. **2.** Ratio between fan mass airflow (lb/sec.) and core engine mass airflow (lb/sec.).

bypass turbojet engine — Forerunner of the bypass fan engine, whereby the low pressure compressor discharge is divided in two; one portion of air to enter a bypass duct and the other portion of air to enter the high pressure compressor inlet.

bypass valve — A valve whose function is to maintain a constant system pressure. When the system pressure is exceeded by a predetermined amount, the valve will allow excess pressures to bypass the system, thereby, not allowing the system to rupture from excess pressure.

byte — A computer term used to describe a group of binary digits consisting of eight bits.

C

C battery — A small, low-voltage battery. Also referred to as a C-cell battery or C-cell.

C check — The intermediate level of inspection in a Continuous Airworthiness Maintenance Program. The entire program consists of A, C, and D checks.

cabane — An arrangement of struts used to support a wing above the fuselage of an airplane. Such an airplane wing attachment is called a parasol.

cabin — That portion of an aircraft used for cargo and/or passengers.

cabin altitude — Cabin pressure in terms of equivalent altitude above sea level.

cabin differential pressure — The difference between the pressure inside a cabin and the outside air pressure. The maximum cabin differential pressure is determined by the aircraft structural strength.

cabin pressure regulator — A means of controlling cabin pressure by regulating the outflow of air from the cabin.

cabin pressurization safety valve — A combination pressure and vacuum relief and dump valve used to prevent cabin pressurization exceeding safe limits.

cabin supercharger — Mechanical air pumps used to provide the air pressure for cabin pressurization.

cabinet file — A coarse file that is flat on one face, and rounded on the other face, used for metalworking and woodworking. Sometimes referred to as a half-round file.

cable — 1. In common usage, any heavy conductor. 2. In electronics, two or more conductive paths bound into a single package. 3. A group of insulated electric conductors usually covered with rubber or plastic to form a flexible transmission line. 4. A stranded wire generally composed of a number of wires enclosed in a single bundle or group.

cable control — The system of operating aircraft controls by the use of high-strength flexible steel cables.

cable drum — A cylindrically shaped spool around which a control cable is wound to increase the amount of cable moved each time the handle is turned.

cable guard — A pin installed in the flange of a control cable pulley bracket to prevent the cable from jumping out of the pulley grooves.

cable rigging tension chart — Charts showing the relationship between control cable tension and temperature.

cadmium — A bluish-white, malleable, ductile, toxic, metallic chemical element with a symbol of Cd and an atomic number of 48.

cadmium cell — A basic unit of the nickel-cadmium battery. It consists of positive and negative plates, separators, electrolyte, a cell vent, and a cell container. The positive plates are made from porous plaque on which nickel-hydroxide is deposited. The negative plates are made from similar plaques on which cadmium-hydroxide is deposited. The voltage that is produced by a cadmium cell at 20°C is 1.0186 volts.

cadmium plating — A thin coating of cadmium metal electroplated on a steel part to protect the steel from corrosion. This is accomplished by the cadmium serving as the anode in a corrosive action.

caging device — A mechanism used in a gyroscopic instrument to erect the rotor of a gyro to its normal operating position prior to flight or after tumbling.

caging system — See caging device.

calcium — A hard metallic element with a symbol of Ca and an atomic number of 20. Used as an alloying element with other metals.

calcium carbide — A combination of calcium and carbon. Reacts chemically with water to produce acetylene gas.

calculated landing time — A term that can be used in place of tentative or actual calculated landing time, whichever applies.

calendar month — The measure of time used by the Federal Aviation Administration for inspections and certification purposes. A calendar month ends at midnight of the last day of the month, regardless of the day it began.

calendering — The process of dipping cotton yarn or fabric into a hot solution of caustic soda to shrink the material and give it greater strength and luster.

calibrate — A procedure in which the indication of an instrument is compared with a standard value in order to inspect and correct the graduations of a measuring device.

calibrated airspeed (CAS) — The indicated airspeed of an aircraft, corrected for position and instrument error. Calibrated airspeed is equal to true airspeed in standard atmosphere at sea level.

calibrated orifice — A hole with a specific internal diameter used to measure or control the amount of flow through it.

calibration — Testing the accuracy of a measuring instrument or scale by comparing it with a known standard.

calibration card — A card mounted on an instrument panel near an instrument to show the errors in an instrument in order for the pilot to apply an appropriate correction.

calibration curve, instrument — A curve on a graph plotted to show the instrument errors at different points on the scale of the instrument. The pilot uses the curve to interpolate the error at points between those that have been plotted.

californium — A radioactive, synthetically produced element with a symbol of Cf and an atomic number of 98.

call for release — Wherein the overlying ARTCC requires a terminal facility to initiate verbal coordination to secure ARTCC approval for release of a departure into the enroute environment.

callouts — Numbers or names used to identify components or parts in an aircraft drawing. Callouts are placed near the part being identified and are connected by a thin leader line.

call-up — Initial voice contact between a facility and an aircraft, using the identification of the unit being called and the unit initiating the call.

calm — The absence of wind or of apparent motion of the air.

calorie — The amount of heat required to raise the temperature of one gram of water 1 °C.

calorimeter — An apparatus for measuring specific heat.

cam — An eccentric plate or shaft used to impart motion to a follower riding on its surface or edge.

cam dwell — The cam dwell is the number of degrees the cam rotates between the time the breaker points close and the time they open.

cam lobe — An eccentric used to change rotary motion into linear motion. For example, the cam followers that operate valves ride on the cam lobes and as the cam shaft rotates, the cam lobes move the cam followers up and down in a direction that is perpendicular to the axis of the cam shaft.

cam nose — The peak, or highest point on a cam, that pushes up on the cam follower.

cam pawl — A special device that allows a wheel or gear to turn in one direction but prevents its from turning in the opposite direction.

cam plate — In radial-engines, a disc or plate with lobes machined onto its circumference. Cam followers ride on the lobes and open the engine valves through a system of push rods and rocker arms.

cam ring — An open cam plate driven by teeth around its circumference.

camber — 1. The curvature of an airfoil above and below the chord line surface. An airfoil is often described as having an upper and lower camber. 2. The mean camber of an airfoil section is a line drawn through a series of points located midway between the upper and the lower camber. 3. The amount of angle the wheels of an aircraft are from the vertical. If the top of the wheel tilts outward, the camber is positive; if it tilts inward, the camber is negative.

cambric — A finely woven cotton or linen material.

cam-ground piston — An aircraft engine piston ground in such a way that its diameter parallel to the wrist pin boss is less than its diameter perpendicular to the boss. When the piston reaches its operating temperature, the difference in mass has caused the piston to expand to a perfect circular form.

Camlock fastener — A patented cowling fastener in which a hard steel pin is turned in a special cam-shaped receptacle.

camshaft — A long shaft running parallel to the crankshaft of an inline or horizontally opposed reciprocating engine. Lobes are ground at intervals along its length to operate the valves through push rods and rocker arms.

can tap valve — A valve fastened onto a small can of Freon refrigerant. It punctures the can seal and controls the flow of refrigerant.

Canadian Minimum Navigation Performance Specification Airspace — That portion of Canadian domestic airspace within which MNPS (Minimum Navigation Performance Specification) separation may be applied.

can-annular combustor — Can annular combustion chambers arranged radially around the axis of a gas turbine engine. The combustion chambers are enclosed in a movable steel shroud, which covers the entire burner section, and is designed for more complete cooling and mixing of fuel and air. The burners are interconnected by projecting flame tubes that facilitate the engine-starting process.

canard — 1. The forward wing of a canard configuration and may be a fixed, movable, or variable geometry surface, with or without control surfaces. 2. An aircraft with its horizontal stabilizing and control surfaces in front of the wings. Also used to describe the forward-mounted control surface.

canard configuration — A configuration in which the span of the forward wing is substantially less than that of the main wing.

candela — A unit of incandescent intensity.

candlepower — The luminous intensity of light expressed in candelas.

cannibalize — To remove serviceable parts from a non-flying aircraft for use on another machine.

cannon connector — A connector used to attach the battery to the aircraft electrical system. It is a high-current connector in which the cables are held onto the battery by pressure applied by a hand screw.

canopy — **1.** A transparent cover for the airplane's cockpit. It provides streamlining and protection for the pilot against the elements. Sometimes referred to as a cockpit canopy. **2.** The large, umbrella-shaped material of a parachute.

canted bulkhead — A bulkhead or wall that is slanted (canted) vertically.

cantilever — A beam or other member that is supported at or near only one end, without external bracing.

cantilever wing — A cantilever wing uses no external wing struts. All of its support is obtained inside the wing itself.

can-type combustor — A combustor, or burner section of a gas turbine engine, made up of eight to ten individual burner cans. These cans are long cylinders that consist of an outer case or housing within which there is a perforated stainless steel combustion chamber liner or inner liner. The can-type combustors are arranged radially around the axis of the engine. Compressed air from the compressor flows through the cans and fuel is sprayed into them and burned to add energy to the air. Cooling air flows through holes in the inner liners to keep the temperature of the liners low enough so that they will not be damaged.

canvas — A heavy, woven coarse cloth generally made of cotton.

cap--The longitudinal members at the top and bottom of a beam that resist most of the bending loads because of their strength in compression and tension.

cap cloud — A standing or stationary cap-like cloud forming on mountain or ridge tops due to cooling of moist air rising on the upwind side followed by warming and drying of downdrafts on the lee side. Also referred to as a Foehn cloud.

cap screw — A threaded fastener. The head of the cap screw that, when turned with a wrench, clamps two pieces of material together.

cap strip — Cap strips are extrusions, formed angles, or milled sections to which the web is attached. These members carry the loads caused by wing bending and also provide a foundation for attaching the skin.

capacitance — The ability of an insulator to store electrical energy in the form of electrostatic fields, expressed in farads. The amount of electricity a capacitor can store depends on several factors, including the type of material of the dielectric. It is directly proportional to the plate area and inversely proportional to the distance between the plates. The formula for capacitance is C=Q/E, in which C is the capacitance in farads, Q is the quantity (amount) of charge in coulombs, and E is the electrical pressure in volts.

capacitance box — An electronic device used to select a capacitance to insert into a circuit for testing purposes.

capacitance bridge — A null-type precision measuring instrument used to measure capacitance.

capacitance-type fuel gauging system — A fuel quantity indicating system using the fuel as the dielectric of a capacitor. It measures the weight of the fuel instead of its volume.

capacitive coupling — The use of a capacitor or capacitors to connect electronic circuits. This type coupling allows AC to pass through and blocks DC.

capacitive electrical load — An electrical load that produces more capacitive reactance than inductive reactance in the circuit.

capacitive feedback — A process of feeding a portion of the output of a circuit through a capacitor back into the input section of the circuit. Can allow the circuit to be somewhat self-regulating.

capacitive reactance (X_C) — The opposition to the flow of alternating current electricity caused by the capacitance in a circuit, and is measured in ohms. Capacitive reactance is calculated using the capacitance of the circuit and the frequency of the AC.

capacitive time constant — The amount of time, determined by the resistance of the circuit and by the capacitance of the capacitor, measured in seconds, needed for the voltage across a capacitor to rise to 63% of the source voltage.

capacitor — A device used to store electrical energy in the form of electrostatic fields. A capacitor is essentially two conductors separated by an insulator.

capacitor-discharge ignition system — An ignition system consisting of two identical independent ignition units operating from a common low-voltage DC electrical power source, the aircraft battery. A high voltage, supplied by the ignition exciter unit, charges a storage capacitor with a charge, up to four joules, that generates an arc across a wide igniter spark gap to ignite the fuel.

capacitor-input filter — An electronic filter used to smooth out the pulsations in the output of an electrical power supply. A capacitor-input filter is installed in parallel with the rectifier output, and an inductor is installed in series with the rectifier output.

capacitor-start induction motor — An AC motor whose rotor is excited by voltage induced from the field windings. A second winding whose phase is shifted by a capacitor is used to provide a rotating field for starting. When the motor gets up to speed, a centrifugal switch opens the circuit in which the capacitor is situated.

capacitor-type ignition system — An ignition system consisting of two identical independent ignition units operating from a common low-voltage DC electrical power source, the aircraft battery. A high voltage, supplied by the ignition exciter unit, charges a storage capacitor with a charge, up to four joules, which generates an arc across a wide igniter spark gap to ignite the fuel.

cape chisel — A cold chisel used when cutting square corners or slots.

capillary action — An action causing a liquid to be drawn up into extremely tiny tubes or between close-fitting parts.

capillary tube — A tube with a very small bore used to meter a fluid or transmit pressure of fluid or gas to an indicating gauge.

capping stable layer — The elevated stable layer found on top of a convective boundary layer. Usually marks a sharp transition between smooth air above and turbulent air below.

capstan — A spool-shaped device in the control system of an aircraft similar to a grooved drum-like wheel. A control cable is wound around the capstan, and the ends of the guide are attached to the aileron, the elevator, or the rudder control cable.

capstan screw — A special purpose machine screw with holes across the head to accommodate a bar that can be passed through these holes to turn the screw.

captive balloon — An aerial platform anchored to the ground.

captive screw — A fastener that has the ability to turn in the body in which it is mounted, but which will not drop out when it is unscrewed from the part it is holding.

capture — The process by which small droplets are swept up by faster-falling large droplets. Also called coalescence.

carbide drill — A specially manufactured cutting drill that has a cutting edges surfaced with tungsten carbide, tantalum carbide, or titanium carbide.

carbide tool — A metal-cutting machine tool that has cutting faces surfaced with either tungsten carbide, tantalum carbide, or titanium carbide.

carbo-blast — A grit blast. Field cleaning agent, a lignocellulose material consisting of ground up walnut shells and apricot pits.

Carboloy — A brand name of certain cutting tools and dies having tungsten carbide bonded to their cutting surface.

carbon — Nonmetallic element with a symbol of C and an atomic weight of 6. Carbon is a part of all organic compounds. It ranges in appearance from black, fluffy soot, to hard, transparent diamond.

carbon arc — An electric arc, produced by a welding machine, that jumps between two carbon electrodes or from a carbon electrode to a metal electrode. A carbon arc makes an intensely bright light and produces enough heat (approximately 10,000°F) to melt metals for welding or cutting.

carbon arc lamp — An electrical lamp in which an electric arc between carbon electrodes produces a high-intensity light.

carbon black — A soft and fluffy carbon deposit. Carbon black is produced by the incomplete burning of acetylene gas when the flame does not have enough oxygen for complete combustion.

carbon brake — An aircraft brake required for extremely high energy dissipation. Both the rotating and stationary discs are made of pure carbon.

carbon composition resistor — A resistor formed by embedding wire leads in a cylindrical slug of carbon and filler material. Usually covered with an epoxy or other plastic insulating jacket.

carbon deposits — Residue from overheated oil or incompletely burned gasoline. It forms as a hard, black crust inside the engine.

carbon dioxide (CO_2) — A colorless, odorless, nonflammable gas often used as a fire extinguishing agent in aircraft.

carbon fiber — Produced by placing carbon (an element) in an inert atmosphere at temperatures above 1,800°F. Used as a reinforcing material. Carbon fiber is a lightweight, high strength and high stiffness fiber. The material can be graphitized by heat-treating at a very high temperature.

carbon fouling — A carbon deposit that forms as a result of overly rich, idle fuel/air mixtures. The carbon forms on the inside of combustion chambers and spark plugs.

carbon knock — The pre-ignition of the fuel/air charge inside the cylinder of a reciprocating engine before the engine is ready for ignition to occur. Caused by the glowing of carbon deposits.

carbon microphone — A microphone used in telephones and some types of radio transmitters. It consists of a flexible diaphragm of carbon granules. Sound waves pressing against the microphone change its resistance, thus modulating the voltage and creating the electronic signal that is transmitted to a receiver.

carbon monoxide (CO) — A colorless, odorless, highly toxic gas that forms from incomplete combustion of a hydrocarbon fuel.

carbon monoxide detector — A device used to detect the presence and concentration of carbon monoxide gas.

carbon oil seal — An oil seal used in gas turbine engines. These seals are usually spring loaded and are similar in material and application to the carbon brushes used in electrical motors. Carbon seals rest against a surface to prevent oil from leaking out along the shaft into the compressor airflow or the turbine sections.

carbon pile resistor — A variable resistor used in some electrical equipment. A carbon pile is made of a stack of thin, pure carbon disks. Its resistance is changed by varying the amount of pressure applied to the stack.

carbon pile voltage regulator — A voltage regulator that depends on the resistance of a number of carbon disks arranged in a pile or stack. The resistance of the carbon stack varies inversely with the pressure applied. When the stack is compressed under appreciable pressure, the resistance in the stack is less. Pressure on the carbon pile depends upon two opposing forces: a spring and an electromagnet. The spring compresses the carbon pile, and the electromagnet exerts a pull that decreases the pressure. When the generator voltage varies, the pull of the electromagnet varies, thereby increasing or decreasing the pressure on the disks. This change allows a change in the generator output voltage.

carbon resistor — An electrical component used to put a controlled amount of resistance in an electrical circuit. Carbon resistors are composed of a rod of compressed graphite and binding material with wire leads, called "pigtail" leads, attached to each end of the resistor. Colored bands marked on the resistor indicate its resistance value.

carbon seal — A heat-resistant device used in turbine engines to seal the lubricating oil in the bearing cavity. A ring of carbon material rides on a highly polished metal surface to prevent lubricating oil from seeping into the gas path. Usually located at main bearings.

carbon steel — A group of iron alloys having carbon as the principal alloying agent. Low-carbon steel contains less than 0.20% carbon and is not as strong as high-carbon steel that contains up to 0.95% carbon.

carbon tetrachloride (Halon 104) — Once used as a fire extinguishing agent. A chemical formula CCl_4, a liquid with a UL toxicity rating of 3. When used as a fire extinguishing agent it becomes very toxic and harmful to humans and other animals.

carbon tracking — A fine track of carbon deposited inside the magneto, distributor, or terminal cavity of a spark plug as a result of a flashover. It acts as an electrical conductor to ground or to another electrical lead.

carbon/graphite fiber or fabric — A fiber used in advanced composites comprised of carbon filaments which may be woven together. The terms carbon and graphite have been used interchangeably for years. The Americans prefer the term graphite, while the Europeans prefer carbon. Depending on the manufacturer of the aircraft, different terms may be used. The term Carbon/Graphite is used throughout this book to include both terms.

carbonaceous — Containing carbon.

carbon-film resistor — An electrical resistor composed of a thin film of carbon on a ceramic cylinder. Wires connected to each end of the carbon film allow the resistor to be connected to an electrical circuit.

carbon-zinc cell — A portable primary cell such as a flashlight battery consisting of a carbon rod placed in a can made of zinc and filled with a paste of ammonium chloride. The chemical reaction between the paste and zinc causes electrons to leave the zinc can and travel through an external circuit to the carbon rod.

carborundum — A manufactured aluminum oxide abrasive similar to natural emery. It is used for grinding wheels and abrasive papers.

carburetor — **1.** Pressure: A hydromechanical device employing a closed feed system from the fuel pump to the discharge nozzle. It meters fuel through fixed jets according to the mass airflow through the throttle body and discharges it under a positive pressure. Pressure carburetors are distinctly different from float-type carburetors, as they do not incorporate a vented float chamber or suction pickup from a discharge nozzle located in the venturi tube. **2.** Float-type: Consists essentially of a main air passage through which the engine draws its supply of air, a mechanism to control the quantity of fuel discharged in relation to the flow of air, and a means of regulating the quantity of fuel/air mixture delivered to the engine cylinders.

carburetor air temperature — The temperature of the induction air before it enters the carburetor. The temperature of the air as it enters the carburetor must be controlled to keep the fuel/air mixture temperature high enough to prevent water from condensing out of the air and freezing and low enough to prevent detonation.

carburetor float — A component between the fuel supply and the metering system of a carburetor. The float chamber provides a nearly constant level of fuel to the main discharge nozzle. The float is connected to a needle valve and seat that meters the correct amount of fuel to the induction system according to the demand.

carburetor heater — A heater muff or jacket installed around the exhaust manifold through which induction air is drawn to warm it before it enters the carburetor. This heat prevents the formation of carburetor ice.

carburetor ice — Ice that forms inside the carburetor due to the temperature drop caused by the vaporization of the fuel. Induction system icing is an operational hazard because it can cut off the flow of the fuel/air charge or vary the fuel/air ratio.

carburetor icing — Occurs when moist air is drawn into the carburetor and is cooled to a dewpoint temperature less than 0° C.

carburetor maintenance — Maintenance that may include idle speed adjustment, removal and installation, adjusting idle mixtures, rigging, and inspection of carburetors.

carburizing — A form of case hardening of steel in which carbon is infused into the surface of the steel to increase its hardness.

carburizing flame — A flame used in oxy-acetylene welding in  which there is an excess of acetylene gas. Also called a reducing flame. This type of flame introduces carbon into the steel. It can be recognized by the greenish-white, brush-like second cone at the tip of the first cone. The outer flame is slightly luminous and has about the same appearance as an acetylene flame burning freely in air alone.

cardinal altitudes — Full thousands of feet of altitude or flight level. (3,000 feet/4,000 feet/etc. or FL 330/FL 340/etc.)

cardinal flight levels — See cardinal altitudes.

cardinal headings — Headings along the four main points of a compass: North, South, East, and West.

cardioid microphone — A microphone with the ability to pick up sounds ahead of it, rejecting sounds behind it.

cargo — Freight transported in an airplane.

cargo aircraft — An airplane whose main function is to carry freight.

carrier frequency — The high frequency alternating current that produces the electromagnetic waves that radiate from a radio transmitting antenna.

carrier wave — The high frequency alternating current that can be modulated to carry intelligence by propagation as a radio wave.

cartridge fuse — A fuse used to protect an electrical circuit from an excess of current. It consists of a fusible link held between metal rings, or caps, that screw onto each end of an insulating tube.

cartridge starter — An engine starting device that uses electrically ignited solid fuel pellets. The pressure is used to move a piston to start a reciprocating engine or to turn a turbine wheel to start a turbine engine.

cartridge-pneumatic starter — A combination air-turbine starter and cartridge starter. It can be operated by bleed air or by an explosive charge, both of which exhaust through a turbine wheel connected to a reduction gearbox. Its purpose is to start main engines.

cartridge-type filter — A disposable filter element. Used for both fuel and oil systems.

cascade electrical circuits — A system of connecting multiple levels of electrical circuits so that the output of one level feeds the input of the next level.

cascade thrust reverser — A configuration of thrust reverser used in turbojet engines in which thin airfoils or obstructions are placed in the engine's exhaust stream to duct the high-velocity exhaust gases forward. This decreases the airplane's landing roll. Also referred to as an aerodynamic thrust reverser.

cascade transformer — A device used in an electrical circuit to generate a high voltage. A system of connecting multiple levels of electrical step-up transformers where the output of one level feeds into the next level transformer, cascading until the required high-output voltage is obtained.

cascade vane — An air turning vane. One common use is in thrust reversers.

case hardening — A form of heat treatment of a metal in which the surface is made extremely hard and brittle while the core of the metal retains its toughness.

case pressure — A low pressure maintained inside the case of a hydraulic pump. In the event of a damaged seal, fluid will be forced out of the pump rather than allowing air to be drawn in.

casein glue — A form of powdered glue made from milk. Casein glues are widely used in wood aircraft repair work. For aircraft use, casein glues should contain suitable preservatives such as the chlorinated phenols and their sodium salts, to increase resistance to organic deterioration under high humidity.

casing — The rubber and fabric body of a pneumatic tire. The casing is the same as the carcass of the tire. It is composed of diagonal layers of rubber-coated fabric cord (running at right angles to one another), providing the strength of a tire.

casing plies, aircraft tires — Diagonal, strength-providing layers of rubber-coated nylon cord fabric (running at right angles to one another).

cast iron — Iron that contains 6% to 8% carbon and silicon. Cast iron is a hard unmalleable pig iron made by casting.

cast-aluminum alloy — Aluminum alloy that has been heated to its molten state and poured into a mold to give it a desired shape.

castle nut — A general purpose hexagonal nut for aircraft or engine use Their shape resembles a castle with the slots between the "turrets" used for locking the nut to the bolt with cotter pins.

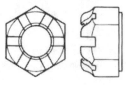

catalyst — A material used to bring about a change, or speeds up the rate of change of a chemical action, but does not actually enter into the change itself.

catalytic cracking — A method of refining petroleum products in which a catalyst is used to change high-boiling-point hydrocarbons into low-boiling-point hydrocarbons.

catalyzed material — A material whose cure is initiated by the addition of a catalyst.

catalyzed resin — A term used to describe the resin mixture after it has been mixed with the catalyst or hardener. It may still be in the workable state.

catapult — A mechanism used to launch an object into the air. Catapults are used to launch heavily loaded aircraft from the decks of aircraft carriers at a high rate of speed.

category — 1. As used with respect to the certification, ratings, privileges, and limitations of airmen, means a broad classification of aircraft. Examples include: airplane; rotorcraft; glider; and lighter-than-air. 2. As used with respect to the certification of aircraft, means a grouping of aircraft based upon intended use or operating limitations. Examples include: transport, normal, utility, acrobatic, limited, restricted, and provisional.

category A — With respect to transport category rotorcraft, means multiengine rotorcraft designed with engine and system isolation features specified in Part 29 and utilizing scheduled takeoff and landing operations under a critical engine failure concept which assures adequate designated surface area and adequate performance capability for continued safe flight in the event of engine failure.

category B — With respect to transport category rotorcraft, means single-engine or multiengine rotorcraft which do not fully meet all Category A standards. Category B rotorcraft have no guaranteed stay-up ability in the event of engine failure and unscheduled landing is assumed.

category II operation — With respect to the operation of aircraft, this is a straight-in ILS approach to the runway of an airport, under a category II ILS instrument approach procedure issued by the Administrator or other appropriate authority.

category II operations — With respect to the operation of aircraft, means a straight-in ILS approach to the runway of an airport under a Category II ILS instrument approach procedure issued by the Administrator or other appropriate authority.

category III operations — With respect to the operation of aircraft, means an ILS approach to, and landing on, the runway of an airport using a Category III ILS instrument approach procedure issued by the Administrator or other appropriate authority.

category IIIA operations — An ILS approach and landing with no decision height (DH), or a DH below 100 feet (30 meters), and controlling runway visual range not less than 700 feet (200 meters).

category IIIB operations — An ILS approach and landing with no DH, or with a DH below 50 feet (15 meters), and controlling runway visual range less than 700 feet (200 meters), but not less than 150 feet (50 meters).

category IIIC operations — An ILS approach and landing with no DH and no runway visual range limitation. Also known as zero-zero landing.

catenary curve — A curve formed by a flexible cord or rope suspended between two points at the same level.

catenary thermal shield — A curved sheet metal section between turbine wheels of a particular set. It serves as a heat barrier between the gas path and the inner portion of a drum-type turbine wheel.

cathedral — The downslope of the wings from the fuselage. It is the opposite of dihedral. Airplanes that employ cathedral have an increase in maneuverability but a decrease in stability.

cathode — 1. The negative terminal of a semiconductor diode or the element in an electron tube from which the electrons are emitted. 2. An active element in an electrochemical cell that loses oxygen in the chemical action that causes electrons to flow.

cathode of a semiconductor diode — The end of a semiconductor diode made of N-type material.

cathode protection — A material more anodic than the material being protected is attached to or plated on the material, which then becomes the cathode and is not corroded. Also referred to as sacrificial corrosion

Cathode-ray oscilloscope — An electrical measuring instrument in which the readout is on the surface of a tube similar to that in a television set. Electrons are made to strike the inside of the tube where they cause the coating of the tube to glow. Recurring voltage changes are displayed on the tube in the form of a green line.

cathode-ray tube (CRT) — An electron tube in which a stream of electrons (cathode rays) from an electron gun impinges upon a fluorescent screen, thus producing a bright spot on the screen. The electron beam is deflected electrically or magnetically to produce patterns on the screen. Also referred to as a display tube.

cation — A positive charged ion that moves toward the cathode in the process of electrolysis.

catwalk — A narrow walkway (as along a bridge or elevated on the side of a building).

caul plates — In composites, smooth plates used during the cure process to apply pressure in a uniform manner.

caustic material — Any substance having the ability of burning, corroding, or eroding other substances by chemical action.

caustic soda — A common name for sodium hydroxide.

cavitation — A partial vacuum of an area of low pressure behind an object that is moving in a fluid.

cavity — A hole or hollow within a body or structure.

C-clamp — A metal clamp in the general shape of the letter C. It is used to exert pressure and to temporarily hold objects together.

C-D inlet or exhaust — **1.** Inlet. The forward section is convergent to increase air pressure and reduce air velocity to subsonic speed. The aft section is divergent to increase air pressure still further and slow airflow to approximately Mach 0.5 before entering the engine. **2.** Exhaust. The forward section is convergent to increase gas pressure. The aft section is divergent to increase gas velocity to supersonic speed. This arrangement is necessary in order for the aircraft to attain supersonic speed.

Ceconite — A fabric woven from polyester fibers.

ceiling — In meteorology in the U.S., (1) the height above the surface of the base of the lowest layer of clouds or obscuring phenomena aloft that hides more than half of the sky, or (2) the vertical visibility into an obscuration.

ceiling balloon — A small balloon used to determine the height of a cloud base or the extent of vertical visibility.

ceiling light — A light used by weather observers to measure the height of the bottom of a layer of clouds at night.

ceilometer — A device used to measure the height of the bottom layer of clouds above a weather station. Consists of a light transmitter and receiver separated by a known horizontal distance. A beam of light shines on the cloud layer and height is calculated using trigonometry.

celestial dome — The hemisphere of the sky as observed from a point on the ground.

celestial navigation — Navigating by use of the stars and sun.

cellular combustor — A combustor, or burner section of a gas turbine engine, that is made up of eight to ten individual burner cans. These cans are long cylinders that consist of an outer case or housing within which there is a perforated stainless steel combustion chamber liner or inner liner. The can-type combustors are arranged radially around the axis of the engine. Compressed air from the compressor flows through the cans and fuel is sprayed into them and burned to add energy to the air. Cooling air flows through holes in the inner liners to keep the temperature of the liners low enough so that they will not be damaged.

Celluloid — The registered trade name of a thermoplastic material consisting essentially of cellulose nitrate and camphor.

cellulose — A material that comes from the cell walls of plants and is the raw material of many manufactured goods.

cellulose acetate butyrate — A compound formed by the action of acetic and butyric acid on cellulose.

cellulose acetate butyrate dope — An aircraft dope having a cellulose acetate butyrate film base and suitable plasticizers, along with the necessary solvents and diluents. Butyrate dope has a better tautening effect on fabric and is less flammable than nitrate dope.

cellulose nitrate — A compound formed by treating cellulose with a mixture of nitric and sulfuric acids.

cellulose nitrate dope — Aircraft dope consisting of a nitrocellulose film base with the appropriate plasticizers, thinners, and solvents. It has excellent encapsulating properties, but its high flammability has caused its decrease in popularity as a finish for fabric covered aircraft.

Celsius (C) — A temperature scale with zero degrees as the melting point of pure ice and 100 degrees as the boiling point of pure water at standard sea level atmospheric pressure.

CENRAP — See center radar ARTS presentation/processing.

CENRAP-plus — See center radar ARTS presentation/processing-plus.

center — 1. A point that is equally distant from the sides or outer boundaries of something. The middle. 2. Air Route Traffic Control Center.

center console — The space between the pilot and copilot where the power lever control system is positioned on most multi-engine airplanes.

center drill — A combination of a twist drill and a 60° countersink. Used to center a hole and a countersink in a piece of metal.

center head — A tool, in a combination measuring set, used for finding the center of circular objects such as a piece of round bar stock.

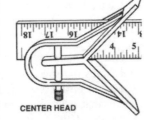

CENTER HEAD

center line — In aircraft drawings, the alternate long and short dashes indicating the center of an object or part of an object.

center of airfoil moments — The point about which the basic airfoil moment coefficients are given, usually the aerodynamic center.

center of gravity (CG) — The theoretical point where the entire weight of the airplane is considered to be concentrated.

center of gravity envelope — A graphic depiction of the fore-and-aft range of center of gravity limits, showing the way these limits vary with the gross weight of the aircraft.

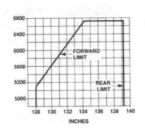

center of gravity limits — The extreme forward and rearward positions at which the center of gravity of an aircraft may be located.

center of gravity range — The distance between the forward and rearward center of gravity limits, as specified on the Type Certificate Data Sheet for the aircraft.

center of lift — The resultant of all of the centers of pressures of an airfoil.

center of mass — The location within an aircraft at which its entire mass can be considered to be in equilibrium.

center of pressure — The point along the wing chord line where lift is considered to be concentrated.

center of pressure coefficient — The ratio of the distance of the center of pressure from the leading edge to the chord length.

center of thrust — The resultant of all of the thrust forces of the propellers or the exhaust jet stream.

center punch — A punch having a somewhat blunt point, used to form an indentation in sheet metal that can be used to start the twist drill.

center radar arts presentation/processing (CENRAP) — A computer program developed to provide a back-up system for airport surveillance radar in the event of a failure or malfunction. The program uses air route traffic control center radar for the processing and presentation of data on the ARTS IIA or IIIA displays.

center radar arts presentation/processing-plus (CENRAP-plus) — A computer program developed to provide a back-up system for airport surveillance radar in the event of a terminal secondary radar system failure. The program uses a combination of Air Route Traffic Control Center Radar and terminal airport surveillance radar primary targets displayed simultaneously for the processing and presentation of data on the ARTS IIA or IIIA displays.

Center Weather Advisory (CWA) — Unscheduled in-flight, flow control, air traffic, and air crew advisory. These can be a supplement to an existing SIGMET/AIRMET, or when conditions are observed or expected to adversely affect air traffic.

center's area — The specified airspace within which an air route traffic control center (ARTCC) provides air traffic control and advisory service.

centering — In gliders, adjusting circles while thermalling to provide the greatest average climb.

centerline thrust — An aircraft design for multi-engine airplanes that eliminates engine-out asymmetrical thrust and asymmetrical drag. Mounting the engines along the fuselage centerline eliminates directional control problems of the conventional twin following an engine failure. However, the loss of an engine can still reduce climb performance significantly. Airplanes incorporating centerline thrust do not have a published V_{mc} speed. Pilots who receive a multi-engine rating in an airplane of this type have the restriction "Limited to Center Thrust" placed on their pilot certificates. This restriction can be removed when they subsequently demonstrate the maneuver during a practical test in a conventional twin.

center-of-rotation line — The line on a drawing about which an object will rotate.

center-tapped winding — A winding on an electrical transformer that has a connection (tap) located in its electrical center. It is used to divide the winding in half.

centervent system — Use of the main rotorshaft as an air-oil separator in place of a driven centrifugal device. After separation, oil is scavenged back to the oil reservoir and air, which was entrained in the oil, is vented through the rotorshaft into the gas path in the area of the turbine wheel.

centigrade — Formerly used for Celsius temperature.

centigrade temperature scale — Same as Celsius temperature scale.

centimeter-gram-second (cgs) system — Metric system of measurement using the centimeter as the basic unit of measurement of length, the gram as the basic unit for weight, and the second as the basic unit for time.

centistoke — A unit of viscosity measurement of both fuels and oils. 1/100 of a "stoke."

Central East Pacific — An organized route system between the U.S. West Coast and Hawaii.

Central Processing Unit (CPU) — The main processing portion of a computer. The CPU stores and operates on data, provides time signals, and performs arithmetic/logic functions.

central refueling system — An aircraft fuel system in which all tanks may be filled from one fueling point. Also referred to as a single-point refueling system or a pressure refueling system.

centrifugal — An apparent force that opposes centripetal force, resulting from the effect of inertia during a turn.

centrifugal brake — A friction brake used to apply friction if the unit rotating turns at a speed that is faster than is permitted.

centrifugal breather — A centrifugal device through which oil laden air from the vent subsystem passes. Oil is returned to the reservoir and air exits back to the atmosphere.

centrifugal clutch — A friction clutch that engages when a drive wheel reaches a predetermined speed. The clutch is engaged by centrifugal force that acts on a flyweight mechanism.

centrifugal filter — A filtering element that separates contaminants from a fluid by centrifugal action. It throws contaminants by rotary motion into traps that hold them until they can be removed.

centrifugal flow compressor — An impeller-shaped device that receives air at its center and slings air outward at high velocity into a diffuser for increased pressure. Also referred to as a radial outflow compressor.

centrifugal force — The outward pull on a body as it rotates or spins.

centrifugal force — The apparent force that an object moving along a circular path exerts on the body constraining the object and that acts outwardly away from the center of rotation.

centrifugal moment — A force that tries to produce a rotation caused by the amount of centrifugal force acting on an object.

centrifugal oil filter — A rotary filtering element used to throw contaminants outward into sediment traps.

centrifugal pump — Any pump that uses a high-speed impeller to throw the fluid outward by centrifugal action.

centrifugal switch — An electrical switch mounted inside of a rotating induction motor of a capacitor-starter. The switch, actuated by centrifugal force, disconnects the starter winding when the rotor is turning at a predetermined speed.

centrifugal tachometer — A mechanical tachometer that measures the speed of a rotating shaft. Flyweights are mounted on a collar around the rotating shaft in such away that centrifugal force pulls the flyweights away from the shaft. As the flyweights move away from the shaft, the collar moves up the shaft causing a pointer to move over a dial registering the shaft speed.

centrifugal twisting force — The centrifugal forces acting on a propeller blade. The twisting force is present in all rotating propellers and always acts to send the blades toward a lower pitch position.

centrifugal twisting moment — The tendency of a propeller blade to twist on its axis due to the centrifugal forces acting on the blade. The twisting moment is present in all rotating propellers and always acts to send the blades toward a lower pitch position.

centrifugal-type pump — A pump that uses a high-speed impeller to throw the fluid outward at a high velocity.

centrifuge — A device used to separate a liquid mixture or a suspension into its various components that have different specific gravities.

centrifuge action — A force that tends to separate particles according to their density, or to pull an object apart by rotating it rapidly about its center.

centrifuging — A method of separating particles of varying density by spinning them in a centrifuge.

centripetal force — The force within a body that opposes the centrifugal force as the body rotates or spins.

centroid — The center of mass of a body or a point about which all of its mass is concentrated.

ceramic — A clay-like material composed primarily of magnesium and aluminum oxide, which may be molded and fired to produce an excellent insulating material.

ceramic magnet — A permanent magnet made by compressing a mixture of ceramic material and sintered magnetic particles.

CERAP — See combined center-RAPCON.

certificate — An official Federal Aviation Administration document authorizing a privilege, fact, or legal concept.

certificated — An object or person that has been granted a certificate of approval, usually by the Federal Aviation Administration.

certificated aircraft — An aircraft designed to meet minimum specifications and requirements specified by the Federal Aviation Administration. When these conditions are met, an Approved Type Certificate is issued for the aircraft. In order for the aircraft to maintain the certificate (to be considered legally airworthy) it must be maintained in such a way that it continues to meet these specifications.

certificated technician — A person who holds a valid technician's certificate issued by the Federal Aviation Administration with either an Airframe or Powerplant rating, or both ratings.

certified tower radar display (CTRD) — A radar display that provides a presentation of primary, beacon radar videos, and alphanumeric data from an Air Traffic Control radar system, which is certified by the FAA to provide radar services. Examples include Digital Bright Radar Indicator Tower Equipment (DBRITE), Tower Display Workstation (TDW) and BRITE.

cesium — A soft, ductile, bluish-gray metallic chemical element with a symbol of Cs and an atomic weight of 55. Used in the manufacture of photoelectric cells.

cesium-barium 137 — A radioactive substance used to coat ignition system air-gap points to synchronize discharge of current to the igniter plug.

CFI renewal — A process that allows CFIs to renew their certificates since flight instructor certificates are only valid for 24 months. CFI renewal is not automatic by any means since it requires a specific certificate action by the FAA. Certificate renewal should be in accordance with FAR Part 61.197 by any of the following methods. A CFI can present the FAA with a record of training that shows endorsement of at least 5 students for a practical test for a certificate or rating, and at least 80% of the students passed on their first attempt. A CFI can also show the FAA a satisfactory record as a Part 121 or 135 check pilot, chief flight instructor, check airman or flight instructor, or that he or she is in a position involving the regular evaluation of pilots. Graduation from an approved flight instructor refresher course (FIRC), consisting of at least 16 hours of ground and/or flight training, also may be the basis of renewing certificates at the discretion of the FAA.

CFR engine — An engine used by Cooperative Fuel Research to determine the octane rating of a hydrocarbon fuel. A CFR engine has a variable compression ratio, and it can cause any of the fuels that are being tested to detonate. When the correct percentages of iso-octane are obtained, an octane number is given to the fuel.

Chadwick balancer — The electronic balancing or tracking of rotor blades. It is manufactured by Chadwick-Helmuth, Inc.

chafe — To wear away by a rubbing action.

chafed surface — A surface resulting from a slight relative movement between two surfaces under high contact pressure. The surface of each part reveals metal removed and metal added.

chafers — In tires, layers of fabric and rubber that protect the tire carcass from damage during mounting and demounting. They insulate the carcass from brake heat and provide a good seal against movement during dynamic operations.

chaff — Thin, narrow metallic reflectors of various lengths and frequency responses, used to reflect radar energy. These reflectors, when dropped from aircraft and allowed to drift downward, result in large targets on the radar display.

chafing — Rubbing action between adjacent or contacting parts under light pressure which results in wear.

chafing strip — See chafing tape.

chafing tape — Cloth or paper tape placed over a metal seam or protruding screw head that is to be covered with fabric. It is used to protect the fabric from wear.

chain gear — A gear or sprocket used to transmit motion from one shaft to another shaft connected by a roller chain similar to that used in a bicycle.

chain hoist — A mechanism used in a shop to lift heavy weights. A chain hoist uses an endless loop of chain to drive a geared wheel that also supports and lifts the load as it is pulled up by the geared wheel.

chain reaction — A self-sustaining action in which one event causes other events to happen.

chaining — Combines behaviors students already know to assemble more complex behaviors.

chamfer — A bevel cut on the edge of a piece of material.

chamfered point of a threaded fastener — The point of a bolt or a screw formed in the shape of a cone with its top cut off. The chamfered point allows easy entry into the hole for starting.

chamfered tooth — The tooth of the gear on the rotating magnet or the distributor gear which is beveled to identify it for use when timing the magneto.

chamois — A piece of soft leather used to filter gasoline. Gasoline will pass through a chamois, but water will not. Gasoline that has been filtered through a chamois can be considered to be free from water.

chandelle — A maximum performance 180° climbing turn. It involves continual changes in pitch, bank, airspeed, and control pressures. During the maneuver, the airspeed gradually decreases from the entry speed to a few knots above stall speed at the completion of the 180° turn.

change of state — The transformation of a substance from one form (state) to another such as a solid changing to a liquid. In meteorology, a change of state usually refers to the change of water from one form to another.
Examples include: condensation – vapor to liquid; evaporation – liquid to vapor; sublimation – solid (ice) to vapor; freezing – liquid to solid; melting – solid to liquid.

changeover point (COP) — The point at which a pilot changes frequencies between navigation aids when other than the midpoint on an airway.

channel — A metal structural member either extruded or bent into a U-shape

channel iron — Extruded steel either extruded or bent into a U-shape.

channel section — A form of structural material that has the cross sectional shape of a channel or the letter U.

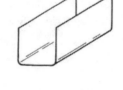

characteristic curves — 1. A series of graphically presented curves that describe in mathematical terms the characteristics of lift and drag produced by an airfoil section. 2. A graph that shows the performance of an electron tube or a transistor under various operating conditions.

characteristic potential difference — The theoretical potential difference produced by a chemical cell using specific pole materials.

characteristics of learning — Effective learning shares several common characteristics. Learning is dynamic and should be purposeful, based on experience, multifaceted, and involve an active process.

charcoal — Black porous carbon.

charge — 1. A quantity of electricity. If the charged material holds a greater number of electrons than normal, it is said to be negatively charged. If the material has a deficiency of electrons, it is positively charged. 2. The physical condition that gives rise to an electric field.

charging a battery — The preparation of battery for service by passing low-voltage DC through the battery in the opposite direction of normal battery output. A lead acid battery should be recharged when a cell has a hydrometer reading of 1.240 or below. A fully charged battery reads about 1.300 on the hydrometer.

charging current — A current passed through a secondary cell that restores the active material on the plates to a condition that allows them to change chemical energy into electrical energy.

charging stand — A handy and compact arrangement of air conditioning service equipment containing a vacuum pump, manifold set, and a method of measuring and dispensing the refrigerant.

Charles' law — A law of physics which states that if a gas is held at a constant pressure, it will expand in direct relationship to the increase in its absolute temperature.

chart — 1. A pictorial presentation of data. 2. A graph. 3. A graphic representation of the operation of engine performance, fuel consumption, horsepower, or limitation of some specific unit. 4. A map.

chart NOTAMS — Jeppesen Chart NOTAMs include significant information changes affecting Enroute, Area, and Terminal charts. Entries are published until the temporary condition no longer exists, or until the permanent change appears on revised charts. Enroute chart numbers / panel numbers / letters and area chart identifiers are included for each entry in the enroute portion of the chart NOTAMs. To avoid duplication of information in combined Enroute and Terminal Chart NOTAMs, navaid conditions, except for ILS components, are listed only in the Enroute portion of the Chart NOTAMs. All times are local unless otherwise indicated. Arrows indicate new or revised information. Chart NOTAMs are only an abbreviated service. Always ask for pertinent NOTAMs prior to flight.

chart, navigation — A special map used for aerial navigation that gives the location and the necessary information about all of the navigation aids. A chart shows the grids of latitude and longitude and provides a surface for plotting courses and locating fixes.

charted VFR flyways — Charted VFR Flyways are flight paths recommended for use to bypass areas heavily traversed by large turbine-powered aircraft. Pilot compliance with recommended flyways and associated altitudes is strictly voluntary. VFR Flyway Planning charts are published on the back of existing VFR Terminal Area charts.

charted visual flight procedure (CVFP) approach — An approach conducted while operating on an instrument flight rules (IFR) flight plan which authorizes the pilot of an aircraft to proceed visually and clear of clouds to the airport via visual landmarks and other information depicted on a charted visual flight procedure. This approach must be authorized and under the control of the appropriate air traffic control facility. Weather minimums required are depicted on the chart.

chase — An aircraft flown in proximity to another aircraft normally to observe its performance during training or testing.

chase aircraft — See Chase.

chasing threads — The process of cutting screw threads by moving a tool along the surface of the work to be threaded.

chassis — An aluminum, copper, or plated steel body around which an electronic unit is built. It serves as the support for the electronic components, power supply, etc., and is often used as a voltage reference point.

chassis ground — In electricity, a ground connected to the case or chassis of electronic equipment. See also earth ground.

chattering brakes — A heavy vibration in the brakes produced by the brake friction as the disks rotate. Chattering can be caused by glazed discs.

check flight — A test flight to check the aircraft performance after major re-work or repairs.

check list — A sequential systematic list of specific procedures to be followed when performing any complex operation. For example, check lists are used in the performance of preflight inspections, 100-hour inspections, and annual inspections of aircraft to ensure all required operations are completed.

check nut — A thin nut jammed against another nut to prevent it from loosening.

check valve — A valve that allows free flow of fluid in one direction but prevents or restricts fluid from flowing in the opposite direction.

checklist — A systematic list of items and equipment on board an aircraft intended for reference, verification, or identification. An essential tool for safely flying the airplane.

checkpoint — A navigation location identified either visually or electronically.

cheesecloth — A lightweight cotton gauze that has no sizing in it. It is used as a polishing cloth or as a straining element to remove lumps and contaminants from liquids.

chemical bond — An adhesive agent that is applied to two or more parts or pieces. Joins them together by molecular attraction

chemical compound — The substance formed by the chemical reaction between two or more chemical elements.

chemical element — A fundamental substance that consists of atoms of only one kind. Examples of chemical elements include oxygen, carbon, gold, silver, and hydrogen.

chemical energy — The energy stored in chemicals due to their attraction to or reaction with other chemicals.

chemical etching — 1. A process in which small cracks in aluminum are detected by application of a caustic soda solution. 2. A chemical process used to etch (roughen) the surface of metal in preparation for priming or painting.

chemical fire extinguisher — A fire extinguisher that extinguishes fire by expelling a chemical fire extinguishing agent.

chemical milling — A chemical etching process used to machine large sheets of metal. Chemical milling economically reduces the weight of the aircraft and produces a lightweight skin that has all of the needed strength and rigidity than can be done with conventional machining or by using riveted-on stiffeners.

chemical reaction — A chemical alteration in a substance to form a chemical compound. This is always accompanied by an energy change.

chemical salt — The result of the combination of an alkali with an acid. Salts are generally porous and powdery in appearance and are the visible evidence of corrosion in a metal.

cherry picker — A hydraulically operated boom with a large basket on its end. A person can be lifted into the basket in order to work at high locations on large airplanes.

Cherry® rivet — A form of blind rivet patented and manufactured by the Cherry division of Textron, Inc. An upset head is formed by pulling the tapered stem through its hollow shank.

chevron seal — A single-direction seal in a hydraulic or pneumatic actuator. It derives its name from its V-shaped (chevron) cross section.

chilled iron — Cast iron that has been cast in a steel mold. The casting is quickly cooled by the steel mold so that it retains most of the carbon as well as a high degree of hardness.

chin — An aircraft structure that sticks out from the bottom of the forward part of the fuselage.

chine — The longitudinal member on the side of a float or seaplane hull where the bottom and the side meet.

chine tire — A nose wheel tire that has a deflector molded into its sidewall. Chine tires are mounted on the nose wheel of jet aircraft and prevent water, ice, snow, and slush from getting into the intake of the engines by throwing the water and slush outward and away from the engines.

chinook — A warm, dry wind that blows down the eastern slopes of the Rocky Mountains into the United States from Canada. The moisture in the air of a chinook is almost completely lost as it blows up the western slopes of the mountains, and it is dry and warm as it blows down the eastern slopes.

chip — 1. A small fragment of metal removed from a surface by cutting with a tool. 2. An electronic component containing an integrated circuit.

chip detector — An electrical metal detection warning system. A magnetic sump or drain plug with an electrode at its center and with ground potential at its casing. When ferrous particles bridge the gap, the current path is completed and a warning light illuminates in the cockpit.

chipping — The breaking away of pieces of material by excessive stress or by careless handling.

chisel — A hard steel cutting tool used to shear metal when it is hammered.

chlorate candle — A chemical oxygen supply used as an emergency oxygen supply in large aircraft and in some smaller aircraft. When the candle is heated, it emits oxygen that is then routed to a mask for breathing by the passenger. Also referred to as an oxygen candle.

chlorine — A gaseous chemical element with a symbol of Cl and an atomic number of 17.

chlorobromomethane (Halon 1011) — A chem.ical formula -CH_2ClBr. A liquefied gas with a UL toxicity rating of 3. Commonly referred to as CB, chlorobromomethane is more toxic than CO_2. It is corrosive to aluminum, magnesium, steel, and brass. It is not recommended for aircraft use.

chock — A block of material wedged under the tires of an aircraft to prevent it from rolling on the ground.

choke — An electrical inductor used to oppose the flow of pulsating DC electricity. Chokes are used with capacitors to make filter circuits that smooth out the voltage changes and make pulsating direct current into smooth flowing DC.

choke bore — A method of boring the cylinder of an aircraft engine in which the top, that portion affected by the mass of the cylinder head, has a diameter slightly less than that of the main bore of the barrel. When the cylinder reaches operating temperature, the increased expansion due to the larger mass of the head has caused the bore to be straight throughout its length.

choke coil — An inductance coil designed to provide a high reactance to certain frequencies and generally used to block or reduce currents at these frequencies.

choked — A condition of a turbojet engine where airflow from a convergent nozzle is at Mn = 1.0 (speed of sound) and cannot be further accelerated regardless of the pressure applied. Choke normally occurs at the turbine nozzle and exhaust nozzle, although it is a cause of stall conditions in the compressor.

choked airflow — In gas turbine engines, an airflow condition from a convergent shaped nozzle, where the gas is traveling at the speed of sound and cannot be further accelerated. Any increase in internal pressure will pass out the nozzle in the form of pressure.

choked cylinder bore — The cylinder of a reciprocating engine whose bore is slightly smaller in the part of the cylinder that is screwed into the cast aluminum head than it is in the center of the cylinder barrel. The cylinder head expands at normal operating temperature enough so that the bore straightens out and has the same diameter throughout.

choked nozzle — A jet engine nozzle whose flow rate has reached the speed of sound.

choke-input filter — A form of filter used with an electronic power supply to change pulsating direct current into smooth DC.

choo-choo — A mild compressor surge condition caused by insufficient compression ratio across the compressor.

chopper — Slang for helicopter.

chord — 1. An imaginary line drawn through an airfoil from its leading edge to its trailing edge. The chord, or chord line, is used as a reference (a datum line) for laying out the curve of the airfoil. 2. A straight line that passes through a circle and touches the circumference at two points. Also referred to as the diameter of the circle.

chord line — An imaginary line drawn through an airfoil from its leading edge to its trailing edge. The chord, or chord line, is used as a reference (a datum line) for laying out the curve of the airfoil.

chordwise — Passing from the leading edge to the trailing edge of an airfoil.

chordwise axis — In rotorcraft, a term used in reference to semirigid rotors describing the flapping or teetering axis of the rotor.

chrome molybdenum steel — A strong, tough and highly weldable alloy steel containing chromium and molybdenum. The most commonly used steel for aircraft structure is the SAE 4100 series.

chrome nickel molybdenum steel — Steel that has been alloyed with chromium, nickel, and molybdenum.

chrome pickling — A method used to convert the surface of magnesium to a hard oxide film in order to protect it from corrosion. This is accomplished by soaking the magnesium in a solution of potassium dichromate.

chrome plated cylinder — Hard chrome plating applied to the inside walls of an aircraft cylinder to form a hard, wear-resistant surface.

chrome plating — 1. An electroplating process transferring chromium to the surface of the steel. Either hard chrome or decorative chrome can be applied. 2. A treatment for cylinder walls of reciprocating engines. It hardens the walls and helps lubricate them. Worn cylinder barrels may be ground so that their bore is straight and round. Then hard chromium is electroplated on the cylinder walls to a depth that brings the diameter of the cylinder bore back to its original dimensions. The surface of the chrome plating on the cylinder walls resembles a maze of spider webs. There are thousands of tiny, interconnected cracks in its surface. The electroplating current is then reversed, and these tiny cracks open up enough so that they can hold oil. Porous chrome plating provides a hard, wear-resistant surface for the piston rings to ride on. The oil that is trapped in the tiny grooves helps seal the rings and lubricate the wall to minimize piston ring and cylinder wall wear.

chrome vanadium steel — A steel alloyed with chromium and vanadium. It is the SAE 6100 series and is used extensively in the manufacturing of technicians' hand tools.

chrome-alumel — Bimetallic strip of metal used in the exhaust temperature indicating system. Alumel contains an excess of free electrons which, when heated, move into the chromel lead. This current flow is read as an indication of temperature.

chromel — An alloy of nickel and chromium that is highly resistant to oxidation and has a high electrical resistance.

chromel-alumel thermocouple — A thermocouple is a device that generates a small current when heated. A chromel-alumel thermocouple consists of a positive lead of chromel and a negative lead of alumel. This device is used primarily to measure high temperatures in reciprocating and jet turbine engines.

chromic acid — An acid similar to sulfuric acid except for the substitution of chromium for sulfur. It is used as an etchant to prepare aluminum alloys for finishing and as a corrosion inhibitor.

chromic acid etch — A solution of sodium dichromate, nitric acid, and water used to etch or roughen a surface.

chromium — A hard, brittle, white metallic chemical element with a symbol of Cr and an atomic number of 24. Chromium is highly resistant to corrosion and is used for plating metal to harden its surface or to protect it from rust or corrosion.

chronometric tachometer — An instrument used to measure the speed in revolutions per minute of a reciprocating aircraft's engine crankshaft. The chronometric tachometer repeatedly counts the number of revolutions in a given period of time and displays the average speed on its dial.

chuck — A special clamp-like device on a lathe used to hold the material being worked or a drill to hold the drill bit. Chucks have three or more jaws that are used to clamp and hold the material or the tool.

chugging — Low frequency airflow oscillations within a turbine engine. Chugging is a mild, audible stall condition that can usually be controlled by proper throttle movement.

chute — An inclined trough or channel used to allow objects or materials to be sent from one level or place to another.

cigarette — A ceramic or synthetic rubber insulator used at the end of an ignition lead to insulate it from the shielded barrel of a spark plug.

circle — A closed-plane curve in which all points along the curve are equidistant from a point within the center of curve.

circle graph — A graph using a circle divided into pieces like a pie to convey data. Also referred to as a pie chart.

circle to runway (runway number) — Used by ATC to inform the pilot that he must circle to land because the runway in use is other than the runway aligned with the instrument approach procedure. When the direction of the circling maneuver in relation to the airport/runway is required, the controller will state the direction (eight cardinal compass points) and specify a left or right downwind or base leg as appropriate; e.g., "Cleared VOR Runway Three Six Approach circle to Runway Two Two," or "Circle northwest of the airport for a right downwind to Runway Two Two."

circle-to-land maneuver — A maneuver initiated by the pilot to align the aircraft with a runway for landing when a straight-in landing from an instrument approach is not possible or is not desirable. At tower controlled airports, this maneuver is made only after ATC authorization has been obtained and the pilot has established required visual reference to the airport.

circling — A maneuver to align the aircraft with a runway for landing when a straight-in landing from an instrument approach is not possible or desirable. This maneuver is made only after ATC authorization and the pilot has established the required visual reference to the airport.

circling approach — See circle-to-land maneuver.

circling maneuver — See circle-to-land maneuver.

circling minima — A statement of MDA and visibility required for the circle-to-land maneuver.
Note: Descent below the established MDA or DH is not authorized during an approach unless the aircraft is in a position from which a normal approach to the runway of intended landing can be made and adequate visual reference to required visual cues is maintained.

circuit — The complete path in which electrical current flows. It must contain a source of electrical energy, a load to absorb this energy, and conductors to carry the electron flow.

circuit breaker — A circuit-protecting device that opens the circuit in case of excess current flow. A circuit breaker differs from a fuse in that it can be reset without having to be replaced.

circuit diagram — An electrical drawing that uses conventional symbols to show how the components in an electrical system are interconnected.

circuit protector — A device that opens an electrical circuit in the event of an excessive current flow.

circular inch — The area of a circle whose diameter is 1."

circular mil (CM) — A measurement of area equal to that of a circle having a diameter of $1/1,000$," 1 mil., or 0.001."

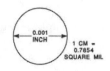

circular motion — The motion of an object along a curved path in which the object stays a constant distance from the center of the motion.

circular saw — A power saw driven by an electric motor that uses a circular blade.

circular slide rule — A slide rule having scales arranged in circles on the surface of a disk. Transparent runners attached at the center of the disk can be moved over the scale to perform various mathematical operations.

circulation — In meteorology, the organized movement of air. Also called an eddy.

circumference of a circle — The linear distance around a circle. The circumference of a circle is always 3.1416 times the length of the diameter of the circle.

circumferential frame — A circular or oval frame. It gives shape to a fuselage or nacelle. Also referred to as a belt or transverse frame.

circumscribed circle — A circle drawn around the outside of another figure in such a way that all of the points touch the circumference of the circle.

cirriform — High altitude clouds made up mostly of small ice crystals.

cirrocumulus (Cc) — Cirroform clouds, elements of which appear as small puffy clouds sometimes aligned in rows.

cirrostratus (Cs) — Layer of cirroform clouds that appear as thin white or light-gray sheets.

cirrus (Ci) — Detached cirroform clouds that appear as white feathers or filaments.

cistern — A container used to store a liquid.

CIT sensor — A device that sends an inlet duct temperature signal to the fuel control as a scheduling parameter.

civil aircraft — Aircraft other than public aircraft.

civil twilight — The period of time before sunrise or after sunset when the sun is not more than 6° below the horizon.

clad aluminum — An aluminum alloy with a coating of pure aluminum rolled onto both sides for corrosion protection.

cladding — A method of protecting aluminum alloys from corrosion by rolling a coating of pure aluminum onto the surface of the alloy. Although cladding protects the alloy, it also reduces its strength.

clamp — Any device used to exert pressure to temporarily hold objects together.

clamp-on ammeter — A hand-held ammeter that clamps around a current-carrying wire that is to be measured. The changing magnetic field around the wire induces a voltage in the jaws of the ammeter proportional to the amount of current flowing in the line.

clamshell doors — Two doors that open on the opposite sides of the center line similar to the way the shell of a clam opens.

clamshell thrust reverser — A thrust reverser, clamshell door system that fits in the exhaust system of a turbojet engine. When the reverser is deployed for thrust reversing, the doors move into position to block the normal tailpipe and duct the exhaust gases around so that they flow forward to oppose the forward movement of the aircraft.

class — **1.** As used with respect to the certification, ratings, privileges, and limitations of airmen, means a classification of aircraft within a category having similar operating characteristics. Examples include: single engine; multiengine; land; water; gyroplane; helicopter; airship; and free balloon. **2.** As used with respect to the certification of aircraft, means a broad grouping of aircraft having similar characteristics of propulsion, flight, or landing. Examples include airplane, rotorcraft, glider, balloon, landplane, and seaplane.

class A airspace — Controlled airspace covering the 48 contiguous United States and Alaska, within 12 nautical miles of the coasts, from 18,000 feet MSL up to and including FL600, but not including airspace less than 1,500 feet AGL.

class B airspace — Controlled airspace designated around certain major airports, extending from the surface or higher to specified altitudes. For operations in Class B airspace, all aircraft must receive an ATC clearance to enter, and are subject to the rules and pilot/equipment requirements listed in FAR Part 91.

class C airspace — Controlled airspace surrounding designated airports where ATC provides radar vectoring and sequencing on a full-time basis for all IFR and VFR aircraft. Participation is mandatory, and all aircraft must establish and maintain radio contact with ATC, and are subject to the rules and pilot/equipment requirements listed in FAR Part 91.

class D airspace — Controlled airspace around at least one primary airport which has an operating control tower. Aircraft operators are subject to the rules and equipment requirements specified in FAR Part 91.

class E airspace — Controlled airspace which covers the 48 contiguous United States and Alaska, within 12 nautical miles of the coasts, from 14,500 feet MSL up to but not including 18,000 feet MSL. Exceptions are restricted and prohibited areas, and airspace less than 1,500 feet AGL. Class E airspace also includes Federal airways, with a floor of 1,200 feet AGL or higher, as well as the airspace from 700 feet or more above the surface designated in conjunction with an airport which has an approved instrument approach procedure.

class G airspace — Airspace that has not been designated as Class A, B, C, D, or E, and within which air traffic control is not exercised.

class of thread — In threaded fasteners, classes of threads are distinguished from each other by the amount of tolerance and/or allowance specified. Classes 1A, 2A, and 3A apply to external threads, whereas classes 1B, 2B, and 3B apply to internal threads. Classes 2 and 3 apply to both external and internal threads.

class-A amplifier — An electronic amplifier that produces current during 100% of the input cycle.

class-B amplifier — An electronic amplifier whose two output devices (vacuum tubes or solid-state) conduct one at a time to produce a composite output signal. One device conducts for the positive portion of an input signal and the other device conducts during the negative portion.

class-C amplifier — An electronic amplifier that produces current during a small part (less than half) of the input cycle.

claw hammer — A hammer used primarily for carpentry. Has a claw-like device opposite the face for removing nails.

clean and true — A term used in valve seat grinding whereby the rough stone is used until the seat is true or exactly matches the valve guide and until all pits, scores, and burned areas are removed.

clean room — A room used to house manufacturing and servicing of high-precision products. Usually has air filtration to prevent contaminating particles from entering the products.

cleanout — The process of cleaning out or cutting away a damaged area to prepare it for a repair.

clear air turbulence — Turbulence that occurs in clear air, and is commonly applied to high-level turbulence associated with wind shear. It is often encountered near the jet stream, and it is not the same as turbulence associated with cumuliform clouds or thunderstorms.

clear air turbulence (CAT)- —Usually, high level (or jet stream) turbulence encountered in air where no clouds are present, may occur in nonconvective clouds.

clear icing (or clear ice) — Generally, the formation of a layer or mass of ice which is relatively transparent because of its homogeneous structure and small number and size of air spaces; used commonly as synonymous with glaze, particularly with respect to aircraft icing. Compare with rime icing. Factors that favor clear icing are large drop size, such as those found in cumuliform clouds, rapid accretion of supercooled water, and slow dissipation of latent heat of fusion.

clear of the runway — 1. A taxiing aircraft, which is approaching a runway, is clear of the runway when all parts of the aircraft are held short of the applicable holding position marking. 2. A pilot or controller may consider an aircraft, which is exiting or crossing a runway, to be clear of the runway when all parts of the aircraft are beyond the runway edge and there is no ATC restriction to its continued movement beyond the applicable holding position marking. 3. Pilots and controllers shall exercise good judgement to ensure that adequate separation exists between all aircraft on runways and taxiways at airports with inadequate runway edge lines or holding position markings.

clearance — 1. The clear space or distance between two mechanical objects or moving parts. 2. An authorization by air traffic control, for the purpose of preventing collision between known aircraft, for an aircraft to proceed under specified traffic conditions within controlled airspace. The pilot-in-command of an aircraft may not deviate from the provisions of a visual flight rules (VFR) or instrument flight rules (IFR) air traffic clearance except in an emergency or unless an amended clearance has been obtained. Additionally, the pilot may request a different clearance from that which has been issued by air traffic control (ATC) if information available to the pilot makes another course of action more practicable or if aircraft equipment limitations or company procedures forbid compliance with the clearance issued. Pilots may also request clarification or amendment, as appropriate, any time a clearance is not fully understood, or considered unacceptable because of safety of flight. Controllers should, in such instances and to the extent of operational practicality and safety, honor the pilot's request. FAR Part 91.3(a) states: "The pilot-in-command of an aircraft is directly responsible for, and is the final authority as to, the operation of that aircraft." The pilot is responsible to request an amended clearance if ATC issues a clearance that would cause a pilot to deviate from a rule or regulation, or in the pilot's opinion, would place the aircraft in jeopardy.

clearance fit — An assembly that leaves a clearance between mating parts. The shaft will be smaller than the hole.

clearance limit — The fix, point, or location to which an aircraft is cleared when issued an air traffic clearance.

clearance void if not off by (time) — Used by ATC to advise an aircraft that the departure clearance is automatically cancelled if takeoff is not made prior to a specified time. The pilot must obtain a new clearance or cancel his IFR flight plan if not off by the specified time.

clearance void time — A time specified by an air traffic control unit at which a clearance ceases to be valid unless the aircraft concerned has already taken action to comply therewith.

clearance volume — The volume of the cylinder of a reciprocating aircraft engine with the piston at the top of its stroke.

cleared (type of) approach — An ATC authorization for an aircraft to execute a specific instrument approach procedure to an airport, i.e. "Cleared ILS Runway Three Six Approach."

cleared approach — An ATC authorization for an aircraft to execute any standard or special instrument approach procedure for that airport. Normally, an aircraft will be cleared for a specific instrument approach procedure.

cleared as filed — A statement that refers to an aircraft being cleared to proceed in accordance with the route of flight filed in the flight plan. This clearance does not include the altitude, DP, or DP Transition.

cleared for takeoff — An ATC authorization for an aircraft to depart. It is predicated on known traffic and known physical airport conditions.

cleared for the option — An ATC authorization for an aircraft to make a touch-and-go, low approach, missed approach, stop and go, or full stop landing at the discretion of the pilot. It is normally used in training so that an instructor can evaluate a student's performance under changing situations.

cleared through — An ATC authorization for an aircraft to make intermediate stops at specified airports without refiling a flight plan while enroute to the clearance limit.

cleared to land — An ATC authorization for an aircraft to land. It is predicated on known traffic and known physical airport conditions.

clearing engine — Purging the combustion chambers of unburned fuel by rotating the engine with the starter. The air flow through the engine carries off dangerous accumulations of fuel vapors and vaporizes any liquid fuel present.

clearing turns — Turns consisting of at least a 180°change in direction, allowing the pilot to see areas blocked by blind spots. A visual check of the airspace around the airplane to avoid conflicts while maneuvering.

clearway — 1. An area beyond the takeoff runway under the control of airport authorities within which terrain or fixed obstacles may not extend above specified limits. These areas may be required for certain turbine-powered operations and the size and upward slope of the clearway will differ depending on when the aircraft was certificated. 2. For turbine engine powered airplanes certificated after August 29, 1959, an area beyond the runway, not less than 500 feet wide, centrally located about the extended centerline of the runway, and under the control of the airport authorities. The clearway is expressed in terms of a clearway plane, extending from the end of the runway with an upward slope not exceeding 1.25 percent, above which no object nor any terrain protrudes. However, threshold lights may protrude above the plane if their height above the end of the runway is 26 inches or less and if they are located to each side of the runway. 3. For turbine engine powered airplanes certificated after September 30, 1958, but before August 30, 1959, an area beyond the takeoff runway extending no less than 300 feet on either side of the extended centerline of the runway, at an elevation no higher than the elevation of the end of the runway, clear of all fixed obstacles, and under the control of the airport authorities.

Cleco fastener — A spring-type fastener used to hold metal sheets together until drilling or riveting procedures are accomplished.

clevis — The forked end of a push-pull tube usually fastened to a bell crank in a control assembly.

clevis bolt — A special-purpose bolt whose round head is slotted or recessed to accept a screwdriver. Used only for shear loads, the threaded portion of the shank is very short and used only to secure the bolt in the clevis.

clevis pin — A high-strength steel pin with a flat head on one end and a hole for a cotter pin on the other end. A clevis pin is used as a hinge for control surfaces or for attaching a clevis to a control horn. Clevis pins are designed to take shear loads only. Also referred to as a flathead pin.

climate — Weather conditions such as temperature, wind, cloud cover, precipitation, etc. that typically exist in a specified area when averaged over a long time, usually decades. .

climatological forecasts — A forecast based on the average weather (climatology) of a particular region.

climatology — The study of the average conditions of the atmosphere.

climb gradient — A minimum climb rate expressed in feet per nautical mile. For example, a climb gradient of 400 feet per nautical mile requires a minimum climb performance of 400 feet in a horizontal distance of one n.m. Climb gradient can be converted mathematically (or by use of a table) to feet per minute if groundspeed is known.

climb indicator — A rate of pressure change indicator used to furnish the pilot with information regarding the rate of vertical ascent or descent.

climb propellers — A fixed-pitch propeller that provides the aircraft with the best performance during takeoff and climb.

climb to VFR — An ATC authorization for an aircraft to climb to VFR conditions within Class B, C, D, and E surface areas when the only weather limitation is restricted visibility. The aircraft must remain clear of clouds while climbing to VFR.

climbing blade — A condition when one or more blades are not operating in the same plane of rotation during flight. This might occur only in flight and not occur during ground operations.

climbout — That portion of flight operation between takeoff and the initial cruising altitude.

climbout speed — With respect to rotorcraft, means a referenced airspeed which results in a flight path clear of the height-velocity envelope during initial climbout.

clinometer — A closed-end, curved glass tube filled with a liquid similar to kerosene and enclosing a round glass ball. It may be used as a leveling device or in a turn and slip indicator to indicate the relationship between the force of gravity and centrifugal force in a turn.

clip — A small attachment device used to join parts in aircraft construction.

clock — In electronic equipment, a pulse generator that allows all components to be synchronized.

clockwise rotation — The direction in which the hands of a clock rotate.

close parallel runways — Two parallel runways whose extended centerlines are separated by less than 4,300 feet, having a Precision Runway Monitoring (PRM) system that permits simultaneous independent ILS approaches.

closed angle — The angle formed in sheet metal after it has been bent more than 90°. For example, if a piece is bent through 135°, it forms a 45° closed angle.

closed loop — An electronic circuit that allows some of the output signal to be fed back into the input section in order to allow the circuit to be self regulating.

closed low — A low-pressure area enclosed in at least one closed isobar or (aloft) one closed contour.

closed runway — A runway that is unusable for aircraft operations. Only the airport management/military operations office can close a runway.

closed traffic — Successive operations involving takeoffs and landings or low approaches where the aircraft does not exit the traffic pattern.

close-grain wood — Wood that has grown with the annual rings very close together. Close-grain wood is very consistent in its cross-sectional strength.

close-tolerance bolt — A hex-head aircraft bolt with a shank that has been ground to a tolerance of + 0.000 — 0.0005". It is identified by a triangle on its head enclosing the material identification mark.

cloud — A visible accumulation of minute water droplets and/or ice particles in the atmosphere above the Earth's surface. Cloud differs from ground fog, fog, or ice fog only in that the latter are, by definition, in contact with the Earth's surface.

cloud amount — The amount of sky covered by each layer of clouds.

cloud bank — A mass of cloud seen from a distance, spreading across an appreciable section of horizon, but not extending directly over the observation location.

cloud detection radar — Radar used to determine bases and tops of clouds (rather than precipitation) above a reporting station.

cloud height — The height of the base of the cloud layer above ground level (AGL).

cloud layer — Clouds with bases at approximately the same level.

cloud point — The temperature of an oil at which its wax content, normally held in solution, begins to solidify and separate into tiny crystals, causing the oil to appear cloudy or hazy.

cloud streets — Parallel rows of cumulus clouds. Each row can be as short as 10 miles or as long as a hundred miles or more.

cloudburst — A sudden and heavy rain shower.

clouds with great vertical development — Cumulus and cumulonimbus clouds.

cloudy convection — The upward movement of saturated air that is warmer than its surroundings.

clove hitch — A knot used for making individual spot ties for securing electrical wire bundles. In this use, the clove hitch is locked with a square knot.

club propeller — A short, stubby propeller used for ground testing of reciprocating engines.

clubhead — A formed rivet head tipped to one side. Tipping can be corrected by rapidly moving the bucking bar across the rivet head in a direction opposite that of the malformed travel. This corrective action can only be accomplished during the forming of the rivet head.

cluster — Part of an airframe structure made up of two or more tubes meeting at one point. They are welded together in a cluster weld.

cluster weld — A welded joint made at the intersection of a number of tubes that meet at a common point.

clutch — A device used to connect and disconnect a driving and driven part of a system such as between the transmission and main rotor of a helicopter.

clutter — Undesirable signals that show up on a radar screen and mask the desired signal return. Topographical and meteorological phenomena are some of the causes of clutter.

CMM — Component Maintenance Manual. A manual developed by the component manufacturer and frequently adopted by an airframe manufacturer. A CMM is most frequently not approved by the FAA. Blanket approval comes through the AMM (Aircraft Maintenance Manual) or SRM Structural Repair Manual).

coalesce — To unite into a single entity. To come together.

coalescent bag — A bag in the water separator of an air-cycle air conditioning system on which the moisture that condenses from the air may coalesce.

coastal fix — A navigation aid or intersection where an aircraft transitions between the domestic route structure and the oceanic route structure.

coast-down check — The amount of time a turbine engine takes to motor down to a complete stop from idle speed after the fuel is shut off. A maintenance test cell check of engine performance.

coated cathode — In a vacuum tube, the cathode emits electrons. Coated cathodes are covered with a substance that enhances the emission of electrons.

coating — The application of material such as a metal or organic compound to a surface.

coaxial — Having a common axis or shaft.

coaxial cable — A transmission line in which the center conductor is surrounded by an insulator and a braided outer conductor. All of this is enclosed in a weatherproof outer insulator.

coaxial propellers — Propellers with concentric shafts that allow them to rotate in opposite directions, canceling torque and p-factors inherent in single propellers. Coaxial propellers can be geared for power by a single engine or by multiple engines.

coaxial rotors — Rotors with concentric shafts that allow them to rotate in opposite directions, canceling torque factors inherent in single rotors. Coaxial rotors can be geared for power by a single engine or by multiple engines.

coaxial shafts — Shafts that are mounted concentrically (having a common center) in order to drive two or more propellers, rotors, compressors, etc.

cobalt — A metallic chemical element with a symbol of Co and an atomic number of 27.

cobalt chloride — An additive to silica gel dehydrator plugs that serves as an indicator of the amount of moisture absorbed by the plug. A dry dehydrator with this additive will be bright blue, but if it has been exposed to excessive moisture, it will turn pink.

cobalt chromium steel — A steel alloy containing cobalt and chromium. Used in exhaust valves.

cobalt-based alloy — A cobalt, tungsten, molybdenum alloy of extreme high temperature strength. Very expensive and used almost exclusively in the hot section of turbine engines.

COBOL — A high-level computer language designed for business applications. Stands for COmmon Business Oriented Language.

cock — A valve (British).

cockpit — The pilot's compartment of an aircraft.

cockpit resource management (CRM) — Effective use of all resources by a flight crew. Emphasis on good communication and interpersonal skills.

code markings — Aircraft fluid lines are often identified by markers of color codes, words, and geometric symbols. These identify each line's function, content, direction of fluid flow, and primary hazard.

codes, transponder — The number assigned to a particular multiple pulse reply signal transmitted by a transponder.

coefficient — A dimensionless number expressing degree of magnitude.

coefficient of expansion — A dimensionless number relating dimensional changes in a material with changes in temperature.

coefficient of friction — The coefficient of friction is the value obtained by dividing the force necessary to move one substance over another at a constant speed by the weight of the substance.

coefficient of linear expansion — A value obtained by measuring the length (inches or centimeters) a material expands per degree Celsius change in temperature, i.e. the thermal expansion properties.

coefficient of thermal expansion — The change in unit of length of volume accompanying a change of temperature.

coercive force — The amount of force (in the form of magnetism with opposite polarity) required to remove magnetism from an object.

cog belt — A drive belt that incorporates teeth on the drive surface in order to prevent slippage. Cog belts are often used in systems where timing is critical.

cognitive domain — A grouping of levels of learning associated with mental activity which range from knowledge through comprehension, application, analysis, and synthesis to evaluation.

cognitive information processing — The method in which information is gathered, processed, and stored by the brain in much the same way as a computer.

cognitive theory — Learning is not just a change in outward behavior but involves changes in thinking, feeling, or understanding. It involves mental processes, such as decision making and problem solving, which are difficult, if not impossible, to observe or measure.

coherent light — Light such as that produced by a laser, made up of waves of the same wavelength and in phase. Ordinary light consists of light of different wavelengths and phase relations.

cohesion — The tendency of a single substance to adhere to itself. The force holding a single substance together.

coil — A conductor consisting of turns of wire in which the magnetic field around one turn cuts across the other turns, increasing the inductive effect of the wire.

coil assembly — The magneto coil assembly consists of a soft iron core around which is wound the primary and secondary coil with the secondary coil wound on top of the primary coil.

coil booster — A transformer coil used with a vibrator to produce a high voltage at the spark plugs during starting.

coil spring — A spiral of hardened steel wire. When loosely wound, it is used in compression applications to provide cushioning, while tightly wound coils are used in tension applications.

coin dimpling — Performed by a special machine that has, in addition to the usual dies, a "coining ram." This ram applies an opposing pressure to the edges of the hole so that the metal is made to flow into all the sharp contours of the die giving the dimple greater accuracy and improving the fit.

coin pressing — A dimpling process using a countersunk rivet as the male dimpling die, placing the female die in the usual position, and backing it with a bucking bar. The rivet is then struck with a pneumatic hammer to form the dimple.

coin tap — In composites, the use of a coin to tap a laminate in different spots to detect a change in sound, which would indicate the presence of a defect.

coke — A solid, carbon-like residue left by mineral oil after the removal of the volatile material by heat.

coking — The carbon buildup from decomposition of oil in vent lines. This buildup can, over a period of time, cause a restriction of flow.

cold — The absence of heat.

cold air funnel — A weak vortex that occasionally develops behind a cold front with rainshowers and nonsevere thunderstorms.

cold airmass — An airmass that is colder than the ground it is passing over.

cold bending — The bending of sheet metal without the use of heat. Thin sheets of metal and some soft metals are bent in this manner. Thick sheets of metal and some harder materials must be heated to allow bending without causing the material to break.

cold chisel — A hardened steel cutting tool. Used to cut metal materials without the need for softening by use of heat.

cold circuit — An electrical circuit without power. As opposed to a "hot circuit" that has electrical power applied.

cold cylinder test — A method of determining an inoperative cylinder on a reciprocating engine by measuring the relative lack of heat on the bad cylinder compared with the normal cylinders.

cold dimpling — Accomplished while the material is at room temperature by either the coin ram or coin dimpling method.

cold downslope winds — A bora type wind.

cold drawing — The process of pulling rod or tubing through progressively smaller dies in order to reduce the diameter of the material. The material is not heated for this process.

cold flow — The deep and permanent impressions or cracks in a hose caused by hose clamp pressure.

cold front — The boundary between two airmasses where cold air is replacing warm air.

cold front occlusion — An occlusion where very cold air behind a cold front lifts the warm front and the cool airmass preceding it.

cold heading — The process of cold forming wire or bar stock into intricate shapes by reducing the diameter of the stock in some areas, while expanding the diameter in others. Cold heading is performed by upsetting or heading the metal into larger diameters or extruding it into smaller diameters.

cold junction — In electricity, the reference junction in a thermocouple. A thermocouple produces current in relation to the difference in temperature between a reference junction, known as the cold junction, and the junction where temperature measurement is taken, known as the hot junction.

cold light — Light that produces very little heat. Used for applications requiring low intensity lights such as formation lights and obstruction lighting.

cold rolling — The process of rolling sheets of metal in order to produce a specific thickness of material. The material is not heated for this process.

cold section — The air compression sections of a turbine engine.

cold soaked — **1.** The condition of a unit of equipment being extremely cold after prolonged exposure to very cold temperatures. Often makes operation extremely difficult. This is caused by the waxes and tars in paraffin-based and asphalt-based engine oils precipitating out after a week or so of very cold temperatures. When they seep out, the viscosity of the engine oil rises much more than that caused by the cold-induced rise in viscosity. **2.** In gliding, the condition of a self-launch or sustainer engine making it difficult or impossible to start in flight due to long-time exposure to cold temperatures. Usually occurs after a long soaring flight at altitudes with cold temperatures, e.g., a wave flight.

cold spark plug — A spark plug in which the nose insulator provides a short path for heat to travel from the center electrode to the shell. Cold spark plugs are used in high-compression engines to minimize the danger of preignition.

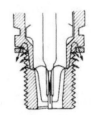

cold swaging process — A method of reducing or forming steel or other material while it is cold by drawing it to a point or reducing its diameter.

cold tank system — A lubrication system wherein the oil cooler is located in the scavenge oil subsystem. The oil passes through the cooler and returns to the tank cooled.

cold weld — The process where two pieces of metal are joined by pressing them together until the materials are fused. No external heat is applied during this process.

cold working — Any mechanical process that will increase the hardness of a metal. This may be done by repeatedly hammering the material, passing it through rollers, or pulling it through dies.

cold-cathode vacuum tube — A vacuum tube that assists a cathode in emitting electrons without the use of a heater.

cold-cranking amps — The capacity of a battery under extreme usage conditions. This gives a rough estimate of the ability of a battery to operate under extreme conditions, and allows for comparison between batteries. Also referred to as a high-rate discharge.

cold-rolled steel — Steel that has been cold worked by passing through a series of compression rollers or dies.

cold-solder joint — An improperly soldered joint. Usually caused by movement of the components before the solder sets. Has a grainy appearance. It is not satisfactory, and must be reheated and allowed to set without movement of the components.

cold-starting oil relief valve — A bypass relief valve in a main oil system that acts as an emergency pop-off valve when cold oil causes excessive system pressure. Used in systems having no oil pressure regulating relief valve.

collapsed surface — A dimensional change with neither removal of material nor an abrupt change of surface and usually affecting large sections of the object. Causes are excessive pressure or forces and improper abusive engine operation. Parts affected usually include valves, piston rings, and springs.

collar — A collar is a raised ring or flange of material on the head or shank of a fastener.

collective pitch control — The control in a helicopter in which the pitch of all the rotor blades are changed at the same time.

collective pitch control — In rotorcraft, the control for changing the pitch of all the rotor blades in the main rotor system equally and simultaneously and, consequently, the amount of lift or thrust being generated.

collector — 1. The electrode in a transistor through which conventional current leaves the transistor. 2. The exhaust cone collector in a turbine engine that collects the exhaust gases discharged from the engine turbine buckets and gradually converts them into a solid jet. In performing this, the velocity of the gases is decreased slightly and the pressure increased. It also helps to direct the flow of hot gases rearward and prevents turbulence and, at the same time, imparts a high final exit velocity to the gases.

collector ring — A corrosion-resistant steel assembly that collects the exhaust gases from the cylinders of a radial engine and routes them overboard.

collimated light — An optical device that causes all light beams passing through to be parallel.

color code — A means of identifying an object by the use of various combinations of colors.

color temperature — The approximate temperature of an object, based on the peak intensity-emitting wavelength (its color). Specified in degrees Kelvin.

color tempering — A method of removing brittleness from a part during heat treatment. The part is heated to a specified color, then quenched (cooled by dipping into a specified liquid).

color wheel — A circular diagram or tool that helps the user visualize colors that result when colors are mixed together.

combination compressor — In turbine engines, a compressor design that utilizes an axial compressor and a centrifugal compressor (usually attached together) to compress incoming air prior to combustion.

combination inertia starter — An inertia starter for reciprocating aircraft engines that can be energized by either an electric motor or a hand crank.

combination set — A set of measuring tools that often includes a tri-square, a 45° miter, a depth gauge, a marking or scribing gauge, and a level.

combination square — A device used in laying out material. It consists of a steel scale attached to a squaring device that is machined with precise 90° and 45° angles.

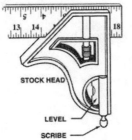

combination wrench — A tool that has a box-end wrench on one end and an open-end wrench on the other.

combined center-RAPCON (CERAP) — An air traffic facility which combines the functions of an ARTCC and a radar approach control facility.

combustibles — Materials capable of burning.

combustion — A chemical process in which a material is united with oxygen at such a rapid rate that light and heat are released.

combustion chamber — The section of the engine into which fuel is injected and burned. See also combustor.

combustion liner — In a turbine engine, the perforated and louvered inner section of the combustor in which fuel burning is controlled.

combustion liner louvers — In a turbine engine, small slots in the liner that direct cooling airflow and provide the inner walls with a cooling air blanket.

combustion section — The section located directly between the compressor and the turbine sections. It contains a casing, a perforated inner liner, a fuel injection system, some means for initial ignition, and a fuel drainage system to drain off unburned fuel after engine shutdown. The combustion section houses the combustion process, which raises the temperature of the air passing though the engine. This process releases energy contained in the air/fuel mixture.

combustion starter — A fuel engine starting accessory that utilizes a combustion section similar to a turbine engine combustor. Combustion products are exhausted through a turbine connected to a reduction gearbox to create starting torque.

combustor — In gas turbine engines, the section of the engine into which fuel is injected and burned to create expansion of the gases.

combustor drain valve — A spring-loaded valve in a turbine engine that drains excess fuel from the combustion section when the engine is not running. It is normally open and is held closed by the pressures of combustion when the engine is operating.

combustor efficiency — A measure of the percentage of fuel burned completely or Btus of heat attained as opposed to the Btu potential of fuel introduced. Typical figures are in the 99% range.

coming-in speed — The speed of a magneto that is just sufficient to produce the voltage required to fire all of the spark plugs consistently.

comma cloud — A cloud mass shaped like a comma as seen in satellite imagery.

commercial fastener — A fastener manufactured to published standards and stocked by manufacturers or distributors. The material, dimensions, and finish of commercial fasteners conform to the quality level generally recognized by manufacturers and users as commercial quality.

commercial maneuvers — Consists of maximum performance takeoffs and landings, steep turns, chandelles, lazy eights, and eights-on-pylons

commercial operator — A person who, for compensation or hire, engages in the carriage by aircraft in air commerce of persons or property, other than as an air carrier or foreign air carrier or under the authority of Part 375. Where it is doubtful that an operation is for "compensation or hire", the test applied is whether the carriage by air is merely incidental to the person's other business or is, in itself, a major enterprise for profit.

commercial pilot — An FAA pilot rating that allows the pilot to operate an aircraft for hire.

common point — A significant point over which two or more aircraft will report passing or have reported passing before proceeding on the same or diverging tracks. To establish/maintain longitudinal separation, a controller may determine a common point not originally in the aircraft's flight plan and then clear the aircraft to fly over the point.

common portion — See common route.

common route — That segment of a North American Route between the inland navigation facility and the coastal fix.

common traffic advisory frequency (CTAF) — A frequency designed for the purpose of carrying out airport advisory practices while operating to or from an uncontrolled airport. The CTAF may be a UNICOM, MULTICOM, FSS, or tower frequency and it is identified in appropriate aeronautical publications.

communication process — Consists of three basic elements: the source, the symbols used to communicate the message, and the receiver. In addition, feedback is essential for effective communication to take place.

communication skills — The skills an instructor must develop to communicate effectively with students. The ability to communicate effectively as an instructor begins with an understanding of the communication process and is enhanced by experience and training.

communications — In electronics, changing information (audible or digital) into a form that can be transmitted and then changed back into information at the receiver.

communications receiver — A device that receives electronic data and converts it into information appropriate for use.

communications satellite — A satellite placed in geostationary orbit used for the relay of communications transmissions from Earth.

community aerodrome radio station (CARS) — An aerodrome radio that provides weather, field conditions, accepts flight plans and position reports.

commutator — The copper bars on the end of a generator armature to which the rotating coils are attached. AC is generated in the armature and the brushes riding on the commutator act as a mechanical rectifier to convert it into DC.

compact disk (CD) — A small plastic optical disk that contains recorded music or computer data. Also, a popular format for storing information digitally. The major advantage of a CD is its capability to store enormous amounts of information.

comparator — A device for inspecting parts by comparing them with a greatly enlarged standard chart.

compartment — A separate and enclosed space in an aircraft structure.

compass — 1. A device for determining direction measured from magnetic north. 2. A drafting instrument used to create circles or arcs.

compass card — In navigation, a card affixed to a magnet and allowed to rotate freely so as to show the direction of travel. Divided into degrees and cardinal compass headings.

compass compensation — The procedure of adjusting the magnetic compass in an aircraft for deviation caused by magnetic forces within the aircraft.

compass correction card — A card mounted near a compass and in full sight of the pilot that indicates the difference between the compass reading and the actual magnetic heading.

compass heading — A compass reading that will make good the desired course. It is the desired course (true course) corrected for variation, deviation, and wind.

compass locator — A low-frequency, non-directional beacon (NDB) co-located with the marker beacons on an ILS approach. Used to help establish the pilot's position on the localizer.

compass locator — A low power, low or medium frequency (L/MF) radio beacon installed at the site of the outer or middle marker of an instrument landing system (ILS). It can be used for navigation at distances of approximately 15 miles or as authorized in the approach procedure.
a. Outer Compass Locator (LOM) — A compass locator installed at the site of the outer marker of an instrument landing system.

b. Middle Compass Locator (LMM) — A compass locator installed at the site of the middle marker of an instrument landing system.

compass north- The north to which a compass actually points. Its field is produced by the combination of the Earth's magnetic field and the local magnetic fields within the aircraft.

compass rose — A circle marked out on a flat part of an airport away from magnetic interference. It is marked every thirty degrees of magnetic direction. The airplane is taxied onto this rose, and the magnetic compass is adjusted to agree with the heading of each of the marks.

compass saw — A small, thin-bladed saw used to cut intricate designs into a material.

compass swinging — The process of aligning the aircraft on a series of known magnetic headings and adjusting the compensating magnets to bring the compass heading as near the magnetic heading as possible.

compass turns — Maintaining heading and making turns with reference to the magnetic compass only.

compensated cam — The magneto cam used on high-performance radial engines. One lobe is provided for each cylinder, and the lobes are ground in such a way that the magneto points will open when the piston is a given linear distance from the top of the cylinder rather than a given angular distance. This compensates for the relationship of the master rod pistons and those connected to the crankshaft through the link rods.

compensated relief valve — An oil pressure relief valve with a thermostatic valve that decreases the regulated oil pressure when the oil warms up. High pressure is allowed to force the cold oil through the engine, but the pressure is automatically decreased when the oil temperature increases.

compensating cam — A cam used in conjunction with the collective pitch control to add the correct amount of engine power for the pitch of the rotor. Used on turbine powered helicopters.

compensating port — A port inside a brake master cylinder that vents the wheel cylinder to the reservoir when the brake is not applied. It prevents fluid expansion from causing the brakes to drag when temperatures increase.

compensating winding — A series winding in a high-output generator wound between the main pole and the interpoles to aid in brushless commutation and overcoming armature reaction.

compensation — A defense mechanism that attempts to disguise a weak or undesirable quality by emphasizing a more positive one. Students may lower the scope of their goals to avoid possible failure in achieving goals that are more difficult.

compiler — A special computer program that converts a high-level computer language used by a programmer into machine language that can be used by the computer.

complement of an angle — The angle that, when added to the original angle, makes a 90° angle.

complex airplane — An airplane that has a retractable landing gear, flaps, and a controllable pitch propeller.

complex circuit — A circuit consisting of a number of components, where some are arranged in series and others in parallel.

complex number — The sum of a real number and an imaginary number.

compliance — Being in accordance with a regulation or directive.

component — Any one of several parts in a combination of parts that make up a unit or whole.

composite — Two or more substances which are combined to produce material properties not present when either substance is used alone.

composite fan blades — An advanced technology blade design not yet in current use. Its composition is an epoxy-resin material and graphite fiber. It is stronger than fiberglass and 20% to 30% lighter than metals of the same strength.

composite flight plan — A flight plan which specifies VFR operation for one portion of flight and IFR for another portion. It is used primarily in military operations.

composite route system — An organized oceanic route structure, incorporating reduced lateral spacing between routes, in which composite separation is authorized.

composite separation — A method of separating aircraft in a composite route system where, by management of route and altitude assignments, a combination of half the lateral minimum specified for the area concerned and half the vertical minimum is applied.

composite structure — A structure made up of two or more substances that, when combined, produce a structure with properties not present when the individual substances are used alone.

composition resistor — An electrical device made up of carbon and some insulator. The amount of resistance to current flow depends on the relative amounts of carbon and the insulator.

compound — A new entity formed by a union of elements or parts. The constituents lose their original identity and assume the characteristics of the compound.

compound curve — A metal surface curved in more than one plane.

compound lever — A series of two or more levers connected to achieve a result. Usually used to multiply force.

compound-wound generator — A generator with both a series and a shunt field.

comprehensiveness — A characteristic of a measuring instrument when it is based on a liberal sampling of the knowledge or skill to be measured. It must be broad enough to ensure that conclusions are representative of the whole.

compressed air — Air compressed until its pressure is above ambient.

compressibility — The capability of a material to be compressed to a smaller volume. Gases are compressible; liquids are not.

compressibility burble — A region of disturbed flow produced by a shock wave. The region is produced aft of the shock wave.

compressibility error — An error introduced to the pitot system due to the compressibility of air. The air entering the pitot tube is compressed slightly as a result of impacting air that has stopped within the system. At altitudes below 10,000 feet and airspeeds below 200 knots the amount of error is negligible, but it becomes significant at higher altitudes and airspeeds.

compressible flow — In aerodynamics, airflow that is inconsistent in the degree of compressibility, that is, it experiences a fractional volume change per unit of pressure change.

compression — The resultant of two forces that act along the same line and also act toward each other.

compression failure — The failure of a material due to compression. Usually causes material to buckle or bulge.

compression fastener — A fastener with the primary function of resisting forces that tend to compress it.

compression ignition — An ignition used in diesel engines. Rather than an electrical discharge causing ignition, the fuel ignites when compression causes the temperature of the fuel-air mixture to rise to the ignition point.

compression member or strut — A heavy member, usually of tubular steel, that separates the spars in a Pratt truss wing and is used to carry only compression loads.

compression molding — The creation of parts by compressing a material into a mold. This ensures the complete filling of all parts of the mold.

compression ratio — **1.** The ratio of the volume of an engine cylinder with the piston at the bottom center to the volume when the piston is at top center. **2.** Sometimes used to refer to compressor pressure ratio; however, this is not entirely correct because compression ratio infers a ratio of volumes as in a piston engine.

compression rib — A heavy-duty rib made with heavy cap strips and extra strength webs. A compression rib is designed to withstand compression loads between the wing spars.

compression rings — In reciprocating engines, the top piston rings. Used to provide a seal for the gases in the cylinder and to transfer heat from the piston into the cylinder walls.

compression spring — A spring used to withstand compression. The coils are loosely wound in order to leave room for compression.

compression stress — Stress applied to a material by a squeezing force.

compression strut — A brace that fits between two structural members. The members tend to move toward each other in compression. One example is the compression struts used between upper and lower spar sections in Pratt truss construction.

compression test — A test used to determine the condition of the cylinders on a reciprocating engine. The test pressurizes the cylinder and then measures excessive loss of pressure, which indicates bad rings or bad exhaust or intake valves.

compressive load — A load or a force that tends to compress or squeeze an object together.

compressive strength — The ability of a body to resist a force that tends to shorten, compress, or squeeze it.

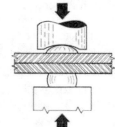

compressive stress — The basic stress that tends to shorten an object by pressing its ends together.

compressor — In gas turbine engines, an impeller or a multi-bladed rotor assembly. A component which is driven by a turbine rotor for the purpose of compressing incoming air.

compressor blade — A rotating airfoil in a turbine engine. Compresses and accelerates the air before arrival at the combustion chamber.

compressor bleed air — Air taken out of the compressor section of a turbine engine to prevent stall and to operate certain components.

compressor case — The outer compressor housing, usually split front to rear or top to bottom. This case provides support for the stator vanes.

compressor discharge pressure (Pt4) — A measurement taken at the compressor exit and sent to the fuel control unit for fuel scheduling purposes.

compressor disk — The inner section of the compressor to which the blades are attached. A disk assembly is made up of one segment per stage and bolted together to form one large rotating piece.

compressor efficiency — A measure of aerodynamic efficiency, one important factor of which is the ability to compress air to the maximum pressure ratio with the minimum temperature rise.

compressor front frame — The compressor inlet case.

compressor hub — The front and rear portion of the compressor to which the compressor shafts attach.

compressor pressure ratio — The ratio of compressor discharge pressure to compressor inlet pressure.

compressor stage — 1. Each section of a compressor in which the air pressure is progressively increased. A stage of compression consists of one row of blades and one row of stator vanes in an axial flow compressor. 2. A rotor blade set followed by a stator vane set. The rotating airfoils create air velocity, which then changes to pressure in the numerous diverging ducts formed by the stator vanes.

compressor stall — In gas turbine engines, a condition in an axial-flow compressor in which one or more stages of rotor blades fail to pass air smoothly to the succeeding stages. A stall condition is caused by a pressure ratio that is incompatible with the engine rpm. Compressor stall will be indicated by a rise in exhaust temperature or rpm fluctuation, and if allowed to continue, may result in flameout and physical damage to the engine.

compressor stall-margin curve — A curve showing a relationship between the compression ratio and mass air flow that must be maintained for a particular engine. If either of the factors goes out of limits, a compressor stall results.

compressor surge — 1. A severe compressor stall across the entire compressor that can result in severe damage if not quickly corrected. This condition occurs with a complete stoppage of airflow or a reversal of airflow. 2. An operating region of violent pulsating air flow usually outside of the operating limits of the engine flow control settings. A primary cause is compressor blade stall. Surge can result in flameout and, in severe cases, structural damage.

compulsory reporting points — Reporting points which must be reported to ATC. They are designated on aeronautical charts by solid triangles or filed in a flight plan as fixes selected to define direct routes. These points are geographical locations which are defined by navigation aids/ fixes. Pilots should discontinue position reporting over compulsory reporting points when informed by ATC that their aircraft is in "radar contact."

computer — A device (usually electronic) that receives information, performs some sort of operation on the data, and then produces an output.

computer assisted instruction (CAI) — Similar to computer-based training, except emphasizes the point that the instructor is responsible for the class and uses the computer to assist in the instruction.

computer-based instruction (CBI) — Synonymous with computer-based training. The use of the computer as a training device.

computer-based training (CBT) — The use of the computer as a training device. CBT is sometimes called computer-based instruction (CBI); the terms and acronyms are synonymous and may be used interchangeably.

concave surface — An inwardly curved surface.

concentration cell corrosion — A corrosion in which the electrode potential difference is caused by a difference in ion concentration of the electroyte instead of a difference in galvanic composition within the metal.

concentric — Having a common center.

concentric circles — Circles having the same center point.

concentric shafts — Two shafts having a common axis, one inside the other.

concurrent lines — Lines that pass through a common point.

condensation — The process of changing state from a vapor into a liquid.

condensation level — The height at which a rising parcel or layer of air would become saturated if lifted adiabatically.

condensation nuclei — Small particles in the air on which water vapor condenses.

condensation trail (or contrail) — A cloud-like streamer frequently observed to form behind aircraft flying in clear, cold, humid air.

condenser — 1. Another name for capacitor. 2. The component in a vapor cycle air conditioning system where heat energy is given up to the air and the refrigerant vapor is changed into a liquid.

condition lever — A turboprop cockpit lever. On some aircraft, it serves as prop control lever for flight (alpha range). On other engines, it serves only as a fuel shut-off lever.

conditional routes (CDR) — (Europe) — Category 1,2,3. Category 1: Permanently plannable CDR during designated times. Category 2: Plannable only during times designated in the Conditional Route Availability Message (CRAM) published at 1500Z for the 24 hour period starting at 0600Z the next day. Category 3: Not plannable. Useable only when directed by ATC.

conditionally unstable air — Unsaturated air that will become unstable on the condition it becomes saturated.

conditions — The second part of a performance-based objective, which describes the framework under which the skill or behavior will be demonstrated.

conductance — The ability of a substance to conduct electricity.

conduction — The transfer of heat through a substance or from one substance in contact with another; transfer is always from warmer to colder temperature.

conductive coating — A coating applied to the surface of a material in order to allow it conduct electricity or to make it able to conduct electricity more efficiently.

conductivity — The characteristic of a material that makes it possible for it to transmit heat or electrical energy by conduction.

conductor — A material whose outer ring electrons are loosely bonded. Therefore, a relatively low voltage will cause a flow of these electrons.

conduit — A duct or tube enclosing electrical wires or cable.

cone — A three-dimensional geometric surface formed by the rotation of a line around a fixed axis such that the line always passes through a fixed point on the axis (called the vertex) and always makes the same angle with the axis.

cone clutch — A drive clutch consisting of nesting cones as the contact surface.

cone of confusion — The cone shaped area above a VOR station in which there is no signal and the TO/FROM flag momentarily flickers to OFF.

cones — The cells concentrated in the center of the retina which provide color vision and fine detail.

conflict alert — A function of certain air traffic control automated systems designed to alert radar controllers to existing or pending situations between tracked targets (known IFR or VFR aircraft) that require his immediate attention/action.

conflict resolution — The resolution of potential conflictions between aircraft that are radar identified and in communication with ATC by ensuring that radar targets do not touch. Pertinent traffic advisories shall be issued when this procedure is applied. Note: This procedure shall not be provided utilizing mosaic radar systems.

conformance — The condition established when an aircraft's actual position is within the conformance region constructed around that aircraft at its position, according to the trajectory associated with the aircraft's Current Plan.

conformance region — A volume, bounded laterally, vertically, and longitudinally, within which an aircraft must be at a given time in order to be in conformance with the Current Plan Trajectory for that aircraft. At a given time, the conformance region is determined by the simultaneous application of the lateral, vertical, and longitudinal conformance bounds for the aircraft at the position defined by time and aircraft's trajectory.

conformity — Meeting all of the requirements of its original conditions as specified in the Type Certificate Data (TCD) Sheets and the manufacturer's specifications, or properly altered per Supplemental Type Certificates (STC).

conformity inspection — An inspection of an aircraft or component to verify that all components installed on the aircraft are approved either in the original Type Certificate Data and manufacturer's specifications or as modified by Supplemental Type Certificates or Field Approvals (337's).

confusion between the symbol and the symbolized object — Results when a word is confused with what it is meant to represent. Words and symbols create confusion when they mean different things to different people.

congeal — Changing from a fluid to a solid state, or from a free flowing liquid state to one that is less free flowing.

congealed oil — Oil that has solidified because of cold or contaminants.

conical — Cone-shaped.

conifer — A softwood cone-bearing tree.

coning — In rotorcraft, an upward sweep of rotor blades as a result of lift and centrifugal force.

coning angle — The angle formed by the rotor blades and the axis of rotation of a helicopter rotor system. The magnitude of the angle is determined by the relationship between the centrifugal force and the lift produced by the blades.

connecting rod — The component in an internal combustion engine that connects the piston to the crankshaft.

connector — A device used to join two pieces of wire, tubing, or hose to a component.

CONSOLAN — A low frequency, long-distance NAVAID used principally for transoceanic navigations.

console — The pedestal or panel in an aircraft cockpit where the operating controls are located.

constant — A value used in a mathematical computation that is the same every time. For instance, the relationship between the circumference of a circle and its diameter is a constant, 3.1416 (pi, Π).

constant current charge — A method of charging a battery in which the voltage is adjusted as the charge progresses in order to keep the current constant.

constant displacement pump — A pump that displaces a constant amount of fluid each time it turns. The faster it turns, the greater the output.

constant pressure chart — A weather chart that represents conditions on a constant pressure surface; may contain analyses of height, wind, temperature, humidity, and/or other elements.

constant pressure cycle — The thermodynamic cycle of a gas turbine engine to produce thrust. This is a varying volume, constant pressure, cycle of events. A more recent term is continuous combustion cycle because of the four continuous and constant events, including the intake, compression, expansion (including power), and exhaust. Also referred to as the Brayton cycle.

constant section — That part of an aircraft's fuselage having a uniform cross-sectional shape.

constant voltage charge — A method of charging a battery in which the voltage across the battery remains constant. The current is high at the start of the charge but tapers off to a low value as the charge progresses.

constantan — A copper-nickel alloy used as a negative lead in thermocouples used for measuring temperatures in reciprocating engines.

constant-force spring — A spring that exerts a constant pressure regardless of how much the spring is compressed or extended.

constant-speed drive (CSD) — A hydraulic transmission that can be controlled either electrically or mechanically. It is used for alternators and enables the alternator to produce the same frequency regardless of the engine's variation of speed from idle to maximum RPM.

constant-speed propeller — A controllable-pitch propeller whose pitch is automatically varied in flight by a governor to maintain a constant RPM in spite of varying air loads.

constant-voltage power supply — A voltage supply that is regulated so that voltage remains constant through a range of loads.

constrained-gap igniter — A turbine igniter plug that has the center electrode recessed in the insulator in order to cause the spark to arc well past the tip of the igniter. This allows it to operate at cooler temperatures than other igniter plugs.

constructivism — Provides a unique way of thinking about how students learn. Constructivism is based upon the idea that learners construct knowledge through the process of discovery as they experience events and actively seek to understand their environment.

contact — **1.** Establish communication with (followed by the name of the facility and, if appropriate, the frequency to be used). **2.** A flight condition wherein the pilot ascertains the attitude of his aircraft and navigates by visual reference to the surface.

contact approach — An approach where an aircraft on an IFR flight plan, having an air traffic control authorization, operating clear of clouds with at least one mile flight visibility, and a reasonable expectation of continuing to the destination airport in those conditions, may deviate from the instrument approach procedure and proceed to the destination airport by visual reference to the surface. This approach will only be authorized when requested by the pilot and the reported ground visibility at the destination airport is at least one statute mile.

contact cement — A syrupy adhesive that bonds on contact.

contact cooling — The process by which heat is conducted away from the warmer air to the colder earth.

contactor — A heavy-duty electrical switch. Used to connect high current loads to the power supply.

contaminant — An impurity or foreign substance present in a material or environment that affects one or more properties of the material.

contaminated runway — A runway is considered contaminated whenever standing water, ice, snow, slush, frost in any form, heavy rubber or other substances are present. A runway is contaminated with respect to rubber deposits or other friction-degrading substances when the average friction value for any 500-foot segment of the runway within the ALD (Available Landing Distance) falls below the recommended minimum friction level and the average friction value in the adjacent 500-foot segments fails below the maintenance planning friction level.

contamination — The entry of foreign materials into the fuel, oil, hydraulic, or other system.

conterminous U.S. — The 48 adjoining States and the District of Columbia.

continental control area — Obsolete term. Formerly a control area consisting of the airspace covering the continental United States above 14,500 feet, not including airspace below 1,500 feet AGL, and most prohibited and restricted areas.

Continental United States — The 49 states located on the continent of North America and the District of Columbia.

continue — When used as a control instruction should be followed by another word or words clarifying what is expected of the pilot. Example: "continue taxi", "continue descent", "continue inbound" etc.

continuity — The condition of being unbroken or uninterrupted.

continuity light — A simple test device in which a light indicates continuity in an electrical circuit while an open circuit prevents the light from illuminating.

continuous airworthiness program — **1.** A maintenance program consisting of the inspection and maintenance necessary to maintain an aircraft or a fleet of aircraft in airworthy condition. Large or turbine powered aircraft typically go through a continuous airworthiness program. **2.** A program of Federal Aviation Administration-approved inspection schedules that allows aircraft to be continually maintained in a condition of airworthiness without being taken out of service for long periods of time. Usually consists of a system of "letter checks."

continuous casting — A manufacturing method that injects a molten substance into a mold, but is then continuously pulled from the mold as it cools; similar to extrusion.

continuous curved-line graphs — A graph utilizing a smooth and even line to convey data.

continuous filament — An individual reinforcement that is flexible and indefinite in length. The fibers used to weave fabric are considered continuous filaments.

continuous gusset — A brace used to strengthen corners in a structure. It runs the full width of the structure.

continuous ignition system — A secondary, lower power, ignition system installed along with the main system. It is used to fire one igniter plug during takeoff, landing, and in bad weather, or for relight purposes in case of flameout.

continuous load — In electrical equipment, a non-intermittent draw of electrical current. Many electrical devices have a continuous-duty rating that specifies the maximum continuous load the device can withstand.

continuous wave (cw) — An RF carrier wave whose successive oscillations are identical in magnitude and frequency. Often used to indicate Morse code.

continuous-duty ignition system — An ignition system used on turbine engines where an igniter plug continues firing even after the engine is operating. Used in situations where there is danger of flameout. One area of use would be in icing conditions where airflow to the turbine could be interrupted.

continuous-duty rating — Many electrical devices are limited to operation only for a specified percentage of time without being turned off for cooling. A device that can run continuously will specify the amount of continuous load.

continuous-element-type detector — A fire detection system consisting of a stainless steel tube containing a discrete element that has been processed to absorb gas in proportion to the operating temperature set point. When the temperature rises due to fire, overheating, etc., the gas is released from the element, causing a pressure increase in the tube. This mechanically actuates a diaphragm switch, activating the warning lights and an alarm bell.

continuous-flow oxygen system — Any oxygen system that provides a continuous flow of oxygen at a rate constant for any given altitude.

continuous-loop fire detector system — A fire detection system utilizing a continuous loop consisting of two conductors separated by a thermistor material. At normal temperatures, the thermistor is an insulator, but in the presence of a fire or overheat condition, the thermistor becomes conductive and signals the presence of a fire.

contour — **1.** The outline of a figure. **2.** In meteorology, a line of equal height on a constant pressure chart; analogous to contours on a relief map.

contour map — A map with lines of equal elevation drawn on it. Closely-spaced lines indicate steeper terrain.

contour template — A tool used to measure or duplicate the contour of a surface.

contouring circuit — An electronic circuit that adjusts output devices to match the following input stage. Often used to match amplifier characteristics to speakers.

contract — To become reduced in size by squeezing or drawing together.

contrail — Condensation trail. Cloud-like streamer that frequently forms behind an aircraft. There are two types of contrails. Aerodynamic contrails are caused by sudden lowering of pressure over wings and propellers. Exhaust contrails are caused by the hot, moist air from turbine engines mixing with the cold air at high altitudes.

contrarotating propellers — Two propellers mounted on concentric shafts that turn in opposite directions. This type of rotation cancels the torque forces caused by the rotation of propellers.

control — The act of regulating, directing, or coordinating any device or activity.

control and performance concept — A method of teaching attitude instrument flying, which focuses on controlling attitude and power as necessary to produce the desired performance. This method divides the instruments into three categories: control, performance, and navigation. It is used predominantly with high-performance turbine aircraft.

control area — A controlled airspace extending upwards from a specified limit above the Earth.

control cable — Specially designed steel cable connected to linkages used in flight control systems and engine controls.

control circuit — Any one of a variety of circuits designed to exercise control of an operating device, to perform counting, timing, switching, and other operations.

control column — A vertical column in the cockpit on which a yoke is mounted. The ailerons are controlled by rotation of the yoke and the elevators are controlled by its fore-and-aft movement.

control grid — The electrode in a vacuum tube to which the signal is applied.

control locking devices — Devices used to secure control surfaces in their neutral positions when the aircraft is parked. Prevents damage from strong winds that could force the controls violently against the stops.

control rod — A rigid, tubular rod used to actuate control surfaces. Also referred to as a push-pull rod or torque tube.

control sector — An airspace area of defined horizontal and vertical dimensions for which a controller, or group of controllers, has air traffic control responsibility, normally within an air route traffic control center or an approach control facility. Sectors are established based on predominant traffic flows, altitude strata, and controller workload. Pilot-communications during operations within a sector are normally maintained on discrete frequencies assigned to the sector.

control slash — A radar beacon slash representing the actual position of the associated aircraft. Normally, the control slash is the one closest to the interrogating radar beacon site. When ARTCC radar is operating in narrow-band (digitized) mode, the control slash is converted to a target symbol.

control snubber — A method of protecting control surfaces equipped with a hydraulic booster unit.

control stick — A vertical stick in the cockpit of an airplane used to move the elevators by fore-and-aft movement or the ailerons by side-to-side movement.

control surface — Any of the major flight controls such as the ailerons, elevator, and rudder.

control wheel — The hand-operated wheel in the cockpit of an airplane used to actuate the elevators by in-and-out movement and the ailerons by rotation of the wheel. Also referred to as the yoke.

control zone — Obsolete term. Formerly controlled airspace extending upward from the surface of the Earth to a specified upper limit.

control zone (ICAO) — A controlled airspace extending upwards from the surface of the earth to a specified upper limit.

controllability — A measure of the response of an aircraft relative to the pilot's flight control inputs.

controllable-pitch propeller — A propeller with a pitch that can be changed in flight.

controlled airport — An airport which has a control tower, sometimes called a tower airport.

controlled airspace — An airspace of defined dimensions within which air traffic control service is provided to IFR flights and to VFR flights in accordance with the airspace classification.
a. Controlled airspace is a generic term that covers Class A, Class B, Class C, Class D, and Class E airspace.
b. Controlled airspace is also that airspace within which all aircraft operators are subject to certain pilot qualifications, operating rules, and equipment requirements in FAR Part 91. For IFR operations in any class of controlled airspace, a pilot must file an IFR flight plan and receive an appropriate ATC clearance. Each Class B, Class C, and Class D airspace area designated for an airport contains at least one primary airport around which the airspace is designated.

Controlled airspace in the United States is designated as follows:
1. CLASS A: Generally, that airspace from 18,000 feet MSL up to and including FL600, including the airspace overlying the waters within 12 nautical miles of the coast of the 48 contiguous States and Alaska. Unless otherwise authorized, all persons must operate their aircraft under IFR.
2. CLASS B: Generally, that airspace from the surface to 10,000 feet MSL surrounding the nation's busiest airports in terms of airport operations or passenger enplanements. The configuration of each Class B

airspace area is individually tailored and consists of a surface area and two or more layers (some Class B airspaces areas resemble upside down wedding cakes), and is designed to contain all published instrument procedures once an aircraft enters the airspace. An ATC clearance is required for all aircraft to operate in the area, and all aircraft that are so cleared receive separation services within the airspace. The cloud clearance requirement for VFR operations is "clear of clouds".

3. CLASS C: Generally, that airspace from the surface to 4,000 feet MSL above the airport elevation surrounding those airports that have an operational control tower, are serviced by a radar approach control, and that have a certain number of IFR operations or passenger enplanements. Although the configuration of each Class C airspace area is individually tailored, the airspace usually consists of a surface area with a 5 nautical mile (NM) radius, an outer circle with a 10 nm radius that extends from 1,200 feet to 4,000 feet above the airport elevation and an outer area. Each person must establish two-way radio communications with the ATC facility providing air traffic services prior to entering the airspace and thereafter maintain those communications while within the airspace. VFR aircraft are only separated from IFR aircraft within the airspace.

4. CLASS D: Generally, that airspace from the surface to 2,500 feet MSL above the airport elevation surrounding those airports that have an operational control tower. The configuration of each Class D airspace area is individually tailored and, when instrument procedures are published, the airspace will normally be designed to contain the procedures. Arrival extensions for instrument approach procedures may be Class D or Class E airspace. Unless otherwise authorized, each person must establish two-way radio communications with the ATC facility providing air traffic services prior to entering the airspace and thereafter maintain those communications while in the airspace. No separation services are provided to VFR aircraft.

5. CLASS E: Generally, if the airspace is not Class A, Class B, Class C, or Class D, and it is controlled airspace, it is Class E airspace. Class E airspace extends upward from either the surface or a designated altitude to the overlying or adjacent controlled airspace. When designated as a surface area, the airspace will be configured to contain all instrument procedures. Also in this class are Federal airways, airspace beginning at either 700 or 1,200 feet AGL used to transition to/from the terminal or enroute environment, enroute domestic, and offshore airspace areas designated below 18,000 feet MSL. Unless designated at a lower altitude, Class E airspace begins at 14,500 MSL over the United States, including that airspace overlying the waters within 12 nautical miles of the coast of the 48 contiguous States and Alaska, up to, but not including 18,000 feet MSL, and the airspace above FL 600.

controlled airspace [ICAO] — An airspace of defined dimensions within which air traffic control service is provided to IFR flights and to VFR flights in accordance with the airspace classification. Note: Controlled airspace is a generic term which covers ATS (Air Traffic Services) airspace Classes A, B, C, D, and E.

controlled departure time programs — These programs are the flow control process whereby aircraft are held on the ground at the departure airport when delays are projected to occur in either the enroute system or the terminal of intended landing. The purpose of these programs is to reduce congestion in the air traffic system or to limit the duration of airborne holding in the arrival center or terminal area. A CDT is a specific departure slot shown on the flight plan as an expected departure clearance time (EDCT).

controlled firing area — A controlled firing area is established to contain activities, which if not conducted in a controlled environment, would be hazardous to nonparticipating aircraft.

controlled flight into terrain (CFIT) — An accident where an aircraft is flown into terrain or water with no prior awareness by the crew that the crash is imminent.

controlled time of arrival — The original estimated time of arrival adjusted by the ATCSCC ground delay factor.

controller — A person authorized to provide air traffic control services.

convection — 1. In general, mass motions within a fluid resulting in transport and mixing of the properties of that fluid. 2. In meteorology, the circular motion of air that results when warm air rises and is replaced by cooler air. These motions are predominantly vertical, resulting in vertical transport and mixing of atmospheric properties. On a global scale, convection causes warm air to move from the warmer latitudes to the cooler latitudes.

convection cooling — The internal cooling air that escapes through small holes and slots, as opposed to transpiration cooling through porous walls.

convection current — Transport and mixing of a fluid caused by mass movement in the fluid. For example, as a fluid is heated, warmer portions of the fluid tend to rise in a current while cooler portions descend.

convective cloud — A cloud of vertical development that forms in an unstable environment (stratocumulus, cumulus, cumulonimbus, altocumulus, cirrocumulus).

convective condensation level (CCL) — The lowest level at which condensation will occur as a result of convection due to surface heating. When condensation occurs at this level, the layer between the surface and the CCL will be thoroughly mixed, temperature lapse rate will be dry adiabatic, and mixing ratio will be constant.

convective lifting — In unstable atmospheric conditions, a parcel of air warmer than its surroundings rises.

convective sigmet — A weather advisory concerning convective weather significant to the safety of all aircraft. Convective SIGMET's are issued for tornadoes, lines of thunderstorms, embedded thunderstorms of any intensity level, areas of thunderstorms greater than or equal to VIP (video integrator and processor thunderstorm intensity standard) level 4 with an area coverage of 4/10 (40%) or more, and hail 3/4 inch or greater.

convective significant meteorological information — See Convective SIGMET.

conventional — Conforming to formal or accepted standards in drawings or rules.

conventional current — Current flowing in an electrical circuit from positive to negative, outside the power source. This current flow theory is not in common use today.

conventional landing gear — A landing gear with wheels attached to a strut assembly located forward of the center of gravity and either a skid or wheel assembly at the tail.

conventional tail — An aircraft design with the horizontal stabilizer mounted at the bottom of the vertical stabilizer.

converge — The process of two objects moving closer to each other. Two lines forming an angle are said to converge at that angle.

convergence — In meteorology, the condition that exists when the distribution of winds within a given area is such that there is a net horizontal inflow of air into the area.

convergence zone — In meteorology, an area of convergence, sometimes several miles wide, at other times very narrow. In soaring these zones often provide organized lift for many miles along the convergence, for instance, a sea-breeze front.

convergent duct — In gas turbine engines, a cone-shaped passage or channel in which a gas may be made to flow from its largest area to its smallest area, resulting in an increase in velocity and a decrease in pressure. Referred to as nozzle shaped. With this relationship present, the weight of airflow will remain constant.

convergent-divergent exhaust — An afterburner design, a supersonic exhaust duct. The forward section is convergent to increase gas pressure. The aft section is divergent to increase gas velocity to supersonic speed.

convergent-divergent inlet — A supersonic engine inlet duct. The forward section is convergent to increase air pressure and reduce air velocity to subsonic speed. The aft section is divergent to increase air pressure still further and slow airflow to approximately Mach 0.5 before entering the engine.

conversion coating — A chemical solution used to form a dense, nonporous oxide or phosphate film on the surface of aluminum or magnesium alloys.

converter — A circuit in the control box of an anti-skid system using AC wheel speed sensors. It converts changes in AC frequency into changes in DC voltage.

convex — Having a surface that curves outward.

convey — To communicate or transmit.

convoluted — Involved, intricate. Often used to describe routing of tubing or conduit that winds circuitously around and through other components.

coolant — A fluid used to remove heat from a system or component. Anti-freeze is used to remove heat from a reciprocating engine and a coolant oil is used to remove heat from cutting tools.

cooling fins — Ribs projecting from the surface of a component to increase its area so that heat may be more easily transferred into the airstream flowing over the fins.

cooperative or group learning — An instructional strategy which organizes students into small groups so that they can work together to maximize their own and each other's learning.

coordinated universal time — Greenwich Mean Time. Referred to as coordinated universal time because the time is the average of the times from a network of atomic clocks around the world. Abbreviated UTC. (Formerly abbreviated Z.)

coordinates — The intersection of lines of reference, usually expressed in degrees/minutes/ seconds of latitude and longitude, used to determine position or location.

coordination fix — The fix in relation to which facilities will handoff, transfer control of an aircraft, or coordinate flight progress data. For terminal facilities, it may also serve as a clearance for arriving aircraft.

copal resins — A natural resin that is used in some aircraft finishes.

copilot — A pilot on a flight crew who assists in flying the aircraft, but who is not the pilot-in-command.

coping saw — A small handsaw with a U-shaped frame and a thin blade. Used for making curved cuts.

copper — A metallic element with a symbol of Cu and an atomic number of 29.

copper brazing — A method of joining metals where molten copper is used to stick to and flow between two or more metallic objects. Differs from welding in that the object metals are not melted and fused together.

copper crush gasket — A copper gasket with a fiber core that crushes when tightened between mating surfaces. Takes the shape of the mating surfaces for the purpose of creating a leakage-free seal.

copper steel — When any minimum copper content is specified, the steel is classed as copper steel. The copper is added to enhance corrosion resistance of the steel.

copper-constantan thermocouple — A low-temperature thermocouple that has a copper positive element and a constantan reference element.

copper-oxide rectifier — A rectifier that utilizes a copper-oxide coating to restrict the flow of electrons to one direction and not the other. Such a device is used to convert alternating current to direct current.

copperweld — An electrical conductor that uses a steel core for strength and copper coating for good conductance of electricity.

copter — Slang expression for helicopter

cord body — The diagonal layers of rubber-coated nylon cord fabric (running at right angles to one another), which provide the strength to a tire.

core — In composites, the central member of a sandwich part (usually foam or honeycomb). Produces a lightweight, high strength component when laminated with face sheets.

core crush — In composites, compression damage of the core.

core depression — In composites, a gouge or indentation in the core material.

core engine — The basic portion of a turbine engine consisting of the compressor that supplies air for combustion, the combustor section, and the turbine section.

core orientation — In composites, the placement of the honeycomb core to line up the ribbon direction, thickness of the cell depth, cell size, and transverse direction.

core separation — In composites, a breaking of the honeycomb core cells.

core speed sensor — Same as tachometer generator speed (N_2 speed on a dual spool engine).

core splicing — In composites, joining two core segments by bonding them together, usually with a foaming adhesive.

coriolis effect — In rotorcraft, the tendency of a rotor blade to increase or decrease its velocity in its plane of rotation when the center of mass moves closer or further from the axis of rotation.

Coriolis force — The apparent force that is produced when an object moves in relation to a surface that is rotating beneath it. Used to explain why the wind moves in a counter-clockwise direction around a low-pressure area (in the Northern Hemisphere).

cork — A lightweight wood used for gaskets and sometimes as an insulator.

cornice brake — A large sheet metal forming tool used to make straight bends. Also referred to as a leaf brake.

corona — The discharge of electricity from a wire when it has a high potential.

corrected altitude — Altitude corrected for temperature. Approximately the same as true altitude.

correction — An error has been made in the transmission and the correct version follows.

correlation — A basic level of learning where the student can associate what has been learned, understood, and applied with previous or subsequent learning.

corrolation box — Correlator. A device to automatically add power to a helicopter engine as the collective pitch control is raised.

corrosion — An electrochemical process in which a metal is transformed into chemical compounds that are powdery and have little mechanical strength.

corrosion inhibiting primer — A primer formulated to stop, slow down, or prevent corrosion on the surface of the material to which the primer is applied..

corrosion inhibitor — A substance added to a coating that slows down, stops, or prevents corrosion on the surface of the material to which the coating is applied.

corrosion prevention compound — An oil that has special corrosion prevention properties that is added to a reciprocating engine to prevent corrosion while the engine is in storage.

corrosion resistant steel — Stainless steel. A steel that has resistance to corrosion.

corrugated fastener — A fastener made up of wrinkles or folds that is used to hold two pieces of wood together in a butt joint.

corrugated sheet metal — Metal sheets that have been formed into wrinkles or folds in order to increase rigidity.

corrugation — Parallel wrinkles or folds in a material that increase the material's rigidity.

cosecant — In trigonometry, the ratio of the hypotenuse of a right triangle of which the angle is considered part and the leg opposite the angle.

cosine — In trigonometry, the ratio between the leg adjacent to the angle when it is considered part of a right triangle and the hypotenuse.

cotter pin — A metal fastener with a shank that splits into two halves. It is used to lock castellated nuts onto drilled bolts by passing the cotter pin through the hole and then spreading the ends of the cotter pin.

cotton braid — A loosely woven fabric used to encase rubber hoses and shock cords to protect them from wear and ultraviolet radiation.

coulomb — A measure of electrical charge. Equal to 6.28×10^{18} electrons. One ampere of electricity is equal to the flow of one coulomb per second.

counter bore — A tool used to cut a counterbore. A counterbore is cut into the surface of a material at the top of a smaller hole to allow the head of the bolt or the nut to be recessed below the surface.

counterboring — The act of cutting a counterbore into a material.

counterelectromotive force — When alternating current is conducted through a winding, a small current is generated that is 180° out of phase with the primary current. This is a counterelectromotive force (CEMF).

counterrotating engine — An engine designed to allow the crankshaft and other moving parts to rotate in the direction opposite the manner of a conventional engine. As a result, the propeller rotates in a counterclockwise fashion from the pilot's perspective.

countersink — A cutting tool used to cut a recess into material so the rivet or screw head is flush with the surface of the material.

countersunk-head rivet — A rivet designed with a cone shaped head that allows the rivet to fit flush with the skin of the material being riveted.

counterweight (control surface) — A weight attached to a control surface in order to reduce the effort required to move the surface and to eliminate flutter in flight.

counterweight (crankshaft) — A weight attached to the crankshaft of a reciprocating engine to balance the piston/connecting rod/crankshaft assembly.

coupled approach — A coupled approach is an instrument approach performed by the aircraft autopilot which is receiving position information and/or steering commands from onboard navigation equipment. In general, coupled nonprecision approaches must be discontinued and flown manually at altitudes lower than 50 feet below the minimum descent altitude, and coupled precision approaches must be flown manually below 50 feet AGL (See Autoland Approach). Note: Coupled and autoland approaches are flown in VFR and IFR. It is common for carriers to require their crews to fly coupled approaches and autoland approaches (if certified) when the weather conditions are less than approximately 4,000 RVR.

course — **1.** The intended direction of flight in the horizontal plane measured in degrees from north. **2.** The ILS localizer signal pattern usually specified as front course or back course. 3. The intended track along a straight, curved, or segmented MLS path.

course deviation indicator (CDI) — The instrument used for flying along a VOR-defined course. Also referred to as a left-right indicator.

course of training — A complete series of studies leading to attainment of a specific goal, such as a certificate of completion, graduation, or an academic degree.

course reversal — A method of reversing course, which is depicted on an instrument approach procedure. Some procedures do not provide for straight-in approaches unless the airplane is being radar vectored. In these situations, the pilot is required to complete a course reversal, generally within 10 nautical miles of the primary navaid or fix designated on the approach chart, to establish the aircraft inbound on the intermediate or final approach segments.

covalent bond — A bond between two atoms that comes about when valence electrons are "shared" by atoms.

cowl flaps — Movable doors on the air exit of an aircraft engine cowling. The cylinder head temperature can be controlled by varying the amount the flaps are opened.

cowl panels — The detachable coverings that cover areas requiring regular access.

cowl support ring — A large ring attached to a radial engine mount to provide firm support for cowl panels and also for attachment of cowl flaps.

cowling — A removable cover or housing placed over or around an aircraft component or section, especially an engine.

cowling, NACA — A cowling enclosing a radial air-cooled engine, consisting of a hood, O-ring, and a portion of the body behind the engine, so arranged that the cooling air smoothly enters the hood at the front and leaves through a smooth annular slot between the body and the rear of the hood; the whole forming a relatively low-drag body with a passage through a portion of it for the cooling air.

crab — The procedure of altering the heading of an aircraft into the wind in order to maintain a desired course across the ground.

crack — A partial separation of material usually caused by vibration, overloading, internal stresses, defective

assemblies, fatigue, or too rapid changes in temperature.

crack arrester — A hole drilled into a material in order to stop the spread of a crack. This hole, created by "stop drilling," distributes stresses over a larger area and keeps the crack from continuing. Commonly used in both sheet metal and acrylic plastics.

cradle — A support with pads used for cradling fuselage and wings during assembly, disassembly, or repairs.

crankcase — The housing that serves as an engine's foundation and encloses the different mechanisms of the engine.

crankpin — That part of a crankshaft to which the connecting rods attach.

crankshaft — A shaft with a series of throws used for transforming the reciprocating motion of the pistons into rotary motion used to turn the propeller of an aircraft.

crankshaft runout — A measurement of how much a crankshaft is bent. The crankshaft is rotated and the amount of "out-of-round" is measured using a dial indicator.

crankshaft throw — The distance from the crankshaft centerline to the centerline of the rod journal. Crankshaft throw is equal to half the stroke.

crater — A small pool of molten metal in the flame or arc during the process of welding.

craze — To produce minute cracks on the surface of a material.

crazing — Region of ultrafine cracks, which may extend in a network on or under the surface of a resin or plastic material.

creep — A condition of permanent elongation in a material, often due to stretching and high heat.

creeper — A platform with small wheels that allows a person to lie prone and work on the undersurface of a machine or vehicle being repaired.

crepe masking tape — A paper tape with a crinkled surface and adhesive on one side. Used to shield a surface from paint spray.

Crescent wrench — An adjustable open-end wrench manufactured by Crescent®. Often used incorrectly to describe any adjustable wrench.

crest — The surface of a thread that joins the flanks of the thread and is farthest from the cylinder or cone from which the thread projects.

crest clearance — As in a thread assembly, the distance, measured perpendicular to the axis, between the crest of a thread and the root of its mating thread.

crew member — A person assigned to perform duty in an aircraft during flight.

crew resource management (CRM) — The application of team management concepts in the flight deck environment. It was initially known as cockpit resource management, but as CRM programs evolved to include cabin crews, maintenance personnel and others, the phrase "crew resource management" has been adopted. This includes single pilots, as in most general aviation aircraft. Pilots of small aircraft, as well as crews of larger aircraft, must make effective use of all available resources; human resources, hardware, and information. A current definition includes all groups routinely working with the cockpit crew who are involved in decisions required to operate a flight safely. These groups include, but are not limited to, pilots, dispatchers, cabin crewmembers, maintenance personnel, and air traffic controllers. CRM is one way of addressing the challenge of optimizing the human/machine interface and accompanying interpersonal activities.

crewmember — A person assigned to perform duty in an aircraft during flight time.

crimping — Forming a series of small bends into a piece of sheet metal in order to shorten its length. Also used to create a bend in angle stock by crimping the sock on one leg of the angle.

crimping tool — Device used to create small bends in sheet metal for the purpose of shortening the sheet or bending angle stock.

crimp-on terminals — A solderless terminal placed on the end of a wire and then squeezed around the wire to make a solid electrical connection.

crinkle finish — A protective finish that dries to a wrinkled surface. Often used on surfaces that should not reflect light.

criteria — Part of a performance-based objective. The standards against which the desired outcome is measured.

criterion reference testing (CRT) — System of testing where students are graded against a measurable standard or criterion rather than against each other.

critical altitude — The maximum altitude at which a turbocharged reciprocating engine can deliver its rated horsepower.

critical angle of attack — The angle of attack at which maximum lift is generated. Above this angle of attack, an aircraft will stall. Also referred to as the stalling angle of attack.

critical compression ratio — On a diesel engine, the lowest compression ratio that will still allow ignition of a particular fuel.

critical coupling — In electricity, the relationship between primary and secondary windings of a transformer that provides maximum transfer of electrical energy.

critical engine — The engine that would have the most adverse effect on controllability and climb performance if it were to fail. Multi-engine airplanes with propellers turning in the same direction are designed so the descending blade of one engine is further from the centerline of the aircraft than the descending blade of the engine mounted on the other side. This makes the failure of one engine more critical than the other because there is a greater yawing moment created by the engine that is producing thrust further from the aircraft centerline. To eliminate the critical engine, manufacturers use counter-rotating engines. In this case, the descending propeller blades of both engines are the same distance from the aircraft centerline, and neither engine would affect climb performance or controllability more than the other if it were to fail.

critical height — Lowest height in relation to an aerodrome specified level below which an approach procedure cannot be continued in a safe manner solely by the aid of instruments.

critical Mach number — That airspeed for a specific airframe where supersonic airflow is first encountered at any point on the aircraft.

critical part — That part in an assembly that would cause total failure of the assembly.

critical pressure — The highest compression of a fuel-air mixture in a reciprocating engine that allows normal ignition of the mixture without premature spontaneous detonation.

critical strain — The strain at the yield point.

critical stress area — The area of a structure that would first cause total failure of the structure when overstressed.

critical temperature — The highest temperature of a fuel-air mixture in a reciprocating engine that allows normal ignition of the mixture without premature spontaneous detonation.

critique — Informal appraisals of student performance, designed to quickly convey feedback. Use critiques to summarize and complete a lesson, as well as to prepare students for the next lesson.

crocus cloth — A fabric embedded with abrasive compound, used to polish metal surfaces.

cross (fix) at (altitude) — Used by ATC when a specific altitude restriction at a specified fix is required.

cross (fix) at or above (altitude) — Used by ATC when an altitude restriction at a specified fix is required. It does not prohibit the aircraft from crossing the fix at a higher altitude than specified; however, the higher altitude may not be one that will violate a succeeding altitude restriction or altitude assignment.

cross (fix) at or below (altitude) — Used by ATC when a maximum crossing altitude at a specific fix is required. It does not prohibit the aircraft from crossing the fix at a lower altitude; however, it must be at or above the minimum IFR altitude.

cross coat — A technique used in painting where a second coat is sprayed on at a right angle to the first coat. This method is used to ensure full coverage of material.

cross country — In soaring, any flight out of gliding range of the take-off airfield. Note that this is different than the definition in the FARs for meeting the experience requirements for pilot certificates and/or ratings.

cross feed — An assembly of valves installed in a fuel system to allow any engine to receive fuel from any fuel tank, or to move fuel from one tank to another for fuel balance.

cross firing — High voltage discharge from one ignition lead to another, causing a spark plug to fire at the incorrect time.

cross linking — In composites, with thermosetting and certain thermoplastic polymers, the setting up of chemical links between the molecular chains.

cross member — A structural member that joins two longerons or other lengthwise structural members. It carries loads other than the primary loads.

cross modulation — The modulation of a desired signal by an unwanted signal resulting in two signals in the output.

cross section — The representation on a drawing of a part's interior cut at right angles to the viewing axis.

cross sectional area — The area of the plane section of an object cut at right angles to its length.

cross talk — In electronics, where signals in one portion of the equipment bleed over to another section where the signal is not desired.

cross-bleed — An assembly of valves installed in an air ducting system of a multi-engine turbine powered aircraft that allows bleed air from one engine to be used in starting another engine.

cross-brace wires — Wires arranged in an "X" fashion across rectangular sections in order to stiffen the structures by triangulation.

cross-check — A systematic way of observing instrument indications during attitude instrument flying. Also called scanning, it requires logical and systematic observation of the instrument panel. It saves time and reduces the workload of instrument flying because the pertinent instruments are observed as needed.

cross-country flying — Flying from one airport to another over a distance that requires some sort of navigation.

crosscut saw — A saw with closely spaced teeth used for cutting across the grain of wood.

crossed-control stall — A demonstration stall that a flight instructor shows a student pilot. This type of stall can occur during a skidding turn, and is most likely to occur when a pilot tries to compensate for overshooting a runway during a turn from base to final while on landing approach.

cross-hatching — A series of parallel lines on a mechanical drawing. Denotes a section of the drawing has been removed. The style of the lines indicates the type of material depicted in the drawing

crossover — A condition that exists in a helicopter rotor system in which the climbing and diving blades cross.

crossover tube — A small tube connecting multiple burner cans together for the purpose of flame propagation during starting.

cross-ply laminate — In composites, a laminate with plies usually oriented at 0° and 90° only.

cross-sectional view — A view in a mechanical drawing that depicts the shape of a cross-sectional area of the object at an indicated point.

crosswind — Wind that is blowing across the flight or taxi path of the aircraft. When used in conjunction with traffic patterns, the term is short for 'crosswind leg.'

crosswind component — Component of the wind (in knots) perpendicular to flight or taxi path of aircraft.

crosswind landing — A landing where the wind is blowing from other than parallel to the runway. Landings with small crosswind angles or velocities are of little consequence and are not usually considered to be crosswind landings.

crosswind landing gear — Landing gear on an aircraft that is specially adapted to landing in a crosswind. They are spring loaded on pivots to allow the gear to align with the runway when the aircraft is landed in a crab.

crowfoot wrench — An open or box-end wrench with a short extension and a square drive to be driven with a ratchet and an extension bar. A crowfoot wrench allows a fastener in a recess to be driven when a normal ratchet or wrench would not work.

crucible steel — Steel made from heating iron and steel in a crucible until impurities have been boiled off. Alloying elements are then added and the resulting high-grade steel is cast into ingots.

crude petroleum — Unrefined petroleum, in a raw or natural condition, before being refined for use.

cruise — **1.** A moderate speed of travel at optimum speed for sustained flight. **2.** Used in an ATC clearance to authorize a pilot to conduct flight at any altitude from the minimum IFR altitude up to and including the altitude specified in the clearance. The pilot may level off at any intermediate altitude within this block of airspace. Climb/descent within the block is to be made at the discretion of the pilot. However, once the pilot starts descent and verbally reports leaving an altitude in the block, he may not return to that altitude without additional ATC clearance. Further, it is approval for the pilot to proceed to and make an approach at destination airport and can be used in conjunction with:
a. An airport clearance limit at locations with a standard/special instrument approach procedure. The FARs require that if an instrument letdown to an airport is necessary, the pilot shall make the letdown in accordance with a standard/special instrument approach procedure for that airport, or
b. An airport clearance limit at locations that are within/below/outside controlled airspace and without a standard/special instrument approach procedure. Such a clearance is NOT AUTHORIZATION for the pilot to descend under IFR conditions below the applicable minimum IFR altitude nor does it imply that ATC is exercising control over aircraft in Class G airspace; however, it provides a means for the aircraft to proceed to destination airport, descend and land in accordance with applicable FARs governing VFR flight operations. Also, this provides search and rescue protection until such time as the IFR flight plan is closed.

cruise climb — A climb technique employed by aircraft, usually at a constant power setting, resulting in an increase of altitude as the aircraft weight decreases.

cruise control — Engine operation procedures that allow the best efficiency for power and fuel consumption during cruising.

cruise power — Sixty percent to 70 percent of maximum continuous power; used for fuel economy and engine life during cruising.

cruise propellers — A fixed-pitch propeller that provides the aircraft with the best performance during cruise flight.

cruising altitude — An altitude or flight level maintained during enroute level flight. This is a constant altitude and should not be confused with a cruise clearance.

cruising level — See cruising altitude.

crush gasket — A gasket usually formed of a sandwich of copper with an asbestos core. When the gasket is crushed between mating surfaces, it conforms to the surfaces and creates a leak-proof seal.

cryogenic liquid — A gas that has been cooled to a temperature low enough for it to be liquefied. Liquid oxygen and liquid nitrogen are commonly found cryogenic liquids.

crystal — 1. A thin piece of piezoelectric material having a specific resonant frequency used to control the frequency of an oscillator. 2. A small piece of galena or lead sulfide that allows electron flow in only one direction.

crystal diode — An electronic component that passes electrons in one direction, but not the other. A crystal diode is constructed from a germanium or silicon crystal.

crystal earphones — Earphones that reproduce sound by use of piezoelectric crystals. Audio-frequency alternating voltage is applied to the crystal and the crystal vibrates in accord with the frequency of the voltage. The crystal is attached to a sound diaphragm that causes the pressure of the air to change in phase with the voltage. These changes in air pressure are perceived by the ear as sound.

crystal filter — A crystal filter vibrates at a certain frequency. This absorbs vibrations at that frequency, but passes frequencies above and below the frequency of the crystal.

crystal lattice — The basic pattern in which atoms are arranged, repeated throughout the solid.

crystal microphone — A microphone making use of the piezoelectric properties of a crystal, acted on by the pressure of sound waves.

crystal oscillator — An electronic device that produces an alternating current. A crystal oscillator produces a frequency determined by a crystal.

crystal transducer — An electronic device that converts vibrations into electronic signals. A crystal transducer uses a piezoelectric crystal that converts vibrations to a minute alternating current.

crystal-controlled oscillator — See crystal oscillator.

crystalline — Arranged in a definite pattern, with the atoms or molecules tending to be developed in precisely oriented plane surfaces.

C-stage — In composites, the final stage in the curing of the mixed thermoset resin system. It cannot be softened by heat and is insoluble at this stage.

cube — A three dimensional geometric figure consisting of six congruent square sides.

cube root — In mathematics, the cube root is the number that, when multiplied by itself twice, is the given number. Example: The cube root of 27 is 3 (3 X 3 X 3).

cumuliform — A term descriptive of all convective clouds exhibiting vertical development in contrast to the horizontally extended stratiform types.

cumulonimbus (Cb) — Deep convective clouds with a cirrus anvil and may contain any of the characteristics of a thunderstorm: thunder, lightning, heavy rain, hail, strong winds, turbulence, and even tornadoes. Also referred to as thunderclouds.

cumulous congestus — A cumulous cloud of significant vertical extent and usually displaying sharp edges. In warm climates, these sometimes produce precipitation. Also called towering cumulous, these clouds indicate that thunderstorm activity may occur soon.

cumulus (Cu) — A principle cloud type having a flat base and sharp rounded tops and sides. Clouds develop vertically and tops often resemble a cauliflower.

cumulus stage — The initial stage of a thunderstorm. The cloud grows from cumulus to towering cumulus and usually lasts 10 or 15 minutes.

Cuno filter — The proprietary name of a fluid filter made up of a stack of discs separated by scraper blades. Contaminants collect on the edge of the discs and are periodically scraped out and collected in the bottom of the filter case.

cup washer — A cup-shaped, spring steel washer. When compressed, the spring action prevents the nut on a bolt from loosening.

cure — In composites, to change the physical properties of a material by chemical reaction, by the application of catalysts, heat and pressure, alone or in combination.

cure temperature — In composites, the temperature that the resin system attains for its final cure. Does not include the ramp up or down.

cure time — The time required for a resin to complete its solidification.

curing agent — A catalytic or reactive agent that causes polymerization when added to resin. Also referred to as a hardener.

current — The flow of electricity. Technically, it is electrons that flow, but more commonly, this is called current-flow. Current-flow is measured in amperes.

current amplifier — An electronic device whose output current is greater than its input current.

current density — The amount of electrical current per cross-sectional area of a conductor. Expressed in amperes per square inch.

current flight plan [ICAO] — The flight plan, including changes, if any, brought about by subsequent clearances.

current limiter — A device that limits the generator output to a level within that rated by the generator manufacturer.

current plan — The ATC clearance the aircraft has received and is expected to fly.

current regulator — An electronic device in a circuit that maintains a constant current.

current switch — A switch used to activate a separate circuit in response to a specified current. When the AC current being monitored reaches or exceeds the switch setting, the switch will conduct to activate another AC or DC control circuit connected to that switch.

current transformer — An electronic device used to indicate the amount of current in an alternating current transmission line. The primary winding is connected in series with the line current and the secondary winding is connected to an indicator that reads out in volts, which are proportionate to the amperage in the primary.

current-fed antenna — A half-wave antenna fed in its center.

current-limiting resistor — A resistor placed in an electronic circuit to limit the amount of amperage flowing in the circuit.

curriculum — A set of courses in an area of specialization offered by an educational institution. A curriculum for a pilot school usually includes courses for the various pilot certificates and ratings.

curvature — The shape of a surface that has a smooth, bending surface.

curvic coupling — A circular set of gear-like teeth on each of two mating flanges that provide a positive engagement when meshed together and bolted. Used to attach together turbine wheels, compressor disks, etc.

cusp — The indentation on each side at the floor level when a fuselage shape has a "figure eight" shape. The cusp design is used to avoid unneeded width in this area.

cut off — To sever an object or to stop a flow.

cut thread — A thread produced by removing material from the surface with a form cutting tool.

cut-away — Model of an object that is built in sections so it can be taken apart to reveal the inner structure.

cutaway drawing — A mechanical drawing with a portion of the outside not drawn in order to show what the inside of the object looks like.

cut-aways — Model of an object that is built in sections so it can be taken apart to reveal the inner structure.

cutout switch — An electrical switch that interrupts the power to a motor or actuator when the limit of its desired travel is reached.

cuts out — The intermittent operation of a magneto or ignition system.

cutting edge — The edge of a tool or device used to remove material when some type of force is applied.

cutting fluid — A fluid, often consisting of oil and water that is flowed over the cutting surface of a cutting tool. Its purpose is to cool the surfaces, lubricate the cutting device, flush chips away, and inhibit corrosion.

cutting plane — A line on an aircraft drawing that indicates the surface of an auxiliary view.

cutting speed — The speed at which a cutting tool moves across or through a material. The speed must be set according to the material being cut and the cutting tool type in order to reduce wear and optimize the speed of the cutting procedure.

cutting torch — A torch used to cut metals. It consists of pre-heating jets to heat the work to near melting and a cutting jet to burn through the metal.

CVFP (Charted Visual Flight Procedure) approach — An approach conducted while operating on an instrument flight rules (IFR) flight plan which authorizes the pilot of an aircraft to proceed visually and clear of clouds to the airport via visual landmarks and other information depicted on a charted visual flight procedure. This approach must be authorized and under the control of the appropriate air traffic control facility. Weather minimums required are depicted on the chart.

cyaniding — A hardening metal where the heated metal is introduced to a cyanide bath. The metal absorbs nitrogen and carbon from the cyanide and creates a hardened surface.

cyanoacrylate — A single-component, polyester-type resin that hardens by exposure to ambient moisture and surface alkalinity.

cybernetics — The comparative study of the interaction of biological, mechanical, and electrical systems.

cycle — A complete series of events or operations that recur regularly. The series ends at the same condition as it started so that the next series of events can immediately take place.

cycle — A period of time during which a sequence of recurring events is completed. In aviation, a cycle can be one takeoff and one landing, one extension and retraction of the flaps, one retraction and extension of the landing gear, etc.

cycles per second — A measure of how often something is repeated or oscillates (cycles) per second. One cycle per second is called one hertz (Hz).

cyclic feathering — In rotorcraft, the mechanical change of the angle of incidence, or pitch, of individual rotor blades independently of other blades in the system.

cyclic pitch control — The control in the cockpit of a helicopter with which the pilot can change the pitch of the rotor blades at a specific point in their rotation. The resulting change imparts lateral, forward, or backward movement to the helicopter.

cycling — The operation of a unit such as the landing gear retraction system through its full range of operation.

cycling switch — A switch that opens and closes the circuit, permitting a unit to cycle on and off.

cyclogenesis — Any development or strengthening of cyclonic circulation in the atmosphere.

cyclone — In meteorology, the closed circulation around a low pressure area. The flow is counterclockwise in the Northern Hemisphere and clockwise in the Southern Hemisphere. Large tropical cyclones are referred to as typhoons or hurricanes.

cyclonic flow — In the Northern Hemisphere the counterclockwise flow of air around an area of low pressure and a clockwise flow in the Southern Hemisphere.

cyclostrophic wind — A wind in a circular path, in which Coriolis acceleration is negligible as compared to centrifugal forces and pressure gradient acceleration.

cylinder — 1. A geometric shape having ends of a circular form and parallel sides. 2. That component of a reciprocating engine in which the fuel is burned and in which the piston moves up and down.

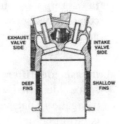

cylinder barrel — That portion of the cylinder or a reciprocating engine in which the piston moves up and down.

cylinder bore — The diameter of the cylinder barrel.

cylinder fins — Metal fins on a cylinder that increase the area of metal exposed to air. Allows the heat to radiate out into the air for cooling.

cylinder flange — The base of a cylinder that incorporates a machined mounting flange by which the cylinder assembly is attached to the crankcase.

cylinder head — The closed end of the combustion chamber of a reciprocating engine.

cylinder head temperature — The temperature of the cast-aluminum head of an air-cooled reciprocating aircraft engine cylinder.

cylinder honing — A machining process that trues the inside surface of a cylinder. The process also prepares the inner surface of a reciprocating engine cylinder to receive proper lubrication.

cylinder pads — The machined surfaces on the crankcase of an aircraft engine on which the cylinders are mounted.

cylinder skirt — That portion of a reciprocating engine's cylinder that extends below the mounting flange.

cylinder taper — Many reciprocating engine cylinders are tapered so that the top portion of the cylinder (nearest the combustion chamber) is slightly smaller than the opposite end. This allows for increased expansion due to higher temperatures and increased mass at the combustion end of the cylinder. Also referred to as a choke-bore cylinder.

D

D check — The highest and most thorough level of inspection in a Continuous Airworthiness Maintenance Program. Entails a complete teardown of the aircraft for inspection, repair, and replacement of worn components. the entire program consists of A, C, and D checks.

Dacron® — Polyester fibers made by DuPont, Inc.

dado head — A woodworking tool used to cut a wide, flat-bottomed groove. Usually consists of two normal blades separated by a wood-chipping blade.

dado plane — A narrow, woodworking plane used to cut a wide, flat-bottomed groove in wood.

Dalton's law — One of the basic gas laws that explains partial pressures. The law states that the total pressure of a gaseous mixture made up of two or more separate gases is the sum of the partial pressures of each of the gases. This explains why, at high altitudes, the partial pressure of oxygen is not enough to allow the oxygen to enter the blood stream without assistance from supplemental oxygen or pressurization.

dampen — To deaden, depress, reduce, or lessen.

damper — A device used to limit movement.

damper valve — A plate in the flue of a furnace that regulates the draft. The plate fills the flue when turned perpendicular, shutting off the flow of gas. When the plate is turned parallel to the flue, gases can freely flow. In aviation, similar valves control the flow of fluid in a tube.

damper vane — A vane in a fuel flow meter used to dampen fluctuations caused by erratic flow.

damper-type combustion air fuel valve — An automatically actuated damper-type valve located in the combustion air inlet of an aircraft heater. The valve is held open by fusible links that melt and allow the valve to shut off the combustion air in case of a fire or over-temperature condition.

damping action — An action that smoothes out pulsations in the flow of an oscillation.

damping tube — A short length of tubing with an extremely small inside diameter inside a manifold pressure gauge. It prevents a pressure surge (caused by the engine backfiring) from damaging the mechanism of the instrument.

danger area — A specified or specific area designated as dangerous. Constitutes a potentially hazardous situation to persons or property.

danger area [ICAO] — An airspace of defined dimensions within which activities dangerous to the flight of aircraft may exist at specified times. Note: The term "Danger Area" is not used in reference to areas within the United States or any of its possessions or territories.

d'Arsonval meter movement — Most commonly used meter movement in DC measuring instruments. A movable coil on which a pointer is mounted rotates in the field of a permanent magnet. The amount of current in the coil determines the strength of the coil's electromagnetic field and the amount the pointer is deflected by the magnetic field of the permanent magnet.

dart leaders — An individual event in a return stroke that the eye cannot distinguish that occurs following the initial discharge. See lightning.

dash numbers — Numbers following, and separated from, a part number by a dash identifying the components of the part.

dashpot — A mechanical damper used to cushion or slow down movement by restricting the flow of a viscous fluid.

data — Information available on a subject that is processed for the purpose of reasoning, discussion, calculations, etc. Computers process data and provide a desired output.

data base — A grouping of data that can be easily searched for specific information and then processed or acted upon.

data block — See alphanumeric display.

data plate — A permanent identification plate affixed to an aircraft, engine, or component.

data stamp — Information stamped on units or components providing information on the correct name, part number, date of manufacture, and cure date, as applicable.

database identifier — A specific geographic point in space identified on an aeronautical chart and in a navigation database, officially designated by the controlling state authority or derived by Jeppesen. It has no ATC function and should not be used in filing flight plans nor used when communicating with ATC.

data-plate speed — The speed at which the manufacturer determines "rated power" of a gas turbine engine. Stamped on a data plate affixed to the engine. The engine is required to perform within a certain range of this value throughout its service life.

datum — An arbitrary reference line from which all measurements are made when determining the moments used for weight and balance computations. Also called datum line.

Davis wing — A wing with a narrow chord and a thick cross-section.

DC amplifier — An electronic amplifier whose output is greater than and proportional to its DC input signal.

dead — In electricity, having no potential difference or current flow.

dead center — Either of the two positions at the ends of a stroke in a crank and connecting rod when the crank throws and rod lie in a straight line. See bottom dead center and top dead center.

dead engine — An engine that is not running during flight.

dead reckoning navigation — A method of navigating from one location to another without the use of outside navigation aids. Dead reckoning navigation relies on calculating the effects of predicted winds on the course and speed of the aircraft.

dead reckoning — Dead reckoning, as applied to flying, is the navigation of an airplane solely by means of computations based on airspeed, course, heading, wind direction, and speed, groundspeed, and elapsed time.

dead short — An electrical path with very low resistance, usually disastrous in nature.

dead-stick landing — An aircraft landing accomplished without an operative engine. There is no capability to go around from an inaccurate approach.

dead-weight tester — A calibration device used to test pressure gauges for accuracy.

de-aeration — The process of removing air from a liquid.

de-aerator chamber — In hydraulics, an area of a reservoir where de-aeration takes place.

de-aerator tray — A container that collects the return oil from the oil system of a turbine engine and allows the air bubbles to separate out of the oil before it returns to the system.

debarkation — The unloading of passengers and cargo.

debooster — A unit used in the brake system to reduce system pressure and give faster application and release of the brakes.

debug — To troubleshoot electronic equipment or computer programs and return to proper operation.

Deca (Deka) — A metric prefix indicating ten times the term shown. Example: decameter or decaliter.

decade resistance box — An electrical test device that provides the capability to insert selected values of resistance into a circuit.

decades — A series of quantities in multiples of 10.

decalage — The difference in the angle of incidence of two aerodynamic surfaces of an aircraft. (Between wings of a biplane or between the wing and the horizontal stabilizer).

decarbonizers — Potent solvents used to soften the bond of carbon to a metal part.

decay — The gradual decrease in the amplitude of a vibration or oscillation.

decelerate — To slow down the velocity of an object. Opposite of accelerate.

deceleration — The rate of decrease in velocity.

deceleration check — A check made on an engine while retarding the throttle from the acceleration check. The RPM should decrease smoothly and evenly with little or no tendency for the engine to after fire.

deci — A metric prefix indicating one tenth the term shown. Example: decimeter or deciliter.

decibel (dB) — 1. A measure of sound intensity equal to $1/10$ of a bel. 2. A unit used to express the ratio between two amounts of electrical or acoustical power and equal to 10 times the logarithm of this ratio.

DECIDE model — To assist in teaching pilots the elements of the decision-making process, a six-step model has been developed using the acronym "DECIDE."
Detect the fact that a change has occurred.
Estimate the need to counter or react to the change.
Choose a desirable outcome for the success of the flight.
Identify actions, which could successfully control the change.
Do the necessary action to adapt to the change.
Evaluate the effect of the action.

decimal — A proper fraction in which the denominator is a power of ten. Usually expressed as a number divided by a decimal (period). Example: 12.345

decimal digit — One of the ten Arabic numerals 0, 1, 2, 3, 4, 5, 6, 7, 8, or 9.

decimal number system — A number system that uses the base of ten. Each number is expressed with one of the first nine integers or 0 in each place and each place value represents a power of 10.

decimal point — The dot that separates a whole number from its decimal fraction. Example: 19.44.

decision altitude/decision height [ICAO] — A specified altitude or height (A/H) in the precision approach at which a missed approach must be initiated if the required visual reference to continue the approach has not been established. Note 1: Decision altitude (DA) is referenced to mean sea level (MSL) and decision height (DH) is referenced to the threshold elevation. Note 2: The required visual reference means that section of the visual aids or of the approach area which should have been in view for sufficient time for the pilot to have made an assessment of the aircraft

position and rate of change of position, in relation to the desired flight path.

decision height (DH) (USA) — With respect to the operation of aircraft, means the height at which a decision must be made, during an ILS or PAR instrument approach, to either continue the approach or to execute a missed approach.

NOTE: Jeppesen approach charts use the abbreviation DA(H). The decision altitude "DA" is referenced to mean sea level (MSL) and the parenthetical decision height (DH) is referenced to the TDZE or threshold elevation. A DA(H) of 1,440 ft.. (200 ft.) is a Decision Altitude of 1,440 ft. and a Decision Height of 200 ft.

decision-making process — Involves an evaluation of risk elements to achieve an accurate perception of the flight situation. The risk elements include the pilot, the aircraft, the environment, the operation, and the situation.

declination — The angular difference between magnetic north and true north caused by the magnetic north pole not being located at the same place as the true north pole. In navigation for aircraft this is called variation.

decoder — The device used to decipher signals received from ATCRBS transponders to effect their display as select codes.

decomposing — The process of a material being broken down into its basic elements. Example: water breaking down into oxygen and hydrogen by the process of electrolysis.

decontamination — The removal or neutralization of undesired material from an area, a piece of equipment, a building, or a person.

decouple — To release or disconnect a unit.

dedicated computer — A computer that is used only for one function.

deductible — The amount that the policyholder is responsible for in the event of a claim.

de-energize — To turn off a piece of electrical equipment.

deep cycling — A treatment of nickel cadmium batteries in which the battery is completely discharged, the cells shorted out and allowed to "rest." The battery is then recharged to 140% of its ampere-hour capacity.

deep discharge — The procedure of removing all electrical energy from a battery. After discharging a battery down to a low level, the cells are shorted out with shorting clips until all cells are completely discharged. This allows all cells to be recharged from the same starting condition.

deepening — A decrease in the central pressure of a pressure system; usually applied to a low rather than to a high, although technically, it is acceptable in either sense.

defect — Any imperfection, fault, flaw, or blemish that may require repair or replacement of a part.

defective — Faulty. Not operating normally.

defense mechanisms — Subconscious ego-protecting reactions to unpleasant situations.

defense visual flight rules (DVFR) — Rules applicable to flights within an ADIZ conducted under the visual flight rules in FAR 91.

deferred item — An item on a Minimum Equipment List (MEL) that can be inoperative under specified conditions. When an item is deferred, it is often accompanied by some sort of flight limitation.

deflate — To decrease the amount of air or gas held by an object.

deflecting-beam torque wrench — A hand-operated torque wrench in which the amount of torque applied to a bolt is indicated by the amount the beam is bent. The indication is read against a fixed scale on the handle of the wrench.

deflection — The movement of an electron beam up and down or sideways in response to an electric or magnetic field in a cathode-ray tube.

degauss — To remove magnetism or magnetic field from an object or piece of equipment.

degeneration — Feedback of a portion of the output of a circuit to the input in such a direction that it reduces the magnitude of the input; also called negative feedback. Degeneration reduces distortion, increases stability, and improves frequency response.

degreaser — A solvent used for removing oil or grease from a part.

dehumidify — To reduce the amount of water vapor in the air.

dehydrator — A piece of equipment or substance used to remove water vapor from the air. Silica gel is often used to absorb water within a closed space.

dehydrator plug — A plastic plug with threads screwed into a spark plug opening of an aircraft engine cylinder. These plugs are filled with silica-gel and an indicator to remove moisture from the air inside the cylinder and indicate the condition of preservation of the cylinder.

deicer — A system or substance that removes ice from an aircraft structure.

deicer boots — Inflatable rubber boots attached to the leading edge of an airfoil. They can be sequentially inflated and deflated to break away ice that has formed over their surface.

deicer tubes — The inflatable tubes in the deicer boot.

deicing — Removing ice after it has formed.

de-icing equipment — Aircraft equipment that is actuated to remove ice from the structure of the aircraft that has already formed.

deicing fluid — A liquid (usually heated) that is sprayed on an aircraft to remove ice. Deicing fluid prevents the return of ice for a specified period of time.

delaminate — In composites, the separation of layers due to adhesive failure. This also includes the separation of the layers of fabric to a core structure. A delamination may be associated with bridging, drilling, and trimming.

delaminated — **1**. A condition caused by exfoliation corrosion in which the layers of grain structure in an extrusion separate from one another. **2**. Separation of the core and face sheets of a bonded structure along a bond line.

delay indefinite (reason if known) expect further clearance (time) — Used by ATC to inform a pilot when an accurate estimate of the delay time and the reason for the delay cannot immediately be determined; e.g., a disabled aircraft on the runway, terminal or center area saturation, weather below landing minimums, etc.

delay time — The amount of time that the arrival must lose to cross the meter fix at the assigned meter fix time. This is the difference between ACLT and VTA.

delta — Greek letter (Δ) used in weight and balance computations to indicate amount of change.

delta connection — A method of connecting three components to form a three-sided circuit, usually drawn as a triangle, hence the term delta.

delta hinge — The hinge located at the root end of the rotor blade with its axis parallel to the plane of rotation of the rotor that allows the blade to flap. The flapping blade equalizes lift between the upwind and downwind sides of the rotor disc.

delta winding — The connection of the windings of three-phase AC machines. The three windings are connected together to form a loop or a single path through the three windings.

delta wing — The triangular wing planform of an aircraft.

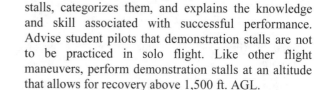

demagnetize — To remove magnetic properties from an object.

demand oxygen system — Any oxygen system in which the delivery of oxygen is metered according to the rate and depth of a user's breathing.

demodulation — The recovery of the signal from a radio frequency carrier wave. Also referred to as detection.

demonstration stalls — Stalls that the instructor demonstrates mainly as an in-flight portion of stall/spin awareness training. The flight instructor PTS lists these stalls, categorizes them, and explains the knowledge and skill associated with successful performance. Advise student pilots that demonstration stalls are not to be practiced in solo flight. Like other flight maneuvers, perform demonstration stalls at an altitude that allows for recovery above 1,500 ft. AGL.

demonstration-performance method — An educational presentation in which an instructor first shows the student the correct way to perform an activity and then has the student attempt the same activity.

demulsibility — The measure of an oil's ability to separate from water.

demulsifier — A system or substance that causes an emulsion (mixture) of two or more substances to break down into component materials.

denatured alcohol — Grain alcohol (ethyl alcohol or ethanol) with substances added to make it unfit for human consumption without affecting its use for other purposes.

denial of reality — A psychological defense mechanism where students may ignore or refuse to acknowledge disagreeable realities. They may turn away from unpleasant sights, refuse to discuss unpopular topics, or reject criticism.

denier — A numbering system for filaments in the yarn used for weaving. The number is equal to the weight in grams of 9,000 meters of yarn.

denominate number — A number associated with a unit of measurement.

denominator — The part of a fraction below the line indicating division. It is an indication of the number of parts into which a number is divided.

density — The mass of a substance per unit of its volume. The weight per unit volume expressed in pounds per cubic foot.

density altitude — Pressure altitude corrected for nonstandard temperature variations.

dent — A depression in a surface usually caused by the surface being struck by another object.

dented surface — A depression on a surface that usually affects a small area. Cams, tappet rollers, and ball and roller bearings are the parts most often involved.

departure center — The ARTCC having jurisdiction for the airspace that generates a flight to the impacted airport.

departure control — A function of an approach control facility providing air traffic control service for departing IFR and, under certain conditions, VFR aircraft.

departure sequencing program — A program designed to assist in achieving a specified interval over a common point for departures.

departure time — The time an aircraft becomes airborne.

depletion area — That area on both sides of the junction of a semiconductor that varies between acting as a conductor and an insulator.

depolarization — The absorption of generated gases in a chemical cell especially during the "rest" periods. This may cause an apparent rejuvenation of the cell.

deposition — The direct transformation of a gas to a solid state, where the liquid state is bypassed.

depreservation — A procedure that includes a special inspection and cleaning of aircraft parts removed from storage prior to being installed on the aircraft or engine.

depressants — Drugs that reduce the body's functioning usually by lowering blood pressure, reducing mental processing, and slowing motor and reaction responses.

depression — In meteorology, an area of low pressure, as in a trough. Often used to describe the stage of development of a cyclone, as in "tropical depression."

depth gauge — A device used to measure the depth of a hole, groove, or depression.

depth micrometer — A form of micrometer caliper used to measure the depth of a recess.

derate — To reduce the allowable power output of an engine or motor in order to extend the life of the equipment.

derichment — An automatic leaning of the fuel-air mixture ratio to a ratio that will produce maximum power regardless of the heat released. Derichment occurs when the anti-detonation injection system injects liquid into the cylinders to remove this excess heat.

descent — A reduction in altitude.

descent speed adjustments — Speed deceleration calculations made to determine an accurate VTA. These calculations start at the transition point and use arrival speed segments to the vertex.

description of the skill or behavior — The first part of a performance-based objective, which explains the desired outcome of instruction in measurable concrete terms.

desiccant — 1. Any form of absorbent material. 2. A material used in a receiver-dryer to absorb moisture from the refrigerant.

desiccant bags — Cloth containers of a silica gel desiccant packed with an engine or component that is placed in long-term storage.

design load — The load for which a member is designed. It is usually obtained by multiplying a basic load by a specified design load factor.

design maneuvering speed (VA) — The maximum speed at which you can use full, abrupt control movement without over stressing the airframe.

design size — That size from which the limits of size are derived by the application of tolerances. When there is no allowance, the design size is the same as the basic size.

designated — Being given the legal right and authority to perform certain specified functions by the Federal Aviation Administration.

designated examiner — Any person authorized by the Administrator to conduct a pilot proficiency test or a practical test for an airman certificate or rating issued under FAR Part 61. Also, a person authorized to conduct a knowledge test under FAR Part 61.

desired course — 1. True desired course — A predetermined desired course direction to be followed (measured in degrees from true north). 2. Magnetic desired course — A predetermined desired course direction to be followed (measured in degrees from local magnetic north).

desired track — The planned or intended track between two waypoints. It is measured in degrees from either magnetic or true north. The instantaneous angle may change from point to point along the great circle track between waypoints.

destructive testing — Testing that results in the destruction of the item being tested. For example, as a part of the certification process, aircraft wing structures are often loaded until they fail.

detail drawing — A drawing that describes a single part in detail.

detail view — An auxiliary view incorporated into a drawing to show additional details of a part.

detailed inspection item — An item of a progressive inspection that requires close and careful inspection. May require disassembly or complete overhaul of a component or part.

detector — That portion of an electronic circuit that demodulates or detects the signal.

detents — The points along a line of movement of a control where distinct resistance to movement can be felt. These points are usually found at specific levels of importance to the operation of the controlled unit. An example would be the flap control lever, where detents could be assigned either to the number of degrees of travel (15°, 30°, and 45°) or at specific performance points such as takeoff flaps and landing flaps.

detergent — A cleansing material. An oil-soluble substance that holds insoluble matter in suspension. Detergent is sometimes referred to as soap, though detergent is chemically different than soap.

detergent oil — A mineral oil to which ash-forming additives have been added to increase its resistance to oxidation. Because of its tendency to loosen carbon deposits, it is not used in aircraft engines.

deteriorate — To become worse.

determiners — In test items, words that give a clue to the answer. Words such as "always" and "never" are determiners in true-false questions. Since absolutes are rare, such words usually make the statement false.

detonation — The sudden release of heat energy from fuel in an aircraft engine caused by the fuel-air mixture reaching its critical pressure and temperature. Detonation occurs as a violent explosion rather than a smooth burning process.

DETRESFA (distress phase) [ICAO] — The code word used to designate an emergency phase wherein there is reasonable certainty that an aircraft and its occupants are threatened by grave and imminent danger or require immediate assistance.

Deutsch rivet — A type of high-strength blind rivet.

developed width — Width of the flat layout of a sheet metal part.

developer — A powder that has been treated with a penetrating dye. When sprayed on a surface, the powder acts as a blotter, pulling penetrant out of any crack and exposing the crack.

development — The main body of an instructional lesson that contains a detailed listing of the subject matter. Developing material in a structured way speeds up the process and makes it easier to follow a logical progression.

deviation — A compass error caused by magnetic disturbances from electrical and metal components in the airplane. The correction for this error is displayed on a compass correction card placed near the magnetic compass in the airplane.

deviations — 1. A departure from a current clearance, such as an off course maneuver to avoid weather or turbulence. 2. Where specifically authorized in the FARs and requested by the pilot, ATC may permit pilots to deviate from certain regulations.

dew — Moisture that has condensed from water vapor. Usually found on cooler objects near the ground, such as grass, as the near-surface layer of air cools faster than the layers of air above it.

dew point — The temperature at which air becomes saturated and produces dew or moisture.

Dewar flask — A double-wall vacuum chamber used for the storage of liquid oxygen. Similar in principle to a Thermos® bottle.

dewaxed oil — A pure lubricating oil that has had waxy products removed during the refining process.

dewpoint — The temperature to which air must be cooled to become saturated with water vapor.

DF approach procedure — Used under emergency conditions where another instrument approach procedure cannot be executed. DF guidance for an instrument approach is given by ATC facilities with DF capability.

DF fix — The geographical location of an aircraft obtained by one or more direction finders.

DF guidance — Headings provided to aircraft by facilities equipped with direction finding equipment. These headings, if followed, will lead the aircraft to a predetermined point such as the DF station or an airport. DF guidance is given to aircraft in distress or to other aircraft which request the service. Practice DF guidance is provided when workload permits.

DF steer — See DF guidance.

diagonal cutting pliers — A wire cutting tool that cuts by chiseling action rather than by shearing.

diagram — A graphic representation of an assembly or system.

dial — An analog style instrument face.

dial indicator — A precision linear measuring instrument whose indication is amplified and read on a circular dial.

dial-indicating torque wrench — A hand-operated, deflecting beam-type torque wrench that uses a dial indicator to measure the deflection of the beam and reads directly in foot-pounds, inch-pounds, or meter-kilograms of torque.

diamagnetic material — A material having extremely low magnetic permeability and considered to be nonmagnetic.

diameter — The length of a chord passing through the center of a circular body.

diamond — A hard substance consisting of nearly pure carbon. Some are of gem quality while others, of industrial quality, are used as cutting surfaces.

diamond chisel — A cutting tool with a diamond shaped cutting face.

diamond dressing tool — An industrial diamond mounted in a tool and used to true a grinding wheel.

diamond-point cutting tool — A machine tool used for machining very hard metals. The cutting face consists of an industrial grade diamond.

diaphragm switch — A switch whose position is controlled by movement of a diaphragm.

diaphragm type pump — A pump that uses a cam-operated flexible diaphragm to move fuel past spring-loaded valves. The pump can have a single diaphragm that only pumps on one direction of movement of the cam, or it can have a double diaphragm that operates on both the powered stroke and the spring return.

diaphragm-controlled — A mechanical movement controlled by the action of a pressure or suction applied to a diaphragm.

dibromodifluoromethane — A fire extinguishing agent. Noncorrosive to aluminum, brass, and steel, it is less toxic than CO_2. It is one of the more effective fire extinguishing agents available. Halon 1202.

dichromate solution — Cr_2O_7. Used with zinc as zinc dichromate, a protective coating on metal parts. Sometimes referred to as gold iridite and found on many brake components. Also found in sodium dichromate ($Na_2Cr_2O_7$) and Potassium dichromate ($K_2Cr_2O_7$). Sodium dichromate is used to protect magnesium parts.

die — 1. A tool used to shape or form metal or other materials. 2. A cutting tool used to cut threads on the outside of round stock.

die casting — Method by which molten metal is forced into suitable permanent molds by hydraulic pressure in order to improve the grain structure of the resulting casting.

dielectric — A material that will not conduct electricity.

dielectric constant — Symbol: k. The characteristic of an insulator that determines the amount of electrical energy that can be stored in electrostatic fields.

dielectric qualities — Insulating characteristics of a material.

dielectric strength — A measure of a dielectric's ability to withstand puncture by electrical stresses.

diesel engine — An internal combustion reciprocating engine whose ignition is achieved by the heat of compression rather than with an electrical ignition.

diesel fuel — A fuel used in diesel engines. Very close in grade to jet-fuel and kerosene.

dieseling — The continued firing of a reciprocating engine after the ignition has been turned off. Ignition is caused by incandescent particles in the combustion chamber.

difference between the symbol and the symbolized object — The result of a word being confused with an unintended meaning. Words and symbols do not always represent the same thing to every person. Confusion results when the name of an object is not differentiated from the characteristics of the object itself.

differences training — Training given to pilots who wish to transition between similar makes and models of a given manufacturer. For example, transitioning from a C-210 to a P-210 would require differences training for pressurization and turbocharging.

differential aileron travel — The increased travel of the aileron moving up over that of the aileron moving down. The up aileron produces extra parasite drag to compensate for the additional induced drag caused by the down aileron. This balancing of the drag forces helps minimize adverse yaw.

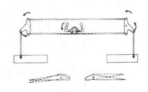

differential compression check — A test of the condition of an engine cylinder in which the amount of leakage past the piston rings and valves is determined by measuring the pressure drop across a calibrated orifice in the tester.

differential heating — The heating of objects that have dissimilar heat capacities. See heat capacity.

differential pressure — A difference between two pressures. The measurement of airspeed is an example of the use of a differential pressure.

differential pressure range of pressurization — The range of cabin pressurization in which a constant differential is maintained between cabin pressure and the outside air pressure.

differential pressure switch — A diaphragm and electrical microswitch arrangement that senses two pressures. If the difference in the two pressures exceeds a certain value, the microswitch illuminates a warning light.

differential-voltage reverse-current relay — Removes a battery from and connects a generator to a load when the generator voltage is higher than the battery voltage. If the generator voltage falls below the battery voltage, the battery will be reconnected.

differentiating circuit — A circuit that produces an output voltage proportional to the rate of change in input voltage.

diffuser — **1.** A duct used on a centrifugal flow turbine engine to reduce the velocity of the air and increase its pressure. **2.** The divergent section of a gas turbine engine used to convert velocity energy of compressor discharge air into pressure energy.

diffuser vane — A turning or cascading vane in a centrifugal flow engine diffuser used to change air from a radial direction as it leaves the impeller to an axial direction as it moves into the combustor.

diffusion — **1.** A process where particles of liquids, gases, or solids move from a region of higher concentration to one of lower concentration. **2.** The scattering of light by a rough surface or by transmission through a translucent material.

diffusion welding — A method of fusing materials to each other that relies on the two materials being pressed together under high pressure and the joint heated to less than melting temperature. The two pieces fuse together without the strength degradation found with normal welding.

digital readouts — The presentation of information by an instrument in a digital form such as light emitting diodes or drums, rather than by the movement of a pointer over a numbered dial.

digital voltmeter — A voltage measuring device that reads out in digital form rather than an analog dial.

digitize — To change an analog reading to digital format (binary) that can be used by a computer.

dihedral — The positive acute angle between the lateral axis of an airplane and a line through the center of a wing or horizontal stabilizer.

dikes — A common expression for diagonal cutting pliers.

diluent — A material used to change the concentration of some other substance without changing the characteristics of the base material. Common diluents include water- used to dilute salt water solutions and water-based paints and paint thinners used to dilute oil-based paints.

diluter-demand oxygen system — An oxygen system that delivers oxygen mixed or diluted with air in order to maintain a constant oxygen partial pressure as the altitude changes.

dilution air — The portion of combustion secondary air used to control the gas temperature immediately prior to its entry into the turbine nozzle area.

dimension — A measurement of length, width, thickness, size, or degree listed within a drawing.

dimension line — A light solid line broken at its midpoint for insertion of measurement indications. Dimension lines typically have outward pointing arrowheads at each end to show origin and termination of a measurement.

dimensional inspection — The physical measurement of a part against a recognized standard to determine the amount of wear or deformation of the part.

dimmer — A device that controls illumination. Can be either a simple rheostat or a variable pulse control.

dimming relay — A relay that allows a light or lights in a circuit to be dimmed.

dimming rheostat — A rheostat used to control the degree of brilliance of a lighting circuit.

dimpling — A process used to indent the hole into which a flush rivet is to be installed. Some metals  such as the harder aluminum alloys cannot be dimpled while the metal is cold because it is likely to crack. This type of metal must be hot dimpled. Hot dimpling equipment consists of a pair of electrically heated dies with a pilot that is inserted into the rivet hole. The pilot is passed through the hole and the heated dies are pressed together. The dies heat the metal enough to soften it and force it into the shape of the die.

diode — A device that allows a flow of electrons in one direction but not the opposite. Also referred to as an electron check valve.

dip — A vertical attraction between a compass needle and the magnetic poles. The closer the aircraft is to the pole, the more severe the effect. In the Northern Hemisphere a weight is placed on the south-facing end of the compass needle; in the Southern Hemisphere a weight is placed on the north-facing end of the compass needle to somewhat compensate for this effect.

DIP (Dual Inline Packaging) — A standard configuration used for chips, DIP switches, etc., used on computer circuit boards.

dip coating — A process of coating various products with a soft, rubber-like, plastic coating applied by dipping the product into a container of plastic coating. Often used to insulate the handles of hand tools.

dip soldering — A process of lowering a printed circuit board onto the top of a container of molten solder in order to solder the leads of components that extend through the board.

dipole antenna — A center-fed, half-wave antenna.

dipping — A method of paint application in which a part is dipped into a tank of finishing material.

dipstick — A bayonet-type gauge used for measuring the quantity of fluid in a reservoir.

direct — Straight line flight between two navigational aids, fixes, points or any combination thereof. When used by pilots in describing off-airway routes, points defining direct route segments become compulsory reporting points unless the aircraft is under radar contact.

direct altitude and identity readout — The DAIR System is a modification to the AN/ TPX-42 Interrogator System. The Navy has two adaptations of the DAIR System - Carrier Air Traffic Control Direct Altitude and Identification Readout System for Aircraft Carriers and Radar Air Traffic Control Facility Direct Altitude and Identity Readout System for land based terminal operations. The DAIR detects, tracks, and predicts secondary radar aircraft targets. Targets are displayed by means of computer generated symbols and alphanumeric characters depicting flight identification, altitude, ground speed, and flight plan data. The DAIR System is capable of interfacing with ARTCCs.

direct control — In rotorcraft, the ability to maneuver a rotorcraft by tilting the rotor disc and changing the pitch of the rotor blades.

direct current (DC) — A flow of electrons in one direction throughout a circuit.

direct question — A question used for follow-up purposes, but directed at a specific individual.

direct route — A requested route published on a Jeppesen Enroute or Area chart to assist pilots who have previous knowledge of acceptance of these routes by ATC. Use of a direct route may require prior ATC approval and may not provide ATC or Advisory services, or be acceptable in flight plans.

direct shaft turbine — A turbine engine in which the compressor and power section are mounted on a common driveshaft.

direct user access terminal service (DUATS) — A computer-based program providing NWS and FAA weather products that are normally used in pilot weather briefings.

direct-cranking electric starter — A high-torque, direct-current electric motor used to rotate a reciprocating aircraft engine for starting.

direct-current amplifier — An electronic amplifier whose output is greater than and proportional to its DC input signal.

direct-current electricity — Electricity that flows in one direction only. The current from solar cells, chemical cells (batteries), and thermocouples are direct-current (DC).

direct-current generator — Electrical generator that has a DC output rather than an AC output. Since generators naturally produce AC, an electronic rectifier or a commutator (mechanical rectifier) must convert the AC output to DC.

direct-current motor — A motor that operates on direct current (DC).

direction finder — A radio receiver equipped with a directional sensing antenna used to take bearings on a radio transmitter. Specialized radio direction finders are used in aircraft as air navigation aids. Others are ground based primarily to obtain a "fix" on a pilot requesting orientation assistance, or to locate downed aircraft. A location "fix" is established by the intersection of two or more bearing lines plotted on a navigational chart using either two separately located Direction Finders to obtain a fix on an aircraft or by a pilot plotting the bearing indications of his DF on two separately located ground-based transmitters, both of which can be identified on his chart. UDF's receive signals in the ultra high frequency radio broadcast band; VDF's in the very high frequency band; and UVDF's in both bands. ATC provides DF service at those air traffic control towers and flight service stations listed in the Airport/Facility Directory and the DOD Flip IFR En Route Supplement and the Jeppesen Radio Aids Section.

directional antenna — An antenna that either transmits or receives signals in a field pattern other than omnidirectional. Usual patterns for waves that are longer than microwaves are either cardioid or figure-eight. Microwave antennae usually have narrow field patterns.

directional gyro — An instrument used to determine heading in an aircraft. It does not indicate magnetic direction, but is gyro-stabilized to indicate direction after being set with reference to a magnetic compass.

directional reference — In gas turbine engines, an industry standard to describe engine locations. The orientation is to look from the rear towards the front of the engine and use standard twelve hour clock reference points. Right Side and Left Side are also determined in this manner.

directional stability — Stability about the vertical axis of an aircraft, whereby an aircraft tends to return, on its own, to flight aligned with the relative wind when disturbed from that equilibrium state. The vertical tail is the primary contributor to directional stability, causing an airplane in flight to align with the relative wind.

dirigible — A lighter-than-air aircraft. Dirigibles are constructed of a lightweight metal framework supporting a fabric gas envelope filled with helium (early dirigibles used flammable hydrogen). Dirigibles are propelled by engine-driven propellers and are capable of being maneuvered as opposed to balloons that drift with the prevailing wind.

disbond — In composites, the separation of a bond from one structure to another. Many times this term is used for referring to the separation of the laminate skin to the core structure. It is also used for a separation from a fitting to the skin.

disc area — In rotorcraft, the area swept by the blades of the rotor. It is a circle with its center at the hub and has a radius of one blade length.

disc loading — The total helicopter weight divided by the rotor disc area.

discharge indicator disk — An indicator disk on the exterior of an aircraft that indicates the status of the onboard fire protection system. If the system has been discharged normally, a yellow disk is blown out, and if the system has discharged because the system overheats, a red disk is blown out.

discharge nozzle — The portion of a carburetor that sprays the fuel into the intake airstream. Serves to atomize the fuel for best combustion.

disconnect — To remove a load from a source of power or the removal of one electrical device from another by the use of a manually-operated switch or an automatic relay.

discontinuity — In meteorology, a condition where there is a rapid change from one type of weather phenomena or air mass to another. A cold front is an example of a discontinuity.

discreet — Showing good judgment in conduct or speech.

discrete — Not linear. A single action rather than action changing through a range.

discrete beacon code — See discrete code.

discrete code — As used in the Air Traffic Control Radar Beacon System (ATCRBS), any one of the 4096 selectable Mode 3/A aircraft transponder codes except those ending in zero zero; e.g., discrete codes: 0010, 1201, 2317, 7777; non-discrete codes: 0100, 1200, 7700. Non-discrete codes are normally reserved for radar facilities that are not equipped with discrete decoding capability and for other purposes such as emergencies (7700), VFR aircraft (1200), etc.

discrete frequency — A separate radio frequency for use in direct pilot-controller communications in air traffic control which reduces frequency congestion by controlling the number of aircraft operating on a particular frequency at one time. Discrete frequencies are normally designated for each control sector in enroute/terminal ATC facilities. Discrete frequencies are listed on Jeppesen charts.

discrimination — **1**. In knowledge testing, able to detect small differences in understanding of material tested. **2**. In a measuring instrument, means being able to detect small differences in understanding of material between individuals.

discriminator — In a radio receiver, a demodulator that derives an audio signal from an incoming frequency- or pulse-modulated RF signal.

disengage — To break the connection between mechanical or electronic components.

dish antenna — A parabolic-shaped antenna used for transmitting and receiving microwave signals.

disk area — A description of an area swept by the blades of a helicopter.

disk brake — A brake that achieves brake action with a disk or disks attached to the rotating wheel being squeezed between brake pucks and caliper attached to the aircraft gear structure.

disk loading — A ratio of the gross weight of a helicopter to the disk area.

dispersant — A substance that keeps particles suspended in solution. An example is ashless dispersant (AD) oil. It contains a substance that causes ash and other contaminants suspended in the oil to be filtered out.

displaced threshold — When the landing area begins at a point on the runway other than the beginning of the runway.

display tube — An electron tube in which a stream of electrons (cathode rays) from an electron gun impinges upon a fluorescent screen, thus producing a bright spot on the screen. The electron beam is deflected electrically or magnetically to produce patterns on the screen. Also referred to as a cathode ray tube (CRT).

dissimilar metal corrosion — Corrosion caused by the different atomic structures of metals. All metals are listed on a nobility chart from the most vulnerable to corrosion (least noble) to the least vulnerable (most noble). Magnesium and Zinc are the most vulnerable to corrosion, while gold and platinum are the least. When two metals are in contact with each other, the farther apart they are on the nobility chart, the more likely dissimilar metal corrosion will occur. All that is necessary is for an electrolyte (usually water) to contact the area where they touch for corrosion will occur.

dissipate — To lessen in concentration or intensity. Smoke can dissipate due to mixing with surrounding cleaner air. Electrical power can be dissipated by being changed to heat by a motor, transformer, or resistor.

dissipating stage — Thirty minutes or so after a single-cell airmass thunderstorm begins, downdrafts spread throughout the lower levels of the cell. Without the necessary source of energy, (heat and moisture), the end of the thunderstorm is near and the clouds take on a strataform appearance.

dissipation contrail (distrail) — A streak of clearing that occurs behind an aircraft as it flies near the top of, or just within a thin cloud layer.

dissolve — To change form from a solid to become part of a liquid into which the solid is placed, so as to pass into solution. A crystal of salt dissolves when placed in water to become a new liquid, salt water.

dissymmetry of lift — On a helicopter in forward flight, the advancing blade develops more lift at a given angle of attack than does the retreating blade. This is because the effective velocity of the relative wind is greater on the advancing blade. The blades of a helicopter are designed to flap in order to compensate for this.

distance measuring equipment (DME) — Equipment (airborne and ground) to measure, in nautical miles, the slant range distance of an aircraft from the DME navigation aid.

Distant Early Warning Identification Zone (DEWIZ) — An ADIZ (Air Defense Identification Zone) located over the coastal waters of Alaska.

distillate fuel — A liquid hydrocarbon fuel that has been condensed from vapors of crude petroleum.

distillation — A process whereby a liquid is heated until gases boil off, and the gases are then cooled until they condense back into a liquid. The process is used to purify water or to select certain petroleum products from a base liquid of crude oil. Distillation of individual petroleum products is possible since each boils off at a different temperature.

distilled water — Water that has been heated until it boils and is then condensed back into a liquid. The effect purifies the water, leaving the non-water products behind.

distortion — An undesired change in the waveform of the output of a circuit –as compared with the input.

distractions — During training flights, an instructor should interject realistic distractions to determine if students can maintain aircraft control while their attention is diverted.

distractors — Incorrect responses to a multiple-choice test item.

distress — A condition of being threatened by serious and/or imminent danger and of requiring immediate assistance.

distress frequencies — Radio frequencies used for broadcasting distress signals. In aviation, the frequencies are 121.5 MHz, 243.0 MHz, and 406.0 MHz.

distributed pole motor — An electric motor that has the stator windings wound into slots in the motor frame rather than on discrete pole shoes.

distributor — That part of a high-tension magneto that distributes the high voltage to each spark plug at the proper time. Distributors for low-tension ignition systems distribute the low voltage to the transformers at each spark plug at the proper time for air-fuel ignition.

distributor block — A dielectric block in a magneto that contains stationary electrodes to pick up the voltage from a rotating distributor brush or finger and deliver it to the proper ignition lead.

distributor brush — A carbon brush used on a low-tension magneto to distribute the voltage to the distributor block as it rotates.

distributor finger — A rotating conductor in the distributor of a high-tension magneto that delivers voltage to the distributor electrodes.

distributor valve — A device that controls the inflation sequence of deicer tubes.

disturbance — In meteorology, another name for an area that shows indications of cyclonic development, along with associated weather such as clouds and precipitation.

disuse — A theory of forgetting that suggests a person forgets those things which are not used.

dither signal — A varying signal that is mixed into a command signal and increases sensitivity of control. Dithering the commands to a stepper motor enables response with lessened hysteresis. The least significant bit of an analog-to-digital signal converter can be dithered to lessen ambiguity.

diurnal effects — A daily variation (may be in temperature, moisture, wind, cloud cover, etc.) especially pertaining to a cycle completed within a 24-hour period, and which recurs every 24 hours.

diurnal variation — Daily, especially pertaining to a cycle completed within a 24-hour period, and which recurs every 24 hours.

dive — A steep, nose-down descent.

dive brake — An auxiliary control that slows an aircraft during descent. Used to limit speed or to minimize "shock cooling" on reciprocating engines by allowing higher power settings during descent. Also referred to as a speed brake.

dive flaps — Devices used on an airplane to produce drag without an attendant increase in lift. Also referred to as speed brakes.

divergence — In meteorology, the condition that exists when the distribution of winds within a given area is such that there is a net horizontal flow of air outward from the region. The opposite of convergence.

divergent duct — A cone-shaped passage or channel in which a gas can be forced to flow from its smallest area to its largest area resulting in decreased velocity and increased pressure.

diverse vector area — In a radar environment, that area in which a prescribed departure route is not required as the only suitable route to avoid obstacles. The area in which random radar vectors below the MVA/MIA, established in accordance with the TERPS criteria for diverse departures obstacles and terrain avoidance, may be issued to departing aircraft.

dividend — A number to be divided.

dividers — A measuring tool having two movable legs, each with sharp points. Dividers can be used to transfer measurements or to divide straight or uniformly curved lines into an equal number of parts.

diving blade — A blade track of a helicopter's main rotor that lowers with an increase in revolutions per minute (RPM).

divisor — The number by which a dividend is divided.

DME fix — A geographical position determined by reference to a navigational aid which provides distance and azimuth information. It is defined by a specific distance in nautical miles and a radial, azimuth, or course (i.e., localizer) in degrees magnetic from that aid.

DME separation — Spacing of aircraft in terms of distances (nautical miles) determined by reference to distance measuring equipment (DME).

dock — An enclosed work area where airplanes can be placed for repairs.

docking — Placing an airplane in a hangar where dock platforms are used to facilitate maintenance.

DOD FLIP — Department of Defense Flight Information Publications used for flight planning, enroute, and terminal operations. FLIP is produced by the National Imagery and Mapping Agency (NIMA) for worldwide use. United States Government Flight Information Publications (enroute charts and instrument approach procedure charts) are incorporated in DOD FLIP for use in the National Airspace System (NAS).

doghouse — A mark on a turn and slip indicator that resembles a doghouse. It is located one needle width away from the center and when the pointer aligns with it a standard rate of turn is being made.

dolly — A low mobile platform on wheels or casters used for moving heavy aircraft components.

dolly block — Variously shaped anvils used to form and finish sheet metal parts.

dolphin flight — In gliders, straight flight following speed-to-fly theory. Glides can often be extended and average cross-country speeds increased by flying faster in sink and slower in lift without stopping to circle.

domain — Spheres of magnetic influence around molecules of metals containing iron. Magnetic fields.

domains of learning — In addition to the four basic levels of learning, psychologists have developed three domains of learning: cognitive, psychomotor, and affective. These domains represent what is to be gained during the learning process, either knowledge, skills, or attitudes.

domestic airspace — Airspace which overlies the continental land mass of the United States plus Hawaii and U.S. possessions. Domestic airspace extends to 12 miles offshore.

donor — An impurity used in a semiconductor to provide free electrons as current carriers. A semiconductor with a donor impurity is said to be type N.

donor atom — An atom of a material that has more electrons than needed for normal covalent bonding. Used with germanium or silicon in semiconductors to create diodes and transistors.

donor impurity — The elements added to germanium or silicon to create diodes and transistors.

dope — The finishing material used on fabric surfaces that tautens, strengthens, and weatherproofs the fabric.

dope proofing — Coating the structural elements of a fabric-covered aircraft to protect them from the solvents in the dope.

dope roping — A condition in the application of dope in which the surface dries while the dope is being brushed. This results in a stringy, uneven surface.

doped-in panel — An entire panel between ribs and from the trailing edge to the leading edge that is doped in place, but includes rib stitching on sections over the ribs.

doped-on fabric repair — The repair of small damage to a fabric covered aircraft by doping a patch directly to the fabric covering using no other attachment

doped-on fabric repair — Repair of a small section of a fabric structure. Typically applied by doping on a patch over a small hole or tear in the fabric.

dope-proof paint — A paint applied to protect structure from being damaged by the solvents in the dope.

Doppler effect — The effect where any oscillating frequency, whether sound or radar beams, are compressed (increased in frequency) as the source and the receiver move closer to each other and decompressed (lowered in frequency) when moving apart.

Doppler radar — A radar system that indicates speed by measuring the amount of Doppler shift and equating it to the speed that the measured object is moving toward or away from the measuring point.

dorsal — Situated near the rear or on the back of an object.

dorsal fin — A fixed vertical control surface on the upper surface of an aircraft. Usually on the rear of an aircraft and tapering into the vertical stabilizer, it increases the directional stability of the aircraft.

double flare — On rigid fluid lines, a connecting flare where the tubing is bent back on itself creating a double thickness. Only allowed on soft materials in small diameters.

double magneto — A single magneto housing that holds one rotating permanent magnet and one cam with two sets of breaker points, two condensers, two coils, and two distributors. For all practical purposes, this constitutes two ignition systems. Also called a dual magneto.

double spread — -The spreading of an adhesive equally divided between the two surfaces to be joined.

double-acting actuator — An actuator that uses hydraulic or pneumatic power to move the piston in both directions.

double-acting hand pump — A hydraulic hand pump that moves fluid with both the forward and rearward movement of the handle.

double-backed tape — Adhesive tape that is sticky on both sides.

double-cut file — A file with two sets of parallel grooves, cut at an angle to each other.

double-cut saw — A saw with teeth shaped so that it cuts in both the down and return strokes.

double-loop rib-stitching — Attachment of fabric to the aircraft structure using a double loop of rib-stitch cord at each stitch.

double-pole, double-throw (DPDT) switch — A switch that controls two circuits, selecting either of two positions (and sometimes a third, off position) for each circuit.

double-pole, single-throw (DPST) switch — A switch that controls two circuits, selecting either off or on.

doubler — In aircraft sheet metal, a thickness of metal attached to the skin of an aircraft to strengthen it. Often used on the inside of the fuselage to strengthen the skin where an antenna is attached.

doubler plies — In composites, a patch that extends over the sanded out area to the existing structure which strengthens the repair. A doubler can also be used where fasteners are used or where there are abrupt load transfers.

double-row radial engine — A radial engine having two rows of cylinders and using two master rods attached to a single crankshaft having two throws.

double-sided tape — Tape with adhesive on both sides.

double-tapered wing — A wing where both the chord and the thickness ratio vary along the span.

double-throw switch — A switch that selects one circuit from two possible circuits. Can have a center "off" position.

dovetail — A method of joining two materials where one has a base shaped like a widened, inverted, triangle that fits into a similarly shaped cutout in the adjoining piece. Often found in attaching turbine and compressor blades to rotor disks.

dovetail fit — A shape similar to a cabinet maker's interlocking "dovetail joint." Primarily used to fit compressor blades into a compressor disk.

Dow 19 treatment — An acid treatment for magnesium alloy parts, which produces an oxide film that inhibits the formation of harmful corrosion.

Dow metal — A series of magnesium alloys produced by the Dow Chemical Corporation.

dowel — A short wood or metal rod used to hold objects together.

downburst — A strong downdraft which induces an outburst of damaging winds on or near the ground. Damaging winds, either straight or curved, are highly divergent. The sizes of downbursts vary from 1/2 mile or less to more than 10 miles. An intense downburst often causes widespread damage. Damaging winds, lasting 5 to 30 minutes, could reach speeds as high as 120 knots.

downdraft — Any downward flow of air, for example, the flow downwind of buildings or hills, or the sinking air near thunderstorms especially in areas of precipitation.

downdraft carburetor — A carburetor mounted on top of the engine in which the flow of air into the engine is downward through the venturi.

downlock pin — A landing gear safety device consisting of a pin that, when placed into a hole in the landing gear apparatus, prevents the gear from retracting or collapsing while on the ground. The downlock pin must be removed before flight.

downlocks — Mechanical locks that hold a retractable landing gear in the "on" or "down" position, preventing its retracting when the hydraulic pressure is released.

downslope wind — Wind moving down a slope, the wind can be either a cold downslope wind or a warm downslope wind.

downtime — The time an airplane is out of commission.

downwash — 1. Air deflected perpendicular to the motion of the airfoil. 2. Air that has been accelerated downward by the action of the main rotor of a helicopter.

downwash angle — The angle the air is deflected downward by an airfoil. It is the difference between the angle of air approaching the airfoil and the air leaving it.

downwind — The direction in which the wind is blowing.

downwind landing — Landing in a direction so that the wind is coming from behind the aircraft.

downwind leg — A flight path parallel to the landing runway in the direction opposite to landing. The downwind leg normally extends between the crosswind leg and the base leg.

draftsman — A person who makes mechanical drawings.

drag — An aerodynamic force on a body acting parallel and opposite to the relative wind. The resistance of the atmosphere to the relative motion of an aircraft. Drag opposes thrust and limits the speed of the airplane.

drag brace — 1. In main landing gears, a device that acts as side support for shock strut. 2. An adjustable brace used to position the main rotor in a fixed position preventing movement of the blade at the attached point on semi-rigid rotors.

drag chute — A relatively small parachute attached to the rear of an aircraft. Drag chutes are deployed during landing rolls to help slow an aircraft and sometimes in flight to increase drag.

drag coefficient — One of the aerodynamic characteristics of an airfoil section that illustrates the increase in induced drag as the angle of attack is increased.

drag demonstration — The demonstration of the effects of drag on a multi-engine airplane from landing gear, flaps, and windmilling propellers.

drag hinge — The hinge on a helicopter rotor blade parallel to the axis of rotation of the blade. It allows the blade to move back and forth on a horizontal plane, minimizing the blade vibrations.

drag wire — A diagonal, load-carrying member of a Pratt truss wing. It runs from the front spar inboard to the rear spar outboard and carries tensile loads that tend to drag back on the wing.

dragging brakes — Brakes that have not fully released and that maintain some friction as the wheel rolls. Dragging brakes cause serious overheating.

drain — The electrode in a field effect transistor (FET) that corresponds to a collector of the ordinary transistor.

drain can — A container to catch fuel drained from the main fuel manifold after shutdown of a turbine engine.

drain hole — A hole placed in the lower surface of a wing or other sealed component to provide ventilation and allow the drainage of any accumulated moisture.

drain plug — A removable plug located at the lowest point of a system used for drainage purposes.

drain valve — A spring-opened and burner pressure-closed mechanical valve located in the lower portion of the combustor outer case, installed to drain off puddled fuel after an aborted start or after shutdown. Also referred to as a drip valve.

drainage wind — A shallow, small scale current of cold dense air accelerated down a slope by gravity.

drape — In composites, the ability of a fabric or pre-preg to conform to a contoured surface.

draw filing — A method of hand filing in which the file is grasped with both hands and moved crosswise over the work. Draw filing produces an exceptionally smooth surface.

draw knife — A knife with handles on either end, designed to be pulled (drawn) toward the user.

draw set — A riveting tool used to force sheets of metal together before they are riveted.

drawing — A graphic method of conveying information.

drawing number — The number assigned to each drawing in a set of drawings and located in the lower right-hand corner of the title block. It identifies the drawing and is usually associated with the part number of the component or part depicted in the drawing.

drift angle — The horizontal angle between the longitudinal axis of an aircraft and its path relative to the ground.

drift magnet — A small permanent magnet in a fixed-coil ratiometer indicator. Used to drift or pull the pointer off scale when the instrument is not energized.

drift punch — A pin punch with a long straight shank.

driftdown — The unavoidable descent due to the loss of an engine when above the engine-out absolute ceiling of an airplane. If the airplane is above its engine-out altitude limits, it is incapable of maintaining altitude with one engine inoperative, and the airplane will drift down to the engine-out absolute ceiling.

drill — 1. A rotary cutting tool driven with a drill motor or a drill press. 2. To sink a hole with a drill, usually a twist drill. 3. A pointed, rotating cutting tool.

drill bushing — Hardened steel sleeves inserted in jigs, fixtures, or templates to provide a guide for drills so holes will be straight and in the proper location.

drill chuck — The clamp on the spindle of a drill motor or drill press into which the drill bit is fastened.

drill jig — A device that holds parts or units in the proper position while holes are being drilled.

drill press — A power-driven drilling device that includes a table for holding the material, a chuck for holding the drill bit, a motor for driving the chuck, and a means of feeding the drill into the material.

drill rod — A high carbon tool steel. It has a combination of hardness and toughness for good wear characteristics.

drilling burrs — Sharp ragged particles of metal left by the drill when a hole is made.

drip pan — A shallow pan placed beneath an engine to catch dripping fluids.

drip valve — A spring-opened and burner pressure-closed mechanical valve located in the lower portion of the combustor outer case, installed to drain off puddled fuel after an aborted start or after shutdown. Also referred to as a drain valve.

drip-stick gauge — A stick used as a visual means of checking the fuel level from beneath the wing of a large jet transport aircraft. The stick is released and pulled downward until fuel drips from its end, signifying that the inside is even with the top of the fuel in the tank. The fuel quantity is read where the drip-stick enters the wing.

drive coupling — A coupling between the accessory section of an engine and the component that is driven. It is used to absorb torsional shock or to serve as a safety link that will shear in case the component seizes.

drive fit — A fit between mating parts in which the part to be inserted into the hole is larger than the hole and therefore must be driven or forced together with the other part. Also referred to as an interference fit.

drive gear — In a gear train, the gear nearest the power source. Meshes with and drives (turns) the driven gear.

drive screws — Plain-headed, self-tapping screws used for attaching name plates to castings or to plug holes in tubular structures through which rust-preventative oil has been forced.

driven gear — In a gear train, the gear being driven (turned) by the drive gear.

driver — In electronics, a device that supplies input power to an output device.

driver head — A head on a bolt or screw designed for driving the fastener by means of a tool such as a screwdriver or Allen wrench rather than a conventional wrench.

drizzle — A form of precipitation. Very small water drops that appear to float with the air currents while falling in an irregular path (unlike rain, which falls in a comparatively straight path, and unlike fog droplets which remain suspended in the air).

drogue — A device that provides drag to a moving body. For boats, a sea anchor provides drag that stabilizes the position of the boat in relation to current. In aircraft a drogue chute stabilizes or slows the airplane. A drogue receptacle for refueling is a basket-shaped device into which another aircraft can insert a probe to receive fuel from an aerial refueling tanker.

drone — An unmanned aircraft guided by remote control. Newest terminology refers to these vehicles as UAVs (Unmanned Aerial Vehicles) and RPVs (Remote Piloted Vehicles).

droop — Refers to the RPM loss that occurs when a fuel control flyweight governor speeder spring is extended and weakened. It takes less flyweight force to come to equilibrium with the weaker spring force and consequently slightly less speed results. The inability of the engine power to increase as the rotor pitch is increased causes the rotor to slow down.

droop compensator — In helicopters, a device that automatically adds power when collective pitch is increased in order to compensate for the increased load placed on the engine.

droop restraint — A device used to limit the droop of the main rotor blades at low RPM.

drop cloth — A plastic or fabric sheet used to protect floors, furniture, or other objects from paint drips or overspray from painting operations.

drop forging — A process of forcing semi-molten metal to flow into a mold or die under the pressure of repeated hammer blows.

drop hammer — A large, heavy, hammer-type, metal-forming machine that uses sets of matched dies to form compound curved sheet metal parts. The metal is placed over the female die and the male die is dropped into it, forcing the metal to conform to the shape of the two dies.

drop tank — An externally mounted fuel tank designed to be dropped in flight.

drop-forged part — A steel part that has been formed by the drop-forging process.

droplets — Tiny drops of liquid.

drop-out voltage — In electricity, the minimum voltage that can be applied to a device without it dropping from the circuit. Usually used in referring to relays that de-energize when control voltage drops below a certain drop-out voltage.

dropping resistor — A resistor used to decrease the voltage in a circuit.

drum brake — A friction device used on some aircraft landing gear. Consists of a cylindrical metal drum attached to the wheel and shoe-shaped friction pads attached to the landing gear. This device is applied against the inner side of the drum when braking action is desired.

dry adiabat — A line on a thermodynamic chart representing a rate of temperature change at the dry adiabatic lapse rate.

dry adiabatic lapse rate (DALR) — The rate of decrease of temperature with height of unsaturated air lifted adiabatically (not heat exchange). Numerically the value is 3C or 5.4F per 1000'.

dry adiabatic process — The cooling of an unsaturated parcel of air by expansion and the warming of a parcel of air by compression.

dry air — Air that contains no water vapor. Dry air weighs 0.07651 pounds per cubic foot under standard sea level atmospheric conditions of 59°F (15°C) and a barometric pressure of 14.69 PSI, or 29.92 inches of mercury.

dry air pump — An engine-driven air pump using carbon vanes and that does not require any lubricating oil in the pump for sealing or cooling.

dry bulb — A name given to an ordinary thermometer used to determine temperature of the air; also used as a contraction for dry-bulb temperature. Compare wet bulb.

dry clearance — Clearance adjusted in adjusting valve tappets without any oil in the tappet body.

dry fiber — In composites, a condition in which fibers are not fully encapsulated by resin during pultrusion.

dry ice — Frozen carbon dioxide (CO_2). Dry ice sublimates from solid to gas without becoming a liquid. It is used for non-mechanical cooling of perishables or for cooling parts for an interference fit.

dry laminate — In composites, a laminate containing insufficient resin for complete bonding of the reinforcement.

dry line — The moisture boundary, where the moisture content of the air changes rapidly from one side to the other.

dry operation — The operation of an aircraft engine equipped with a water injection system, but operating without the benefit of water injection.

dry rot — The condition of wood attacked by fungus. Causes brittleness and decay.

dry wash — Aircraft cleaning method in which cleaning material is applied by spray, mop, or cloth and removed by mopping or wiping with a clean, dry cloth. It is used to remove airport film, dust, and small amounts of dirt and soil.

dry-bulb temperature — The temperature of the air without the effect of water evaporation.

dry-cell battery — Common name for a carbon-zinc, single-cell battery. D-, C-, AA-, and AAA-cells are all dry-cell batteries. They are not rechargeable or serviceable.

dry-charge battery — The common way of shipping a lead-acid battery. The battery is fully charged, drained, and the cells are washed and dried. The battery is sealed until it is ready to be put into service. Electrolyte is added and the battery is given a freshening charge, making it ready for service.

dry-chemical fire extinguisher — An extinguishing agent such as sodium bicarbonate used as a compressed, non-flammable gas as a propellant. Dry-chemical fire extinguishers are usually rated for multiple purpose use.

dry-sump engine — An engine in which most of the lubricating oil is carried in an external tank and is fed to the pressure pump by gravity. After it has lubricated the engine, the oil is pumped back into the tank by an engine-driven scavenger pump.

dry-sump system — An oil system in which the oil is contained in a separate tank and circulated through the engine by pumps.

dual controls — Two sets of flight controls for an aircraft that allow the airplane to be flown from either of two positions.

dual indicator — An aircraft instrument that provides two sets of indications on one dial. For example, the oil pressure of both engines can be shown on one indicator using one dial and two pointers.

dual magneto — A single magneto housing that holds one rotating permanent magnet and one cam with two sets of breaker points, two condensers, two coils, and two distributors. For all practical purposes, this constitutes two ignition systems.

dual rotor system — The rotor system of a helicopter in which there are two separate main rotors spinning in such a direction that they tend to cancel the torque of each other.

dual-spool compressor — A turbine engine with two separate compressors, each with its own stage of turbine. The low-pressure compressor is N_1 and the high-pressure compressor is N_2.

duckbill pliers — Flat-nosed pliers used extensively in safety wiring.

duct — A hollow tube used to transmit and direct the flow of air through an aircraft.

duct support systems — Methods and apparatus used to support cabin air supply ducts.

ducted-fan engine — An engine-propeller combination that has the propeller enclosed in a radial shroud. Enclosing the propeller improves the efficiency of the propeller.

ductility — The property that allows metal to be drawn into thinner sections without breaking.

ductwork — The channels or tubing through which the cabin air supply is distributed.

due regard — A phase of flight wherein an aircraft commander of a State-operated aircraft assumes responsibility to separate his aircraft from all other aircraft.

dump chute — In an aircraft fuel system, a device designed to carry dumped fuel away from the aircraft to prevent it from being ignited by static electricity or engine exhaust.

dump valve — The valve that allows the fuel in a tank to be dumped in flight in order to decrease the landing weight of the aircraft.

duo-servo brakes — Brakes that use the momentum of the aircraft to wedge the lining against the drum and assist in braking when the aircraft is rolling either forward or backward.

duplex bearing — A matched pair of bearings with a surface ground on each to make contact with the other matched surface. When three bearings are used they are called triplex bearings; when four bearings are used they are called quadplex bearings, and so on. These are usually ball bearings.

duplex fuel nozzle — A turbine engine fuel nozzle that has two different spray patterns: one for low airflow and one for high airflow. This keeps the flame pattern centered in the combustor section.

duplex operation — In communications, the capability of all stations to transmit and receive simultaneously. Telephone communication is a duplex operation. For radio communication to be a duplex operation, two frequencies are required.

duplexer — A circuit that makes it possible to use the same antenna for transmitting and receiving without allowing excessive power to flow to the receiver.

durability — The ability to withstand hard wear.

Dural® — A high strength, low weight alloy consisting of 95% aluminum, 4% copper, and 1% magnesium.

Duralumin — The original name of the aluminum alloy now known as 2017. First produced in Germany and used in its Zeppelin fleet of WWI.

dust — Small soil particles suspended in the atmosphere.

dust devil — An unusual, frequently severe weather condition characterized by strong winds and dust-filled air over an extensive area.. A small vigorous circulation that can pick up dust or other debris near the surface to form a column hundreds or even thousands of feet deep. At the ground, winds can be strong enough to flip an unattended aircraft. Dust devils mark the location where a thermal is leaving the ground.

Dutch roll — A combination of rolling and yawing oscillations that normally occurs when the dihedral effects of an aircraft are more powerful than the directional stability.

duty cycle — The comparison of the time a piece of equipment can operate to how long it must cool before being operated again. A device that can operate one minute before having to cool off for four minutes would have a 25% duty cycle.

duty runway — See runway in use and active runway.

D-value — Departure of true altitude from pressure altitude; obtained by algebraically subtracting true altitude from pressure altitude.

DVFR flight plan — A flight plan filed for a VFR aircraft which intends to operate in airspace within which the ready identification, location, and control of aircraft are required in the interest of national security.

DVOR (Doppler VOR) — Provides increased accuracy compared to a conventional VOR. Employed in areas such as mountainous terrain.

Dwell angle — In ignition systems, the time that the ignition points remain closed. During this time, the magnetic field builds in the primary coil. When the points open, the magnetic field collapses, creating a much higher voltage in the secondary coil, which is connected to the spark plugs.

dye — A material added to a substance to add or change its color. In aviation, dye is added to aviation fuels to identify the grade of fuel.

dye penetrant inspection — An inspection method for surface cracks in which a penetrating dye is allowed to enter any cracks present and is pulled out of the crack by an absorbent developer. A crack appears as a line on the surface of the developer.

dynafocal engine mount — A mount that attaches an aircraft engine onto the airframe in which the extended center line of all of the mounting bolts would cross at the center of gravity of the engine and propeller combination.

dynamic — Continuous review, evaluation, and change to meet demands.

dynamic balance — The condition that exists in a rotating body in which all of the rotating forces are balanced within themselves and no vibration is produced by the body in motion.

dynamic braking — A method of slowing the rotation of equipment driven by an electric motor. Braking is accomplished by disconnecting electrical power from the motor and replacing it with a resistance load. The inertia of the motor/equipment causes the motor to become a generator and the resistance absorbs the electrical output of the motor and thus resists and slows the rotation.

dynamic damper — A counterweight on the crankshaft of an aircraft engine. It is attached in such a way that it can rock back and forth while the shaft is spinning and absorb dynamic vibrations. In essence, it changes the resonant frequency of the engine/propeller combination.

dynamic factor — The ratio between the load carried by any part of an aircraft when accelerating and the corresponding basic load.

dynamic load — The effective weight of an aircraft. It is the actual weight of the aircraft multiplied by the load factor (G-load). When an aircraft is sitting on the ramp, the load factor is one and the dynamic load equals the weight of the aircraft. In flight, if the aircraft is experiencing a load factor of two (as it would be if in a level 60° bank turn) the dynamic load would be twice the actual weight of the aircraft.

dynamic microphone — A device used to convert acoustic pressure waves into electrical waves. Incoming sound impinges a movable diaphragm that has a coil of wire on it. The coil is in a magnetic field and its movement produces an electrical signal that correlates to the sound.

dynamic pressure — The product $\frac{1}{2} p V^2$, where p is the density of the air and V is the relative speed of the air.

dynamic restrictions — Those restrictions imposed by the local facility on an "as needed" basis to manage unpredictable fluctuations in traffic demands.

dynamic rollover — The tendency of a helicopter to continue rolling when the critical angle is exceeded, if one gear is on the ground, and the helicopter is pivoting around that point.

dynamic stability — The property of an aircraft that causes it, when disturbed from straight and level flight, to develop forces or moments that restore the original condition of straight and level.

dynamometer — An instrument used to measure torque force or power.

dynamotor — A machine with two windings on a single armature that simultaneously operates as a motor with one of the windings and as a generator with the other. The armature windings are usually different so that the voltage on the generator side is different from the voltage on the motor side and the machine acts as a rotary transformer.

dynatron effect — The area of operation in a tetrode electron tube where plate current decreases as plate voltage increases. This effect is caused by secondary electrons to the screen grid.

dyne — A unit of force. Dyne is the amount of force required to accelerate one gram of mass one centimeter per second squared. In science, the term is slowly becoming obsolete.

dynode — The elements in a multiplier tube that emit secondary electrons.

Dzus fastener — A patented form of cowling fastener in which a slotted stud is forced over a spring steel wire and rotated to lock the wire in a cam.

E

early warning radar — Long-range radar used to detect incoming aircraft or missiles soon enough to be intercepted by defensive aircraft or missiles.

earplug — A rubber, wax, or soft plastic device worn in the canal of the ear to keep loud noises from damaging the ear and to help prevent hearing loss.

earth connection — An electrical connection with a ground. The United Kingdom uses "earth" for the electrical term "ground."

earth ground — In electricity, a ground connected directly to Earth. This connection is usually achieved by attaching a grounding clamp to a metal water pipe or to a grounding rod driven into the ground. The third plug in a three-wire system is usually connected to earth ground.

Earth induction compass — A direction indicator in an aircraft that derives its signal from the lines of flux of the Earth cutting across the windings of the flux valve.

Earth's magnetic field — The magnetic lines of flux that surround the Earth. These lines enter and exit the Earth at the magnetic north and south poles, close to, but not at, the geographic poles.

easy-out — A screw extractor used to remove broken screws or studs. It is made of hard steel and has a point with a tapered, left-hand spiral-like thread. A hole is drilled in the shank of the broken screw and the easy-out is screwed into it by turning it counter-clockwise.

eccentric — A disk or wheel having its axis of revolution displaced from its center so that it is capable of imparting reciprocating motion.

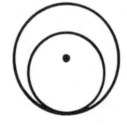

eccentric cam — A circular cam with a displaced axis. As the shaft rotates about its axis, the outside of the eccentric cam rises and falls, changing rotary motion into reciprocating motion.

echo — In radar terminology, the energy reflected or scattered by a target and the radar scope presentation of the return from a target.

economizer system — A power compensator or a power enrichment system in a carburetor or fuel injection system that adds additional fuel. It enriches the fuel mixture at high power engine operations. The economizer is closed during cruising speeds.

E-core — The laminated core of an electric transformer cut in the shape of the letter "E." The coil windings are mounted on the core.

eddy — In meteorology, an organized movement of air in a circulation of a particular size. The organization is more obvious in the larger scale circulations.

eddy current — A current induced into the core of a coil, transformer, or the armature core of a motor or generator by current flowing in the winding. Eddy currents cause power loss and are minimized by laminating the iron cores.

eddy current inspection — A nondestructive inspection used to locate surface or subsurface defects in a metal part. This is a comparison-type inspection, based on the difference in conductivity of a sound and a defective part.

eddy current losses — The electrical losses in the core of a transformer or other electrical machine. The induction of eddy currents into the core robs the machine of some of its power.

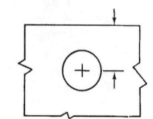

edge distance — The distance from the center of a bolt or rivet hole to the edge of the material.

edge thickness - The thickness of the edge of a material.

edge-grain wood — Wood that has been sawed from the tree in such a way that the edges of its grain are visible in the wide part of the plank. Also referred to as quarter-sawed wood.

Edison effect — The emission of electrons from a heated filament attached to an electrode placed in an evacuated tube. The discovery of Thomas A. Edison in 1883.

eductor — A jet pump used in some aircraft fuel systems to remove fuel from a vent-drain tank and return it to the main tank.

effect — A principle of learning, which states that learning is strengthened when accompanied by a pleasant or satisfying feeling, and that learning is weakened when associated with an unpleasant feeling.

effective pitch — The actual distance a propeller moves through the air in one revolution. It is the difference between the geometric pitch of the propeller and the propeller slip.

effective thread — The effective (or useful) thread includes the complete thread and that portion of the incomplete thread having crests not fully formed.

effective value — The root mean square (RMS) value of sine wave AC. It is equivalent to 0.707 times the peak value and corresponds to the DC value that will produce the same amount of heat.

effective voltage — The equivalent direct current (DC) voltage in an alternating current system. It is calculated as .707 times the peak sinusoidal waveform voltage value. In a non-sinusoidal waveform, it is the root mean square voltage value.

efficiency — A measure of a system's effectiveness or a mechanism found by dividing the output of the mechanism by its input. It is usually expressed as a percentage.

effort arm — The distance from the input to the fulcrum on a lever. The distance from the fulcrum to the output point is referred to as the load arm.

E-gap angle — The number of degrees of magnet rotation beyond its neutral position at which the primary magneto breaker points open. It is at this point that the primary current flow is the greatest and, therefore, the rate of collapse of the primary field will induce the greatest voltage into the secondary winding.

E-glass — In composites, a type of fiberglass. The E stands for electrical. It is used primarily when there could be interference to radio signals such as with a radome.

egoistic needs — Basic personal needs that relate to a student's self-esteem and are directly linked to self-confidence, independence, achievement, competence, and knowledge. Another type relates to a student's reputation, such as status, recognition, appreciation, and respect of associates.

eight-harness satin — In composites, a type of fabric weave. The fabric has a seven-by-one weave pattern in which a filling thread floats over seven warp threads and then under one. Like the crowfoot weave, it looks different on one side than on the other. This weave is more pliable than any of the others and is especially adaptable to forming around compound curves, such as on radomes.

ejection seat — An emergency escape seat found in military aircraft that, when used, is propelled or shot from the aircraft and parachutes to the ground.

ejector pump — A pump that produces a low hydraulic pressure. The low pressure is used to eject or move fluid from one place to another. See also eductor.

elastic limit — The maximum load (in PSI) a metal can withstand without causing a permanent deformation. If a metal hasn't reached its elastic limit, it will return to its original dimensions when the load is removed no matter how many times the load is applied.

elastic stop nut — A self-locking nut with a collar of elastic material and an inside diameter slightly smaller than the outside diameter of the bolt or stud it fits. The collar fills the bolt threads and prevents the nut from backing out inadvertently.

elasticity — The capability of an object or material to be stretched and to recover its size and shape after deformation.

elastomeric bearing — A metal and rubber composite bearing used to carry oscillating loads where complete rotation is not needed. The bearing is made of alternate layers of an elastomer and metal bonded together. Elastomeric bearings can be designed to take radial, axial, and torsional loads.

elastomers — A rubber or synthetic rubber used in the layers between the metal in an elastomeric bearing. Elastomers can be stretched to twice their original length and can return to their original size and shape when released.

E-layer — A layer of ionized air in the ionosphere found approximately 55 to 90 miles above Earth.

elbow — A fluid line fitting used to join two pieces of tubing at an angle of 90°.

Elcon connector — A slip-on type terminal, held in place with a hand screw, for aircraft battery installation. It is similar to the Cannon connector.

electret — A permanently polarized dielectric material.

electret microphone — A device that changes sound pressure into an electrical signal. It consists of a diaphragm made of a thin foil of electret placed next to a metal coated plate. Sound pressure that is picked up by the microphone vibrates the diaphragm causing a voltage. The waveform of this voltage is a copy of the waveform of the sound that vibrated the diaphragm.

electric bonding — The connecting of metal structural parts together with electrical conductors in order to keep them at the same electrical potential. Bonding eliminates static electricity build-up, which causes radio interference.

electric discharge machining — A process of machining complex metal shapes by a controlled electric arc that erodes the metal.

electric drill motor — An electric motor, usually of the universal type, which is geared down to provide additional torque and is equipped with a chuck to hold a twist drill.

electric inertia starter — An electric starter motor for aircraft reciprocating engines that spins a small flywheel to a high speed. The energy in the flywheel, when coupled to the engine crankshaft, turns or "cranks" the engine.

electric strain gauge — A device used to measure the amount of physical strain placed on a piece of conductive material. A strain gauge is made of a piece of very fine wire that is bonded to the material in which the strain is to be measured, and the two ends of the wire are connected into a sensitive, resistance-measuring bridge circuit. When the material on which the strain gauge is mounted is strained, the wire in the strain gauge is stretched. It becomes longer and thinner. When its length and its cross sectional area changes, its resistance changes. The change in resistance of the strain gauge is proportional to the amount of strain in the material, and the amount of strain is proportional to the amount of stress that caused it.

electric wave — One of the components of a radio wave produced along the length of the antenna.

electrical bus — An electrical distribution point to which many circuits can be connected.

electrical charge — An excess (negative charge) or deficiency (positive charge) of electrons in a body.

electrical diagram — A diagram or drawing showing the relationship of electrical components within a system.

electrical energy — The energy possessed by a substance or device because of a difference in electrical potential. This can exist because of electromagnetic or electrostatic forces.

electrical equipment — An electrical unit or combination of units that make up the electrical system.

electrical filter — An arrangement of a choke coil and condenser used in an electrical circuit to create a steady current flow.

electrical generator — A mechanical device that converts mechanical energy into electrical energy.

electrical insulator — A material that prevents the passage of electricity. The material's outer electrons are so forcibly held to the nucleus that they cannot be dislodged to flow in a circuit. Glass is an example of a good insulating material.

electrical lines — The wiring used for transporting electrical energy to electrical equipment.

electrical resistance welding — The fusion of metals by clamping them together and passing a high amperage electrical current through the joint. The resulting heat melts the metal, and the pressure causes the two pieces to fuse together. Spot and seam welding are forms of electrical resistance welding.

electrical shield — A housing made of a conductive material that encloses an electrical circuit. The shield picks up any electrical energy radiated from the circuit and carries it to ground so it cannot interfere with any other electrical or electronic equipment.

electrical short — An unintentional electrical system fault connection that provides a low-resistance path across an electrical circuit. Electrons can flow through the short to ground without passing through the load.

electrical steel — A low-carbon steel alloy that contains up to 5% silicon and is used in the form of thin laminations for the cores of transformers and the armatures of electrical motors and generators.

electrical strength — The maximum amount of voltage that can be placed on an insulator before the insulator breaks down and allows electrons to flow through it.

electrical symbols — The graphic symbols used in aircraft drawings to represent electrical wiring or components.

electrical zero — A designated rotor position in synchro systems. It is a reference position for meshing gears and for installing indicators.

electrically detonated squib — An explosive charge, usually installed in fire extinguisher systems, that is ignited by electrical methods.

electrically suspended gyroscope — A gyroscope with a rotor supported in an electromagnetic field. This allows the rotor to spin with an absolute minimum of friction.

electricity — The physical phenomena arising from the existence of positive and negative electrical charges. Electricity can be observed in the attracting and repelling of objects electrified by friction and in natural phenomena (lightning and aurora borealis). Usually employed in the form of electrical currents generated by a mechanical generator device that forces the flow of electrons from an area having an excess of electrons to an area with a shortage of electrons. In the process, heat is produced, and a magnetic field surrounds the conductor.

electroacoustic transducer — A device that converts variations in sound pressure into variations of voltage, or vice versa.

electrochemical action — The corrosive results of the potential difference of two different metals in contact with each other in the presence of an electrolyte.

electrochemical series — A list of metallic elements that ranks them according to the ease with which they give up electrons. All metals are listed on an electrochemical series or nobility chart from the most vulnerable to corrosion (least noble) to the least vulnerable (most noble). See also dissimilar metal corrosion.

electrochemistry — The branch of chemistry that deals with the electrical voltages existing within a substance because of its chemical composition.

electrode — A terminal element in an electric device or circuit. Examples are the plates in a storage battery, the elements in an electron tube, and the carbon rods in an arc light.

electrode potential — A voltage that exists between different metals and alloys because of their chemical composition. An electrical current will flow between these materials when a conductive path is provided.

electrodynamic damping — The diminishing of oscillations of the pointer of an electrical meter by the generation of electromagnetic fields in the frame of the moving coil.

electrogalvanizing — -The process of coating metal with zinc by electroplating.

electro-hydraulic control — A hydraulic control that is electrically actuated.

electrolysis — A chemical change produced in an electrolyte by an electric current.

electrolyte — A nonmetallic conductor, such as a liquid or a gas, in which current is carried by the movement of ions.

electrolytic — The action of conducting electrical current through a nonmetallic conductor by the movement of ions.

electrolytic capacitor — A capacitor that uses metal foil for the electrodes and a thin film of metallic oxide as the dielectric. The sheets of metal foil are separated by a piece of porous paper that is impregnated with an electrolyte. The capacity is affected by the thickness of the dielectric.

electromagnet — A magnet produced by electrical current flowing through a coil of wire. The coil is often wound around a soft iron core that concentrates the lines of flux and intensifies the magnetic field.

electromagnetic emission — The radiation of electromagnetic energy that is produced when electricity flows through a conductor.

electromagnetic induction — A transfer of electrical energy from one conductor to another by means of a moving electromagnetic field. A voltage is produced in a conductor as the magnetic lines of force cut or link with the conductor. The value of the voltage produced by electromagnetic induction is proportional to the number of lines of force cut per second. An emf (electromagnetic force) of 1 volt will be induced when 100,000,000 lines of force are cut per second.

electromagnetic radiation — An electrical energy of extremely high frequency and short wavelength that will penetrate solid objects and expose photographic film.

electromagnetic vibrator — A device that interrupts the flow of DC through a set of contacts and changes it into pulsating DC. The contacts will vibrate between open and closed as long as the vibrator is connected to a source of DC electricity.

electromagnetic waves — A resonance of electric and magnetic fields that move at the speed of light.

electromagnetism — The magnetic field emanating from a conductor carrying electrical current. Its strength is determined by the amount of current flowing in the conductor.

electromechanical frequency meter — An instrument that uses the resonant frequency of a vibrating metal reed to measure the frequency of alternating current. Also referred to as a vibrating-reed frequency meter.

electron — 1. A subatomic particle with a negative electric charge. It spins around the nucleus of an atom, and under certain conditions, can be caused to move from one atom to another. Electrons that travel in this manner are called free electrons. 2. The negatively charged part of an atom of which all matter is made. Electrons circle around the nucleus of an atom in orbits or shells.

electron beam — A narrow beam of free electrons in a vacuum. In cathode-ray tubes (CRTs), the electron beam strikes luminescent phosphors, causing them to emit light. CRTs are used in TV receivers, radar, computer monitors, and some instrument displays.

electron beam welding — A process of welding metal by the heat that is produced when a high-speed stream of electrons strike the metal.

electron current flow — The flow of electrons from negative to positive in a circuit outside of the source.

electron drift — The relatively slow natural movement of individual electrons that move from atom to atom within the conductor.

electron force — The force causing electrons to move through a conductor (electromagnetic force). The symbol "E" is used in calculations until the actual number of volts is determined.

electron gun — The combination of an electron-emitting cathode together with accelerating anodes and beam-forming electrodes to produce the electron beam in a cathode-ray tube.

electron spin — The rotation of an electron about its own axis.

electron tube — 1. A device consisting of an evacuated or gas-filled envelope containing electrodes for the purpose of controlling electron flow. The electrodes are usually a cathode "electron emitter," a plate "anode," and one or more grids. 2. A vacuum tube. It consists of a cathode and its heater, the grids, and plate that are usually housed in a glass envelope. The tube may also be filled with an inert gas.

electron-flow — The current-flow in a circuit is actually the flow of electrons. Electrons flow from negative to positive in the external circuit.

electronic countermeasures — Methods of decreasing the effectiveness of enemy communications or radar.

electronic counter-countermeasures — Methods used by the military to counter the ECM (Electronic Countermeasures) of the enemy and decrease the effectiveness of the enemy's countermeasures. See also electronic countermeasures.

electronic emission — The freeing of electrons from the surface of a material usually produced by heat.

electronic leak detector — An electronic oscillator device that emits an audible tone if any refrigerant gas is picked up in its sensor tube. When a refrigerant leak is detected, the tone changes.

electronic moisture indicator — A device for checking moisture in a material. It operates on the principle of measuring the conductivity of the material.

electronic oscillator — 1. An electronic device used in a leak detector that emits a changing audible tone when a leak is detected. 2. An electronic circuit that converts DC into AC electricity.

electronic voltmeter — An electronic instrument used to measure voltage.

electronics — 1. The branch of science that deals with electron flow and its control. 2. In physics, the study and use of the movement and effects of free electrons and with electronic devices.

electroplating — An electrochemical method of depositing a thin layer of metal on some object. The object to be plated is the cathode, the metal that will be deposited is the anode, and the electrolyte is a nonmetallic conductor that will form ions of the plating metal.

electrostatic charge — A stationary electrical charge on an object caused by an accumulation of electrons or by a depletion of electrons.

electrostatic deflection control — A method of controlling the position of a beam of electrons on the face of a cathode-ray tube. The beam of electrons that forms the trace or picture on a cathode-ray tube can be deflected to the correct position on the screen by electrostatic charges on plates that are placed above, below, and on each side of the beam.

electrostatic energy — In a capacitor, the energy stored when two opposing electrical charges act across the dielectric.

electrostatic field — A field of force that exists around a charged body. Also referred to as a dielectric field.

electrostatic stress — The electrical force that tends to puncture an insulator. It is caused by an accumulation of electrical charges on a body.

electrostatics — The branch of physics that deals with the attraction and repulsion of static electrical charges.

electrovalent or ionic bond — The bond formed by two atoms when one atom gives up one or more valence electrons to the other. The bond is based on the attraction between the positive and negative ions thus formed.

element — The basic chemical substances that cannot be divided into simpler substances by chemical means.

element of threat — A perception factor that describes how a person is unlikely to easily comprehend an event if that person is feeling threatened since most of a person's effort is focused on whatever is threatening them.

elevator — A horizontal, movable control surface on the tail of an airplane. It is used to rotate the airplane about its lateral axis.

elevator angle — The angular displacement of the elevator from its neutral position.

elevator control tab — A metal tab located on the elevator that helps the pilot control the elevator.

elevator trim stall — A demonstration stall that a flight instructor shows a student pilot. It simulates the danger zone defined by a rejected landing or go-around, and is demonstrated at altitude with the airplane configured and trimmed for a typical final approach to landing.

elevon — A control surface that combines the functions of both ailerons and elevators. Movement of the control wheel to the right or left causes the elevons to move differentially. (When the left elevon moves up, the right elevon moves down.) The differential movement of the elevons causes the airplane to rotate about its longitudinal (roll) axis.

ellipse — An oval. A curve generated by a point moving such that the sum of its distances from two fixed points (foci) is a constant. It is the plane cross-section of a cylinder at other than a right angle or a right cone at other than parallel to the base.

elliptical — Of or having the form of an ellipse. An elongated oval shape.

elongate — To stretch or lengthen.

e-mail — Electronic mail. Mail sent and received between computers.

embarkation — The loading of passengers and cargo onto the airplane.

embedded circulation — A relatively small scale circulation embedded in, and driven by, a larger scale circulation.

embossing — The process of raising a boss or protuberance on the surface.

emergency — A distress or an urgency condition.

emergency air pressure — The compressed air stored in high-strength steel cylinders used to provide emergency landing gear extension and emergency braking in the event of the failure of the main power system.

emergency descent — A maneuver designed to allow the aircraft to descend to a lower altitude at the fastest, most practical airspeed, in the event of an engine fire, loss of cabin pressure, or other emergency.

emergency locator transmitter (ELT) — A radio transmitter attached to the aircraft structure which operates from its own power source on 121.5 MHz, 243.0 MHz, and 406 MHz. It aids in locating downed aircraft by radiating a downward sweeping audio tone, 2-4 times per second. It is designed to function without human action after an accident. The 406 MHz ELT broadcasts an identification signal and is capable of broadcasting location if linked to a GPS receiver.

emery paper — A fine abrasive paper composed of pulverized corundum or aluminum oxide and used for polishing.

emery wheel — A wheel consisting of a fine abrasive material composed of pulverized corundum or aluminum oxide and used for grinding and polishing.

emitter — The electrode of a transistor that corresponds to the cathode of a vacuum tube. Conventional current enters a transistor through the emitter.

empennage — The rear or tail portion of an airplane.

empty field myopia — The normal tendency of the eye to focus at only 10 to 30 feet when looking into a field devoid of objects, contrasting colors, or patterns.

empty weight — The weight of an aircraft, its power plant, and all of the fixed equipment. It includes unusable fuel and undrainable oil for aircraft not certified under FAR Part 23 (aircraft certified under FAR Part 23 include full oil as part of empty weight).

empty weight center of gravity (EWCG) — The center of gravity of an airplane that includes all fixed equipment, the unusable fuel, and undrainable oil for aircraft not certified under FAR Part 23 (aircraft certified under FAR Part 23 include full oil as part of empty weight).

empty weight center of gravity range — The range determined so that the empty weight center of gravity limits will not be exceeded under standard specifications loading arrangements.

empty weight moment — The moment of an aircraft at its empty weight.

emulsion-type cleaner — A chemical cleaner that mixes with water or petroleum solvent to form an emulsion. It is used to loosen dirt, soot, or oxide films from the surface of an aircraft.

En Route Automated Radar Tracking System (EARTS) — An automated radar tracking system that combines inputs from multiple short- and long-range radars into one display.

enamel — A material whose pigments are dispersed in a varnish base. The finish cures by chemical changes within the base.

encapsulate — To completely surround or cover something.

encased — Enclosed in a housing.

enclosed relay — An electrical relay in which both the coil and the contacts are enclosed in a protective housing.

encode — In electronics, to put an analog signal into digital format.

encoding altimeter — A pneumatic altimeter that provides a signal to the transponder that indicates the altitude on the radar operator's screen.

end spanner — A socket wrench that has a series of raised lugs around its end rather than splines broached inside.

end voltage — The voltage across a chemical cell when the cell should be discarded or recharged. The end voltage of a particular type of cell is usually defined by the manufacturer. Also referred to as an end-of-life voltage.

endurance — The length of time an aircraft can remain in the air. The power produced by the engines can be regulated to give the aircraft the greatest speed, the greatest range, or the greatest endurance.

energy — Inherent power or the capacity for performing work. When a portion of matter is stationary, it often has energy due to its position in relation to other portions of matter. This is called potential energy. If the matter is moving, it is said to have kinetic energy, or energy due to motion.

engaging solenoid — A solenoid used to engage an inertia starter with the engine.

engine — A machine that converts energy into mechanical power.

engine analyzer — An electronic instrument using a cathode ray oscilloscope as an indicator to analyze the condition of the ignition system and to visually display the vibrations in the engine.

engine breather — The vent for the crankcase of a reciprocating engine. It allows fumes to escape from the crankcase and prevents pressure build-up inside the engine.

engine compartment — The area of an aircraft in which the engine and its components and accessories are located.

engine compressor — The section of a turbine engine in which the air is compressed before it enters the burner section.

engine conditioning — An integrated system of engine checks and tests whereby engines can be brought up to or kept in top operating condition. Two of the most important checks of engine conditioning are the compression test and the cold cylinder check.

engine controls — The controls required for the proper operation of an aircraft engine. They include the throttle, mixture control, propeller pitch control, carburetor heat, engine cowl flap control, and the ignition switch.

engine cycle — **1.** The cycle of events that must be accomplished in the transformation of chemical energy into mechanical energy. The two most common cycles of events are the Otto cycle, which describes the events of the reciprocating engine, and the Brayton cycle, which describes the transformation taking place in a turbine engine. **2.** One takeoff and landing as recorded by an airline. Also described as one start and one stop of the engine or sometimes as one full advance and retard of the throttle. The latter two situations require special recording procedures for maintenance runs if cycle times are needed.

engine gauge unit — A three-in-one instrument used to show the operating condition of an engine. It houses a fuel pressure, oil pressure, and oil temperature gauge in one case.

engine history recorder — An electronic data collection device on some newer engines which records the number of times certain normal operating parameters such as speed and temperature are reached.

engine inoperative loss of directional control demonstration — This demonstration is required during a multi-engine practical test to show the control pressures necessary to maintain directional control with one engine inoperative. This demonstration should be accomplished within a safe distance of a suitable airport, and the entry altitude should allow completion no lower than 3,000 feet AGL. Since actual VMC varies with existing condition, the pilot should not try to duplicate the published VMC, which was established during initial certification. Pilots should expect a loss of directional control at a speed that may be higher than the published VMC. Remember, as altitude increases, actual VMC decreases, and under some weight and altitude combinations, VMC and stall speed are the same. This means that the loss of directional control demonstration cannot be accomplished safely. Pilots should be prepared to recover at the first indication of stall or loss of directional control, whichever occurs first. The bottom line is that the intent of the engine-out loss of directional control demonstration is to demonstrate the onset of control limits. Normally, this occurs when the nose begins to move even though full rudder is applied.

engine logbook — A record book of an aircraft engine's time in service, maintenance performed, inspections, etc.

engine mount — The structure used to attach an engine in the airframe. It normally includes shock mounts.

engine mounting pads — The shock absorbing units connected between the engine and the engine mount.

engine nacelle — The streamlined, enclosed housing on a wing or fuselage in which the engine is mounted.

engine performance — The relationship between power, RPM, fuel consumption, and manifold pressure of an engine.

engine pressure ratio (EPR) — In gas turbine engines, the ratio of turbine discharge pressure divided by compressor inlet pressure. Displayed in the cockpit as an indication of engine thrust.

engine ratings — The engine power ratings as type certificated by the FAA. These ratings list thrust or shaft horsepower at takeoff, cruise, etc.

engine ring cowl — The ring-shaped covering over the cylinders of a radial engine for the purpose of streamlining and improving the airflow through the engine.

engine seizure — The locking-up or stopping of an engine because of some internal malfunction.

engine stations — In gas turbine engines, numbered locations along the engine length, or along the gas path used for the purpose of identifying pressure and temperature points, component locations and the like.

engine stroke — In a reciprocating engine, the distance a piston travels from bottom dead center (BDC) to top dead center (TDC). Engine stroke is equal to two times the crankshaft throw.

engine sump — The lowest point in the engine from which the oil may be drained.

engine trimming — The adjustment of the fuel control unit of a gas-turbine engine.

engine-driven air pump — An air pump driven from an accessory drive on the engine. Also referred to as a vacuum pump.

engineer — A person who practices the profession of engineering. In the United Kingdom an engineer is an aircraft maintenance technician.

engineered performance standards — A mathematically derived runway capacity standard.

EPS's are calculated for each airport on an individual basis and reflect that airport's aircraft mix, operating procedures, runway layout, and specific weather conditions. EPS's do not give consideration to staffing, experience levels, equipment outages, and intrail restrictions as does the AAR.

enhanced training materials — While aviation instructors are expected to be familiar with all regulatory training requirements, use of instructor-oriented training materials, which are enhanced for regulatory compliance, are beneficial for ensuring that required training is accomplished, endorsed, and properly documented. Examples of these materials may include training syllabi, maneuver guides or handbooks, and computer-based training.

enrich — To make a fuel-air mixture ratio richer. When the amount of fuel metered into the engine is increased without increasing the amount of air, the mixture is enriched.

enroute air traffic control services — Air traffic control service provided aircraft on IFR flight plans, generally by centers, when these aircraft are operating between departure and destination terminal areas. When equipment, capabilities, and controller workload permit, certain advisory/assistance services may be provided to VFR aircraft.

enroute charts — a. Enroute Low Altitude Charts — Provide aeronautical information for enroute instrument navigation (IFR) in the low altitude stratum. Information includes the portrayal of airways, limits of controlled airspace, position identification and frequencies of radio aids, selected airports, minimum enroute and minimum obstruction clearance altitudes, airway distances, reporting points, restricted areas, and related data. Area charts, which are a part of this series, furnish terminal data at a larger scale in congested areas. b. Enroute High Altitude Charts — Provide aeronautical information for enroute instrument navigation (IFR) in the high altitude stratum. Information includes the portrayal of jet routes, identification and frequencies of radio aids, selected airports, distances, time zones, special use airspace, and related information.

enroute descent — Descent from the enroute cruising altitude which takes place along the route of flight.

enroute flight advisory service — A service specifically designed to provide, upon pilot request, timely weather information pertinent to his type of flight, intended route of flight and altitude. The FSSs providing this service are listed in the Airport/Facility Directory.

enroute high altitude charts — Provide aeronautical information for enroute instrument navigation (IFR) in the high altitude stratum. Information includes the portrayal of jet routes, identification and frequencies of radio aids, selected airports, distances, time zones, special use airspace, and related information.

enroute low altitude charts — Provide aeronautical information for enroute instrument navigation (IFR) in the low altitude stratum. Information includes the portrayal of airways, limits of controlled airspace, position identification and frequencies of radio aids, selected airports, minimum enroute and minimum obstruction clearance altitudes, airway distances, reporting points, restricted areas, and related data. Area charts, which are a part of this series, furnish terminal data at a larger scale in congested areas.

enroute minimum safe altitude warning — A function of the NAS Stage A enroute computer that aids the controller by alerting him when a tracked aircraft is below or predicted by the computer to go below a predetermined minimum IFR altitude (MIA).

enroute spacing program — A program designed to assist the exit sector in achieving the required in-trail spacing.

entrained air — The foam or bubbles in the scavenged oil caused by heat and the centrifugal action of the oil-wetted parts. Oil with large quantities of entrained air is a poor lubricant. This air has to be removed.

entrained water — The water held in suspension in aircraft fuel. It is in such tiny droplets that it passes through filters and will do no damage until the temperature of the fuel drops to the point that these tiny particles accumulate or coalesce to form free water in the tank.

envelope — A pre-sewn cover made of aircraft fabric that is slipped over the structure and attached.

envelope method of recovering — A method of recovering an aircraft structure in which a pre-sewn fabric envelope is slipped over the structure and attached. The opening is closed either by cementing the fabric to the structure or by hand sewing.

envelope power — In electricity, a measurement of average power supplied to an antenna by a radio transmitter.

environmental — The conditions surrounding an object.

environmental control systems — In an aircraft, the systems, including the supplemental oxygen systems, air conditioning systems, heaters, and pressurization systems, which make it possible for an occupant to function at high altitude.

environmental stress cracking (ESC) — In composites, the susceptibility of a resin to cracking or crazing when in the presence of surface-active chemicals.

Eonnex — A fabric woven from polyester fibers.

epicyclic gear train — An arrangement of gears in which one or more gears travel around the circumference of another gear.

epoxy — A flexible, thermosetting resin made by the polymerization of an epoxide. It is noted for its durability and chemical resistance.

epoxy primer — A two-part catalyzed paint material used to provide a good bond between a surface and the topcoat.

epoxy resin — A common thermoset material used in aircraft construction. Used as the bonding matrix to distribute the stresses to the fibers, and hold the fibers together. When mixed with a catalyst, they are adhesive, resistant to chemicals, are water resistant, and are unaffected by heat or cold. One part of a two-part system which combines the resin and the catalyst to form the bonding matrix. In composites, the term "resin" is often used to describe the two parts mixed together.

EPR-rated gas turbine — A method of expressing the thrust of a gas turbine engine in terms of engine pressure ratio (EPR).

equalization — The process of restoring all of the cells of a nickel-cadmium battery to a condition of equal capacity. All of the cells are discharged, shorted out, and allowed to "rest." The battery is then said to be equalized and ready to receive a fresh charge.

equalizer circuit — A circuit in a multiple-generator voltage-regulator system that tends to equalize the current output of the generators by controlling the field currents of all the generators in the system.

equilibrium — A condition that exists within a body when the sum of the moments of all of the forces acting on the body is equal to zero.

equilibrium level — The altitude where the updraft temperature is equal to its surroundings.

equinox — Noon on the day when the sun's rays are perpendicular to the earth's surface at the equator.

equipment — Any item that is secured in a fixed location to the aircraft and is to be utilized in the aircraft.

equipment ground — In electricity, a ground connected to the case or chassis of electronic equipment. The chassis is in turn, connected to an earth ground. See also earth ground.

equipment list — A comprehensive list of equipment installed on a particular aircraft. This includes the required and optional equipment.

equivalent airspeed (EAS) — The calibrated airspeed, shown on the airspeed indicator, corrected for errors that are caused by the compressibility of the air inside, the pitot tube, or by the installation of the instrument.

equivalent circuit — A circuit containing only one or two components that has the same properties as a more complex circuit. Used to more easily analyze the characteristics of the circuit.

equivalent flat plate area — The area of a square flat plate, normal to the direction of motion, which offers the same amount of resistance to motion as the body or combination of bodies under consideration.

equivalent monoplane — A monoplane wing with equal lift and drag properties as a combination of two or more wings.

equivalent shaft horsepower (ESHP) — A unit of measured power output of turboprops and some turboshaft engines. Where ESHP equals SHP plus HP from jet thrust (HP from jet thrust equals static thrust divided by 2.6).

erosion — The removal of material by abrasion, dissolution, and/or corrosion.

escape velocity — The speed an aircraft or missile must reach in order for it to escape from the gravitational field of the Earth.

escutcheon — A reinforcement around a hole or opening in a material that helps prevent tearing out of the base material.

established — To be stable or fixed on a route, route segment, altitude, heading, etc.

estimated ceiling — A ceiling value that derives from an estimate made by an observer or when the method of determining ceiling does not fall definitively in any other category.

estimated elapsed time [ICAO] — The estimated time required to proceed from one significant point to another.

estimated off-block time [ICAO] — The estimated time at which the aircraft will commence movement associated with departure.

estimated position error (EPE) — A measure of the current estimated navigational performance. Also referred to as Actual Navigation Performance (ANP).

estimated time enroute — The estimated flying time from departure point to destination (lift-off to touchdown).

estimated time of arrival — The time the flight is estimated to arrive at the gate (scheduled operators) or the actual runway on times for nonscheduled operators.

etch — To chemically remove a part of a material. Clad aluminum alloy sheets are etched before painting to microscopically roughen them so that the primer can bond tightly to their surface.

etching — A process of detecting defects in aluminum alloy by use of a caustic soda and nitric acid solution.

ethylene dibromide — A chemical compound of bromine that is added to aviation gasoline. Used to convert the lead deposits from the tetraethyl lead into lead bromides that are volatile enough to vaporize and pass out the exhaust rather than foul the spark plugs.

ethylene glycol — A viscous form of liquid alcohol (C_2HSO_2) used as a coolant for high-powered, liquid-cooled engines.

eutectic metal — A metal alloy whose melting point, due to the proportion of its components, is lower than would be possible before the mixture of the components.

evacuated bellows — A set of bellows from which most of the air has been removed and the bellows sealed. They serve as the sensitive element in an aneroid barometer for measuring atmospheric pressure.

evaluation — Measures a demonstrated performance against a criteria or standard, such as a grade of at least 70% to pass a written test. Formal evaluations are typically in the form of written tests, oral quizzing, or check flights, and are used to measure performance and document whether the course objectives have been met.

evaporation — Change of state from liquid to vapor.

evaporator — The unit in a vapor-cycle air conditioning system in which liquid refrigerant absorbs heat from the cabin to change the refrigerant into a vapor. Air blown over the evaporator loses its heat and is cooled.

execute missed approach — Instructions issued to a pilot making an instrument approach which means continue inbound to the missed approach point and execute the missed approach procedure as described on the Instrument Approach Procedure Chart or as previously assigned by ATC. The pilot may climb immediately to the altitude specified in the missed approach procedure upon making a missed approach. No turns should be initiated prior to reaching the missed approach point. When conducting an ASR or PAR approach, execute the assigned missed approach procedure immediately upon receiving instructions to "execute missed approach." (Refer to AIM)

exercise — A principle of learning that those things most often repeated are best remembered.

exhaust contrails — Forms when the water vapor added from an aircraft exhaust is sufficient to saturate the atmosphere.

exhaust gas temperature (EGT) — In gas turbine engines, temperature taken at the turbine exit. Often referred to as T_t7.

exhaust manifold — A pipe with several apertures used to collect exhaust gases from reciprocating engines. The manifold attaches to the individual exhaust ports and carries the exhaust gases overboard through a common discharge.

exhaust nozzle — The rear opening of a turbine engine exhaust duct. The nozzle acts as an orifice, the size of which determines the density and velocity of the gases as they emerge from the engine.

exhaust port — The hole in the cylinder of a reciprocating engine through which the exhaust gases are expelled.

exhaust stacks — The short, individual pipes attached to the exhaust ports of the cylinders of reciprocating engines, through which the exhaust gases are discharged overboard.

exhaust stroke — The stroke of the Otto cycle where the exhaust gases are forced out of the cylinder as the piston is moving away from the crankshaft and the exhaust valve is open.

exhaust valve — The valve in an aircraft engine cylinder through which the burned gases leave the combustion chamber.

exit guide vanes — The fixed airfoils at the discharge end of an axial flow compressor that straighten out the swirling air caused by the rotating rotors so that the air leaves the engine in an axial direction.

expand — To increase the dimensions of.

expanded plastic — The increase in the volume of plastic resin generated when the materials that make up the plastic are mixed. The volume is increased by gas bubbles.

expander-tube brake — A nonservo brake in which the composition blocks are forced out against a rotating drum by hydraulic fluid, expanding a synthetic rubber tube on which they rest.

expansion boots — Inflatable deicer boots.

expansion coefficient — A number that describes the change in linear dimensions of a material with a specified change in its temperature.

expansion reamers — A precision cutting tool used to enlarge and smooth the inside circumference of a drilled hole. The diameter of the reamer can be changed by an adjustable wedge inside the blades.

expansion turbine — A turbine wheel in an air-cycle air conditioning system used to extract some of the energy from the bleed air. The energy of the bleed air drives the turbine and the air is further cooled by expansion.

expansion wave — The change in velocity and density of the air as it passes over the thickest part of an airfoil moving through the air at speeds greater than the speed of sound.

expect (altitude) at (time) or (fix) — Used under certain conditions to provide a pilot with an altitude to be used in the event of two-way communications failure. It also provides altitude information to assist the pilot in planning.

expect further clearance (time) — The time a pilot can expect to receive clearance beyond a clearance limit.

expect further clearance via (airways, routes or fixes) — Used to inform a pilot of the routing he can expect if any part of the route beyond a short range clearance limit differs from that filed.

expected departure clearance time — The runway release time assigned to an aircraft in a controlled departure time program and shown on the flight progress strip as an EDCT.

expedite — Used by ATC when prompt compliance is required to avoid the development of an imminent situation. Expedite climb/descent normally indicates to a pilot that the approximate best rate of climb/descent should be used without requiring an exceptional change in aircraft handling characteristics.

expel — To force or drive out.

expendable weight — The weight that is decreased in flight. The fuel on board is an expendable weight as it is used in flight.

experimental category — A grouping for aircraft that do not have type certificates or do not conform to their type certificates. Special airworthiness certificates may be issued in the experimental category for the following purposes: research and development, showing compliance with regulations, crew training, exhibition, air racing, market surveys, operating amateur-built aircraft, or operating kit-built aircraft.

explode — To make a violent change in chemical composition. To violently release chemical, mechanical, or nuclear energy. Usually accompanied by loud noise, a flash of light and a great deal of heat.

explosion-proof motor — An electric motor sealed in a way that prevents explosive gases from being ignited by electrical sparks within the motor.

explosion-proof switch — An electric switch sealed in a way that prevents explosive gases from being ignited by electrical sparks within the switch.

explosive atmosphere — A gaseous environment containing explosive vapors or explosive concentrations of fine dust that can be ignited.

explosive bolt — A special bolt that contains an explosive charge that can be ignited when the bolt must instantly release. Used to allow instant release of expended rocket stages.

explosive charge — A quantity of explosive that can be used to break a seal and discharge a substance.

explosive rivet — A patented blind rivet manufactured by the DuPont Company. Its hollow end is filled with an explosive and sealed with a cap. When the rivet is heated, it explodes, swelling its end and clamping the metal together.

exponent — In mathematics, a number superscripted above and to the right of a base number. It indicates the power to which the base number is to be multiplied. In the example 2^3, the base number is 2 while the exponent is 3. This is stated as two cubed (or two to the third power) and is shown as 2 X 2 X 2 = 8.

extend — To move away from the normal or closed position. When landing gear is extended, it is moved from its retracted position to the "gear down" position. Flaps can be extended from their stowed position to full extension or to some point in between.

extended over-water operation — **1**. With respect to aircraft other than helicopters, an operation over water at a horizontal distance of more than 50 nautical miles from the nearest shoreline. **2**. With respect to helicopters, an operation over water at a horizontal distance of more than 50 nautical miles from the nearest shoreline and more than 50 nautical miles from an off-shore heliport structure.

extension lines — The lines on a technical drawing that extend from a view for the purpose of identifying a dimension.

extent of damage — An amount of damage sustained by a unit of equipment.

exterior angle — The angle between one of the sides of a polygon and an extension to an adjacent side.

exterior view — A view of an object showing only its outer or visible surfaces.

external combustion engine — A heat engine in which the chemical energy in the fuel is converted into heat energy released to the outside of the engine. Heating water to produce steam that is put to mechanical use is a form of an external combustion engine.

external inspection — A visual inspection done externally to the airframe, engine, or unit component without having to inspect the internal mechanism by disassembly.

external load — A load that is carried or extends outside of an aircraft.

external resources — Many potential resources exist outside the cockpit such as air traffic controllers, maintenance technicians, and flight service personnel.

external tooth lock washer — A thin, spring steel, shake-proof lock washer with twisted teeth around its outside circumference that holds pressure between the head of a screw or bolt and the metal surface to prevent the fastener from loosening.

external-control surface locks — The locks applied on the exterior of the control surfaces of a parked aircraft to prevent movement in windy conditions.

external-load attaching — The structural components used to attach an external load to an aircraft, including external-load containers, the backup structure at the attachment points, and any quick-release device used to jettison the external load.

extinguishing agent — The agent used in a fire extinguishing system to either cool the fuel below its kindling point or to exclude oxygen from the surface of the fire.

extra-flexible control cable — A special metal cable flexible enough to pass around pulleys. It consists of seven strands of wire with 19 wires in each strand.

extratropical cyclone — A macroscale low-pressure disturbance that develops outside the tropics.

extrude — To form by forcing through a die of the desired shape.

extrusion — A strip of metal, usually of aluminum or magnesium, that has been forced through a die in its plastic state. This can produce complex cross-sectional shapes required for modern aircraft construction.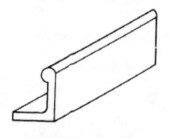

eye — **1**. The hole of an eyebolt. **2**. In meteorology, the roughly circular area of calm or relatively light winds and comparatively fair weather at the center of a well-developed tropical cyclone. A wall cloud marks the outer boundary of the eye.

eye wall — The cloudy region embedded with cumulonimbus (CB) clouds immediately adjacent to the eye of an intense tropical cyclone.

eyebolt — A bolt that has a flattened head with a hole in it. An eyebolt is used to attach a cable to a structure.

eyebrow lights — Small shielded lights positioned over the top corners of an instrument on an aircraft instrument panel. The lights illuminate the dials of the instrument, but they do not shine in the eyes of the pilot.

eyelet — A flanged tubular fastener designed for securing by curling or splaying the end.

F

FAA air carrier operations specifications — Document issued to users operating under Federal Aviation Administration Regulations (FAR) Parts 121, 125, 127, 129, and 135. Operations Specifications are established and formalized by FARs. The primary purpose of FAA Air Carrier Operations Specifications is to provide a legally enforceable means of prescribing an authorization, limitation and/or procedures for a specific operator. Operations Specifications are subject to expeditious changes. These changes are usually too time critical to adopt through the regulatory process.

FAA Form 337 — Major Repair and Alteration Form.

FAA Form 8500-8 — Application for airman medical certificate, or airman medical and student pilot certificate.

FAA Form 8710-1 — Application for an airman certificate and/or rating.

FAA-approved data — Data that can be used as authorization for the techniques or procedures necessary to make a repair or an alteration to a certificated aircraft. Approved data can consist of documents such as Manufacturer's Service Bulletins, Manufacturer's kit instructions, Airworthiness Directives, or specific details of a repair issued by the engineering department of the manufacturer.

FAA-PMA — The identifying letters required on an aircraft part or component to signify it as being manufactured under a Federal Aviation Administration Parts Manufacturing Approval.

fabric — Individual fibers woven together to produce cloth. Unidirectional or matted fibers may be included in this classification.

fabric material — A cloth used to cover aircraft structures. The basic fabric is Grade-A long staple cotton, but Irish linen is sometimes used interchangeably. Synthetic fabrics made of polyester resins and specially treated glass fibers also can be used in place of cotton.

fabric punch test — A test performed to measure the condition of fabric covering material on fabric-covered aircraft. The tester determines fabric condition by measuring the resistance of the fabric to a sharp punch pressed against the fabric until it pierces.

fabric punch tester — A hand tester used to give an indication of the relative strength of aircraft fabric. It measures the force required to press a specially shaped pointed plunger a specific distance into the fabric.

fabric repair — The repair made to a fabric-covered structure that produces the same strength and tautness in the fabric as it originally had.

fabric warp face — That side of a woven fabric on which the greatest number of yarns are parallel to the selvage.

fabricate — To construct or build something.

fabrication — The assembling of parts to make a complete unit or a structure.

face of a propeller — The flat side of a propeller blade.

face of the drawing — The surface of an object as seen from the front view.

face shield — A transparent protective guard covering the entire face to protect against flying objects or liquid spray.

face-end spanner — A type of semicircular, open-end wrench having short pins extending from its face and used to turn special circular type nuts.

faceplate, lathe — Used for turning metal. A heavy, steel disk with a smooth face mounted on the headstock of the lathe and is turned by it.

face-to-face bearings — Bearing sets installed in such a way that one set carries thrust loads in one direction while the other bearing set carries thrust loads in the opposite direction.

facing — A machining operation on the end, flat face, or shoulder of a part.

facsimile — A copy of a document that is transmitted over telephone lines or radio waves by a facsimile machine to be recreated by a facsimile machine at the other end. Also referred to as a fax.

factor of safety — The ratio of the ultimate strength of a member to the probable maximum load. This ratio is larger than one.

fading — **1.** A decrease in the friction applied by a drum-type brake when it is hot. As the drum is heated, it expands in a bellmouth fashion and part of it moves away from the lining. This decreases the friction area and causes a loss of breaking action. **2.** A decrease in strength of a received radio signal.

fahnstock clip — An electrical type of spring clip connector used to temporarily connect a wire to an electrical circuit.

Fahrenheit (F) — A temperature scale on which pure water freezes at 32° and boils under standard atmospheric pressure conditions at 212°.

fail hardover — A failure of an automatic flight control system in which a steady signal is produced that drives the controls to the extreme end of their travel and holds them there.

fail-safe — A design feature that transmits the loads into a secondary portion of the structure in the event of the primary structure failure.

fail-safe control — A type of control that automatically puts the controlled device in a safe condition if the control system should fail.

faired curve — A smoothly curved object.

fairing — A smooth covering over a joint or a junction in an aircraft structure to provide a smooth surface for the airflow. Its primary purpose is to reduce drag.

fairlead — Wood or plastic guides for aircraft control cable, used to hold straight runs of cable away from the structure.

fallstreaks — Ice crystals that descend from cirrus clouds.

false rib — Short, semi-rib extending from the spar to the leading edge of the wing. It is used to give rigidity and better shape to the leading edge of fabric covered wings.

false spar — A partial spar in an aircraft wing to which the aileron hinges attach.

false start — A condition in starting a turbine engine in which ignition occurs but the RPM will not increase. This condition is usually caused by the engine not being turned fast enough by the starter when ignition occurs.

fan --A feature of some turbojet engines. It is, in effect, a ducted, multi-bladed propeller driven by a gas turbine engine. It can be constructed as an extension of some of the turbine blades (aft fan), but it is more commonly an extension of compressor blades or powered by gears from the compressor or turbine sections (forward fan). These fans pull large volumes of air around the outside of the gas generator portion of the engine.

fan air — The portion of airflow through a turbofan engine that is acted upon by the fan stages of the compressor.

fan marker — An aircraft directional radio signal transmitted vertically upward from a transmitter located along a navigational radio range. It is heard only when the aircraft is directly over the transmitter.

Farad (f) — The basic unit of capacitance. A capacitor of one farad will hold one coulomb (6.28×10^{18} electrons) under a pressure of one volt.

Faraday's law of electrolysis — The amount of chemical change produced by current during electrolysis is proportional to the quantity of electricity used.

Faraday's law of electromagnetic induction — When there is relative motion between a conductor and a magnetic field, or when there is a change in magnitude of the magnetic field, electrons are induced to flow through the conductor.

fast file — A system whereby a pilot files a flight plan via telephone that is tape recorded and then transcribed for transmission to the appropriate air traffic facility. Locations having a fast file capability are contained in the Airport/Facility Directory. (Refer to AIM)

fastener — A device such as a rivet, screw, or bolt used to fasten two objects together.

fatigue — The weakening and eventual failure of a metal due to continued reversal or repeated stresses beyond the fatigue limit.

fatigue crack — A crack in a structural member caused by flexing or vibration.

fatigue failure — The failure of a material due to flexing or vibration.

fatigue limit — The amount of flexing or vibration a body can withstand before fatigue failure occurs.

fatigue resistance — The property that enables a metal to withstand repeated loads or reversals of loads and vibrations.

fatigue strength — The ability of a material to withstand vibration and flexing.

fault — A defect in an electrical circuit.

fault tree — A chart used to determine the possible causes of some undesired event. The chart includes the undesired occurrence with the possible causes listed below it. A fault tree is also referred to as a logic chart since the branches of the "tree" are connected through the use of logic gates.

faying strip — The strip along the edge of a sheet metal skin where a lap joint is formed. This inaccessible area is highly susceptible to the formation of corrosion.

faying surface — The overlapping area of adjoining surfaces.

FDC NOTAM — A NOTAM issued by the National Flight Data Center which contains information of a regulatory nature such as temporary flight restrictions or amendments to instrument approach procedures and other current aeronautical charts.

feather — To change the angle of propeller blades so that the chords become approximately parallel to the line of flight.

feather edge — A very thin, sharp edge of a material. Susceptible to damage as a result of bending or breaking.

feathered propeller — A propeller whose blades have been rotated so that the leading and trailing edges are nearly parallel with the aircraft flight path to stop or minimize drag and engine rotation. Normally used to indicate shutdown of a reciprocating or turboprop engine due to malfunction.

feathering — In rotorcraft, the action that changes the pitch angle of the rotor blades by rotating them around their feathering (spanwise) axis.

feathering axis — In rotorcraft, the axis about which the pitch angle of a rotor blade is varied. Sometimes referred to as the spanwise axis.

feathering propeller — A controllable pitch propeller with a pitch range sufficient to allow the blades to be turned parallel to the line of flight. This reduces drag and helps prevent further damage to an engine that has been shut down after a malfunction.

feathering solenoid — A locking, electrical solenoid used with a Hamilton Standard Hydromatic propeller. The solenoid keeps the feathering pump running after the feathering button has been momentarily depressed. The solenoid is de-energized when the propeller becomes fully feathered.

feathering switch — Hydromatic propellers are standard equipment on certain models of airplanes, i.e., the D18C and D18C-T airplanes. These propellers are controlled by individual, single-acting, Hamilton Standard propeller governors, and feathering or unfeathering action is accomplished by individual, electric motor-driven pump feathering systems. The pumps are controlled by a feathering switch.

Federal Airways — See Low Altitude Airway Structure.

Federal Aviation Administration (FAA) — An organization within the Department of Transportation. The FAA establishes aviation rules and regulations as well as enforces those policies. The purpose of the FAA is to set the standards for civil aircraft in the interest of public safety.

Federal Aviation Regulations (FAR) — The rules, regulations, and guidelines established by the FAA to govern the operation of aircraft, airways, airmen, and the safe operation of civil aircraft.

Federal Communications Commission (FCC) — A government board, made up of seven commissioners, responsible for regulating all interstate electrical communications, including foreign communications that originate in the United States.

feedback — **1.** Another way to gauge whether students are receiving the correct message. Students must interpret and evaluate the information received and then respond. The transmission of evaluative or corrective information to the original or controlling source about an action, event, or process. **2.** In rotorcraft, the transmittal of forces, which are initiated by aerodynamic action on rotor blades, to the cockpit controls. **3.** A portion of the output signal of a circuit that is returned to the input. Positive feedback occurs when the feedback signal is in phase with the input signal. Negative feedback occurs when the feedback signal is out of phase with the input signal.

feeder fix — The fix depicted on instrument approach procedure charts which establishes the starting point of the feeder route.

feeder route — A route depicted on instrument approach procedure charts to designate routes for aircraft to proceed from the enroute structure to the initial approach fix (IAF).

feedthrough capacitor — A capacitor used to block low frequencies while passing radio frequencies. Some magnetos use a feed-through capacitor to serve as the normal capacitor to minimize arcing of the points. They also decrease the amount of radio interference caused by electrical energy being radiated from the ignition switch lead.

feedthrough connector — A connector used to carry a group of conductors through a bulkhead.

feel — The feedback from a power-controlled flight control or brake system. This "feel" allows the pilot to sense how much pressure is needed for the operation of the system.

feeler gauge — A measuring tool consisting of a series of precision ground steel blades of various thicknesses. It is used to determine the clearance or separation between parts.

fence — A fixed vane that extends chordwise across the wing of an airplane. Fences prevent air from flowing along the span of the wing.

Fenwal spot-type fire detection system — A fire detection system utilizing bimetallic thermal switches. A fire or overheat condition closes the switches and signals the presence of the fire or overheat.

ferrite — A magnetic substance that consists of ferric oxide combined with the oxides of one or more metals. Ferrite has high magnetic permeability and high electrical resistance.

ferritic stainless steel — Straight chromium carbon and low-alloy steels that are strongly magnetic.

ferromagnetic materials — Magnetic materials composed largely of iron.

ferrous metal — Iron or any alloy containing iron.

ferrule resistor — A group of resistors that have metal bands (ferrules) around each end so that they can be mounted in standard fuse clips.

ferrule terminals — The terminals on each end of a tubular fuse. Used for making connections with the circuit.

ferry — The movement of an aircraft from one location to another.

ferry flight — A flight for the purpose of returning an aircraft to base; delivering an aircraft from one location to another; or moving an aircraft to and from a maintenance base. Ferry flights, under certain conditions may be conducted under terms of a special flight permit.

ferry permit — Commonly used name for Special Flight Permit. A Special Flight Permit issued by the FAA allows an unlicensed aircraft to be flown from one location to another location.

fiber — A single strand of material, used as a reinforcement because of its high strength and stiffness.

fiber bridging — In composites, reinforcing fiber material that bridges an inside-radius of a pultruded product. This condition is caused by shrinkage stresses around such a radius during cure.

fiber content — The amount of fiber in a composite expressed as a ratio to the matrix. The most desirable fiber content is a 60:40 ratio. This means there is 60% fiber and 40% matrix material.

fiber direction or orientation — In composites, the orientation of the fibers in a laminate to the 0° reference designated by the manufacturer.

fiber locknut — A type of self-locking fastener with a fiber insert that puts pressure on the threads to lock the nut in place. This prevents the nut from turning when installed in areas subject to vibration.

fiber optics — The transmission of light through a bundle of fiber rods. Often used in areas where light must be transmitted, but where it is inconvenient or dangerous to use electrical transmission. An optical device called a Borescope uses tiny glass rods that conduct light and vision to make inspection around corners practicable. Fiber optics are also used to transmit information from one electronic device to another.

fiber reinforced plastics (FRP) — Term used interchangeably for advanced composites

fiberglass — Extremely thin fibers of glass. May be woven into a cloth or lightly packed into a mat. Used to reinforce epoxy or polyester resin for aircraft structure.

fiberglass reinforcement — In composites, fiberglass used as a reinforcement in a plastic matrix.

fidelity — The degree of similarity between the input and output waveforms of an electronic circuit.

field — A space in which magnetic or electric lines of force exist.

field coil — Coil or winding used to produce a magnetic field.

field effect transistor (FET) — A special form of semiconductor device with high input impedance. Electron flow between its source and drain is controlled by a voltage applied to the gate.

field elevation — See Airport Elevation

field excitation — DC supplied to the field of an alternator or generator to produce magnetic flux, which is cut by the conductors in the armature or stator.

field frame — The main structure of a generator or motor on which field poles and windings are mounted.

field maintenance — The maintenance performed on aircraft remotely or semi-remotely from the home station. Typically, there are few tools or equipment to implement normal maintenance procedures.

field strength — **1.** The intensity of the magnetic strength of a magnet or electromagnet. **2.** The intensity of the electromagnetic field emanating from a radio transmitter antenna.

field strength meter — An electrical instrument that measures the strength of an electromagnetic field radiating from a radio transmitting antenna.

filament — 1. The heated element in a light bulb or electron tube. In an electron tube, the heat speeds up the molecular movement in the cathode that emits electrons. 2. The smallest unit of a fibrous material.

filament winding — In composites, a manufacturing method in which long continuous fiber is wound around a mandrel to produce a structure.

file — A hand-operated cutting tool made of high-carbon steel and fitted with rows of very shallow teeth extending diagonally across the width of the tool.

filed — Normally used in conjunction with flight plans, meaning a flight plan has been submitted to ATC.

filed enroute delay — Any of the following preplanned delays at points/areas along the route of flight which require special flight plan filing and handling techniques.
a. Terminal Area Delay. A delay within a terminal area for touch-and-go, low approach, or other terminal area activity.
b. Special Use Airspace Delay. A delay within a Military Operating Area, Restricted Area, Warning Area, or ATC Assigned Airspace.
c. Aerial Refueling Delay. A delay within an Aerial Refueling Track or Anchor.

filed flight plan — The flight plan as filed with an ATS unit by the pilot or his designated representative without any subsequent changes or clearances.

files, classification — The teeth on a file vary from very fine to coarse in the following sequence:
a. Dead-smooth cut
b. Smooth cut
c. Second cut
d. Bastard cut
e. Coarse cut
Files with a single set of cutting teeth are called single-cut files. A file that has a second set of cutting teeth crossing the first set is called a double-cut.

filiform corrosion — A thread- or filament-like corrosion that forms on aluminum skins beneath a covering of protective coating.

fill — The direction across the width of fabric.

fill threads — Threads running across the width of a piece of fabric from one selvage edge to the other. Also referred to as weft or woof threads.

filler — In composites, material added to the mixed resin to increase viscosity, improve appearance, and lower the density and cost.

filler material — Any material mixed or added to a base material in order to give body to the base material.

filler metal — A metal used to increase the area of a weld. Normally supplied in the form of the welding rod or the electrode used in arc welding.

filler neck — Usually, a cylinder-shaped neck or tube leading into a reservoir for replenishing fluids.

filler plug — The plug installed in a sheet metal or wood structural repair to make the surface of the repair coincide with the original skin contour. The filler plug is used only to make the surface aerodynamically smooth. The strength of the repair comes from the doubler inside the structure.

filler ply — In composites, an additional patch to fill in a depression in the repair, or to buildup an edge.

filler rod — A thin metal rod or wire used in welding to provide the necessary filler metal. Used to provide additional strength to the weld.

filler valve — A readily accessible valve that provides a means of servicing an installed oxygen, air, or fluid system.

fillet — A rounded-out addition at the intersection of two plane surfaces to produce a smooth junction where the two surfaces meet. Fillets produce a smooth aerodynamic junction between the wing and the fuselage of an airplane.

filling — An increase in the central pressure of a pressure system; opposite of deepening; more commonly applied to a low rather than a high.

fillister-head screw — A machine screw whose shape consists of a rounded top surface, cylindrical sides, and a flat bearing surface.

film adhesive — In composites, a synthetic resin adhesive, usually of the thermosetting type, in the form of a thin, dry film of resin.

film resistor — A resistor formed by coating a ceramic, glass, or other insulating cylinder with a metal oxide or other thin resistive film.

film strength — A lubricant's ability to maintain a continuous lubricating film under mechanical pressure without breaking down.

filter — 1. A device for straining out unwanted solid particles in a fluid. 2. An electrical circuit arranged to pass certain frequencies while blocking all others. A high pass filter passes high frequencies and blocks low frequencies.

filter capacitor — A low-impedance capacitor attached across the output of a DC power supply to filter any ripple to ground, leaving the DC untouched.

filter choke — An inductor placed in series with the output of a DC power supply to reduce the amount of ripple while allowing the DC to pass.

filtering — The separation of unwanted components from either a fluid flow or an electrical flow.

fin — 1. The vertical stabilizer of an airplane to which the rudder is hinged. It produces directional stability. 2. A key under the head of a fastener that serves to keep the fastener from turning during assembly and use.

final — Commonly used to mean that an aircraft is on the final approach course or is aligned with a landing area. (See Final Approach Course) (See Final Approach - IFR) (See Traffic Pattern) (See Segments of an Instrument Approach Procedure)

final approach — A flight path of a landing aircraft in the direction of landing along the extended runway centerline from the base leg or straight in to the runway.

final approach - IFR (USA) — The flight path of an aircraft which is inbound to an airport on a final instrument approach course, beginning at the final approach fix or point and extending to the airport or the point where a circle-to-land maneuver or a missed approach is executed.

final approach (ICAO) — That part of an instrument approach procedure which commences at the specified final approach fix or point, or where such a fix or point is not specified, 1. at the end of the last procedure turn, base turn or inbound turn of a racetrack procedure, if specified; or 2. at the point of interception of the last track specified in the approach procedure; and ends at a point in the vicinity of an aerodrome from which: a. a landing can be made; or b. a missed approach procedure is initiated.

final approach course — A bearing/radial/track of an instrument approach leading to a runway or an extended runway centerline all without regard to distance.

final approach fix (FAF) — The fix from which the final approach (IFR) to an airport is executed and which identifies the beginning of the final approach segment. It is designated in the profile view of Jeppesen

Terminal charts by the Maltese Cross symbol for nonprecision approaches and by the glide slope/path intercept point on precision approaches. The glide slope/path symbol starts at the FAF. When ATC directs a lower-than-published Glide Slope/Path Intercept Altitude, it is the resultant actual point of the glide slope/path intercept.

final approach fix (FAF) (Australia) — A specified point on a non-precision approach which identifies the commencement of the final segment. The FAF is designated in the profile view of Jeppesen Terminal charts by the Maltese Cross symbol.

final approach point — The point, applicable only to a nonprecision approach with no depicted FAF (such as an on-airport VOR), where the aircraft is established inbound on the final approach course from the procedure turn and where the final approach descent may be commenced. The FAP serves as the FAF and identifies the beginning of the final approach segment.

final approach point (FAP) (Australia) — A specified point on the glide path of a precision instrument approach which identifies the commencement of the final segment. NOTE: The FAP is co-incident with the FAF of a localizer based non-precision approach.

final approach segment — That segment of an instrument approach procedure in which alignment and descent for landing are accomplished.

final controller — The controller providing information and final approach guidance during PAR and ASR approaches utilizing radar equipment. (See Radar Approach)

final monitor aid — A high resolution color display that is equipped with the controller alert system hardware/software which is used in the precision runway monitor (PRM) system. The display includes alert algorithms providing the target predictors, a color change alert when a target penetrates or is predicted to penetrate the no transgression zone (NTZ), a color change alert if the aircraft transponder becomes inoperative, synthesized voice alerts, digital mapping, and like features contained in the PRM system. (See Radar Approach)

final monitor controller — Air Traffic Control Specialist assigned to radar monitor the flight path of aircraft during simultaneous parallel and simultaneous close parallel ILS approach operations. Each runway is assigned a final monitor controller during simultaneous parallel and simultaneous close parallel ILS approaches. Final monitor controllers shall utilize the Precision Runway Monitor (PRM) system during simultaneous close parallel ILS approaches.

fineness ratio — The ratio of the length to the maximum diameter of a streamlined body such as an airship hull.

fine-wire spark plug — A spark plug using platinum or iridium electrodes. The small electrodes allow the firing end cavity to be open resulting in better scavenging of the lead oxides from the plug. The heat transfer characteristics of the fine wires prevent their overheating.

finger brake — A metal-forming machine similar to a leaf (cornice) brake. It is used to form all four sides of a box. The sides that have been bent up fit between the fingers of the clamp while the last bends are being made. Also referred to as a box brake.

finger patch — A form of welded patch to go over a cluster in a steel tube fuselage. Fingers extend along all of the tubes in the cluster.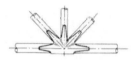

finger screen — A finger-shaped filter screen used on fuel tank standpipes. Used to screen or filter out large pieces of contamination.

finish — In composites, a material that is applied to the fabric after it is woven to improve the bond of the fiber to the resin system.

finish turning — A final smoothing process in the machining of a metal part in which the part is turned to its correct dimension.

finite life — The length of usefulness of a limited-life part. The part should not be used past the end of its predetermined operating life.

fire alarm relay — A relay actuated by a fire detection system. Provides energy for the fire warning bell and the fire warning lights in the cockpit.

fire detection system — A system in an aircraft that informs the pilot of a fire on board the aircraft.

fire extinguisher — Any device containing an extinguishing agent used to either cool a material below its kindling point or to exclude oxygen from its surface.

fire extinguishing agent — Any approved chemical used to extinguish a fire by either reducing the temperature of the fuel to a temperature that is below its kindling point or by excluding oxygen from the fire.

fire point — The temperature at which the vapors given off by a substance will ignite and continue to burn when a flame is passed above it.

fire valve — A valve that automatically shuts off the supply of combustion air to a combustion-type cabin heater in the event of a fire or overheat condition.

fire zone — An area or region of an aircraft designated by the manufacturer to require fire detection, fire extinguishing equipment, and a high degree of inherent fire resistance.

fireproof — The capacity to withstand the heat associated with fire without being destroyed.

fireproof structure — A structure constructed of nonflammable materials.

fire-resistant structure — A structure capable of resisting fire or exposure to high temperature for a specified period of time without being destroyed or structurally damaged.

firewall — A fire-resistant bulkhead that must be installed between an engine compartment and the rest of the aircraft structure.

firewall shutoff valve — A valve located on the airframe side of a firewall that will completely shut off the flow of fuel, oil, or hydraulic fluid to the engine during an engine fire.

firing order — The order or sequence in which the cylinders of an internal combustion engine fire in a normal cycle of operation.

firing position — The position of the piston in the cylinder of a reciprocating engine at the time ignition should occur. Igniting the mixture at this position allows the peak cylinder pressure to occur shortly after the piston passes top center.

firmer chisel — A woodworking chisel that has a thin, flat blade.

firmware — A computer program contained permanently in a computer. Usually contained in read-only memory (ROM).

first solo — A student pilot may not operate an aircraft in solo flight unless that student has met the requirements of FAR Part 61. "Solo flight," refers to the flight time that a student pilot is the solo occupant of the aircraft, or when the student performs the duties of a pilot in command of a gas balloon or airship requiring more than one pilot flight crewmember. "First solo" refers to the very first time a student embarks on a solo flight.

first tier center — The ARTCC immediately adjacent to the impacted center.

first-class lever — A lever in which the fulcrum or pivot point is positioned between two forces that act in opposite directions.

fisheyes — Isolated areas on a surface that have rejected the material's finish because of wax or silicone contamination.

fishmouth splice — A welded splice in steel tube structure in which one tube telescopes over the other. The outside tube is cut into a "V," resembling an open fishmouth, to provide additional area for the weld.

fishtail — A method no longer used to decrease the speed of an airplane on its approach for landing. Fishtailing consists of alternately skidding the airplane to the left and right by using the rudder while keeping the wings level with the control stick.

fissure — A scratch or crack.

fit — The range of tightness in the design of mating parts. Tightness of fit is the result of applying specific combinations of allowances and tolerances.

fitting — A part used to join or attach assemblies together.

five-hour rating — The ampere-hour rating of a battery that will discharge the battery in five hours. This is the most commonly used rating for aircraft batteries.

five-minute rating — A rating of the ampere-hour capacity of batteries normally used to indicate the capacity of a battery for high current drains such as starting current. The five-minute rating of a battery is an indication of the way the battery will function under the severe loads required by the engine starter.

fix — A geographical position determined by visual reference to the surface, by reference to one or more radio navaids, by celestial plotting, or by another navigational device. (Note, fix is a generic name for a geographical position and is referred to as a fix, waypoint, intersection, reporting point, etc.)

fix balancing — A process whereby aircraft are evenly distributed over several available arrival fixes reducing delays and controller workload.

fixed — The state of a permanently installed system in contrast to any type of portable equipment system.

fixed base operator (FBO) — An entrepreneur on an airfield that conducts business involved with general aviation. Often sells aviation fuel, aircraft and aircraft parts, aviation related goods, repairs aircraft, and conducts flying training.

fixed cowl flap — Fixed or ground adjustable doors on the air exit of an aircraft engine cowling. The cylinder head temperature can be controlled by varying the amount the flaps are opened if ground adjustable.

fixed equipment — Non-movable, attached equipment.

fixed landing gear — Landing gear that is not retractable.

fixed tail surfaces — Surfaces, mounted rigidly to the fuselage, that stabilize aircraft during takeoff, flights, and landing. Fixed tail surfaces also provide anchorage for the rudder and elevators. Often referred to as stabilizers.

fixed-pitch propellers — Propellers with fixed blade angles. Fixed-pitch propellers are designed as climb propellers, cruise propellers, or standard propellers.

fixed-wing aircraft — An airplane with rigidly attached wings, as distinguished from a helicopter or autogyro.

fixture — A small jig or device for holding parts in the proper position for assembly.

flag — A warning device incorporated in certain airborne navigation and flight instruments indicating that: **1.** instruments are inoperative or otherwise not operating satisfactorily, or **2.** Signal strength or quality of the received signal falls below acceptable values.

flag alarm — See flag.

flame out — In gas turbine engines, an unintentional loss of combustion due to a blowout (too much fuel) or die-out (too little fuel).

flame spraying — In flame spraying, molten metal is sprayed onto a base material. The combustion of a fuel gas is used to melt a metal. Melted particles are propelled toward the base material by a high-pressure gas where they are quenched very rapidly. This process allows a high wear resistant or high temperature resistant coating to be applied to a surface that normally does not have these characteristics.

flameout — A condition in the operation of a gas turbine engine in which the fire in the engine goes out due to either too much or too little fuel sprayed into the combustors.

flameout pattern — An approach normally conducted by a single-engine military aircraft experiencing loss or anticipating loss of engine power or control. The standard overhead approach starts at a relatively high altitude over a runway ("high key") followed by a continuous 180 degree turn to a high, wide position ("low key") followed by a continuous 180 degree turn final. The standard straight-in approach with a high rate of descent to the runway. Flameout approaches terminate in the type approach requested by the pilot (normally fullstop).

flammable — Any material that will burn or support combustion.

flammable liquid — Any liquid that gives off easily ignited, combustible vapors.

flange — Any design of a machine, motor, or other mechanism having a ridge that sticks out from the device. It is generally used for attaching something to the device or for connecting two or more devices together.

flaperon — A type of control used on airplanes that serves as both aileron and wing flap.

flapper valve — A type of check valve that allows fluid to flow through it in the direction of flow, forcing the valve off of its seat. It does not allow fluid to flow in the opposite direction since this causes the flapper valve to close.

flapping — In rotorcraft, the vertical movement of a blade about a flapping hinge.

flapping hinge — In rotorcraft, the hinge that permits the rotor blade to flap and thus balance the lift generated by the advancing and retreating blades.

flaps — Hinged portion of the trailing edge between the ailerons and fuselage. In some aircraft ailerons and flaps are interconnected to produce full-span 'flaperons.' In either case, flaps change the lift and drag on the wing.

flare — **1.** A flight maneuver made by pulling back on the control wheel just before touchdown to reduce speed and settle the airplane onto the runway with the least amount of vertical speed. **2.** A signal device that was, at one time, carried in most airplanes in the event of a crash landing. The flare was usually fired into the air from a specially designed gun and signaled the general location of the downed aircraft's position. **3.** A 37° cone-shaped expansion at the end of a piece of tubing. A sleeve and nut are slipped over the tubing prior to using a special cone-shaped flaring tool to form the flare. 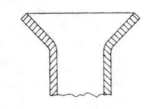 **4.** In rotorcraft, a maneuver accomplished prior to landing to slow down the aircraft.

flareless fitting — A form of fluid line fitting used on some hydraulic lines. Instead of using a formed flare on the end of the tube, a compression sleeve is forced into the tube. A fluid-tight sealing surface is formed when the compression sleeve is tightened onto a recess in the attachment fitting.

flaring — An operation used to expand the end of a length of tubing designed to produce a tight seal when coupled to another unit.

flaring block — A split clamp, usually made of hardened steel, used to hold tubing while it is being flared.

flaring tool — A split block with chamfered holes to clamp the various sizes of tubing while a hardened and polished cone is forced into the end of the tubing to form it against the chamfer.

flash — The thin fin of metal along the sides or around the edges of a forged or upset section. It is caused when metal flows out between the edges of the forging dies.

flash line — A raised line along the boundary of a cast part.

flash plating — A very thin deposit of metal sufficient to give a solid color.

flash point — The temperature at which a fluid will momentarily ignite (flash), but not sustain combustion when a small flame is passed above its surface.

flashback — A malfunction in an oxyacetylene torch in which the gases burn inside the mixing head. Flashback is very dangerous and can cause an explosion unless the gases are immediately shut off at the regulator. This stops the fire inside the torch from burning back through the hoses to the supply tanks.

flasher mechanism — An automatic electrical switching device used for the flashing operation of lights.

flashing off — The drying process of a finish to which solvents have been added for proper spray paint viscosity. Although the surface feels dry to the touch, the film is not completely dry until the proper cure time has been established.

flashing the field — A procedure in which a battery is momentarily connected to the field coil of an aircraft DC generator. Current flows through it for a few seconds to make a permanent magnet of the field frame. This process restores the residual magnetism.

flashover — A condition inside the distributor of a high-tension magneto in which the spark jumps the air gap to the wrong electrode. This may be caused by moisture inside the distributor or by a dirty distributor block.

flash-resistant — Not susceptible to burning violently when ignited.

flat file — A file slightly tapered toward the point in both width and thickness. Cuts on all sides. Double-cut on both sides and single-cut on both edges.

flat lacquer — Any lacquer that dries with a non-glossy or flat finish.

flat machine tip — Compressor or turbine blade tips that have a constant cross section, as opposed to a squeeler-tip or a shrouded-tip configuration.

flat rating — A current means of referring to "rated thrust" at a specific temperature above Standard Day value.

flat spin — A dangerous flight condition or flight maneuver in which the aircraft is yawing around the vertical axis with a pitch attitude approximately level with the horizon.

flat washer — A flat, thin ring used under the head of a bolt or nut in order to protect the surface of the material from damage. Also referred to as a plain washer.

flat-compounded generator — A generator that has both a series and a parallel winding. The series field is adjusted by a regulator to keep the output voltage of the generator constant from a no-load condition to the maximum load the generator can produce.

flathead pin — A high-strength steel pin with a flat head on one end and a hole for a cotter pin on the other end. Used as a hinge for control surfaces or for attaching a cable to a control horn. Also referred to as a clevis pin. Clevis pins are designed to take shear loads only.

flathead rivet — An AN442 rivet used for internal structure where the head of the rivet will not be exposed to the airstream. Flathead rivets are usually driven with an automatic riveting machine.

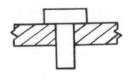

flatnose pliers — Pliers with deep, square jaws and a firm hinge, used to produce a sharp, neat bend in sheet metal and to make flanges along the edge of a part.

fleet weight — The average weight of several aircraft of the same model and with the same equipment. This weight may be used for weight and balance calculations by FAR Part 121 and 135 operators.

flex hose — The colloquial term for flexible tubing used in an aircraft plumbing system to allow relative movement between the two ends of the hose.

flexibility — A material characteristic that allows it to be repeatedly bent, stretched, or twisted within its elastic limits and still return to its original condition each time the bending, stretching, or twisting force is removed.

flexible control cable — A steel aircraft control cable consisting of seven strands of steel wire, each strand having seven separate wires. Also referred to as seven-by-seven cable.

flight — **1.** Travel through the air. An aircraft is considered to be in flight from the time it airplane departs from the ground until it lands. **2.** To take flight physically, students may develop symptoms or ailments that give them acceptable excuses for avoiding lessons. More frequent than physical flights are mental flights, or daydreaming. Mental flight provides a simple escape from problems.

flight assist — Help given to a pilot when the pilot takes the precaution of requesting assistance from Air Traffic Control (ATC). A flight assist report (FAA Form 7230-6) is filed by ATC personnel in order to help other pilots learn from the incident. This form is not used to initiate enforcement action.

flight check — A call-sign prefix used by FAA aircraft engaged in flight inspection/certification of navigational aids and flight procedures. The word "recorded" may be added as a suffix; e. g, "Flight Check 320 recorded" to indicate that an automated flight inspection is in progress in terminal areas. (See Flight Inspection) (Refer to AIM)

flight control surfaces — The movable airfoils used to change the attitude of the aircraft in flight.

flight controller — The command unit of an autopilot system. It is manually operated to generate signals that cause the aircraft to climb, descend, or perform coordinated turns.

flight deck — The area in an aircraft that houses all of the occupants who fly the aircraft, along with all of the controls used in flight. It includes the flight stations for the pilot, copilot, flight engineer, navigator, and radio operator as required.

flight director system — A form of automatic flight control in which all of the information is displayed to the pilot rather than being used to actuate control servos.

flight engineer — The member of the flight crew responsible for the mechanical operation of the aircraft in flight.

flight following — A radar traffic information service which routinely provides traffic information for IFR aircraft. However, when available, VFR aircraft may request flight following and be alerted, by ATC, to air traffic which is relevant to the flight.

flight following — See traffic advisories.

flight idle — Engine speed, usually in the 70% to 80% range, for minimum flight thrust.

flight information region — An airspace of defined dimensions within which Flight Information Service and Alerting Service are provided.
a. Flight Information Service — A service provided for the purpose of giving advice and information useful for the safe and efficient conduct of flights.
b. Alerting Service — A service provided to notify appropriate organizations regarding aircraft in need of search and rescue aid, and to assist such organizations as required.

flight information service — A service provided for the purpose of giving advice and information useful for the safe and efficient conduct of flights.

flight inspection — Inflight investigation and evaluation of a navigational aid to determine whether it meets established tolerances. (See Navigational Aid) (See Flight Check)

flight instructor refresher course (FIRC) — An educational seminar for flight instructors, which consists of ground training or flight training or a combination of both. The FIRC must be completed within the 3 calendar months preceding the expiration month of the current flight instructor certificate, and usually consists of at least 16 hours of ground and/or flight training.

flight level — A level of constant atmospheric pressure related to a reference datum of 29.92 inches of mercury. Flight levels are stated in three digits, representing hundreds of feet. For example, flight level 250 represents a pressure altitude of 25,000 ft.

flight line — 1. The area of an airfield where airplanes are parked. Also referred to as a ramp or tarmac. 2. A term used to describe the precise movement of a civil photogrammetric aircraft along a predetermined course(s), at a predetermined altitude, during the actual photographic run.

flight management system — A computer system that uses a large data base to allow routes to be pre-programmed and fed into the system by means of a data loader. The system is constantly updated with respect to position accuracy by reference to conventional navigation aids. The sophisticated program and its associated data base insures that the most appropriate aids are automatically selected during the information update cycle.

flight management system procedure — An arrival, departure, or approach procedure developed for use by aircraft with a slant E (/E) or slant F(/ F) equipment suffix.

flight manual — Approved information that must be carried in an airplane. This includes the speeds, engine operating limits, and any other information vital to the pilot.

flight path — A line, course, or track along which an aircraft is flying or intended to be flown.

flight plan — Specified information relating to the intended flight of an aircraft that is filed orally or in writing with an FSS or an ATC Facility.

flight plan area — The geographical area assigned by regional air traffic divisions to a flight service station for the purpose of search and rescue for VFR aircraft, issuance of notams, pilot briefing, in-flight services, broadcast, emergency services, flight data processing, international operations, and aviation weather services. Three letter identifiers are assigned to every flight service station.

flight recorder — A general term applied to any instrument or device that records information about the performance of an aircraft in flight or about conditions encountered in flight. Flight recorders may make records of airspeed, outside air temperature, vertical acceleration, engine RPM, manifold pressure, and other pertinent variables for a given flight Housed in a crash-proof container, the flight recorder is used to determine the probable cause of any accident the aircraft should be involved in..

flight review — An industry-managed, FAA monitored currency program designed to assess and update a pilot's knowledge and skills.

flight service stations — Air traffic facilities which provide pilot briefing, enroute communications and

VFR search and rescue services, assist lost aircraft and aircraft in emergency situations, relay ATC clearances, originate Notices to Airmen, broadcast aviation weather and NAS information, receive and process IFR flight plans, and monitor NAVAIDs. In addition, at selected locations, FSSs provide Enroute Flight Advisory Service (Flight Watch), take weather observations, issue airport advisories, and advise Customs and Immigration of transborder flights.

flight simulator — A device that is a full-size aircraft cockpit replica of a specific type of aircraft, or make, model, and series of aircraft; includes the hardware and software necessary to represent the aircraft in ground operations and flight operations; uses a force cueing system that provides cues at least equivalent to those cues provided by a 3 degree freedom of motion system; uses a visual system that provides at least a 45 degree horizontal field of view and a 30 degree vertical field of view simultaneously for each pilot; and has been evaluated, qualified, and approved by the Administrator.

flight standards district office (FSDO) — An FAA field office serving an assigned geographical area. Its staff of flight standards personnel serves the aviation industry and the general public on matters relating to the certification and operation of air carrier and general aviation aircraft. Activities include general surveillance of operational safety, certification of airmen and aircraft, accident prevention, investigation, and enforcement action, among other duties.

flight test — A flight for the purpose of:
a. Investigating the operation/flight characteristics of an aircraft or aircraft component.
b. Evaluating an applicant for a pilot certificate or rating.

flight time — The time from the moment an aircraft first moves under its own power for the purpose of flight until the moment it comes to rest at the next point of landing.

flight training devices (FTD) — A full-size replica of the instruments, equipment, panels, and controls of an aircraft, or set of aircraft, in an open flight deck area or in an enclosed cockpit. A force (motion) cueing system or visual system is not required.

flight visibility — The average forward horizontal distance, from the cockpit of an aircraft in flight, at which prominent unlighted objects can be seen and identified by day and prominent lighted objects can be seen and identified by night.

flight watch — A shortened term for use in air-ground contacts to identify the flight service station providing Enroute Flight Advisory Service; e.g., "Oakland Flight Watch."

flight-path angle — The angle between the flight path of the aircraft and the horizontal.

flint lighter — A flint and steel friction lighter used to ignite an oxyacetylene torch for welding.

FLIP — Department of Defense Flight Information Publications used for flight planning, enroute, and terminal operations. FLIP is produced by the National Imagery and Mapping Agency (NIMA) for worldwide use. United States Government Flight Information Publications (enroute charts and instrument approach procedure charts) are incorporated in DOD FLIP for use in the National Airspace System (NAS).

flip-flop — An electronic device capable of assuming either of two stable states. Term used to describe a communications radio set up with two frequencies entered and the pilot is able to switch back and forth (flip-flop) between them.

float charging potential — A charging potential that can be left connected across the poles of a chemical cell on standby service without damage or destructive overcharge. For a lead-acid battery, this potential is about 14.2 volts for a 12 volt battery.

floated battery — A permanently installed storage battery positioned across the output of a generator. The generator carries the normal electrical load and keeps the battery fully charged at all times.

floating ground — A ground not attached directly to an earth ground. See also chassis ground or equipment ground.

floating waypoint — An airspace fix at a point in space not directly associated with a conventional airway. In many cases, it may be established for such purposes as an ATC metering fix, holding point, RNAV-direct routing, gateway waypoint, STAR origination point leaving the enroute structure, and SID terminating point joining the enroute structure.

float-type carburetor — A fuel metering device which that a float-operated needle valve to maintain a constant fuel metering pressure or head.

flock — Pulverized wool or cotton fibers attached to screen wire used as an air filter. The flock-covered screen is lightly oiled to prevent dirt and dust from entering the engine.

flood valve — A control valve used to direct the flow of extinguishing agent in a CO_2 fire extinguisher system.

flow control — Measures designed to adjust the flow of traffic into a given airspace, along a given route, or bound for a given aerodrome (airport) so as to ensure the most effective utilization of the airspace. (See Quota Flow Control)

flow control valve — A valve that controls the direction or amount of fluid flow.

flow indicator — A device in an oxygen system that provides users with a positive indication that oxygen is flowing.

flow meter — An autosyn electrical transmitter that provides a signal to a cockpit instrument. This indicator shows pounds per hour of fuel flow being consumed by an operating engine.

flow reverser — A mechanical device placed in the tail pipe of a turbojet engine to deflect the exhaust gases forward. This decreases the aircraft landing roll.

flowchart — A diagram that uses symbols connected by lines to indicate the sequence of steps that must be followed in order to achieve a desired end result.

fluctuate — **1.** The swing or oscillation of a dial from low to high. **2.** To continually change or vary in an irregular way.

fluctuating arc — A malfunction in an inert-gas arc welding system caused by improper grounding.

fluid — **1.** A substance, either gaseous or liquid, that will conform to the shape of the container that holds it. A gaseous fluid will expand to fill the entire container, while a liquid fluid will fill only the lower part of the container. **2.** Any material whose molecules are able to flow past one another without destroying itself. Gases and liquids are both fluids.

fluid mechanics — The science and technology of forces produced by fluids.

fluid ounce — A liquid volume equal to $1/16$ liquid pint or 1.8 cu. in.

fluid power — The transmission of force by the movement of a fluid. The best examples are hydraulics and pneumatics.

fluidics — The branch of science that studies the various shapes of ducts to sense, measure, and control physical conditions.

fluidity — The ability of a liquid or gas to flow easily and smoothly.

fluorescent — A substance that will glow or fluoresce when excited. Some types of dye penetrant material use fluorescent dyes, which can be observed under ultraviolet light.

fluorescent finish — A highly light-reflecting aircraft finish.

fluorescent lamp — A lamp that emits light as the result of the glowing of a fluorescent coating on the inside surface of a tube rather than from the glowing of an incandescent filament.

fluorescent penetrant inspection — A form of nondestructive inspection in which a part is thoroughly cleaned and immersed in a vat of penetrating oil. When the part has soaked for a sufficient time, it is removed and the oil is washed from the surface and the part is dried. The part is then covered with a developer that draws the oil out from any crack into which it may have seeped. The part is inspected under ultraviolet light, which causes the crack to appear as a vivid green line.

fluorescent pigment — A paint pigment that can absorb visible or non-visible electromagnetic radiation and release it as energy in a wavelength.

fluorine — A gaseous element with the symbol F and an atomic number of 9.

fluoroscope — An instrument used for observing the internal structure of an object by means of X rays.

flush — To clean, wash, or empty out with a sudden flow of solvent or other cleaning agent.

flush patch — A type of sheet metal repair that leaves a smooth surface maintaining the skin's original contour. The repair is reinforced on the inside and the damaged area is filled with a plug patch.

flush repairs — Metal repairs designed to maintain an aircraft's skin original contour.

flush rivet — A countersunk rivet in which the manufactured head is flush with the surface of the metal when it is properly driven.

flush riveting — Riveting using countersunk-head rivets. Produces a perfectly smooth outside skin.

flute — A groove that is cut or formed in a material.

flutes — Spiral-cut grooves in a drill bit that extend from the point to the shank and allow chips to move from the hole being drilled and for lubrication to flow to the cutting edges.

flutter — The rapid and uncontrolled oscillation of a flight control resulting from an unbalanced surface. Flutter normally leads to a catastrophic failure of the structure.

flux — **1.** A material used in soldering, brazing, or welding to clean the surface of oxides and prevent oxides from forming. Helps ensure good adhesion or fusion of surfaces. **2.** Magnetic lines of force.

flux density — The number of lines of magnetic force per unit area.

flux valve — A special transformer that develops a signal whose characteristics are determined by the unit's position in relation to the Earth's magnetic field. It is part of an Earth inductor compass system.

flux valve spider — The framework around which the three pick-up coils of a flux valve are wound. The spider's highly permeable material accepts the lines of flux from the Earth's magnetic field.

fly — To travel through the air as a pilot or passenger in an aircraft.

fly cutter — A cutting tool used to cut round holes in sheet metal. It is turned by a drill press and the cutting is done by a tool bit held in an adjustable arm.

fly heading (degrees) — Informs the pilot of the heading he should fly. The pilot may have to turn to, or continue on, a specific compass direction in order to comply with the instructions. The pilot is expected to turn in the shorter direction to the heading, unless otherwise instructed by ATC.

fly-by waypoint — A fly-by waypoint requires the use of turn anticipation to avoid overshoot of the next flight segment.

fly-by-wire — A system that moves aircraft controls by use of electro-mechanical servos and controls them by electrical signals from control inputs such as yoke, rudders, or autopilot. Control devices and servos are connected only by electrical wires with no mechanical linkage between them.

flying boat — A form of seaplane whose fuselage serves as the boat hull.

flying wires — Wires used to hold the wings in position during flight and landing loads. There are three types of flying wires: Standard flying wires keep the wing from moving upward during flight and go from the upper outer wing strut fitting to the lower portion of the fuselage. Landing wires go from the top of the inner cabin strut down to the lower outer wing fitting near the location of the bottom attach fittings of the "N" strut and keep the wing from flexing downward on landing loads. Drag wires go from the firewall to the top of the wing strut area and keep the wings from pivoting rearward due to the air drag. Flying wires are also used to reinforce the attachment of the vertical fin and stabilizer to the fuselage.

fly-over waypoint — A fly-over waypoint precludes any turn until the waypoint is overflown and is followed by an intercept maneuver of the next flight segment.

fly-over waypoint — A fly-over waypoint precludes any turn until the waypoint is overflown and is followed by an intercept maneuver of the next flight segment.

flyweights — The L-shaped speed sensing units pivoted on the outer edges of a rotating disc. When rotational speed is high enough, centrifugal force moves them to an angular position. This motion is utilized for various applications, including propeller governors and mechanical tachometers.

flywheel — A heavy wheel or weight used to smooth out the pulsations in a drive system.

flywheel effect — In electronics, a parallel LC circuit that permits a continuing flow of current even though only small pulses of energy are applied to the circuit.

foam rubber — A form of rubber containing millions of tiny air bubbles beaten into the latex before being vulcanized

foamed plastic — A synthetic resin filled with millions of tiny bubbles. Foamed plastics are characteristically light weight and resilient. Also referred to as expanded plastic.

foaming — An undesirable condition in a lubrication system in which oil passing through the engine picks up air, causing tiny air bubbles to form in the oil. Oil foaming reduces the ability of the oil to lubricate and to absorb heat as it should.

Foehn — A warm, dry, downslope wind, the warmness and dryness being due to adiabatic compression upon descent; characteristic of mountainous regions. See adiabatic process, Chinook, Santa Ana.

fog — Cloud consisting of numerous minute water droplets and based at the surface; droplets are small enough to be suspended in the earth's atmosphere indefinitely. (Unlike drizzle, it does not fall to the surface; differs from cloud only in that a cloud is not based at the surface; distinguished from haze by its wetness and gray color.)

fogger oil jet — An air and oil spray mist device used on some engines for lubricating main bearings as opposed to a fluid stream-type oil jet.

foil — A form of metal such as that used in common household aluminum foil that has been rolled out into very thin sheets.

folded fell seam — A type of machine-sewn seam recommended for use in sewing aircraft fabric.

folding — To make sharp, angular bends in sheets of material.

follow-up question — In the guided discussion method, a question used by an instructor to get the discussion back on track or to get the students to explain something more thoroughly.

foot- pound (ft-lb) — **1.** A unit of work: One pound of force moved through a distance of one foot. **2.** A unit of torque: The amount of torque produced when a force of one pound is applied one foot from the pivot point.

foot-candle — The unit of luminance on a surface that is one foot from a uniform point source of light of one candle. Equal to one lumen per square foot. In science, the lux, one lumen per square meter, is in more common usage.

force — The energy applied to an object that attempts to cause the object to change its direction, speed, or motion.

forced exhaust mixer — A long duct design for a turbofan that causes fan air and hot exhaust streams to mix. Used for sound attenuation primarily. Also referred to as mixed exhaust.

forced landing — Any landing necessitated by a malfunction of the aircraft, engine, or improper flight planning.

forceps — A small tool used to grasp or hold things.

foreflap — The first flap in a triple-slotted segmented flap.

foreign object damage (FOD) — Internal gas turbine engine damage that occurs from the injection of foreign objects into the engine. FOD includes ground debris or objects in the air such as birds or flying debris.

foreign particle — A material or particle that can cause serious damage or contamination if it enters a fluid system.

forge — A method of forming metal parts by heating the metal to a plastic state (nearly, but not quite melted) and hammering it to shape.

forge welding — The joining of metal by forging.

forging — The process of forming a product by hammering or pressing. When the material is forged below the recrystallization temperature, it is said to be cold forged. When worked above the recrystallization temperature, it is said to be hot forged.

fork lift — A steerable machine with two long steel fingers that can be positioned under a pallet for lifting and moving heavy loads.

form drag — Skin friction caused by turbulence induced by the shape of the aircraft.

form factor — The ratio of the length of a wire coil to its diameter.

form of thread — In threaded fasteners, the profile of a thread in an axial plane for a length of one pitch.

formal lecture — An oral presentation where the purpose is to inform, persuade, or entertain with little or no verbal participation by the listeners.

formation flight — More than one aircraft which, by prior arrangement between the pilots, operate as a single aircraft with regard to navigation and position reporting. Separation between aircraft within the formation is the responsibility of the flight leader and the pilots of the other aircraft in the flight. This includes transition periods when aircraft within the formation are maneuvering to attain separation from each other to effect individual control and during join-up and breakaway.

a. A standard formation is one in which a proximity of no more than 1 mile laterally or longitudinally and within 100 feet vertically from the flight leader is maintained by each wingman.

b. Nonstandard formations are those operating under any of the following conditions:
1) When the flight leader has requested and ATC has approved other than standard formation dimensions.
2) When operating within an authorized altitude reservation (ALTRV) or under the provisions of a letter of agreement.
3) When the operations are conducted in airspace specifically designed for a special activity. (See Altitude Reservation) (Refer to FAR 91)

former — A frame of light wood or metal that attaches to the truss of the fuselage or wing in order to provide the required aerodynamic shape.

forming — The process of shaping a part.

forming block — A block, usually made of hardwood, around which metal parts are formed.

forming machine — A hand-operated or power-driven machine used to shape sheet metal.

forward bias — The polarity relationship between a power supply and a semiconductor that allows conduction.

forward center of gravity limit — The most forward location allowed for the center of gravity of an aircraft in its loaded condition.

forward current — The amount of current that flows through a semiconductor device when it is forward-biased.

forward fan — Turbofan with the fan located at the front of the compressor. It can be a part of the compressor or a separate rotor.

forward slip — A slide used to dissipate altitude without increasing the glider's speed, particularly in gliders without flaps or with inoperative spoilers.

fossil fuels — Lubricants, fuels, and other petroleum products derived from oil extracted from the Earth.

fouled spark plug — The condition of the spark plug electrodes when they are contaminated with foreign matter. This condition provides a conductive path for the high voltage to leak off to ground rather than building up enough potential to jump the electrode gap.

four-harness satin — A fabric weave. Also called crowfoot satin because the weaving pattern resembles the imprint of a crow's foot. In this type of weave there is a three-by-one interlacing.

four-stroke engine — The four-stroke, five-event cycle consists of five separate mechanical processes occurring in the following order: First, the intake stroke, in which the piston moves inward with the intake valve open. Second, the compression stroke when the piston moves outward with both valves closed. Third, ignition occurs near the top of the compression stroke. Fourth, the power stroke is an inward stroke of the piston with both valves closed, and finally, the exhaust stroke occurs when the piston moves outward with the exhaust valve open. At this point, the cycle begins again.

Fowler flap — Wing flaps that are lowered by sliding from the trailing edge of the wing on a track. Fowler flaps modify the shape of the airfoil and increase the area of the wing.

fractional distillation — A process of oil refining in which the crude oil is heated to the boiling point. As each type of hydrocarbon rises through the distillation column, the heavier components begin to cool and condense into liquids. Lighter fractions such as propane and butane continue to rise and condense later. As a result, different fractions are separated by a process of condensing, or distilling, at different temperatures.

fractions — The various components of a hydrocarbon fuel separated by the distillation process.

frame — A former ring that provides shape and rigidity to a semi-monocoque or monocoque structure.

free balloon — A lighter-than-air, helium-filled device used in weather observations to find the height of the base of the lower layer of clouds.

free electrons — Those electrons so loosely bound in the outer shells of some atoms that they are able to move from atom to atom when an electro motive force is applied to the material.

free fit — A loose fit between moving parts such as a nut that turns easily on the threads of a screw or bolt.

free power turbine — In gas turbine engines, a turbine wheel which drives a power output gearbox rather than a compressor. Found in Turboprop and Turboshaft engines.

free turbine — A turboshaft engine with no physical connection between the compressor and power output shaft.

free water — Liquid water that has condensed out and is no longer entrained in a turbine engine fuel.

free-air wire — A wire completely open to air circulation and not in a bundle.

free-running multivibrator — An oscillator that alternates between two different output voltage levels during the time it is on. The output remains at each voltage level for a definite period of time. Produces a continuous square or rectangular waveform. The free-running multivibrator has two outputs, but no inputs. Also referred to as an astable multivibrator.

freewheeling unit — In rotorcraft, a component of the transmission or power train that automatically disconnects the main rotor from the engine when the engine stops or slows below the equivalent rotor r.p.m.

freeze — **1.** The process in which a liquid changes into a solid due to the removal of heat energy. **2.** The stoppage of a mechanical device due to binding of the mechanical mechanism.

freeze calculated landing time — A dynamic parameter number of minutes prior to the meter fix calculated time of arrival for each aircraft when the TCLT is frozen and becomes an ACLT (i.e., the VTA is updated and consequently the TCLT is modified as appropriate until FCLT minutes prior to meter fix calculated time of arrival, at which time updating is suspended and an ACLT and a frozen meter fix crossing time (MFT) is assigned).

freeze speed parameter — A speed adapted for each aircraft to determine fast and slow aircraft. Fast aircraft freeze on parameter FCLT and slow aircraft freeze on parameter MLDI.

freeze/frozen — Terms used in referring to arrivals which have been assigned ACLT's and to the lists in which they are displayed.

freezing — Change of state from liquid to solid.

freezing drizzle — Drizzle that freezes on contact.

freezing level — A level in the atmosphere at which the temperature is 32° F (0° C).

freezing level chart — A chart depiction of the freezing levels, reported in hundreds of feet.

freezing point — The temperature at which a liquid will change into a solid.

freezing rain — Rain that freezes upon contact with the ground or other objects, such as trees, power lines and aircraft.

French fell seam — A type of machine-sewn seam. Recommended for sewing together sheets of aircraft fabric.

Freon — A fluorinated hydrocarbon compound used as a fire extinguishing agent or a refrigerant for vapor-cycle air conditioning systems. A registered trademark of E.I. DuPont de Nemours & Company.

frequency — The number of waves that pass some fixed point in a given time interval, measured in cycles per second (cps) or Hertz (Hz).

frequency converter — A circuit device that changes the frequency of an alternating current.

frequency meter — An electronic instrument that measures and indicates the frequency of an alternating current on a dial or digital display.

frequency modulation (FM) — A radio wave transmission method where information is transmitted by varying the modulation signal in proportion to the audio signal.

frequency multiplier — Circuit designed to double, triple, or quadruple the frequency of a signal by harmonic conversion.

frequency synthesizer — An electronic circuit used to produce AC with an accurately controlled frequency.

fresh annual inspection — An annual inspection recently performed on an airplane. Sometimes this is used as a selling point for an airplane. The skill and integrity of the person performing the inspection determines the value in a purchase situation.

freshening charge — The charge given a dry-charged battery to bring it up to its rated capacity.

fretting — Surface erosion caused by a slight movement between two overlapping parts.

fretting corrosion — Corrosion damage between parts that are allowed to rub together. The rubbing removes the protective oxide films and allows the metals to corrode.

friction — The force that resists the relative motion of two bodies in contact.

friction brake — Any of a number of different mechanisms used with a rotating wheel or shaft in which friction is used to slow its rotation.

friction clutch — A mechanism used to connect a motor to a mechanical load.

friction damper — A rubber insert used to limit excessive movement in a pedestal-type dynafocal engine mount.

friction error, instrument — The error caused by friction in an instrument mechanism.

friction horsepower — The amount of horsepower required to turn the engine against the friction of the moving parts and to compress the charges in the cylinders.

friction loss — The loss of mechanical energy in a device caused by the friction that is incidental to changing mechanical energy into heat.

friction mean effective pressure (FMEP) — The average working pressure within an engine used to overcome friction. IMEP - BMEP = FMEP (IMEP is Indicated Mean Effective Pressure and BMEP is Brake Mean Effective Pressure).

friction measurement — A measurement of the friction characteristics of the runway pavement surface using continuous self-watering friction measurement equipment in accordance with the specifications, procedures and schedules contained in AC 150/5320-12, Measurement, Construction, and Maintenance of Skid-Resistant Airport Pavement Surfaces.

friction stir welding — A new technology welding method that joins materials by plasticizing and consolidating them along the weld line. The process involves a rotating pin forced down into the seam between two sheets of metal. The pin continues rotating and moves forward in the direction of welding. As the pin proceeds, the friction heats the surrounding material producing a plasticized area around the pin As the pin moves along the seam, the plasticized material consolidates behind the pin and cools to form a bond. No melting occurs and the weld is left in a fine-grained condition with none of the oxide and gas entrapment problems of more conventional welds.

friction tape — Cloth, electrical, insulating tape impregnated with a black tar-like material.

friction welding — A method of joining materials by vigorously rubbing one mating surface against another while forcing them together with a large amount of pressure.

friction-lock Cherry rivet — A patented blind rivet made by Cherry Rivet, a division of Textron Inc., in which the stem locks in the hollow shank by friction.

Frise-type ailerons — An aileron having the nose portion projecting ahead of the hinge line. When the trailing edge of the aileron moves up, the nose projects below the wing's lower surface and produces some parasite drag, decreasing the amount of adverse yaw.

front — The boundary between two different airmasses.

front spar — The foremost spar of a multispar wing.

frontal cyclone — A low pressure area and associated counterclockwise winds (Northern Hemisphere) that develops on the polar front and moves west to east as a macroscale eddy embedded in the prevailing westerlies. Also called a frontal low or wave cyclone.

frontal lifting — The lifting of a warm airmass over a relative cold airmass.

frontal wind shear — The change of wind speed or direction per unit distance across a frontal zone.

frontal zone — A narrow region of transition between two airmasses.

frost — Ice crystal deposits formed by sublimation when temperature and dewpoint are below freezing.

frustrum — The portion of a cone from which the top has been removed.

fuel — A substance that, when combined with an oxidizer, will burn and produce heat.

fuel boost pump — An auxiliary electrically operated pump located within a fuel tank to force the fuel from the tank to the engine. Usually a centrifugal-type pump. Provides vapor-free fuel with a slight head pressure to the main pump.

fuel cell — The compartment in an aircraft where engine fuel is stored.

fuel consumption — The actual amount of fuel consumed by an engine under a specified set of conditions. Fuel consumption can be expressed in either pounds per hour or gallons per hour.

fuel control unit — In gas turbine engines, the main fuel scheduling device which receives a mechanical input signal from the power lever and various other signals, such as Pt2, Tt2, etc. These signals provide for automatic scheduling of fuel at all ambient conditions of ground and flight operation.

fuel dump system — A portion of the fuel system of large jet transport aircraft that allows fuel to be dumped in flight. This is sometimes necessary to reduce the weight of the aircraft to below maximum landing weight or to minimize fuel in the event of an emergency landing.

fuel evaporation ice — Ice formed due to the cooling effect of the fuel evaporating after it is sprayed into the induction system of reciprocating engines. This evaporation process causes carburetor and induction system parts to become very cold and allows moisture in the air to condense, collect and freeze on them. This type of ice is most troublesome in float-type carburetors.

fuel flow — The rate at which fuel is consumed by the engine in pounds per hour (PPH) or gallons per hour (GPH).

fuel flowmeter — A cockpit instrument used to indicate the rate of the fuel consumed by the engine during flight.

fuel grade — A classification of aviation gasoline according to its anti-detonation characteristics.

fuel heater — A radiator-like device that has fuel passing through its core. A heat exchange occurs to keep the fuel temperature above the freezing point of water. This keeps entrained water from forming ice crystals and blocking fuel flow.

fuel injection manifold valve — A valve used in a fuel injection system that distributes fuel from the fuel control unit to the various injection nozzles. The valve provides a metering force for conditions of low fuel flow and a positive fuel shut off when the engine is shut down.

fuel injection system — A fuel metering system used on some aircraft reciprocating engines utilizing a constant flow of fuel to the injection nozzles. Fuel injectors are located in the heads of all cylinders just outside of the intake valve. It differs from sequential fuel injection that uses a timed charge of high-pressure fuel sprayed directly into the combustion chamber of the cylinder.

fuel load — That part of the useful load of an aircraft consisting of the usable fuel on board.

fuel manifold — A pipe-like fitting that distributes fuel flow to the individual fuel injection nozzles. The manifold contains one single fuel line when used with single-line duplex nozzles and two lines for dual-line duplex nozzles.

fuel metering device — Any apparatus such as a carburetor, fuel injector, or fuel control unit, that mixes fuel with intake air in the correct proportions and delivers the mixture to the engine.

fuel nozzle — In gas turbine engines, a device used to spray fuel into a combustion liner. The two most common types are the atomizing nozzle and the vaporizing nozzle.

fuel nozzle ferrule — The receptacle in the combustion liner of a gas-turbine engine where the fuel nozzle tip is inserted.

fuel pressure — The pressure of fuel within a fuel system. The most common measurement location is at the point where fuel is delivered to the fuel control unit.

fuel pressure gauge — A gauge that indicates the fuel pressure delivered to the carburetor.

fuel pump — An electrical or engine-driven pump used to provide a positive volume of fuel under pressure to the engine.

fuel remaining — A phrase used by either pilots or controllers when relating to the fuel remaining on board until actual fuel exhaustion. When transmitting such information in response to either a controller question or pilot initiated cautionary advisory to air traffic control, pilots will state the APPROXIMATE NUMBER OF MINUTES the flight can continue with the fuel remaining. All reserve fuel SHOULD BE INCLUDED in the time stated, as should an allowance for established fuel gauge system error.

fuel shut-off valve — A valve in an aircraft fuel system that shuts off all fuel flow to the engine.

fuel siphoning — Unintentional release of fuel caused by overflow, puncture, loose cap, etc.

fuel system — The system that stores fuel and delivers the proper amount of clean fuel at the right pressure to meet the demands of the engine.

fuel tank vent — A vent in a fuel tank that allows the air pressure above the fuel to be the same as the surrounding air.

fuel venting — See fuel siphoning

fuel-air combustion starter — A fuel engine-starting accessory that utilizes a combustion section similar to a turbine engine. Combustion products are exhausted through a turbine connected to a reduction gearbox to create starting torque.

fuel-air mixture ratio — The weight ratio in pounds of the fuel and air that are mixed together creating a combustible mixture to be burned in an engine.

fuel-oil cooler — A heat exchange device that heats the fuel and cools the oil. It is a radiator-like unit. Fuel passes through the cores and oil passes around the cores. The oil flow is controlled by a thermostatic valve that routes the oil through the cooler only when a certain oil temperature is reached. On some engines, no fuel heater is required due to the exchange rate of this oil cooler.

fuel-oil heat exchanger — A heat exchanging device used on turbine engines to take heat from the engine oil and put it into the fuel. It is a radiator-like unit directing fuel through the tubes that pass through the hot engine oil. Heat from the oil raises the fuel temperature, and at the same time, lowers the temperature of the oil.

fulcrum — A point on which a lever is supported, balanced, or about which it turns.

full annealing — A process used to produce a fine grained, soft, ductile metal without internal stresses or strains. To fully anneal a metal, the temperature of the metal is raised to its critical temperature followed by controlled cooling.

full fuel — The amount of fuel in an aircraft when all fuel tanks are filled to the quantity called for in the aircraft specifications.

full oil — The quantity of oil shown as oil capacity in aircraft specifications.

full rudder — The movement of the rudder to its extreme limit.

full throw — The full range of control surface travel.

full-register position — The position of the rotating magnet in a magneto when the poles are fully aligned with the pole shoes of the magneto frame. At this point, the maximum number of lines of flux flow in the frame.

full-rich — That position of the mixture control that allows the maximum amount of fuel to flow to the engine relative to air flow.

full-scale drawing — A drawing of a part that is the same size as the part.

full-wave rectifier — A form of rectifier that inverts one half of the input AC signal and provides a pulsating DC output having twice the frequency of the input alternating current.

fully articulated rotor — A rotor that is attached to a helicopter rotor hub in such a way that the pitch angle of each blade can change, and each blade is free to move up and down and back and forth in its plane of rotation.

fumes — Vaporized liquids.

functional check — A check for proper operation of the aircraft and systems. Required before returning an aircraft to service after an annual or 100-hour inspection.

functional test — A method of testing a system through its normal operating range to determine whether or not it functions properly.

fundamentals of instruction — Includes the learning process, elements of effective teaching, student evaluation and testing, course development, lesson planning, and classroom training techniques.

fungicidal paste — A paste mixed with clear dope and applied as a first coat on cotton. The fungicidal agent soaks into the fibers and prevents the formation of mold or fungus.

fungus spores — The seed of certain fungi that attach to organic materials such as cotton or linen and cause the material to rot.

funnel cloud — A tornado cloud extending downward from the parent cloud but not reaching the ground.

funneling effect — An increase in winds due to airflow through a narrow mountain pass.

furrow — A deep groove.

fuse — An aircraft electrical circuit protection device. It consists of a link of low-melting-point metal that melts and opens the circuit when an excessive amount of current flows through it.

fuse holder — A device mounted to an electrical fuse panel that holds tubular fuses and makes connections to both ends.

fuse link — A strip of low-melting-point metal used in an electrical circuit fuse device to protect a circuit. When excessive current flows through the circuit the fuse link melts and opens the circuit.

fuselage — The area of an airplane aft of the firewall and forward of the empennage. The cabin or cockpit, is located in the fuselage. It may also provide room for cargo and attachment points for other major airplane components.

fuselage stations — Distances measured along the longitudinal axis of an airplane. Represents distances from the datum in inches.

fusible alloy — A filler material that melts at approximately 160°F. Used to prevent kinking during tube bending. The alloy is heated in hot water and poured inside the tubing. Once the alloy has cooled and set, the tube can be bent by hand around a forming block or bender. After the tube is bent, it is reheated in hot water to remove the fusible alloy.

fusible plug — A hollow plug in an aircraft wheel filled with a material having a specific melting point. If the melting point is reached due to brake heat, the filler will melt out and deflate the tire rather than allow the pressure to increase to the point that would cause a blowout.

fusion — The melting together of metal parts.

G

gain — The increase in signal power through a circuit.

galling — Fretting or chafing of a mating surface by sliding contact with another surface or body. The heat friction causes the material from one surface to be welded or deposited onto the other surface, ultimately damaging the surface area.

gallon — **1.** Imperial: A unit of liquid measurement used outside of the United States. One Imperial gallon is equal to 277.4 cu. in., 4.55 liters, or 1.201 U.S. gallons. **2.** U.S.: A unit of liquid measurement used in the United States. One U.S. gallon is equal to 231 cu. in., 128 fl. oz., or 3.785 liters.

galvanic action — Electron flow because of a difference of electrode potential between different substances.

galvanic corrosion — Corrosion due to the presence of dissimilar metals in contact with each other in the presence of an electrolyte such as water.

galvanic couple — Dissimilar metals that produce an electrical voltage when they are both in contact with the same electrolyte. When dissimilar metals are in this condition, one metal forms the anode and the other metal forms the cathode, thereby producing a current between the two metals.

galvanic electricity — The electricity produced by chemical action such as that produced in a dry-cell battery or storage battery.

galvanic grouping — An arrangement of metals in a series according to their electrode potential difference.

galvanic metal electrical series — The hierarchical arrangement of metals in order of their chemical activity. The following list of metals indicates their hierarchical chemical activity, and acts as the anode in any electrolytic action to those that follow:
1. Zinc
2. Cadmium
3. Iron and steel
4. Cast iron
5. Chromium iron
6. Lead-tin solder
7. Lead
8. Tin
9. Nickel
10. Brass
11. Copper
12. Bronze
13. Copper-nickel alloys
14. Monel
15. Silver
16. Graphite
17. Gold
18. Platinum

galvanizing — A method of protecting steel parts from corrosion by dipping them in a vat of molten zinc or by electroplating them. The protection actually comes from sacrificial corrosion of the zinc.

galvanometer — An electrical measuring instrument in which electrical current is measured by the reaction of its electromagnetic field to the field of a permanent magnet.

gamma rays — Electromagnetic radiation that results from nuclear fission.

ganged tuning — A mechanical arrangement that permits the simultaneous tuning of two or more electronic circuits.

gap — The distance between two objects.

garnet paper — An abrasive polishing paper consisting of a sheet of flexible paper with a layer of finely crushed garnet.

gas — **1.** A fluid that assumes the shape of the container it is placed in and fills the container. **2.** The physical condition of matter in which a material takes the shape of its container and expands to fill the entire container. Oxygen and nitrogen are two chemical elements that are gases at normal room temperature and pressure. The air we breathe is a physical mixture of gases, primarily nitrogen and oxygen.

gas generator — The basic power producing portion of a gas turbine engine excluding sections such as the inlet duct, the fan section, free power turbines, and the tailpipe. Each manufacturer designates what is included in the gas generator, but generally it consists of the compressor, diffuser, combustor, and turbine.

gas generator turbine — In gas turbine engines, high pressure turbine wheel(s) which drive the compressor of a turboshaft or turboprop engine.

gas path — The airflow or open portion of the engine front to back where air is compressed, combusted, and exhausted.

gas path analysis (GPA) — A computer analysis of engine parameters on some airliners. It is designed to assist the modular maintenance and on-condition maintenance concepts by giving continuous on-condition data. Also used for predicting engine component airworthiness.

gas storage cylinders — Long bottles of high-strength steel used to store compressed gases.

gas turbine — Engine consisting of a compressor, combustor and turbine, using a gaseous fluid as a working medium and producing either shaft horsepower, jet thrust, or both. The four common types of Gas Turbine Engines are Turbojet, Turbofan, Turboprop and Turboshaft.

gas turbine engine — A heat engine in which burning fuel adds energy to compressed air and accelerates the air through the remainder of the engine. Some of the energy is extracted to turn the air compressor, and the remainder accelerates the air to produce thrust. A portion of this energy can be converted into torque to drive a propeller or a system of rotors for a helicopter.

gas welding — The method of fusing metals together by a flame using gas as fuel. The most common types of gas welding use oxygen and acetylene gas (oxyacetylene) and oxygen and hydrogen (oxyhydrogen).

gaseous — Having the nature or form of gas.

gaseous breathing oxygen — Oxygen that is 99.5% pure and contains practically no water vapor.

gaseous fuel — Any mixture of flammable gases used for fuel.

gas-filled tube — An electron tube with gas introduced into the envelope to produce certain desired operating characteristics.

gasket — The static, stationary seal between two flat surfaces.

gasoline — A volatile, highly flammable liquid mixture of hydrocarbons produced by the fractional distillation of petroleum and used as fuel in internal-combustion engines.

gasoline combustion heaters — Aircraft cabin heaters that burn gasoline from the aircraft fuel tanks to produce the required heat.

gassing, battery — The release of hydrogen and oxygen as a free gas during the charging cycle of lead-acid storage batteries.

gate — **1.** A logic device having one or more inputs and/or outputs. The condition of the inputs determines whether or not a voltage is present at the outputs. **2.** The electrode of a silicon-controlled rectifier or a triac through which the trigger pulse is applied.

gate hold procedures — Procedures at selected airports that require aircraft to hold at the gate or other ground locations whenever departure delays exceed or are anticipated to exceed 15 minutes. Departure sequence is maintained in accordance with initial call-up unless modified by flow control restrictions. Pilots should monitor the ground control/clearance delivery frequency for engine start/taxi advisories or new proposed start/taxi times if the delay changes.

gate-type check valve — A one-way flow valve having a swinging gate or flapper. The gate-type check valve can isolate one of the vacuum pumps in a multi-engine aircraft from the rest of the system in the event of a failure of the pump.

gateway fix — A navigation aid or fix where an aircraft transitions between the domestic route structure and the oceanic route airspace.

gauge — Any of a variety of measuring instruments. Some are used to indicate the amount of air pressure, the depth of a hole or a groove, or possibly the thickness or clearance between close-fitting parts of a machine, etc.

gauge pressure — Pressure measured relative to the existing atmospheric pressure. Engine oil pressure and hydraulic pressure are normally measured as gauge pressure. If gauge pressure is measured in pounds per square inch, it is referred to as PSIG (pounds per square inch, gauge). Two other types of pressure that are often used are differential pressure (PSID) and absolute pressure (PSIA). Differential pressure is the difference between two pressures and absolute pressure is a value compared to zero pressure.

gauss — The unit of magnetomotive force. It is equal to one maxwell per cm^2.

gear — A toothed wheel or disc that meshes with another toothed wheel or disc to transmit motion.

gear and pinion mechanism — A mechanical amplifying mechanism consisting of two gears, one being a pinion, which is much smaller than the other. The mechanical advantage of the mechanism is determined by the ratio between the number of teeth on the pinion and the number of teeth on the large gear. Often the large gear is only a portion of the wheel and is called a sector gear.

gear backlash — The measured clearance between the teeth of meshed gears.

gear indicators — Indicators in the cockpit of an airplane having retractable landing gear to inform the pilot of the position of the gear. It will indicate whether they are down and locked, in transit, or up and locked.

gear preload — The pressure with which two gears mate or mesh together.

gear-driven supercharger — An internal, engine-driven supercharger on a reciprocating engine driven from the crankshaft through a gear arrangement.

geared fan — A design that allows a fan to rotate at a different speed than the compressor rotor. The fan being geared down allows for higher speeds in the compressor without creating excessive tip speeds on the fan.

geared fan gas turbine engine — A turbofan engine that uses a set of reduction gears between the first stage of the gas generator compressor and the fan.

geared propeller — A propeller driven from the crankshaft through a series of reduction gears. This allows the engine to operate at an efficient speed while holding the propeller RPM in its efficient range.

gear-type pump — A power-driven fluid pump, usually a constant displacement-type pump, driven by the engine accessory drive and used to pump fluid under pressure. The gear-type pump is made up of two meshed spur gears mounted in a close-fitting housing. Fluid is taken into the inlet side of the housing and fills the space between the teeth of both gears. As the fluid is carried around the housing of the rotating gears to the discharge side of the pump, the gear teeth mesh and the fluid is forced out of the outlet side of the pump.

gel — A jelly-like substance formed by the coagulation of a solution into a solid phase.

gel coat — In composites, a coating of resin, generally pigmented, applied to the mold or part to produce a smooth finish. Considered a nonstructural finish.

gel time — In composites, the period of time from the initial mixing of the reactants of a liquid material composition to the point in time when gelation occurs, as defined by a specific test method.

gelatinous — Having the consistency of gelatin or jelly.

gelled cell battery — A lead-acid battery that has a gelling agent added to the electrolyte to make it non-spillable and retain a high level of electrolyte in the battery.

genemotor — A machine with two windings on a single armature that operates simultaneously as a motor with one of the windings and as a generator with the other. The armature windings are usually different so that the voltage on the generator side is different from the voltage on the motor side and the machine acts as a rotary transformer. Also referred to as a dynamotor.

general aviation — The portion of the aviation industry that covers all of aviation with the exception of military, public-use aircraft, and the airlines.

General Aviation District Office (GADO) — Designated FAA Field Offices staffed to serve the general public and aviation industry on matters pertaining to the certification, maintenance, and operation of general aviation.

general circulation — The wind system that extends over the entire globe; it is a macroscale phenomena with a typical horizontal dimension of 10,000 nautical miles.

generator — A mechanical device consisting of a conductor being turned within a magnetic field. Produces electricity by electromagnetic induction.

generator current limiter — A special high-current fuse capable of carrying momentary current overloads. However, it will melt and open the circuit under current flows that might damage the generator. These are used in generator installations that are not protected by automatic current limiters.

GEO map — The digitized map markings associated with the ASR-9 Radar System.

geodetic construction — A form of aircraft construction in which the stress carrying portion of the skin is made up of a lattice work of thin metal or wood strips.

geographic poles — The poles of the axis about which the Earth rotates. They form true north and true south.

geometric pitch — The distance a propeller should advance in one revolution without any slip.

geostrophic balance — The balance of forces that exists when Coriolis force and pressure gradient force are equal in magnitude but in the opposite direction.

geostrophic wind — The wind that occurs when Coriolis force and pressure gradient are equal and opposite.

German silver — Metals of the copper/nickel/zinc alloy family. Also referred to as nickel silver.

germanium — A grayish-white metallic chemical with a symbol of Ge and an atomic number of 32. Often it is used in the manufacturing of semiconductor devices such as diodes and transistors.

gerotor pump — A form of constant displacement pump using a spur gear driven by the engine and turning inside of an internal tooth gear. The internal tooth gear has one more space than teeth on the drive gear. As the pump rotates, the volume at the inlet port increases while the volume at the outlet decreases, moving fluid through the pump.

Giga — Billion.

Gigacycle — Gigahertz.

Gigahertz — One billion hertz.

gilbert — The unit of magnetomotive force equal to approximately 0.7968 amp-turn.

gill-type cowl flap — A cowl flap used on the trailing edge of each cowling of a horizontally opposed engine. Its purpose is to regulate the flow of air through the engine for cooling.

gimbal — The frame in which a gyro spins. It is designed in such a way that it allows a gyroscope to remain in an upright condition while the base is tilted. Rate gyros use a single gimbal, while attitude gyros are mounted in a double gimbal.

gimlet point — A threaded cone point usually having a point angle of 45° to 50°. It is used on thread-forming screws such as Type A tapping screws, wood screws, lag bolts, etc.

girt bar — Part of the emergency evacuation slide/raft assembly on transport category aircraft. The girt bar enables the slide to deploy when the door is "armed" and allows the door to be opened without deploying when the door is "disarmed."

glacier winds — One of the cold downslope winds. A shallow layer of cold, dense air that rapidly flows down the surface of a glacier.

glareshield — Dark, non-reflective cover of the space between the instrument panel and the windshield. Also used to describe the area surrounding the sides and top of the instrument panel.

glass cloth — An aircraft fabric made from fine spun glass filaments woven into a strong, tough fabric. These fabrics are used for reinforcing plastic resins to mold various types of products.

glass fiber — Filaments of fine spun glass woven into cloth or packed together into a mat used for thermal and acoustical insulation.

glaze — 1. The hard, smooth surface of a finishing system. Glaze must normally be "broken" or roughened before another coat of material will adhere to it. 2. A hard, glass-like surface that forms on the rotating disks of a multiple-disk brake. Glaze forms when the sintered material surface is locally overheated. This slick surface does not produce uniform friction and will cause the brakes to chatter or squeal if not removed.

glide — 1. A slow descent of an aircraft without the aid of the engine. 2. To descend at a normal angle of attack with little or no engine power.

glide path — The path of an aircraft relative to the ground while approaching a landing.

glide path (ICAO) — A descent profile determined for vertical guidance during a final approach.

glide ratio — The ratio of the forward distance the airplane travels to the vertical distance the aircraft descends when operating at low power or without power.

glide slope (GS) — Provides vertical guidance for aircraft during approach and landing. The glide slope/glidepath is based on the following:
a. Electronic components emitting signals which provide vertical guidance by reference to airborne instruments during instrument approaches such as ILS/MLS; or
b. Visual ground aids, such as VASI, which provide vertical guidance for a VFR approach or for the visual portion of an instrument approach and landing.
c. PAR, used by ATC to inform an aircraft making a PAR approach of its vertical position (elevation) relative to the descent profile.

glide slope / glide path intercept altitude — The minimum altitude to intercept the glide slope/path on a precision approach. The intersection of the published intercept altitude with the glide slope/path, designated on Jeppesen Terminal charts by the start of the glide slope/path symbol, is the precision FAF; however, when ATC directs a lower altitude, the resultant lower intercept position is then the FAF.

glidepath — See glideslope.

glidepath [ICAO] — A descent profile determined for vertical guidance during a final approach.

glidepath intercept altitude — See glideslope intercept altitude.

glider — A heavier-than-air aircraft supported in flight by the dynamic reaction of the air against its lifting surfaces. Its free flight does not depend on power generated from an engine.

glideslope — Provides vertical guidance for aircraft during approach and landing. The glide slope/glide path is based on the following:
a. Electronic components emitting signals which provide vertical guidance by reference to airborne instruments during instrument approaches such as ILS/MLS, or
b. Visual ground aids, such as VASI, which provide vertical guidance for a VFR approach or for the visual portion of an instrument approach and landing.
c. PAR. Used by ATC to inform an aircraft making a PAR approach of its vertical position (elevation) relative to the descent profile.

glideslope intercept altitude — The minimum altitude to intercept the glideslope/path on a precision approach. The intersection of the published intercept altitude with the glideslope/path, designated on Government charts by the lightning bolt symbol, is the precision FAF; however, when the approach chart shows an alternative lower glideslope intercept altitude, and ATC directs a lower altitude, the resultant lower intercept position is then the FAF.

gliding angle — The angle between the flight path during a glide and a horizontal axis relative to the ground.

gliding ratio — The ratio of the horizontal distance an aircraft travels while gliding for every unit of vertical distance it descends.

g-load — Gust load, the incremental change in vertical acceleration of an aircraft.

global circulation system — In meteorology, the combination of the general and monsoon circulations.

global navigation satellite systems (GNSS) — An "umbrella" term adopted by the International Civil

Aviation Organization (ICAO) to% encompass any independent satellite navigation system used by a pilot to perform onboard position determinations from the satellite data.

global positioning system (GPS) — A space-base radio positioning, navigation, and time-transfer system. The system provides highly accurate position and velocity information, and precise time, on a continuous global basis, to an unlimited number of properly equipped users. The system is unaffected by weather, and provides a worldwide common grid reference system. The GPS concept is predicated upon accurate and continuous knowledge of the spatial position of each satellite in the system with respect to time and distance from a transmitting satellite to the user. The GPS receiver automatically selects appropriate signals from the satellites in view and translates these into three dimensional position, velocity, and time. System accuracy for civil users is normally 100 meters horizontally.

globules — A tiny ball of liquid often found in suspension with some other liquid.

glow coil igniter — An ignition igniter. Around a pin extending from the body of the igniter is a resistance wire wound into a coil. Direct current causes the coil to become red hot, igniting the fuel/air mixture until the device is operating at a temperature sufficient to maintain the flame. At this point current to the glow coil is automatically turned off.

glow discharge tube — A glass tube with a gas such as neon under low pressure. Two electrodes are embedded in opposite ends of the tube. When a sufficiently high potential difference is applied between the electrodes, the gas will ionize and glow.

glow plug igniter — An igniter that uses a coil of wire heated by high-voltage DC electricity. Air and fuel blowing through the coil is ignited to initiate combustion in the combustor section.

glue — A liquid adhesive capable of holding materials together after drying.

glue blocks — Wood blocks used as backing support when making repairs to a wooden structure. They distribute clamp pressure evenly over the area being glued.

glue joint — Glued wood joints. Two pieces of wood joined using glue rather than a mechanical fastener.

Glyptal — A registered trade name of an insulating varnish used in electrical machinery.

go ahead — Proceed with your message. Not to be used for any other purpose.

go around — **1.** To abort a landing. **2.** Instructions for a pilot to abandon his approach to landing. Additional instructions may follow. Unless otherwise advised by ATC, a VFR aircraft or an aircraft conducting visual approach should overfly the runway while climbing to traffic pattern altitude and enter the traffic pattern via the crosswind leg. A pilot on an IFR flight plan making an instrument approach should execute the published missed approach procedure or proceed as instructed by ATC, e.g., "Go around" (additional instructions, if required).

go/no-go gauge — A measuring gauge consisting of a part having two dimensions: the minimum size and the maximum size. An opening of the correct dimension will allow one side to go, or pass through, and the other dimension will not.

goals and values — A perception factor that describes how a person's perception of an event depends on beliefs. Motivation toward learning is affected by how much value a person puts on education. Instructors who have some idea of the goals and values of their students will be more successful at teaching them.

go-around power or thrust setting — The maximum allowable in-flight power or thrust setting identified in the performance data.

gold — A malleable, ductile, yellow corrosion-resistant chemical element with a symbol of Au and an atomic number of 79. Used on critical electrical contacts because of its resistance to corrosion.

gold leaf — Pure gold rolled into extremely thin sheets.

Gold Seal Flight Instructor Certificate — A flight instructor certificate printed with a distinctive gold seal to recognize excellence in flight training based on a CFI's record of performance. To obtain a gold seal certificate, a CFI must have trained and recommended at least 10 students for practical tests within the previous 24 months, and at least 8 of these students must have passed on their first attempt. A CFI must also hold a ground instructor certificate with an advanced or instrument rating.

goniometer antenna — A fixed-loop antenna used by automatic direction finding equipment, consisting of two coils oriented 90° to each other. It measures the angle between a known reference and the direction from which the radio signal is being received.

gouge — A cut, groove, or hole in a material. Considered to be a defect.

gouging — A furrowing condition in which surface material is displaced or damaged. Usually caused by foreign material between tight fitting, moving parts.

governor — A control that limits the maximum rotational speed of a device.

grade-A cotton — Long-staple cotton fabric with 80 threads per inch in both the warp and fill directions. It is the standard material for covering aircraft structures.

gradient — A consistent rate of change, both increasing and decreasing. This term can be applied to the ascent or descent of an aircraft or changes in temperature, pressure, or concentration levels.

gradient system — A device used to give "artificial feel" to hydraulically boosted flight controls.

grain boundary — The lines in metal that are formed by the surfaces of the grains in the metal.

grains — The individual crystals of a material.

gram (g) — The unit of weight or mass in the metric system. One gram equals $^1/_{1,000}$ kg or about 0.035 ounce.

granular — Containing or consisting of grains or granules.

graph — A pictorial presentation of data, equations, and formulas.

graphic plan display (GPD) — A view available with URET CCLD (User Request Evaluation Tool Core Capability Limited Deployment) that provides a graphic display of aircraft, traffic, and notification of predicted conflicts. Graphic routes for Current Plans and Trial Plans are displayed upon controller request.

graphite — **1.** A soft, black form of carbon that usually has a greasy feel. Graphite is commonly used as a dry lubricant. Graphite is also known as black lead and is used in making pencils. **2.** A carbonized fiber used as a reinforcement. The graphitization is accomplished by heating the carbon fiber to temperatures up to 5400°F. See also carbon fiber and carbon/graphite fiber.

graupel — White, round or conical ice particles 1/8" to 1/4" in diameter. They often form as a thunderstorm matures and indicate the likelihood of lightning. Also referred to as soft hail or snow pellets.

gravitational acceleration — The acceleration of a free-falling object caused by the Earth's gravitational pull. The acceleration rate of a freely falling object is 32.2 ft./second^2, or 980.7 cm./second^2.

gravity — **1.** The force of attraction between any two objects containing mass. This force is proportional to the mass of the objects. Large objects exhibiting forces of gravity are the Earth, moon, and planets. **2.** One of the four main forces acting on an aircraft. Equivalent to the actual weight of the rotorcraft. It acts downward toward the center of the earth.

gravity waves — A small-scale wave of air moving in vertical oscillations caused by gravity. Occurring in a stable atmosphere gravity plays the major role in forcing the air parcels to return to their equilibrium level.

great circle — The largest circle which can be drawn on the earth's surface. The shortest distance between two points on a sphere.

green run — The first run of a new or freshly overhauled engine.

greenhouse effect — The capture of terrestrial radiation by certain atmospheric gases. These gases are commonly called greenhouse gases.

Greenwich mean time (GMT) — The time at the 0° meridian located at the Royal Observatory, London, England. Also referred to as Zulu time.

greige Dacron — A synthetic, polyester fabric in its natural condition as it comes from the loom.

grid — **1.** The electrode of a vacuum tube where the signal is applied. **2.** The framework of a plate in a lead-acid battery cell. It is made of lead and antimony with the actual plate material (spongy lead or lead peroxide) attached. **3.** The electrode in an electron tube between the cathode and the anode. It is used to control the amount, shape, and velocity of the electron stream between the cathode and the anode.

grid minimum off-route altitude (grid MORA) — An altitude derived by Jeppesen or provided by State Authorities. The Grid MORA altitude provides terrain and manmade structure clearance within the section outlined by latitude and longitude lines. MORA does not provide for NAVAID signal coverage or communication coverage. 1.)Grid MORA values derived by Jeppesen clear all terrain and manmade structures by 1000 feet in areas where the highest elevations are 5000 feet MSL or lower. MORA values clear all terrain and manmade structures by 2000 feet in areas where the highest elevations are 5001 feet MSL or higher. When a Grid MORA is shown as "Unsurveyed" it is due to incomplete or insufficient information. Grid MORA values followed by ą denote doubtful accuracy, but are believed to provide sufficient reference point clearance. 2.) Grid MORA (State) altitude supplied by the State Authority provides 2000 feet clearance in mountainous areas and 1000 feet in nonmountainous areas.

grind — The process of removing metal from a part with an abrasive stone or wheel.

grinder — A machine with an abrasive wheel used to remove material from a part.

grinding wheel — An abrasive wheel used on grinders to remove excess material.

grip length — The length of the unthreaded shank of a bolt. Also, the length of a blind rivet between the manufactured head and the maximum extent of the pulled head. It is the maximum thickness of material that can be joined by a fastener.

grip range — The difference between the maximum and minimum thickness of material that may be joined by a fastener.

grit blasting — A process for cleaning metal in which abrasive materials such as sand, rice, baked wheat, plaster pellets, glass beads, or crushed walnut shells are forcefully blown onto the part's surface.

grommet — **1.** A metal or plastic eyelet used for reinforcing holes in aircraft fabric. **2.** A small ring of metal, rubber, or plastic used as a fairlead and protector for tubing or wire going though a hole in a metal structure.

grooved surface — A shallow, smooth channel wider than a scratch resulting from wear caused by concentrated contact stress. Abnormal relative movement between contact surfaces or by foreign material on contact surfaces creates the contact stress. The parts usually affected include cylinder barrels, valve faces, and oil seal outer sleeves.

gross thrust — The thrust developed by an engine, not taking into consideration any pressure of initial air mass momentum. Also referred to as static thrust (Fg).

gross weight — The total weight of a fully loaded aircraft including the fuel, oil, and cargo it is carrying.

ground — **1.** A reference point for voltage measurement in an electrical circuit. **2.** To connect a part or component to the electrical ground (normally the airframe).

ground clutter — A pattern produced on the radar scope by ground returns which may degrade other radar returns in the affected area. The effect of ground clutter is minimized by the use of moving target indicator (MTI) circuits in the radar equipment resulting in a radar presentation which displays only targets which are in motion.

ground communication outlet (gco) — An unstaffed, remotely controlled, ground/ground communications facility. Pilots at uncontrolled airports may contact ATC and FSS via VHF to a telephone connection to obtain an instrument clearance or close a VFR or IFR flight plan. They may also get an updated weather briefing prior to takeoff. Pilots will use four "key clicks" on the VHF radio to contact the appropriate ATC facility or six "key clicks" to contact the FSS. The GCO system is intended to be used only on the ground.

ground controlled approach — A radar approach system operated from the ground by air traffic control personnel transmitting instructions to the pilot by radio. The approach may be conducted with surveillance radar (ASR) only or with both surveillance and precision approach radar (PAR). Usage of the term "GCA" by pilots is discouraged except when referring to a GCA facility. Pilots should specifically request a "PAR" approach when a precision radar approach is desired or request an "ASR" or "surveillance" approach when a nonprecision radar approach is desired.

ground crew — The people who maintain, service, and prepare the aircraft before and after flight.

ground delay — The amount of delay attributed to ATC, encountered prior to departure, usually associated with a CDT program.

ground effect — The condition of slightly increased air pressure below an airplane wing or helicopter rotor system that increases the amount of lift produced. It exists within approximately one-half wing span or one-half of the rotor diameter from the ground.

ground fog — In the United States, a fog that is generally less than 20 feet deep

ground idle — A gas turbine engine speed usually in the 60% to 70% of the maximum RPM range, used as a minimum thrust setting for ground operations.

ground loop — **1.** The sudden reversal of direction of travel on the ground of an airplane having a tailwheel-type landing gear. The center of gravity swings around ahead of the wheels. **2.** An undesirable flow of electrical current through the braid around a shielded wire.

ground plane — The reflector used in a quarter-wave radio antenna. Serves as an additional quarter-wave element.

ground potential — The zero potential (no voltage difference) of electrical circuits.

ground power unit — A small gas turbine whose purpose is to provide either electrical power, air pressure for starting aircraft engines, or both. A ground unit is connected to the aircraft when needed. Similar to an aircraft installed auxiliary power unit.

ground resonance — On rotorcraft, an aerodynamic phenomenon associated with fully-articulated rotor systems. Ground resonance develops when the rotor blades move out of phase with each other and cause the rotor disc to become unbalanced. This condition only occurs when the rotorcraft is in contact with the ground and can cause the craft to self-destruct in a matter of seconds. At high RPM, immediate lift-off will stop the resonance. At low RPM, immediate closing of the throttle and lowering of the collective will stop the resonance.

ground return electrical circuit — An electrical circuit that uses the structure of the aircraft as one of the conductors in the circuit.

ground speed — The speed of an aircraft relative to the surface of the earth.

ground support equipment (GSE) — Equipment separate from the aircraft but used in direct support to facilitate maintenance. GSE can include such items as engine hoist, auxiliary power units, testing equipment, compressed air units, etc.

ground visibility — Prevailing horizontal visibility near the earth's surface as reported by the United States National Weather Service or an accredited observer.

ground wave — The portion of a radio wave that travels to the receiver along the surface of the Earth.

ground-adjustable propeller — A propeller with a pitch that can be adjusted and locked on the ground when the engine is not operating but cannot be changed during flight.

ground-boost engine — An engine that, because of supercharging (including turbocharging), can develop more power at sea level than it could without the supercharging.

ground-controlled landing approach — A directed approach to landing through instructions provided by a ground controller The controller watches the aircraft on a radar scope showing the relative position of the aircraft to the glide slope and its horizontal position. The controller gives the pilot instructions necessary for keeping the aircraft on its intended path until it lands.

grounded — To declare an aircraft or airman unfit for flight.

groundspeed — Speed of the aircraft in relation to the ground.

group task — Part of cooperative, or group learning. Each activity your students engage in is known as a group task.

growler — Test equipment used to check generator and starter armatures for shorts. The growler forms the primary of a transformer and the armature forms the secondary. Shorts show up since they cause vibration in a thin piece of metal, such as a hacksaw blade, held over the armature.

guarded switch — A switch protected against accidental movement by having a guard or shroud located directly over the switch. The guard must be raised before the switch can be actuated.

gudgeon pin — British for wrist pin. See wrist pin.

guide vanes — In a turbine engine, stationary airfoil sections positioned radially around the inside of the engine. The airfoils direct the flow of air or gases from one major part to another.

guided discussion — An educational presentation where the topic to be covered by a group is introduced and the instructor participates only as necessary to keep the group focused on the subject.

guided discussion method — An educational presentation typically used in the classroom where the topic to be covered by a group is introduced and the instructor participates only as necessary to keep the group focused on the subject.

guncotton — Nitrocellulose. A nitrated cotton fiber used in nitrate dope.

G-unit — The unit of acceleration as a measure of the force of gravity. One G-unit is the attraction of gravity for a body.

Gunk — A chemical degreaser used for loosening grease and soft carbon from the surface of metal parts.

gusset — A small reinforcing member used to support the corners of a structure.

gust — A sudden brief increase in wind; according to U.S. weather observing practices, gusts are reported when the variation in wind speed between peaks and lulls is at least 10 knots.

gust front — The sharp boundary found on the edge of a pool of cold air that is fed by the downdrafts and spreads out below the thunderstorm. A gust front is the key to the long life of a multicell thunderstorm.

gust lock — Locks used to prevent controls from being damaged by wind gusts while an airplane is parked on the ground. External locks are positioned between the movable surfaces and the fixed aircraft structure. They are usually painted red and will have a warning streamer to indicate their installation. Internal control surface locks are set in the cockpit.

gustnadoes — A tornado-like vortex that sometimes occurs near gust fronts and the edge of a downburst.

guttered surface — Severe erosion confined to narrow areas in the direction of the gas flow. One cause is improper valve seating that allows escape of combustion gases through a narrow area. Areas affected are valve seats, cylinder heads, pistons, and valve and spark plug inserts.

Guy-Lussac's law — The volume of gas is directly proportional to the absolute temperature, provided the pressure is held constant.

gyro — Short for gyroscope. A wheel or disk mounted to spin rapidly about a specific axis while rotating freely about all axes. The gyroscope exhibits rigidity in space; therefore, it remains in the same relative position to space even as the vehicle it is mounted in changes position.

gyro horizon — An attitude gyroscopic instrument that indicates rotation about the pitch and roll axes.

gyrocompass — A navigational instrument that uses a gyroscope as a stable reference to keep the compass from oscillating.

Gyrocopter — Trademark applied to gyroplanes designed and produced by the Bensen Aircraft Company.

gyrodyne — A rotorcraft whose rotors are normally engine-driven for takeoff, hovering, and landing, and forward flight through part of its speed range. Its means of propulsion, consisting usually of conventional propellers, is independent of the rotor system.

gyroplane — A rotorcraft whose rotors are not engine-driven, except for initial starting, but are made to rotate by action of the air when the rotorcraft is moving; and whose means of propulsion, consisting usually of conventional propellers, is independent of the rotor system.

gyroscopic precession — An inherent quality of rotating bodies, which causes an applied force to be manifested 90° in the direction of rotation from the point where the force is applied.

H

hacksaw — A hand-operated metal cutting saw with narrow, replaceable blades. The replaceable blade is held in the hacksaw frame under tension.

hailstones — Precipitation formed by drops of water carried by upward currents of wind inside a cumulonimbus cloud to a level where the temperature is low enough to freeze the drops into ice. When the hailstones are too heavy to be carried in the cloud, they fall to the ground.

hairline — A very thin line.

hairline crack — A nearly imperceptible crack on the surface of a piece of material.

hairspring — A flat, coiled spring used in aircraft instruments as either a calibrated restraint or a preloading device for the gears.

halation — A distortion on a cathode-ray tube that shows up as a blurred area around an image.

half hitch — A knot used for lacing wire bundles.

half life — One half-life is the measure of the time a particular radioactive material loses one half of its radioactivity. At each half-life interval, the material loses one half more of its remaining life and so on.

half view — An aircraft drawing that shows only one-half of a symmetrical view. Center lines and break lines are used to show that there is more of the object than is shown in the drawing.

half-duplex communication — Communication in which signals can be sent in one direction at a time, but cannot be sent in both directions at the same time.

half-life inspection — A jet engine inspection required under conditions of warranty, completed at half the time between overhaul (TBO) interval. It includes primarily a hot section disassembly, inspection, and repair as necessary.

half-round file — A hand file that is flat on one side and curved on the other.

half-section — A view in which the cutting plane extends only halfway across the object, leaving the other half of the object as an exterior view.

half-wave radio antenna — An antenna with an electrical length that is approximately one half of the wavelength of the frequency for which the antenna is tuned.

half-wave rectifier — A rectifier that changes AC into pulsating DC using one diode and producing only one half of the AC wave in its output.

Hall-effect generator — An electronic device used to measure the intensity of a magnetic field. Converts magnetic field to a low-level electrical signal.

halo — A circle of light that appears around the moon when seen through a thin layer of cirrostratus clouds. The halo appearance is caused by tiny ice crystals in the cloud that scatter the light passing through it.

halogen — One of the five chemical elements (fluorine, chlorine, bromine, iodine, and astatine) in Group VII of the periodic table of chemical elements. Used in some fire extinguishing systems.

Halon — Low-toxicity, chemically stable compounds used in fire protection systems. In aviation, Halon 1301 is most commonly used. The benefit of using Halon fire extinguishers is that they do not leave liquid or solid residues when discharged; therefore, they are preferred for sensitive areas and can be used in the presence of humans, which is important in closed areas. Halons are being phased out since they are a chlorofluorocarbon (CFC) and damage the ozone layer.

hammer — A hand-tool consisting of a heavy head and a handle. It is used for pounding, driving, or shaping.

hammer welding — A forge welding in which the edges of two pieces of metal are heated red-hot and then fused by beating with a hammer.

hand drill — A hand-operated, eggbeater-type tool used to turn a twist drill.

hand driving — A method of forming rivets in which the head is driven with a hand set and hammer, and the shank is bucked with a bucking bar.

hand file — A hand-operated cutting tool made of high-carbon steel and fitted with rows of very shallow teeth extending diagonally across the width of the tool.

hand forming — The process of shrinking, stretching, or forming sheet metal by using soft-faced mallets or hammers to force the metal down against suitable forming blocks or into dies.

hand inertia starter — An inertia starter for reciprocating engines in which the flywheel is brought up to speed by a hand-operated crank.

hand lay-up — In composites, assembling layers of reinforcement by hand. This includes the working the resin into the fabric, as well as using a pre-preg fabric.

hand pump — A pump operated by hand to create a flow of fluid.

hand rivet set — A rivet set that can be clamped into a vise to hold the manufactured head of the rivet while the shank is upset with a hammer and a flat punch.

hand snips — Compound-action hand shears used for cutting sheet metal. They normally come in sets of three: one that cuts to the left, one that cuts to the right, and one that cuts straight.

hand tools — A general name for all of the hand-operated tools used in the performance of maintenance.

hand-bending tools — A hand-operated, tube-bending tool consisting of a clamp, a radius block, and a sliding bar. It is used to bend thin-wall aluminum alloy or copper tubing in such a way that it does not collapse the tube.

handbook — A manual that describes simple operations or a system of operations. A handbook normally does not contain specific detailed information on the maintenance of such systems.

hand-cranked inertia starter — A starter that uses a hand crank to store energy in a spinning flywheel. The crank is geared to the flywheel through a high-ratio gear system so that the flywheel can be spun at a high speed. The flywheel is coupled by a clutch to the crankshaft of the engine for starting.

handoff — An action taken to transfer the radar identification of an aircraft from one controller to another if the aircraft will enter the receiving controller's airspace and radio communications with the aircraft will be transferred.

hangar — A building used for the purpose of housing and maintaining aircraft.

hangar queen — Slang, for an aircraft that is frequently non-airworthy and spends a lot of time in the hangar. In the military or commercial aviation, these aircraft are frequently robbed of parts to keep other aircraft airworthy, thus prolonging the status of the hangar queen. Also referred to as an orphaned airplane.

hard — The condition of a material when it is compact, solid, and difficult to bend or deform.

hard landing — An improper landing of an aircraft that has transmitted undue stresses into the structure. The degree of a landing's hardness determines the type of special inspection that will be performed to determine if there is structural damage to the aircraft.

hard X-rays — The degree of the penetration power of an X-ray as determined by the amount of voltage that is applied to the anode of the X-ray tube. The higher the voltage, the greater its penetrating power.

hardboard — A wood composition material manufactured by bonding sawdust and chips of wood with an adhesive under heat and pressure.

hard-drawn copper wire — Copper wire that has been pulled through dies to reduce its diameter to a predetermined size. Pulling the wire also hardens the wire and increases its tensile strength.

hardenability — In a ferrous alloy, the property of metal that determines the depth and distribution of hardness induced by heat treatment and quenching or by cold working.

hardened steel — Steel that has been hardened by a process of heating the steel above its critical temperature then quenching it in brine, water, or oil. Although the hardened steel is very strong, it is also brittle.

hardener — A chemical constituent of an adhesive that promotes its setting and hardening.

hardening — **1.** A heat treatment of metal that increases its brittleness as well as its resistance to abrasion while it decreases its ductility and malleability. **2.** Aluminum: The process of increasing the strength and hardness of aluminum after it has been solution-heat treated. Age hardening takes several days at room temperature until the metal reaches its full hard state. **3.** Steel: A process whereby steel is made hard and brittle by heating it to a temperature above its critical temperature and immediately quenching it in water or oil.

hard-facing — A process of welding, plating, or spraying a hard material such as carbide on the surface of a tool to increase its hardness and to keep the tool from wearing.

hardness — The property of a metal that enables it to resist penetration, wear, or cutting actions.

hardness test — An evaluation of the hardness of a material by measuring the depth of penetration of a specially shaped probe under a specified load. The surface hardness of aluminum alloy parts such as brake housings and wheels can be measured to determine whether or not their heat treatment has been affected by overheating.

hardware — The nuts, bolts, screws, rivets, etc. necessary for assembling parts.

hardwood — A wood with compact texture.

harmonic — A frequency of vibration that is an even multiple of the fundamental of another vibration frequency.

harness satin — A weaving pattern producing a satin appearance. See also eight-harness satin and four-harness satin.

Hartley oscillator — An electronic oscillator that produces its feedback through a tapped inductor.

have numbers — Used by pilots to inform ATC that they have received runway, wind, and altimeter information only.

hazardous attitudes — Studies have identified five hazardous attitudes that can interfere with a pilot's ability to make sound decisions and exercise authority properly. The five hazardous attitudes are anti-authority, impulsivity, invulnerability, macho, and resignation.

hazardous inflight weather advisory service (HIWAS) — Continuous recorded hazardous inflight weather forecasts broadcasted to airborne pilots over selected VOR outlets defined as an HIWAS Broadcast Area.

hazardous weather information — Summary of significant meteorological information (SIGMET/WS), convective significant meteorological information, (convective SIGMET/WST), urgent pilot weather reports (urgent PIREP/UUA), center weather advisories (CWA), airmen's meteorological information (AIRMET/WA) and any other weather such as isolated thunderstorms that are rapidly developing and increasing in intensity, or low ceilings and visibilities that are becoming widespread which is considered significant and are not included in a current hazardous weather advisory.

haze — Fine dust or salt particles dispersed through a portion of the atmosphere; particles are so small they cannot be felt or individually seen with the naked eye (as compared with the large particle of dust), but diminish the visibility; distinguished from fog by its bluish or yellowish tinge.

heading — The direction in which the longitudinal axis of the airplane points with respect to true or magnetic north. Heading is equal to course plus or minus any wind correction angle.

heads up display (HUD) — A system that displays flight information on a transparent display in the line-of-sight of the pilot so that information can be read without looking down into the cockpit. Newest terminology is Head-up Guidance System (HGS).

head-up guidance system (HGS) — See heads up display.

headwind component — That portion of the wind which acts straight down the runway toward the airplane on takeoff or parallel to the line of flight when airborne.

headwork — Conscious, rational thought process when making decisions.

heat — Energy associated with the motion of molecules within a material. The more heat energy there is in a material, the faster its molecules move.

heat capacity — The amount of heat energy required to raise the temperature of a substance 1°C.

heat dissipation — The loss of heat or the transfer of heat into another object or substance.

heat energy — Energy associated with the motion of the molecules within a substance.

heat engine — Any mechanical device that converts heat energy into mechanical energy. For example, reciprocating and turbine engines are heat engines.

heat exchanger — Any device used to transfer heat from one body to another.

heat lamp — An incandescent lamp that produces a maximum of infrared radiation with a minimum of visible light rays. Used for drying paint or for applying heat to glued parts to decrease curing time.

heat load — The amount of heat that an air conditioner is required to remove from an airplane cabin in order to maintain a constant cabin temperature.

heat of compression — The heat generated when a gas is compressed.

heat pump — Moves heat from one location to another. A heat pump can generally be reversed so that it either adds heat (acts as a heater) or takes heat away (acts as an air conditioner).

heat sink — **1.** A device on which semiconductors can be mounted to absorb the heat that would normally tend to damage them. **2.** A heavy plate of conductive material that will absorb or carry away heat. Especially useful in welding.

heat treatment of a plastic resin — The operation in which a cemented joint in a thermoplastic resin is held at an elevated temperature so the entrapped solvent can diffuse into a greater volume of the resin. This decreases its concentration and increases the strength of the joint.

heat treatment of metals — Any operation in which the physical characteristics of a metal are changed by heating. This includes annealing, hardening, tempering, and normalizing.

heat value — The heat energy available per unit volume of a fuel.

heater — Any device that produces controlled heating.

heating element — An electrical resistance wire that glows red-hot to produce heat.

heat-shrinkable — A quality in a synthetic fiber that allows a fabric to shrink when heat is applied.

heatshrinkable fabric — An inorganic fabric used to cover light aircraft structures. The fabric is sewn and put on the aircraft structure so that it is taut but not tight. After securing the fabric to the structure with a special adhesive, it is shrunk to the correct tautness by ironing it with an ordinary household electric iron or by heating it with a high wattage hair dryer. The material is then given a coating of non-tautening dope.

Heaviside atmospheric layer — A layer of ionized particles that surrounds the Earth. The ionosphere.

heavy (aircraft) — For the purposes of Wake Turbulence Separation Minima, ATC classifies aircraft as Heavy, Large, and Small. Heavy aircraft are capable of takeoff weights of more than 255,000 pounds whether or not they are operating at this weight during any particular phase of flight. See Aircraft Classes

heavy ends — In the fractional distillation process, the last parts of crude petroleum refining that have the highest boiling points.

heavy snow warning — A warning that snowfall may exceed four inches or more in a 12-hour period or six inches in a 24-hour period.

Hecto — One hundred.

hedge-hopping — An aircraft flying very near the Earth's surface yet avoiding obstructions on the ground.

height above airport (HAA) — The height of the Minimum Descent Altitude (MDA) above the published airport elevation. This is published in conjunction with circling minimums.

height above landing — The height above a designated helicopter landing area used for helicopter instrument approach procedures.

height above touchdown (HAT) — The height of the Decision Height or Minimum Descent Altitude above the highest runway elevation in the touchdown zone (first 3,000 feet of the runway). HAT is published on instrument approach charts in conjunction with all straight-in minimums.

height band — In gliders, the altitude range in which the thermals are strongest on any given day. Remaining with the height band on a cross-country flight should allow the fastest average speed.

height gradient — The rate of change of height per unit of distance on a constant pressure chart.

Heliarc — A welding process used extensively on aircraft parts. It is a gas-shielded process to prevent oxidation of the base metal. The two types of Heliarc welding are tungsten inert-gas (TIG) welding and metal inert-gas (MIG) welding.

helical — A line or form that winds around a cylinder or the line of threads on a bolt.

helical potentiometer — A potentiometer with a resistance element made in the form of a spiral. The wiper is moved over the element by turning a multi-turn screw.

helical spline — A spline that winds around a shaft. Helical splines are used to change the linear motion of the device that rides on the splines into the shaft's rotary motion on which the splines are cut.

helical spring — A spring wound in the form of a spiral.

Helicoil — A special helical steel insert screwed into specially cut threads to restore threads that have been stripped out or to provide durable threads in soft castings.

helicopter — A rotorcraft that, for its horizontal motion, depends principally on its engine-driven rotors.

helipad — A small, designated area, usually with a prepared surface, on a heliport, airport, landing/takeoff area, apron/ramp, or movement area used for takeoff, landing, or parking of helicopters.

heliport — An area of land, water, or structure used or intended to be used for the landing and takeoff of helicopters and includes its buildings and facilities, if any.

heliport reference point (HRP) — The geographic center of a heliport.

helium — An inert, gaseous, chemical element with a symbol of He and an atomic number of 2. Used to inflate lighter-than-air aircraft.

henry — The standard unit of inductance. It is the amount of inductance in which a current change of one ampere per second will induce a voltage of one volt. Named for Joseph Henry, an American physicist.

heptane — A liquid hydrocarbon material (C_7H_{17}) having a low critical pressure and temperature and whose detonation characteristics are used in determining the octane rating of aviation gasoline.

heptode — A vacuum tube that has seven active electrodes, including the anode, cathode, control grid, and four other special-purpose grids such as screen grids, suppressor grids, and beam-forming grids.

hermaphrodite calipers — A tool used to scribe lines equidistant from an edge. Consists of a caliper with one pointed leg and one with a hook on the end as with an outside caliper. The hook is placed on the edge of the material, and the point is used to scribe a line.

hermetically sealed — A method of protecting an aircraft instrument by exhausting all of the air from its case and sealing it so that no moisture can get in.

hermetically sealed integrating gyro (HIG) — A gyro mounted in a sealed case with a viscous damping medium. The output is therefore an indication of the total amount of angular displacement of the vehicle in which the gyro is installed rather than the rate of angular displacement.

hertz (Hz) — A unit used for measuring the frequency of vibrations or of AC electricity. It is used for the frequency of any type of repeating cycles of motion. One hertz is equal to one cycle per second. Named for Heinrich Hertz, a German physicist.

Hertz antenna — A half-wave radio antenna.

heterodyne — To mix or beat together two frequencies in order to produce an intermediate frequency.

heterodyne-type frequency meter — A frequency meter that beats (heterodynes) an unknown-frequency signal against a locally-produced signal from a variable-frequency generator. When the beat is of zero frequency, the frequency of the incoming signal can be read from a dial or a digital display showing the frequency of the local generator.

heterogeneous mixture — A mixture composed of dissimilar ingredients.

hexadecimal number system — A number system that uses the base 16. (The decimal system uses the base ten).

hexagon — A figure that has six sides.

hexagon head bolt — A bolt head shaped with six sides (a hexagon).

hexode — An electron tube having six active elements.

HF communications — See high frequency communications.

hidden surfaces — Any surface represented on an aircraft drawing that cannot be seen in a particular view but is represented in outline form with hidden lines.

hierarchy of human needs — A listing by Abraham Maslow of needs from the most basic to the most fulfilling. These range from physical through safety, social, and ego to self-fulfillment.

high — An area of high barometric pressure, with its attendant system of winds; an anticyclone. Also referred to as a high-pressure system.

high altitude checkout — FAR 61.31(f) requires specific ground and flight training for a pilot to act as PIC of a pressurized airplane that has a service ceiling or maximum operating altitude, whichever is lower, above 25,000 MSL. High-altitude checkouts require both ground and flight training. Included in the ground training is a thorough review of the physiological aspects of high-altitude flight. An overview of these effects is contained in Chapter 8 of the Aeronautical Information Manual (AIM) and AC 61-107, Operations of Aircraft at Altitudes Above 25,000 Feet MSL and/or MACH Numbers (Mmo) Greater Than .75. AC 61-107 contains a recommended outline for a high-altitude training program. An instructor needs a logbook endorsement for high-altitude operations to give flight instruction in a pressurized airplane that has a service ceiling or maximum operating altitude above 25,000 feet MSL.

high blower — The high-speed operation of a single-stage, two-speed, internal supercharger system. Usually about 10:1 ratio.

high cycle fatigue — A condition seen as cracking or stretching caused by vibration stresses above the design limit of the engine.

high frequency — The frequency band between 3 and 30 MHz.

high frequency communications — High radio frequencies (HF) between 3 and 30 MHz used for air-to-ground voice communication in overseas operations.

high performance airplane — An airplane with an engine of more than 200 horsepower.

high pressure turbine — In gas turbine engines, the turbine rotor which drives the high pressure compressor in a dual or triple spool axial flow gas turbine engine.

high speed exit — See high speed taxiway.

high speed taxiway — A long radius taxiway designed and provided with lighting or marking to define the path of aircraft, traveling at high speed (up to 60 knots), from the runway center to a point on the center of a taxiway. Also referred to as long radius exit or turn-off taxiway. The high speed taxiway is designed to expedite aircraft turning off the runway after landing, thus reducing runway occupancy time.

high speed taxiway / turnoff (HST) — A long radius taxiway designed and provided with lighting or marking to define the path of an aircraft, traveling at high speed (up to 60 knots), from the runway center to a point on the center of a taxiway. Also referred to as long radius exit or turnoff taxiway. The high speed taxiway is designed to expedite aircraft turning off the runway after landing, thus reducing runway occupancy time.

high speed turnoff — See high speed taxiway.

high strength steel — Steel that has a tensile strength of between 50,000 and 100,000 PSI.

high-bypass turbofan — Turbine engines with 4:1 fan-to-engine bypass ratio or higher. That is, four or more times as much air flows through the fan as through the core engine.

high-bypass turbofan engine — A turbofan engine in which the mass airflow in pounds per second that passes through the fan can be four times or more greater than that which is moved by the gas generator or core of the engine.

high-carbon steel — Steel that contains more than 0.5% carbon.

high-frequency communications — Radio communication at a frequency that is above the upper end of the commercial AM broadcasting band.

high-level language — The language of computer instruction a computer can understand.

high-lift device — Any lift-modifying device such as a slot, slat, or any of the forms of flaps that are used to allow an airfoil to achieve a higher angle of attack before airflow separation occurs.

high-pass filter — An electronic filter that allows AC, above a certain frequency, to pass with little or no opposition.

high-potential ignition lead test — A test performed on the spark plug electrical wires of an aircraft ignition system to see if there is a voltage leak to ground.

high-pressure compressor — The rear section of a dual-spool compressor. Also referred to as an N_2 compressor or high speed compressor.

high-pressure oxygen system — Gaseous oxygen systems whose cylinders carry between 1,000 and 2,000 PSI pressure.

high-pressure system — In gaseous systems, maximum pressure between 1,000 and 2,000 PSI. In liquid systems, it refers to approximately 300 PSI pressures.

high-pressure turbine — The forward most turbine wheels, also called the N_2 turbine or high-speed turbine, that drive the high-speed compressor in a two-spool, axial-flow gas turbine engine.

high-rate discharge — See cold-cranking amps.

high-resistance connection — An electrical connection with excessive resistance. Usually caused by a poor solder joint or loose fastener.

high-speed steel — Alloys of steel that maintain their strength when operating at red-hot temperatures. They are used for metal-cutting tools.

high-speed taxiway exit or turnoff — A wide radius turn-off from the runway to allow an aircraft to turn off at a higher rate of speed than would be possible on a normal right angle turn-off.

high-strength fastener — A fastener with high tensile and shear strengths attained through combinations of materials, work-hardening, and heat treatment.

high-tension magneto — A self-contained magneto ignition system used to provide a high potential voltage to the spark plugs. The magneto consists of a rotating magnet, cam, breaker points, capacitor, and a coil with a primary and a secondary winding. The output of the secondary winding goes to a distributor, then to the spark plugs.

high-voltage igniter plug — An igniter plug utilizing an air gap between its center electrode and its outer casing. Used to start the engine combustion process.

high-voltage ignition system — A main system with a voltage output in the range of approximately 5,000 to 30,000 volts delivered to the igniter plug.

high-wing airplane — A monoplane with a single airfoil mounted on top of the fuselage.

hinge — A fastener that allows one of the connected pieces to pivot with respect to the other.

hinge point — The pivot point about which a control surface or a door hinges.

Hipernik — A magnetic alloy made of 50% iron and 50% nickel.

Hi-Shear rivet — A threadless bolt used for high-speed, high-strength, lightweight construction of an aircraft. A steel pin is held into the structure by an aluminum or mild-steel collar swaged into a groove around the end of the pin.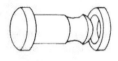

HIWAS area — Hazardous Inflight Weather Advisory Service. Continuous recorded hazardous inflight weather forecasts broadcasted to airborne pilots over selected VOR outlets defined as an HIWAS Broadcast Area.

HIWAS broadcast area — A geographical area of responsibility including one or more HIWAS outlet areas assigned to an AFSS/FSS for hazardous weather advisory broadcasting.

HIWAS outlet area — An area defined as a 150 NM radius of a HIWAS outlet, expanded as necessary to provide coverage.

hold/holding procedure — A pre-determined maneuver which keeps aircraft within a specified airspace while awaiting further clearance from air traffic control. Also used during ground operations to keep aircraft within a specified area or at a specified point while awaiting further clearance from air traffic control.

hold for release — Used by ATC to delay an aircraft for traffic management reasons; i.e., weather, traffic volume, etc. Hold for release instructions (including departure delay information) are used to inform a pilot or a controller (either directly or through an authorized relay) that an IFR departure clearance is not valid until a release time or additional instructions have been received.

hold procedure — A predetermined maneuver which keeps aircraft within a specified airspace while awaiting further clearance from air traffic control. Also used during ground operations to keep aircraft within a specified area or at a specified point while awaiting further clearance from air traffic control.

holding coil — An auxiliary coil in an electrical relay that keeps the relay energized after the current that caused the relay to close has stopped flowing through the main coil.

holding fix — A specified fix identifiable to a pilot by NAVAIDs or visual reference to the ground used as a reference point in establishing and maintaining the position of an aircraft while holding.

holding point [ICAO] — A specified location, identified by visual or other means, in the vicinity of which the position of an aircraft in flight is maintained in accordance with air traffic control clearances

holding procedure — An aircraft maneuver in which a pilot is instructed by air traffic control to maintain a specified air space until given further instruction.

hold-short point — A point on the runway beyond which a landing aircraft with a LAHSO clearance is not authorized to proceed. This point may be located prior to an intersecting runway, taxi-way, predetermined point, or approach/departure flight path.

hold-short position lights — Flashing in-pavement white lights located at specified hold-short points.

hold-short position marking — The painted runway marking located at the hold-short point on all LAHSO runways.

hold-short position signs — Red and White holding position signs located alongside the hold-short point.

hole — 1. The vacancy in the valence structure of an element that will accept an electron from an outside source. 2. The absence of an electron at a point where one might be expected. For most purposes, a hole may be treated as a positive charge. 3. A serious discrepancy, flaw, or weakness. 4. An opening in which something is missing.

hole finder — A tool used in sheet metal work to determine the rivet hole locations to be drilled in a piece of sheet metal so they will match those in the piece of metal being overlapped.

hole punch — A hollow punch resembling a sharp-edged tube. It is used to punch holes in gasket material.

holist — A learning style that focuses on the overall object first and then examines the individual components, using a top to bottom approach.

hollow drill — A drill with a hole through its center making it easier for the drill to be lubricated.

hollow-mill collar cutter — A tool used to remove collar material from pin rivets so the rivet can be tapped out of the work.

home study course — Under FAR 61.35, a home study course may be used to meet the prerequisites for a knowledge test. Home study curriculums may be developed individually by students from material described in the applicable FAA knowledge test guide. Usually, home study courses are designed by pilot schools, colleges and universities, aviation organizations, publishers, or individual ground or flight instructors. The home study course may feature printed material, video, or computer-based training provided on CDs or accessed over the Internet. Regardless of the medium, students must show that the course has been satisfactorily completed and obtain an endorsement from an authorized instructor. Refer to Certification: Pilots and Flight Instructors (AC 61-65), which describes several methods that students can use to show evidence of having satisfactorily completed home study courses.

homebuilt airplane — An aircraft constructed by an amateur builder. Such builders buy kits and/or plans for the aircraft and assemble the airplane over a period of years in their hangars, garages, and even basements. These airplanes are known as homebuilts. An airplane is considered amateur built if the builder constructs at least 51% of the aircraft. Manufacturers of homebuilt kits may not construct more than 49% of the total airplane.

homing — A method of navigating to an NDB by holding a zero relative bearing. The result of homing in a crosswind is a curved course.

hone — A fine abrasive stone used to sharpen cutting tools or create a finished surface.

honeycomb — A material made up of hexagonal-shaped cells. Constructed of thin metal, paper, or plastic and used as a core material for sandwich structures. Named after a bee's honeycomb because of its appearance.

honeycomb shroud ring — Honeycomb material into which a rotating airfoil can cut without degrading its air-sealing function. Usually in the hot section. Also referred to as an abradable shroud.

honing — The process of removing a very small amount of material to produce a smooth finish on a surface or to produce a sharp edge on a cutting tool such as a knife or a chisel.

hook rule — A steel scale with a hook or projection on one end so the rule can be used to measure accurately from the edge of materials.

hook spanner — A semicircular spanner wrench with the handle on one end and a hook on the other. The hook engages notches in the outside circumference of a ring-type nut.

Hooke's law — A law of physics that deals with the relationship between stress and strain in a material. It states that stress in a ductile material is directly proportional to the strain until the limit of elasticity of the material is reached.

hopper — A funnel-shaped container used for storing the abrasive in a sand-blasting machine. The container has an opening in the top for loading and a smaller opening in the bottom for dumping.

hopper-type oil tank — A container within an oil tank used to hold oil diluted with gasoline for cold weather starting. The use of a hopper minimizes the amount of oil that must be diluted.

horizon — The line of sight boundary between the Earth and the sky.

horizontal — Parallel to the Earth's horizon or to the base line of an object.

horizontal needle of the cross-point indicator — The glide slope indicator of the standard ILS indicator (cross point indicator). The horizontal bar is also referred to the horizontal needle and indicates the position of the aircraft on the ILS glideslope.

horizontal pressure gradient force — The force that arises because of a horizontal pressure gradient.

horizontal stabilizer — A fixed horizontal airfoil attached to the rear of the fuselage to provide stability in pitch.

horizontal wind shear — The change in wind direction and/or speed over a horizontal distance.

horizontally opposed engine — A reciprocating engine with the cylinders arranged in two horizontal rows, one on either side of the crankshaft. The cylinders are slightly staggered, with the cylinders in one bank slightly ahead of those in the other bank. Staggering the cylinders allows each piston to be connected to a separate throw of the crankshaft.

horn — A lever or device fastened or connected to a control surface to which an operating cable or rod is attached.

horse latitudes — Approximately 30° to 35° north and south latitudes. An area of high barometric pressure and light, changeable winds.

horsepower (HP) — The standard unit of power used for mechanical measurement. It is equal to 33,000 ft.-lbs. of work done in one minute, or 550 ft.-lbs. of work done in one second. Electrically, it is equal to 746 watts.

horseshoe magnet — A magnet shaped like the letter "U" (or a horseshoe).

hose — A flexible plumbing line used in place of rigid tubing in areas subject to movement or vibration.

hose clamp — A metal clamp used to hold a rubber hose onto a piece of rigid tubing.

hot air muff — A jacket installed around a tail pipe. Air routed through the hot air muff picks up heat by convection through the tail pipe material. This heated air is then routed to the cabin.

hot bond repair — In composites, a repair made using a hot patch bonding machine to cure and monitor the curing operation. Hot bonding equipment typically includes both the heat source and the vacuum source.

hot corrosion — Corrosion occurring in hot sections from a chemical reaction between sulfur in the fuel and salt in the air stream. This condition is more of a problem when operating near salt water.

hot dimpling — A coin dimpling or countersinking of metal for flush rivets or screws. A heating unit heats the metals to prevent cracking around the hole.

hot forming — Working operations such as bending and drawing sheet and plate, forging, pressing, and heading, performed on metal heated to temperatures above room temperature.

hot junction — One end of a thermocouple. When combined with a cold junction, a small current is generated. The same principle is used in many fire detection systems. The cold junction is sometimes referred to as the reference junction.

hot section — The portion of a turbine engine aft of the diffuser where combustion takes place.

hot section distress — Any of the metal deterioration conditions found in the hot section such as warping or creeping.

hot section inspection — An inspection of the hot section of a gas turbine engine.

hot shearing — A method of cutting heavy sheets of magnesium alloys in which the metal is cut while hot. This improves the smoothness of the cut.

hot spark plug — A spark plug with a long-nose insulator in which the heat transferring from the center electrode into the shell has a long path to travel. Hot spark plugs are used in engines that operate relatively cool, and they keep the center insulator hot enough to prevent the accumulation of lead oxides.

hot sparks — Localized areas in the cylinder of an internal combustion engine that are overheated to the point where they become incandescent or glow. They cause pre-ignition.

hot spots — Localized discoloration on hot section parts indicating a breakdown of cooling air or harmful concentration of fuel at that point. This often is the result of a malfunctioning fuel nozzle.

hot stamping — A method of identifying or imprinting plastic materials, cloth, or paper by using heated metal dies.

hot start — In gas turbine engines, a start which occurs with normal engine rotation, but exhaust temperature exceeds prescribed limits. This is usually caused by an excessively rich mixture in the combustor. The fuel to the engine must be terminated immediately to prevent engine damage.

hot streak ignition — An afterburner ignition system in which a stream of raw fuel continues to burn while passing through the turbine section and provides ignition for afterburner fuel supply.

hot valve clearance — The clearance between the valve stem and the rocker arm when all of the engine parts have reached their operating temperature.

hot wire — A wire connected directly to the power source.

hot-tank lubrication system — A gas turbine engine lubrication system in which hot oil returns directly from the engine to the tank without being cooled because the oil cooler is in the pressure portion of the lubrication system.

hot-tank oil system — A lubrication system where the oil cooler is located in the pressure oil subsystem and the scavenge oil returns to the oil tank uncooled.

hot-wire ammeter — A current measuring instrument for measuring high frequency alternating current. The ammeter uses the heating effect of the high frequency current to heat a wire and change its length. As the wire lengthens, it moves a pointer across a dial to show the amount of current flowing through the wire.

hot-wire anemometer — A wind speed indicator that measures the amount of heat the wind removes from a heated wire. The speed of the wind is proportional to the amount of heat removed.

hourmeter — An odometer-type instrument used to measure hours of operating time. When incorporated into a mechanical tachometer, it is accurate only at a specified RPM.

housing — A frame, box or casing that contains a part or mechanism.

hover — The action of a helicopter maintaining a constant position over a selected point. Hovering allows the helicopter to sustain flight with no movement in relation to the ground.

hover check — Used to describe when a helicopter/VTOL aircraft requires a stabilized hover to conduct a performance/power check prior to hover taxi, air taxi, or takeoff. Altitude of the hover will vary based on the purpose of the check.

hover taxi — Used to describe a helicopter/VTOL aircraft movement conducted above the surface and in ground effect at airspeeds less than approximately 20 knots. The actual height may vary and some helicopters may require hover taxi above 25 feet AGL to reduce ground effect turbulence or provide clearance for cargo slingloads.

hovering ceiling — The maximum altitude at which a helicopter can support itself without forward motion.

hovering outside of ground effect — Hovering greater than one rotor diameter distance above the surface. Because induced drag is greater while hovering out of ground effect, it takes more power to achieve a hover out of ground effect.

how do you hear me? — A question relating to the quality of the transmission or to determine how well the transmission is being received.

hub — In helicopters, the part of a propeller or rotor system that attaches to the main driving shaft and to which the blades are fastened.

Huck Lockbolt — A patented, threadless bolt used in the production of aircraft where quickly installed high-strength, lightweight fasteners are required.

HUD — Heads up display. A system that displays flight information on a transparent display in the line-of-sight of the pilot so that information can be read without looking down into the cockpit. Newest terminology is Head-up Guidance System (HGS).

hue — **1.** The gradation of colors. **2.** The characteristics of a color that differentiates between red, blue, or yellow and any of the intermediate colors.

human factors — The study of how people interact with their environments. In the case of general aviation, it is the study of how pilot performance is influenced by such issues as the design of cockpits, the function of the organs of the body, the effects of emotions, and the interaction and communication with other participants in the aviation community, such as other crew members and air traffic control personnel.

human factors related — The phrase "human factors related" more aptly describes an accident since it is not usually a single decision that leads to an accident, but a chain of events triggered by a number of factors. The poor judgment chain, sometimes referred to as the error chain, is a term used to describe this principle of contributing factors in a human factors related accident.

humidifier — A device used to increase the humidity of the air. Humidifiers are primarily used in air-cycle air conditioning systems to increase the comfort within the cabin.

humidity — The amount of water vapor in the air.

hung start — In gas turbine engines, a condition of normal light off but with rpm remaining at some low value rather than increasing to the normal idle rpm. This is often the result of insufficient power to the engine from the starter. In the event of a hung start, the engine should be shut down.

hunting — **1.** An undesirable oscillation above and below a desired value in a control system. **2.** An oscillatory motion of an articulated rotor's blades about the Alpha Hinge caused by coriolis forces.

hurricane — A tropical cyclone in the Western Hemisphere with winds in excess of 65 knots.

hurricane eye — The circular, nearly cloud free region approximately 10 to 20 nautical miles in diameter, located in the center of the storm.

hurricane warning — The warning issued within 24 hours of the arrival of hurricane conditions.

hurricane watch — Issued when hurricane conditions are expected in a particular area within a day or more.

hybrid — In composites, the combination of two or more types of reinforcing materials into the composite structure.

hydraulic booster unit — A unit for moving the flight controls in a large, high-speed aircraft. It is actuated by the normal cockpit controls but greatly amplifies the force the pilot exerts.

hydraulic brake — An aircraft brake operated by means of hydraulic fluid under pressure.

hydraulic filter — A unit that removes foreign particles from the hydraulic system.

hydraulic fluid — A liquid used to transmit and distribute forces to various units being actuated.

hydraulic fuse — A unit designed to stop the flow of hydraulic fluid if a leak occurs downstream of the fuse.

hydraulic lock — A condition that occurs in a reciprocating engine having cylinders below the crankcase. Oil leaks past the piston rings and fills the cylinder with an incompressible fluid. The engine cannot then be rotated without damage.

hydraulic motor — A motor driven by a flow of hydraulic fluid.

hydraulic pump — An engine-driven, electric motor-driven, or hand-operated pump used to move hydraulic fluid through a system.

hydraulic reservoir — A container for the hydraulic fluid supply in an aircraft.

hydraulic system — An aircraft's entire fluid power system, including the reservoir, pump, control valves, actuators, and all of the associated plumbing.

hydraulic valve lifter — The hydraulic units in the valve train of a reciprocating aircraft engine used to automatically adjust for any changes in dimensions of the engine caused by expansion and also to keep the operating clearance in the valve mechanism at zero.

hydraulics — The branch of science that deals with the transmission of power by incompressible fluids under pressure.

hydrocarbon — An organic compound that consists mostly of carbon and hydrogen. The vast majority of our fossil fuels, including gasoline and turbine fuel, are hydrocarbons.

hydrodynamics — The study of forces produced by incompressible fluids in motion.

hydrofoil — An airfoil-shaped plate attached to the bottom of an airplane or boat that lifts the vehicle out of the water by hydrodynamic action when the vehicle is moved through the water at high speed.

hydrogen — A basic element with a symbol of H and an atomic number of 1. In chemical formulas, free hydrogen appears as H_2 because there must be two atoms of hydrogen to form one molecule of free hydrogen gas.

hydrogen bomb — A nuclear weapon that produces heat and light from the fusion of hydrogen atoms.

hydrogen brazing — Braze welding in which hydrogen is used as the fuel gas.

hydrogen embrittlement — A brittle condition caused by the metal absorbing hydrogen while it is being electroplated.

hydrogen fuel — A proposed jet fuel of the future that could be stored as a gas or cryogenic liquid. The present high cost and storage problems prevent its current use.

hydrogen peroxide — H_2O_2. An unstable compound used as an oxidizing agent, antiseptic, and a fuel for rocket engines.

hydrological cycle — The movement of moisture from the earth to the atmosphere and back to the earth again.

hydromechanical fuel control — A fuel control that utilizes hydraulic and mechanical forces to operate its fuel scheduling mechanisms.

hydrometeor — Atmospheric water vapor. Liquid or solid water formation that is suspended in, or falling from, the air, including clouds, fog, ice fog, mist, rain, and hail also any water particles blown by the wind from the Earth's surface.

hydrometer — A device used to measure specific gravity of a liquid. It consists of a weighted float with a long stem in the enlarged glass tube of a syringe. Liquid is pulled up into the tube and the float rides vertically on the surface. The amount the float is submerged is a function of the density of the liquid. The number on the float's stem opposite the liquid level is the specific gravity.

hydroplaning — A condition that exists when landing on a surface with standing water deeper than the tread depth of the tires. When the brakes are applied, there is a possibility that the brake will lock up and the tire will ride on the surface of the water, much like a water ski.

When the tires are hydroplaning, directional control and braking action are virtually impossible. An effective anti-skid system can minimize the effects of hydroplaning.

hydropneumatic — Mechanical equipment that uses both hydraulic and pneumatic forces in order to accomplish its intended purpose.

hydro-ski — A hydrofoil mounted below the hull of a flying boat, which hydrodynamically produces lift by the hydro-ski. As the flying boat begins to move through the water, the hydro-ski helps to lift the hull out of the water.

hydro-sorb — A hydraulic shock absorber used in a bungee shock cord landing gear to prevent rebound.

hydrostatic testing — The method of pressure-testing compressed gas cylinders with high-pressure water rather than a compressible fluid such as air. Water is used for safety reasons.

hydroxide — A chemical compound made up of a metal or a non-metal (acid) base and a negative hydroxyl ion (OH).

hygrometer — An instrument used to determine the amount of moisture in the air.

hygroscopic material — A material such as silica gel that absorbs moisture from the air.

hypergolic — A self-igniting reaction upon contact of the components, without the presence of a spark.

hypersonic — A regime of flight where speeds of Mach 5.00 are exceeded.

hypersonic engine — An engine designed to operate in the speed range above Mach 5. One possibility is the variable cycle engine. This engine functions as a turbojet up to perhaps Mach 3, then as doors close off the compressor inlet, it operates from a ramjet-type duct surrounding the engine.

hypersonic flight — Flight at speeds of Mach 5 or above.

hypersonic flow — Flow at very high supersonic speeds. Mach 5 or above.

hyperventilation — Breathing at such an excess rate that the normal amount of carbon dioxide is depleted from the blood.

hypotenuse — The side of a 90° (right) triangle opposite the right angle.

hypoxia — A lack of sufficient oxygen reaching the body tissues.

hysteresis loop — A graph of the magnetic characteristics of a material.

I

I beam — A structural beam constructed of extruded metal or wood, whose cross section resembles the letter I.

I say again — During communications, this indicates that the message will be repeated.

I'm safe checklist — Personal Checklist — I'm physically and mentally safe to fly; not being impaired by illness, medication, stress, alcohol, fatigue, or eating.

ice — The solid state or condition of water when the temperature is 0°C or 32°F.

ice crystal process — The process by which cloud particles grow to precipitation size. This can only occur where ice crystals and water droplets coexist and the temperature is below 0° C.

ice crystals — Small ice particles that float in the air in clear cold weather. Also referred to as ice needles.

ice fog — Fog composed of minute suspended particles of ice; occurs at very low temperatures.

ice light — A light mounted on an aircraft in such a way that it shines on the leading edge of the wing, allowing the pilot to see the buildup of ice on the wing at night.

ice needles — Small ice particles that float in the air in clear cold weather. Also referred to as ice crystals.

ice pellets — Small, transparent or translucent, round or irregularly shaped pellets of ice. They can be hard grains that rebound on striking a hard surface or pellets of snow encased in ice.

icebox rivet — Rivets made of 2024 or 2017 aluminum alloy, which are too hard to drive unless they are in a softened condition. These rivets must be heat-treated, quenched, and held in a subzero icebox until they are driven.

ice-up — A condition in flight in which ice forms on the aircraft structure.

icing — The accumulation of airframe ice.
a. Types of icing are:
Rime Ice — Rough, milky, opaque ice formed by the instantaneous freezing of small supercooled water droplets.
Clear Ice — A glossy, clear, or translucent ice formed by the relatively slow freezing of large supercooled water droplets.
Mixed — A mixture of clear ice and rime ice.
b. Intensity of ice includes:
Trace — Ice becomes perceptible. Rate of accumulation is slightly greater than the rate of sublimation. Deicing/anti-icing equipment is not utilized unless encountered for an extended period of time (over 1 hour).
Light — The rate of accumulation may create a problem if flight is prolonged in this environment (over 1 hour). Occasional use of deicing/anti-icing equipment removes/prevents accumulation. It does not present a problem if the deicing/anti-icing equipment is used.
Moderate — The rate of accumulation is such that even short encounters become potentially hazardous and use of deicing / anti-icing equipment or flight diversion is necessary.
Severe — The rate of accumulation is such that deicing/anti-icing equipment fails to reduce or control the hazard. Immediate flight diversion is necessary.

ideal cycle — A cycle in which no pressure loss occurs across the combustion section. However, this is not practical for the gas turbine where a slight pressure loss is needed for correct cooling air.

ident — A request for a pilot to activate the aircraft transponder identification feature. This will help the controller to confirm an aircraft identity or to identify an aircraft.

ident feature — The special feature in the Air Traffic Control Radar Beacon System (ATCRBS) equipment. It is used to immediately distinguish one displayed beacon target from other beacon targets. See also ident.

identification symbol — A symbol used in an aircraft drawing to correlate a specific item with its description in the bill of materials or the revision block.

Identification, Friend or Foe (IFF) — A transponder feature that allows a classified code to be set into selected transponders to identify friendly aircraft to interrogating units.

idiot light — A slang term for warning lights that are used instead of digital or analog instruments to indicate the condition or pressure of a system.

idle — In gas turbine engines, a percent rpm setting, the value of which changes from engine to engine. It is the lowest engine operating speed authorized.

idle cut off — That position of the mixture control in which fuel is prevented from flowing from the metering system into the engine.

idle mixture — A fuel-air mixture used by an aircraft engine to provide proper operation at the idle RPM.

idle speed — The RPM of an aircraft engine when the throttle or power control lever is fully closed.

idle thrust — The jet thrust obtained with the engine power control lever set at the stop for the least thrust (idle stop) position.

idler gear — A gear used in a gear train to drive another gear in a reverse direction of rotation without changing the speed of rotation and without adding to or taking away power from the gear train.

idler pulley — The idler pulley changes angular relationships between pulleys but does not change the direction of rotation of either pulley. An idler pulley is also used to adjust the tension on the belt that joins a drive pulley with a driven pulley.

idling current — A low output or operating current that flows in an electronic circuit when there is no input signal or output signal.

if no transmission received for (time) — Used by ATC in radar approaches to prefix procedures that should be followed by the pilot in event of lost communications.

IF/IAWP — Intermediate Fix/Initial Approach Waypoint. The waypoint where the final approach course of a T approach meets the crossbar of the T. When designated (in conjunction with a TAA) this waypoint will be used as an IAWP when approaching the airport from certain directions, and as an IFWP when beginning the approach from another IAWP.

IFR aircraft — An aircraft conducting flight in accordance with instrument flight rules.

IFR conditions — Weather conditions below the minimum for flight under visual flight rules.

IFR departure procedure — See IFR takeoff minimums and departure procedures.

IFR flight — See IFR aircraft.

IFR landing minimums — See landing minimums.

IFR military training routes (IR) — Routes used by the Department of Defense and associated Reserve and Air Guard units for the purpose of conducting low-altitude navigation and tactical training in both IFR and VFR weather conditions below 10,000 feet MSL at airspeeds in excess of 250 knots IAS. [Jeppesen does not chart these routes.]

IFR over-the-top — With respect to the operation of aircraft, means the operation of an aircraft over-the-top on an IFR flight plan when cleared by air traffic control to maintain "VFR conditions" or "VFR conditions on top."

IFR takeoff minimums and departure procedures — Federal Aviation Regulations, Part 91, prescribes standard takeoff rules for certain civil users. At some airports, obstructions or other factors require the establishment of nonstandard takeoff minimums, departure procedures, or both, to assist pilots in avoiding obstacles during climb to the minimum enroute altitude. Those airports are listed in NOS/DOD Instrument Approach Charts (IAPs) under a section entitled "IFR Takeoff Minimums and Departure Procedures." The NOS/DOD IAP chart legend illustrates the symbol used to alert the pilot to nonstandard takeoff minimums and departure procedures. When departing IFR from such airports, or from any airports where there are no departure procedures, DPs, or ATC facilities available, pilots should advise ATC of any departure limitations. Controllers may query a pilot to determine acceptable departure directions, turns, or headings after takeoff. Pilots should be familiar with the departure procedures and must assure that their aircraft can meet or exceed any specified climb gradients.

igniter — The electrical device used to provide the spark for starting combustion in a turbine engine. Some igniters resemble spark plugs while others, called glow plugs, have a coil of resistance wire that glows red hot when electrical current flows through the coil.

igniter plug — In gas turbine engines, an electrical sparking device used to start the burning of the fuel-air mixture in a combustor.

ignition — The process whereby the fuel-air mixture in either a turbine or reciprocating aircraft engine is ignited.

ignition harness — The complete set of wires that carry high-voltage current from the magneto to the spark plugs.

ignition timing — The timing of the fuel-air mixture's ignition in the cylinders of a reciprocating engine that ensures the mixture will be burning before the piston reaches the top of its stroke and the maximum pressure will be produced in the cylinder as the piston starts downward.

illumination — The light output of a light source.

illustrated parts catalog (IPC) — A required document produced by a manufacturer. It has an exploded view of the parts and the part numbers for identification. It does not contain approved data.

illustrated parts list (IPL) — An exploded-view drawing included in a service manual showing every part of a component, along with its proper name, part number, and number required for assembly. This is FAA-approved data and using parts not included in this list jeopardizes the airworthiness of the component.

illustrated talk — An oral presentation where the speaker relies heavily on visual aids to convey ideas to the listeners.

ILS categories —
a. ILS Category I — An ILS approach procedure which provides for approach to a height above touchdown of not less than 200 feet and with runway visual range of not less than 1,800 feet.
b. ILS Category II — An ILS approach procedure which provides for approach to a height above touchdown of not less than 100 feet and with runway visual range of not less than 1,200 feet.

c. ILS Category III —
1) IIIA — An ILS approach procedure which provides for approach without a decision height minimum and with runway visual range of not less than 700 feet.
2) IIIB — An ILS approach procedure which provides for approach without a decision height minimum and with runway visual range of not less than 150 feet.
3) IIIC — An ILS approach procedure which provides for approach without a decision height minimum and without runway visual range minimum.

ILS categories (ICAO) —
a. ILS Category I – An ILS approach procedure which provides for an approach to a decision height not lower than 200 feet (60m) and a visibility not less than 2400 feet (800m) or a runway visual range not less than 1800 feet (550m).
b. ILS Category II (Special authorization required) – An ILS approach procedure which provides for an approach to a decision height lower than 200 feet (60m) but not lower than 100 feet (30m) and a runway visual range not less than 1200 feet (350m).
c. ILS Category III (Special authorization required) –
1) IIIA – An ILS approach procedure which provides for approach with either a decision height lower than 100 feet (30m) or with no decision height and with a runway visual range of not less than 700 feet (200m).
2) IIIB – An ILS approach procedure which provides for approach with either a decision height lower than 50 feet (15m) or with no decision height and with a runway visual range of less than 700 feet (200m) but not less than 150 feet (50m).
3) IIIC – An ILS approach procedure which provides for approach with no decision height and no runway visual range limitations.
d. Some areas require special authorization for ILS Category I approaches. In these areas, an additional category of approach called ILS is available without special authorization. These ILS approaches have minimums higher than a decision height of 200 feet and a runway visual range value of 2600 feet. Jeppesen approach charts, at these locations, will have a notation in the chart heading or in the minimum box titles.

ILS PRM approach — An instrument landing system (ILS) approach conducted to parallel runways whose extended centerlines are separated by less than 4,300 feet and the parallel runways have a Precision Runway Monitoring (PRM) system that permits simultaneous independent ILS approaches.

image frequency — The image produced in a super heterodyne receiver when an unwanted signal of a specific frequency is mixed with the oscillator frequency, it produces an intermediate frequency equal to the intermediate frequency of the desired input frequency.

immediately — Used by ATC or pilots when such action compliance is required to avoid an imminent situation.

immersion heater — An electrical heater used to heat liquids by immersing the heater in the liquid to keep it warm. Insures adequate temperature and flow during cold weather operations.

immersion-type oil heater — An electrical heater immersed in the engine oil reservoir to keep the oil warm when the engine is not operating and ensure an adequate flow of lubricant for starting in extremely cold weather.

immiscible — Liquids that do not mix with each other such as oil and water.

impact area — That portion of a damaged structure that has received the majority of the damage from a collision or other impact.

impact extrusion — Metal forming in which hard metal is forced through a die by striking it with a hard blow.

impact ice — Ice that forms on the wings and control surfaces or on the carburetor heat valve, the walls of the air scoop, or the carburetor units during flight. Impact ice collecting on the metering elements of the carburetor can upset fuel metering or stop carburetor fuel flow.

impact pressure — In pitot-static systems, the difference between pitot pressure and static pressure.

impact test — A test to determine the energy absorbed in fracturing a test bar at high velocity. The test may be in tension or in bending, or it may properly be a notch test if a notch is present, creating multi-axial stresses.

impact wrench — A power wrench, usually air-driven, used to spin nuts onto bolts. Its torque forces are in a series of blows or impacts. Because of the uneven torque it produces, it should not be used for any threaded fastener where the amount of torque is critical.

impedance (Z) — The vector sum of the opposition to the flow of AC caused by circuit resistance, capacitive reactance, and inductive reactance.

impedance coupling — The use in electronics of an impedance matching transformer to connect a power source with a load that has a different impedance.

impedance matching — The matching of the impedance of a source of electrical power with the impedance of the load that uses the power. Allows maximum transfer of power to occur.

impedance matching transformer — An electronic device used to connect a load and a source of electrical power that differ in their impedance. It consists of a transformer with a primary winding that matches the impedance of the source and a secondary winding that matches the load.

impedance triangle — The graphical representation of resistance, reactance, and impedance in an electrical circuit. Resistance in ohms is drawn as the horizontal base of the triangle and reactance in ohms is drawn vertically at 90 degrees to the resistance. Impedance in ohms is the hypotenuse of the triangle. The angle between it and horizontal is called the "phase angle."

impeller — A vaned disc that picks up and accelerates the air outwardly to increase the pressure in a supercharger for a reciprocating engine, or to provide the pressurized air for a centrifugal-type turbine engine.

impingement starting — A turbine engine starting process requiring no engine mounted starter. Air from a source separate from the engine to be started is directed onto the turbine wheel(s) to cause engine rotation for starting and then the air source is removed.

implode — To burst inward. The reverse of explosion.

impregnate — In reinforcing plastics, to saturate the reinforcement with a resin.

impulse — A change in momentum caused by a surge or pulse of energy.

impulse coupling — A spring-loaded coupling between a magneto and its drive gear that causes the magneto to produce a hot and late spark for starting the engine. When the engine is being turned over slowly, the magnet is restrained by stops, and the spring is wound. At the proper time for the starting spark to occur, the spring is released and the magnet is spun, producing a hot, late spark. When the engine starts, centrifugal force holds the coupling engaged so that it acts as a solid unit.

impulse turbine — A stator vane and rotor blade arrangement whereby the vanes form convergent ducts and the blades form straight ducts. The rotor is then turned by impulse as gases impinge on the blades. A design common to turbine driven accessories such as air starters.

impulse-reaction turbine — A stator vane and rotor blade arrangement whereby the base area is an impulse design and the tip is a reaction design. This design is common to flight engines.

impulsive — A learning style where a student makes a quick assessment and then decides to take action. Impulsive students may not read each question or all of the answer choices entirely. As a result, they tend to select the first choice that appears correct.

impurities — 1. Undesired foreign objects in a fluid. 2. A chemical element such as arsenic or phosphorus that is added to silicon or germanium to give them some desired electrical characteristic.

in ground effect (IGE) hover — In rotorcraft, hovering close to the surface (usually less than one rotor diameter distance above the surface) under the influence of ground effect.

in phase — A condition in an electrical circuit where the voltage and current rise and fall together. In an AC circuit, the two pass through 0° and 180° at the same time, going in the same direction.

inactive aircraft — An aircraft that is no longer operational.

inboard — Toward the center of the aircraft.

incandescent — Glowing because of intense heat.

incandescent lamp — An electric lamp that produces light by a white hot filament enclosed in a glass bulb from which the air has been removed and replaced with an inert gas.

incerfa (uncertainty phase) [ICAO] — A situation wherein uncertainty exists as to the safety of an aircraft and its occupants.

inches of mercury (in. Hg) — A measurement of air pressure., normally used for atmospheric pressures. 1 in. Hg. is equal to approximately $\frac{1}{2}$ PSI.

inches per second (ips) — A velocity measurement. Used in electronic balancing.

incidence board — A device used to measure the angle of incidence of a wing.

incipient stage — The time when frontal cyclone development begins, pressure falls at some point along the original stationary front and counterclockwise circulation is generated.

inclined plane — A machine used to gain a mechanical advantage. It consists of a flat surface positioned at an angle with the horizon.

inclinometer — An instrument consisting of a curved glass tube, housing a glass ball, and damped with a fluid similar to kerosene. It may be used to indicate inclination, as a level, or, as used in the turn and slip indicator, to show the relationship between gravity and centrifugal force in a turn.

inclusions — The impurities contained in a material.

incompressible fluids — Any liquid such as oil or water. Liquids cannot be compressed but can be used in a regulated fluid power system such as a hydraulic system to gain a mechanical advantage.

Inconel — A chromium-iron alloy similar to stainless steel, but which cannot be hardened by heat treatment.

increase speed to (speed) — See speed adjustment.

incrementally — Moving in steps rather than in continuous motion.

indefinite ceiling — A ceiling classification denoting vertical visibility into a surface based obscuration.

indicated air temperature (IAT) — Is the temperature of the air as measured by the temperature probe on the outside of the aircraft.

indicated airspeed — The airspeed as indicated on the airspeed indicator with no corrections applied.

indicated altitude — The altitude shown by an altimeter set to the current altimeter setting.

indicated horsepower (IHP) — The total horsepower developed in the engine. It is the sum of the brake horsepower delivered to the propeller shaft and the friction horsepower required to drive the engine.

indicated mean effective pressure (IMEP) — The average measured pressure inside the cylinder of an engine during the power stroke. Expressed in pounds per square inch.

indicating fuse — A fuse assembly that has a small neon light installed in parallel with the fuse that shows when the fuse has blown.

indicating instrument — A device such as a gauge, dial, or pointer that measures or records and also visibly indicates. An apparatus that shows fluid pressures, temperatures, or quantities.

indicator — See indicating instrument.

indirect light — Light reflected onto a surface from another surface.

induced current — The electrical current that is generated in a conductor when it is crossed by magnetic lines of flux.

induced drag — That part of the total drag that is created by the production of lift.

induced flow — In rotorcraft, the component of air flowing vertically through the rotor system resulting from the production of lift.

induced voltage — The voltage generated by a conductor when lines of magnetic flux cut across it.

inducer — The center inlet portion of a centrifugal impeller, sometimes made of a different, harder metal than the impeller for FOD protection.

inductance — The property of a conductor that causes an electromotive force or voltage to be generated when lines of magnetic force cut across it.

inductance bridge — An electronic device used to accurately measure an unknown inductance by comparing it to a known inductance. An inductance bridge is similar to the Wheatstone bridge used to measure resistance.

inductance coil — A coil designed to introduce inductance into a circuit.

induction compass — A direction indicator that derives its signal from the lines of flux of the Earth cutting across the windings of the flux valve mounted in the airplane.

induction furnace — An electric furnace that melts metal by the induction of high-frequency electromagnetic energy.

induction heating — A method of heating a conducting material by passing high frequency alternating current through it.

induction icing — The formation of ice on aircraft air induction ports and air filters.

induction motor — An AC electric motor that has the AC line voltage connected across stationary windings in the motor housing. Current induced into the rotor causes a magnetic field that reacts with the field of the stator, and this reaction causes the rotor to turn.

induction period — The time period after catalyzed material is mixed in which the material is allowed to begin curing before being sprayed onto a surface.

induction system — The complete system of air passages in a reciprocating engine, from the air filter inlet to the intake valve of the cylinder.

induction system fire — A fire in the carburetor or air inlet system of a reciprocating engine usually caused by flooding and a backfire.

induction vibrator — A coil and set of contact points that produce pulsating DC from straight DC. Pulsating DC can be used in the primary winding of a magneto to produce a high voltage in the secondary winding.

induction welding — A method of welding in which the metal is melted by the induction of high-frequency electromagnetic energy.

inductive circuit — An AC circuit in which the capacitive reactance lags behind the inductive reactance.

inductive kick — A slang term used for inductive reactance in a coil of wire. It is a high voltage produced across a coil when current stops flowing through the coil. When current flows through a coil, a magnetic field is set up around each of the turns of wire in the coil. But when the current stops flowing, the magnetic field collapses across the coil producing many times the voltage that was originally in the coil. This high voltage is called an inductive kick.

inductive reactance — An opposition to the flow of AC caused by the generation of an induced voltage whose polarity is opposite to that of the voltage that created it.

inductive time constant — A measurement of the amount of time needed for the current induced into an inductive circuit to reach 63.2% of its maximum value.

inductive tuning — A method of selecting or changing the resonance of a radio frequency circuit by changing the inductance. The circuit is tuned by rotating a tuning coil that increases or decreases the inductance.

inductor — A coil or other device used to introduce inductance into a circuit.

industrial diamond — A diamond used as a cutting tool.

inert agent — A fire extinguishing agent that extinguishes fire by excluding the oxygen from its surface.

inert gas — A gas such as argon or helium that does not form other chemical compounds when it comes into contact with other elements.

inert gas arc welding — A process of arc welding in which the arc is submerged in an envelope of an inert gas such as argon to exclude the oxygen from the molten metal and prevent the formation of oxides.

inertia — The tendency of a mass at rest to remain at rest, or if in motion to remain in motion, unless acted upon by some external force.

inertia anti-icer — A movable vane in the induction air system that, in the extended position, causes the velocity of the incoming air to increase and change direction, thereby discharging the heavier ice overboard while directing the lighter, ice-free air into the engine plenum.

inertia force — A force due to inertia, or the resistance to acceleration or deceleration.

inertia starter — A starter for large reciprocating engines that uses the energy stored in a flywheel, spinning at a high rate of speed, to turn the engine for starting.

inertia switch — An electrical switch built into an emergency locator transmitter (ELT). Designed to close and start the ELT when there is a sudden change in its velocity.

inertia welding — An advanced technology process of welding through use of high speed rubbing friction. Developed to join super alloys that are difficult to weld with traditional methods. In inertia welding, one of the work pieces is connected to a flywheel and the other is restrained from rotating. The flywheel is accelerated to a predetermined rotational speed, storing the required energy. The drive motor is disengaged and the work pieces are forced together by the friction welding force. This causes the faying surfaces to rub together under pressure. The kinetic energy stored in the rotating flywheel is dissipated as heat through friction at the weld interface as the flywheel speed decreases.

inertial navigation — Navigation by means of a self-contained, airborne device that senses changes of direction or acceleration and automatically corrects for deviations in a planned course.

inertial navigation system — An RNAV system which is a form of self-contained navigation.

inflammable — Able to burn and support combustion; easily inflamed. A word that has been replaced with "flammable" to avoid confusion.

inflight refueling — A procedure used by the military to transfer fuel from one aircraft to another during flight.

inflight weather advisory — See weather advisory.

informal lecture — A lecture style that lends itself to active student participation.

information request — A request originated by an FSS for information concerning an overdue VFR aircraft.

infrared (IR) — Electromagnetic radiation having wavelengths longer than red light.

infrared guidance — A guidance system used on heat-seeking missiles to hone in on the infra-red signature of a target.

infrared lamp — An incandescent lamp that produces light energy in the infrared range.

infrared radiation — Electromagnetic radiation having wavelengths longer than red light and shorter than microwaves.

infrasonic frequencies — Frequencies below the audio frequency range. Also referred to as subsonic frequencies.

ingest — To pull in something such as air or to ingest FOD in a gas turbine engine.

ingot — A large cast bar of metal, as poured, with no working.

inherent stability — That built-in characteristic of an aircraft that causes it, when disturbed from straight and level flight, to return to straight and level flight.

inhibited sealer — A material used to exclude moisture and air from a honeycomb repair. In addition to sealing, it inhibits the formation of corrosion.

inhibitive film — A film of material on the surface of a metal that inhibits or retards the formation of corrosion. It does this by providing an ionized surface that will not allow the formation of corrosive salts on the metal.

inhibitor — **1.** An agent added to a resin to retard its curing and increase its shelf life. **2.** Any substance that slows or prevents a reaction.

initial approach fix (IAF) — The fixes depicted on instrument approach procedure charts that identify the beginning of the initial approach segment(s).

initial approach segment [ICAO] — That segment of an instrument approach procedure between the initial approach fix and the intermediate approach fix, or where applicable, the final approach fix or point.

initial approach segment — The segment between the initial approach fix and the intermediate fix or the point where the aircraft is established on the intermediate course or final approach course.

initial lift — One of the two requirements for the production of a thunderstorm, the other is potential instability.

initialization — To facilitate the start-up of a program. In computers it is the start-up of computer language instructions that the computer understands in order to operate as intended.

injection molding — A method of forming thermoplastics by forcing resin, under high pressure, into a mold and allowing it to harden.

injection pump — A high pressure fuel pump used in a reciprocating engine fuel injection system. Fuel is pumped, under high pressure, into the combustion chamber of the engine where it is atomized and ignites as it leaves the injector nozzle.

inland navigation facility — A navigation aid on a North American Route at which the common route and/or the noncommon route begins or ends.

inlet buzz — An audible sound that sometimes occurs in inlets of supersonic aircraft when shock waves alternately move in and out. This condition appears when design speeds are exceeded.

inlet case — The front compressor supporting member, usually one single casting.

inlet duct — That portion of the structure of a turbine-powered aircraft that directs the air into the engine compressor.

inlet gearbox — An auxiliary gearbox driven from and located in front of the compressor in the engine inlet area. Not all engines are configured with this gearbox.

inlet guide vane — In gas turbine engines, stationary airfoil which precedes the first stage compressor rotor blades. These guide vanes form straight through passages and are present to direct air onto the blades at the optimum angle.

inlet particle separator — An inlet device on some turbine powered rotorcraft that prevents sand and other FOD-causing debris from entering the engine. On some separators, the trap has to be cleaned while, on others, the debris is directed overboard.

inlet pressure (Pt2) — The pressure total taken in the engine inlet as a measure of air density, a parameter sent to the fuel control for fuel scheduling purposes.

inlet screen — An anti-FOD screen used on turbine powered rotorcraft and most stationary turbines. Not generally used on other aircraft installations due to icing and other aerodynamic problems that can result.

inlet spike — A moveable inlet device used to control inlet geometry and shock waves. This inlet design diffuses supersonic airflow and reduces it to subsonic speed for entry into the engine.

inlet strut assembly — The spoke-like stationary airfoils that are part of the inlet case. They are used to support the front bearing housing and provide passageways for oil and air line routing from outside the engine to inside.

inlet temperature (Tt2) — The temperature signal taken in the engine inlet to measure air density. Used by the fuel control unit as a fuel scheduling parameter.

in-line engine — An engine with all of the cylinders in a single line. The crankcase may be located either above or below the cylinders. If it is above, it is called an inverted in-line engine.

in-line reciprocating engine — An engine in which all of the cylinders are arranged in a straight line, with each cylinder piston connected to a separate throw of the crankshaft.

inner exhaust cone — The conical-shaped portion of a turbine engine exhaust system that is used to produce the proper area increase for the gases as they leave the engine.

inner liner — Refers to can-annular combustion liner; the innermost section.

inner marker — A marker beacon used with an ILS (CAT II) precision approach located between the middle marker and the end of the ILS runway, transmitting a radiation pattern keyed at six dots per second and indicating to the pilot, both aurally and visually, that he is at the designated decision height (DH), normally 100 feet above the touchdown zone elevation, on the ILS CAT II approach. It also marks progress during a CAT III approach.

inner marker beacon — See inner marker.

inner tube — An air-tight rubber tube that has a stem for inflating it. Used inside a pneumatic tire to hold the air that inflates the tire.

inoperative — Not working.

inoperative components — The lowest landing minimums on an approach are authorized when all components and visual aids are operating. If some components are inoperative, higher landing minimums may be required. If more than one component is inoperative, apply only the greatest increase in altitude and/or visibility required by the failure of a single component.

inoperative equipment — Equipment in the aircraft that is not functional.

input capacitance — Capacitance measured across the input terminals of a circuit.

input circuit — A circuit that provides appropriate power and impedance matching between an input device and the signal source.

input impedance — Impedance measured across the input terminals of a circuit.

input transformer — An electronic device that isolates a signal source from an input device and matches the impedance. See also impedance matching transformer.

inrush current — The high current that flows in an electrical machine or circuit when the switch is first closed.

inside caliper — A measuring instrument with two adjustable legs used to determine an inside measurement. Once the distance has been established, the actual measurement is made with a steel scale, a micrometer, or a vernier caliper.

inside diameter — The diameter measured from one inside surface, through center, and to the opposite inside surface.

inside micrometer — -A micrometer caliper used to measure the inside diameter of a circular object such as a cylinder bore. It measures in increments of $1/1,000$" or smaller. It works on the same principle as an outside micrometer caliper.

insight — The grouping of perceptions into meaningful wholes. Creating insight is one of the instructor's major responsibilities.

insolation — Solar radiation received at the surface of the Earth.

inspect — To determine the condition of something by sight, feel, measurement, or other means.

Inspection Authorization (IA) — An authorization issued by the FAA to experienced A&P technicians meeting certain requirements. This authorization allows them to return aircraft to service after annual inspections or certain major repairs.

inspection door — A small door or hinged plate on the surface of an aircraft structure that can be opened for inspecting the interior of the aircraft.

inspection hole — A hole in the skin of an aircraft, closed with an inspection plate that is held in place with screws on metal skins and with friction clamps on fabric skins. The hole can be opened for inspection or repair inside the structure.

inspection plate — A cover over an inspection hole that is held in place with screws on metal skins and with friction clamps on fabric skins.

instability — **1.** The characteristic of an aircraft that causes it, when disturbed from a condition of level flight, to depart further from this condition. **2.** In meteorology, a general term to indicate various states of the atmosphere in which spontaneous convection will occur when prescribed criteria are met; indicative of turbulence. See also absolute instability, conditionally unstable air.

installation drawing — A drawing that shows all of the parts in their proper relationship for installation.

installation error — An error in pitot static instruments (the airspeed indicator, the altimeter, and the rate of climb indicator) caused by a change in alignment of the static pressure port with the airflow as the aircraft's angle of attack changes.

instantaneous rate of climb indicator (IVSI) — A vertical speed indicator that uses internal accelerometer-type air pumps to overcome the inherent lag of this type of instrument and to provide an instantaneous indication of altitude changes due to pitch attitude changes.

instructional aids — Devices that assist an instructor in the teaching-learning process. They are supplementary training devices and are not self-supporting.

instrument — A device using an internal mechanism to show visually or aurally the attitude, altitude, or operation of an aircraft or aircraft part. It includes electronic devices for automatically controlling an aircraft in flight.

instrument approach procedure — A series of predetermined maneuvers for the orderly transfer of an aircraft under instrument flight conditions from the beginning of the initial approach to a landing or to a point from which a landing may be made visually. It is prescribed and approved for a specific airport by competent authority.
a. U.S. civil standard instrument approach procedures are approved by the FAA as prescribed under FAR 97, and are available for public use.
b. U.S. military standard instrument approach procedures are approved and published by the Department of Defense.
c. Special instrument approach procedures are approved by the FAA for individual operators, but are not published in FAR 97 for public use.

instrument approach — See instrument approach procedure.

instrument approach procedure [ICAO] — A series of predetermined maneuvers by reference to flight instruments with specified protection from obstacles from the initial approach fix, or where applicable, from the beginning of a defined arrival route to a point from which a landing can be completed and thereafter, if a landing is not completed, to a position at which holding or enroute obstacle clearance criteria apply.

instrument approach procedures charts — Portray the aeronautical data which is required to execute an instrument approach to an airport. These charts depict the procedures, including all related data, and the airport diagram. Each procedure is designated for use with a specific type of electronic navigation system including NDB, TACAN, VOR, ILS/ MLS, and RNAV. These charts are identified by the type of navigational aid(s) which provide final approach guidance.

instrument approach waypoint — Fixes used in defining RNAV IAPs, including the feeder waypoint (FWP), the initial approach waypoint (IAWP), the intermediate waypoint (IWP), the final approach waypoint (FAWP), the RWY WP, and the APT WP, when required.

instrument departure procedure (DP) — A preplanned instrument flight rule (IFR) air traffic control departure procedure printed for pilot use in graphic and/or textual form. DP's provide transition from the terminal to the appropriate enroute structure.

instrument departure procedure (DP) charts — Charts designed to expedite clearance delivery and to facilitate transition between takeoff and enroute operations. Each DP is presented as a separate chart and may serve a single airport or more than one airport in a given geographical location.

instrument flight rules (IFR) — Rules that govern the procedure for conducting flight in instrument weather conditions. When weather conditions are below the minimums prescribed for VFR, only instrument-rated pilots may fly in accordance with IFR.

Instrument Flight Rules (IFR) conditions — Weather conditions considered unsafe for flight under visual flight rules.

Instrument Flight Rules (IFR) flight — Aircraft flight conducted entirely by reference to instruments and radio navigation.

instrument ground instructor — A person certificated by the FAA who is authorized to provide the following: ground training in the aeronautical knowledge areas required for issuance of an instrument rating under Part 61; ground training required for an instrument proficiency check; and a recommendation for a knowledge test required for issuance of an instrument rating under Part 61.

instrument interpretation — One of the fundamental skills of basic attitude instrument flying. The three fundamental skills include: instrument cross-check, instrument interpretation, and aircraft control. Interpretation involves an awareness of the instrument indications that represent the desired pitch and bank attitudes for the aircraft.

instrument landing system (ILS) — A precision instrument approach system that consists of the following electronic components and visual aids:
a. Localizer
b. Glide Slope
c. Outer Marker
d. Middle Marker
e. Approach Lights

instrument meteorological conditions (IMC) — Meteorological conditions expressed in terms of visibility, distance from cloud, and ceiling less than the minima specified for visual meteorological conditions.

instrument panel — A panel, typically located in front of the pilot, that holds all of the indicating instruments that show the condition of the aircraft flight and mechanical systems.

instrument proficiency check — An evaluation ride based on the instrument rating practical test standard. Required to regain instrument flying privileges when the privileges have expired due to lack of currency.

instrument runway — A runway equipped with electronic and visual navigation aids for which a precision or nonprecision approach procedure having straight-in landing minimums has been approved.

instrument shunt — An electrical shunt used with an ammeter to make it possible for it to measure current.

instrument training — That time in which instrument training is received from an authorized instructor under actual or simulated instrument conditions.

instrumentation — The installation or use of indicating instruments.

insulated gate field effect transistor (IGFET) — A semiconductor whose gate is insulated from the channel. The IGFET is now called a MOSFET (Metal Oxide Field Effect Transistor). The MOSFET has an extremely large input impedance. Because the insulating oxide layer is extremely thin, the MOSFET is susceptible to destruction by electrostatic charges. Special precautions are necessary when handling or transporting MOS devices.

insulating electrical tape — A flexible, adhesive-backed tape made of a polyvinylchloride material used as insulation over wire terminals and wire splices.

insulation — A heavy material used in an aircraft to prevent the conduction of heat into or out of any of its operating components.

insulation blanket — 1. A layer of fireproof insulating material used to keep the heat of a jet engine tail pipe from radiating into the engine compartment. 2. Any

insulation grip — A plastic-covered, thin metal reinforcing sleeve on a pre-insulated terminal lug that grips the insulation of the wire when the lug is crimped, adding strength and durability to the installation.

insulation resistance — The electrical resistance of an insulating material separating two conductors.

insulation strength — A statement of the electrical insulating property of a substance. Usually stated as the voltage an insulator can withstand without breaking down.

insulator — A material or device used to prevent passage of heat, electricity, or sound from one medium to another.

intake valve — A reciprocating engine valve, located in the head of a cylinder, which provides the passage of the fuel-air mixture into the combustion chamber.

integral fuel tank — A portion of the aircraft structure, usually a wing, which is sealed off and used as a fuel tank. When a wing is used as an integral fuel tank, it is called a "wet wing."

integrated circuit (IC) — A microminiature circuit incorporated on a very small chip of semiconductor material through solid state technology. A number of circuit elements such as transistors, diodes, resistors, and capacitors are build into the semiconductor chip by means of photography, etching, and diffusion.

integrated engine pressure ratio (IEPR) — Used on some turbofans to include fan discharge total pressure and compressor inlet total pressure.

integrated flight instruction — A technique of flight instruction where students are taught to perform flight maneuvers by reference to flight instruments and to outside visual references. Handling of the controls is the same regardless of whether flight instruments or outside references are being used.

integrating circuit — A network circuit whose output is proportional to the sum of its instantaneous inputs.

integrity — The ability of a system to provide timely warnings to users when the system should not be used for navigation.

intensity — A principle of learning where a dramatic or exciting learning experience is likely to be remembered longer than a boring experience. Students experiencing the real thing will learn more than when they are merely told about the real thing.

intensity control — A cathode-ray tube control that meters the quantity of electrons in the beam that strikes the phosphorescent screen inside the cathode=ray tube. The more electrons that strike the screen, the brighter the display.

interactive video — Software that responds quickly to certain choices and commands by the user. A typical system consists of a compact disc, computer, and video technology.

intercom — A communication system within an airplane for the purpose of communicating between flight crew members or to passengers.

interconnector — A small tube connecting multiple burner cans together for the purpose of flame propagation during starting.

intercooler — A devise used to reduce the temperatures of the compressed air before it enters the fuel metering device. The resulting cooler air has a higher density, which permits the engine to be operated with a higher power setting.

intercostal — A longitudinal structure similar to a stringer, but which is attached to a wing rib or fuselage frame and ends at an adjacent rib or frame. Intercostals are usually used to support access doors, equipment, etc.

intercylinder baffles — Sheet metal air deflectors installed between and around air-cooled cylinders to aid in uniform cooling.

interelectrode — The capacitive effect between two elements in an electron tube. At high frequencies, signals can be fed across the interelectrode capacitance between the plate and a grid.

interelectrode capacitance — The capacitance that exists between two electrodes in an electron tube.

interface — A surface that forms the common boundary between two parts of matter such as water interfaces in jet fuel and crystal interfaces in metals.

interference — **1.** A theory of forgetting where a person forgets something because a certain experience overshadows it, or the learning of similar things has intervened. **2.** Barriers to effective communication that are caused by physiological, environmental, and psychological factors outside the direct control of the instructor. The instructor must take these factors into account in order to communicate effectively.

interference fit — A fit between two parts in which the part being put into a hole is larger than the hole itself. In order to fit them together, the hole is expanded by heating and the part is shrunk by chilling. After being united, when the two parts reach the same temperature, they will not separate. The area around the hole is subject to tensile stress and thus vulnerable to stress corrosion.

intergranular corrosion — The formation of corrosion along the grain boundaries within a metal alloy.

interim summary — An interim summary can be made immediately after each topic to bring ideas together, create an efficient transition to the next topic, divert the discussion to another member of the group, or keep students on track.

interlock — An automatic control device that prevents an action until the device that is protected with the interlock is actuated.

intermediate approach segment — The segment of an instrument approach procedure between the intermediate fix or point and the final approach fix.

intermediate approach segment [ICAO] — That segment of an instrument approach procedure between either the intermediate approach fix and the final approach fix or point, or between the end of a reversal, race track or dead reckoning track procedure and the final approach fix or point, as appropriate.

intermediate case — The high-pressure compressor outer case on a turbine engine.

intermediate compressor — On a triple-spool turbine engine, the N_2 compressor.

intermediate fix — The fix that identifies the beginning of the intermediate approach segment of an instrument approach procedure. The fix is not normally identified on the instrument approach chart as an intermediate fix (IF).

intermediate frequency — A frequency generated in a superhetrodyne receiver equal to the difference between the received radio frequency signal and that produced by the local oscillator.

intermediate landing — On the rare occasion that this option is requested, it should be approved. The departure center, however, must advise the ATCSCC so that the appropriate delay is carried over and assigned at the intermediate airport. An intermediate landing airport within the arrival center will not be accepted without coordination with and the approval of the ATCSCC.

intermediate position — The position of some movable unit that lies in between the extreme positions of movement.

Intermediate-Range Ballistic Missile (IRBM) — A ballistic missile with a range generally between 1,500 and 3,437 statute miles.

intermediate turbine temperature (ITT) — The temperature taken usually at a station between the high and low pressure turbine wheels.

intermittent fault — An occasional condition in which a fault in a system does not occur with consistency.

intermittent load — A load that is not continually on the system.

intermittent-duty relay — An electrical relay that cannot be energized over long periods of time. This precludes their use for master relays or for lighting circuits.

internal air pressure — Air pressure within a vessel or container.

internal baffles — The deflector plates installed inside a tank or reservoir to prevent the fluid from sloshing or surging in flight.

internal combustion engine — An engine that obtains its power from heat produced by the combustion of a fuel air mixture within the cylinder of the engine.

internal control lock — A device used to lock a control surface in place when the airplane is parked. It is actuated from a control in the cockpit.

internal damage — Damage that occurs inside a part, component, or mechanism that is not visible externally.

internal resistance — The resistance of the battery to the flow of current. It causes a voltage drop proportional to the amount of current flow.

internal resources — During pilot operations, these are sources of information found within the airplane such as the pilot's operating handbook, checklists, aircraft equipment, aeronautical charts, the instructor, another pilot, and passengers, as well as one's ingenuity, knowledge, and skills.

internal supercharger — A gear-driven centrifugal blower in the accessory section of a reciprocating aircraft engine. Used to increase the pressure of the induction system air.

internal thread — A thread on the internal surface of a hollow cylinder or cone.

internal timing — The timing of the relationship of the E-gap position of the rotating magnet and the opening of the breaker points in a magneto.

internal wrenching bolt — A high-strength steel bolt with its head recessed to allow the insertion of an Allen wrench.

international airport — Relating to international flight, it means: **a.** An airport of entry which has been designated by the Secretary of Treasury or Commissioner of Customs as an international airport for customs service. **b.** A landing rights airport at which specific permission to land must be obtained from customs authorities in advance of contemplated use. **c.** Airports designated under the Convention on International Civil Aviation as an airport for use by international commercial air transport and/or international general aviation.

international airport (ICAO) — Any airport designated by the Contracting State in whose territory it is situated as an airport of entry and departure for international air traffic, where the formalities incident to customs, immigration, public health, animal and plant quarantine and similar procedures are carried out.

International Civil Aviation Organization [ICAO] — A specialized agency of the United Nations whose objective is to develop the principles and techniques of international air navigation and to foster planning and development of international civil air transport.

ICAO Regions include:
1. AFI African-Indian Ocean Region
2. CAR Caribbean Region
3. EUR European Region
4. MID/ASIA Middle East/Asia Region
5. NAM North American Region
6. NAT North Atlantic Region
7. PAC Pacific Region
8. SAM South American Region

international flight information manual — A publication designed primarily as a pilot's preflight planning guide for flights into foreign airspace and for flights returning to the U.S. from foreign locations.

International Morse Code — Dots and dashes used in combination for transmitting messages. Each combination represents a letter of the alphabet or a numeral.

international phonetic alphabet — A list of words used to denote the letters of the alphabet. Agreed upon by all nations, the use minimizes confusion regarding the speaker's intent.

international standard atmosphere (ISA) — A hypothetical atmosphere based on averages in which the surface temperature is 59°F (15°C), the surface pressure is 29.92 in. Hg (1013.2 Mb) at sea level, and the temperature lapse rate is approximately 2°C per 1,000 feet.

international system of units — The system of metric units that includes: meter, kilogram, second, ampere, candela, degrees Kelvin, hertz, radian, newton, joule, watt, coulomb, volt, ohm, farad, tesla, and weber.

Internet — An electronic network which connects computers around the world.

interphone system — A communication system normally carried out between in-flight crew members using microphones and earphones.

interplane struts — Struts that run vertically near the wing tips between the wings of a biplane.

interpolation — The estimation of an intermediate value of a quantity that falls between marked values in a series. Example: In a measurement of length, with a rule that is marked in $1/8$'s of an inch, the value falls between $3/8$ inch and $1/2$ inch. The estimated (interpolated) value might then be said to be $7/16$ inch.

interpole — A field pole in a compound-wound electrical generator used to correct for armature reaction. Armature reaction is the distortion of the generator field flux by the current flowing in the windings of the armature.

interrib bracing — The reinforcing tape of a fabric-covered wing that runs diagonally from the top of one wing rib to the bottom of the next throughout a truss-type wing to hold the ribs upright and in line until the rib-stitching is done.

interrogator — The ground-based surveillance radar beacon transmitter-receiver, which normally scans in synchronism with a primary radar, transmitting discrete radio signals which repetitiously request all transponders on the mode being used to reply. The replies received are mixed with the primary radar returns and displayed on the same plan position indicator (radar scope). Also applied to the airborne element of the TACAN/DME system.

intersecting runways — Two or more runways that cross or meet within their lengths. See also intersection.

intersection — 1. Typically, the point at which two VOR radial position lines cross on a route, usually intersecting at a good angle for positive indication of position, resulting in a VOR/VOR fix. 2. A point defined by any combination of courses, radials or bearings of two or more navigational aids. 3. Used to describe the point where two runways, a runway and a taxiway, or two taxiways cross or meet.

intersection departure — A departure from any runway intersection except the end of the runway.

intersection takeoff — See intersection departure.

interstage transformer — A transformer used to prevent the flow of DC from one stage of a multi-stage transformer system to the other. Provides the correct amount of impedance for the AC output of one stage and for the AC input of the following stage.

interstate air commerce — The carriage by aircraft of persons or property for compensation or hire, or the carriage of mail by aircraft, or the operation ornavigation of aircraft in the conduct or furtherance of a business or vocation, in commerce between a place in any State of the United States, or the District of Columbia, and a place in any other State of the United States, or the District of Columbia; or between places in the same State of the United States through the airspace over any place outside thereof; or between places in the same territory or possession of the United States, or the District of Columbia.

interstate air transportation — The carriage by aircraft of persons or property as a common carrier for compensation or hire, or the carriage of mail by aircraft in commerce: **a**. Between a place in a State or the District of Columbia and another place in another State or the District of Columbia, **b**. Between places in the same State through the airspace over any place outside that State; or **c**. Between places in the same possession of the United States; Whether that commerce moves wholly by aircraft of partly by aircraft and partly by other forms of transportation. Intrastate air transportation means the carriage of persons or property as a common carrier for compensation or hire, by turbojet-powered aircraft capable of carrying thirty or more persons, wholly within the same State of the United States.

intertropical convergence zone — The boundary zone between the trade wind system of the Northern and Southern Hemispheres; it is characterized in maritime climates by showery precipitation with cumulonimbus clouds sometimes extending to great heights.

introduction — The first element of an instructional lesson that sets the stage for the rest of the lesson by relating the coverage of the material to the entire course. The introduction itself is typically composed of three elements: attention, motivation, and an overview of what is to be covered.

Invar — A nickel-iron alloy with an extremely small temperature coefficient of expansion.

inverse peak voltage — The amount of inverse voltage a device can withstand without breakdown.

inverse square law — In physics, a given physical quantity varies inversely with the square of another physical quantity (usually distance). For example, if the distance from a magnet doubles, the strength of magnetism decreases to one fourth of its original value.

inverse voltage — The amount of voltage seen by a rectifier during the half cycle of AC being blocked to create pulsed DC.

inversion — In meteorology, an increase in temperature with height — a reversal of the normal decrease with height in the troposphere; may also be applied to other meteorological properties.

invert — To reverse the position, order, or condition of something.

inverted engine — A reciprocating engine whose crankshaft and crankcase are above the cylinders.

inverted spin — A maneuver having the characteristics of a normal spin except that the airplane is in an inverted attitude.

inverter — An electrical device that changes DC to AC.

investment casting — Casting as in a vacuum furnace or spin chamber to produce a denser, better quality material. Used to produce some steels in turbine engines.

iodine — A poisonous element with a symbol of I and an atomic number of 53.

ion — An atom that has either gained or lost an electron. If an atom has a shortage of electrons, it is a positive ion. If it possesses an excess of electrons, it is a negative ion.

ion engine — A reaction engine that ejects a stream of ionized particles to produce a forward thrust.

ionic charges — The charges in solid-state junctions caused by the dissimilarity of the junction materials. The charges create "barrier voltages." In order for a diode to rectify, it must be presented with an AC voltage whose peak value exceeds its barrier voltage.

ionize — To convert totally or partially into ions.

ionosphere — A series of atmospheric layers of ions that begins approximately 25 miles above the surface of the Earth.

IR — **1**. Voltage. Since E (voltage) = I (current) times R (resistance), IR is another way of stating E, or voltage. **2**. Routes used by the Department of Defense and associated Reserve and Air Guard units for the purpose of conducting low-altitude navigation and tactical training in both IFR and VFR weather conditions below 10,000 feet MSL at airspeeds in excess of 250 knots IAS. [Jeppesen does not chart these routes.]

IR drop — The amount voltage drops in a given conductor due to the resistance of the conductor. The IR drop is found by the formula: I x R=E (voltage), or by multiplying the amount of current (I) in amps, by the amount of resistance (R) in ohms.

IRAN — An acronym used by the military services for a form of maintenance known as Inspect and Repair As Necessary.

iridium — An extremely hard and brittle metallic element of the platinum group, with a symbol of Ir and an atomic number of 77. Used for electrodes of fine-wire spark plugs that must operate in extreme lead-fouling conditions.

iris exhaust nozzle — In turbine engines, a nozzle design similar to a camera shutter. It can be a two-position, partially-open/fully-open or a variable opening type. The widest opening is for afterburner mode. The variable opening type is controlled to continuously seek the optimum position for existing conditions.

Irish linen — A strong fabric made from flax used to cover many of the older aircraft. It is still popular in Europe, but it is no longer readily available in the

United States. It may be used as a direct replacement for grade-A cotton.

iron — A heavy, malleable, ductile, magnetic, silver-white metallic element with a symbol of Fe and an atomic number of 26. It is used in the production of steel and all ferrous metals.

iron-constantan thermocouple — A low temperature thermocouple that operates up to about 800° C. Used in aviation primarily to measure cylinder head temperature.

iron-core coil — An inductor that consists of a soft laminated iron core around which wire is wound.

iron-core transformer — An electrical transformer that has coils wound around soft iron cores. The cores cut down on eddy-current losses but limit the use of the transformer to lower frequencies.

iron-vane movement — An AC electric measuring instrument that depends upon a soft-iron vane or movable core operating with a coil to produce an indication of AC current flow.

irrelevant questions — A question that has no relationship with the subject matter being tested. Serves only to disrupt the orderly learning process.

irreversible controls — Hydraulically-controlled surfaces on an aircraft that do not provide aerodynamic feedback to the cockpit controls.

isobar — A line on a meteorological chart that connects points on the Earth's surface having equal barometric pressure.

isobar — A line of equal or constant barometric pressure.

isobaric metering valve — A metering valve in a cabin pressurization system that maintains a constant cabin altitude.

isobaric range — That range of cabin pressurization in which the cabin maintains a constant pressure, or cabin altitude, as the flight altitude changes.

isogonic lines — Lines on an aeronautical chart connecting points of equal magnetic declination (angle between magnetic North and true North). This angle is known as compass variation.

isogonic lines — Lines on charts that connect points of equal magnetic variation.

isoheight — A line that shows contours of equal height on a constant-pressure weather chart.

isohumes — Lines of equal relative humidity.

isolation mount — A rubber and metal composite used to prevent vibration transfer from one component to another.

isolation transformer — An electrical transformer with equal numbers of turns in the primary and secondary coils. This type transformer is used to isolate a piece of equipment from its power source.

isolation valve — A valve in an aircraft pneumatic system that can be shut off to isolate the components from the source of air pressure so maintenance can be performed without discharging the system.

isoline — A line on a weather chart showing contours of equal values of some quantity. It could be equal height (isoheight), barometric pressure (isobar), temperature (isotherm), wind (isotach), etc.

isometric drawing — The representation of an object in isometric projection in which the lines parallel to the edges of the object are drawn to true length.

isometric projection — A three-dimensional projection in which the three faces of an object are equally inclined to the surface of the drawing, and all of the edges are equally foreshortened.

iso-octane — A hydrocarbon (C_8H_{18}) that has a very high critical pressure and temperature. It is used as a reference for measuring the anti-detonation characteristics of a fuel.

isopleth — A line connecting points of constant or equal value.

isopropyl alcohol — Fluid that prevents the formation of ice on the blades during flight. Used in anti-icing systems for propeller blades.

isostatic forging — A similar process to isothermal forging.

isotach — A line of equal or constant wind speed.

isotherm — A line on a meteorological chart that denotes locations having the same temperature.

isothermal forging — A hot forging method that uses super alloy production. Materials and dies are heated to the same temperature usually in a vacuum or highly controlled atmosphere to prevent oxidation. One such process utilizes an alloy in powder form, which when compressed, results in closer, near-net-shape than older methods. This results in the production of hot section parts with less waste during final machining.

isothermal layers — A layer in the atmosphere where the temperature is the same from the bottom of the layer to the top of the layer.

isotopes — An atom that has the same atomic number as a chemical element, but a different atomic mass and different physical properties.

J

jack pads — Structural locations capable of supporting the weight of an aircraft when being jacked from the ground.

jacket — A metal blanket or shroud used to insulate a portion of a turbine engine.

jacks — The hydraulic or mechanical devices used to lift an aircraft off of the ground for testing or servicing.

jackscrew — A threaded, hardened steel rod that can be rotated to lift an object or to apply a force.

jagged edge — An irregularly shaped edge on a piece of metal, wood, or plastic material.

jam acceleration — The rapid movement of the power control lever of a gas turbine engine. Done when measuring the RPM acceleration rate.

jam nut — A thin check-nut screwed down against a regular nut to lock it in place.

jamming — Electronic or mechanical interference that can disrupt the display of aircraft on radar or the transmission/reception of radio communications/navigation.

J-block — A precision block ground to an accuracy of approximately 0.00001", used as a reference in precision machining operations.

Jeppesen Information Services — A subscription service, for pilots, which provides revisions for several flight information publications including the Jeppesen AIM, Jeppesen FARs for Pilots, the Jeppesen Airport Directory, JeppGuide, and the GPS/LORAN Coordinate Directory

jerry can — A specially designed five gallon container used for carrying fuel.

jet — 1. A calibrated, restricted orifice in the fuel passage of a carburetor used to control the amount of fuel that can flow under a given pressure. The size of the hole (jet) determines the amount of flow through the jet. 2. A forceful stream of fluid discharged from a small nozzle. 3. An aircraft powered by a turbojet engine. 4. The hot, high velocity gas stream issuing from the tailpipe of a gas turbine engine.

Jet A — A kerosene-type turbine engine fuel similar to the military JP-5. It has a very low vapor pressure and a relatively high flash point.

Jet A-1 — A kerosene-type turbine engine fuel similar to the military JP-8 fuel with additives to make it usable at very low temperatures of approximately -58°F.

Jet Assist Takeoff (JATO) — An auxiliary means of assisting a heavily loaded aircraft to takeoff, particularly on a short runway. The JATO consists of small rockets attached to the aircraft and provides the required additional thrust needed for takeoff.

Jet B — A wide-cut blend of hydrocarbon fuels for use in turbine engines. Used primarily in the military as JP-4 fuel.

jet blast — Jet engine exhaust (thrust stream turbulence).

jet efflux — The gas flowing from the exhaust nozzle.

jet engine — A reaction engine that derives its thrust from the acceleration of an air mass through an orifice. There are four common types: rocket, ramjet, pulsejet, and turbojet.

jet fuel control (JFC) — The fuel metering system for a turbine engine. Measures the operating conditions of the engine and meters into the burners the correct amount of fuel for the condition.

jet nozzle — A specially designed device shaped to produce a jet stream.

jet nozzle area — The area in square feet of the opening through which the engine exhaust gases pass to the atmosphere.

jet propulsion — The propulsion produced when a relatively small mass of air is given a large amount of acceleration

jet pump — A pump that operates by producing a low pressure through a venturi. Seen in oil scavenge systems as oil pumps and fuel systems as vapor eliminators.

jet route — A route designed to serve aircraft operations from 18,000 feet MSL up to and including flight level 450. The routes are referred to as "J" routes with numbering to identify the designated route; e.g., J105.

jet silencer — A device used to reduce and modify the lower frequency sound waves emitting from an engine's exhaust nozzle, and thus reducing the noise factor.

jet streak — A portion of the jet stream where wind speeds are greater than in regions up- or downstream. Jet streaks are several hundred to 1,000 miles long.

jet stream — A narrow band of high speed winds (speeds exceed 60 knots). Normally found near the tropopause.

jet stream axis — The line of maximum winds (>60 knots) on a constant pressure chart.

jet stream cirrus — Associated with an extratropical cyclone, these anticyclonically curved bands of cirrus clouds are usually located just downstream of the upper trough.

jet stream front — High-level frontal zone marked by a sloping layer below the jet core.

jet thrust — The thrust produced by a jet.

Jetcal analyzer — A trade name for an electronic test apparatus for checking the calibration of the EGT system, the RPM system, and the accuracy of their associated instruments.

jettison — To cast off or drop from an aircraft in flight.

jettisoning of external stores — Airborne release of external stores; e.g., tiptanks, ordnance.

jewel bearing — A cup-type bearing surface that rides on a hardened steel pivot used extensively in many types of indicating instruments.

jeweler's rouge — A very fine ferric oxide abrasive used for polishing hard metal surfaces.

jig — The framework or alignment structure used in the construction or repair of an aircraft to hold all the parts in proper alignment while they are fastened together.

jigsaw — An electric or pneumatically operated saw tool that uses a variety of narrow blades to cut small curves in wood, metal, or plastic.

Jo-bolts — An internally threaded three-piece rivet.

joggle — A small offset in sheet metal formed to allow one part to overlap another.

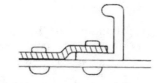

Johnson bar — A brake control found on some early airplanes. A Johnson bar consisted of a long bar mounted between the seats in the cockpit that, when pulled rearward, applied both main wheel brakes and, when pulled diagonally rearward, applied one of the brakes.

joint use restricted area — See restricted area.

jointer — An electrically powered woodworking machine used to smooth the edges of wood.

joule — The international system unit of energy equal to the work done when a current of 1 ampere is passed through a resistance of 1 ohm for 1 second.

joule rating — A turbine engine's ignition system.

journal — The polished surface of a crankshaft on which the bearings ride.

joystick — A slang term used for the control stick that controls an airplanes ailerons and elevator. Moving the joystick fore and aft moves the elevators, and moving the joystick side to side moves the ailerons.

JP-4 — A gas turbine engine kerosene-type fuel made up of approximately 65% gasoline and 35% distillates. Also referred to as Jet B.

JP-5 — A highly refined gas turbine kerosene-type engine fuel. Also referred to as Jet A.

JP-7 — A gas turbine engine kerosene-type fuel with additives for use at extremely high altitudes.

judgment — Process of recognizing and analyzing information, evaluating alternative actions, and making a timely decision on action to take.

jumbo jets — The name given to wide-body airplanes such as the Boeing 747, 757, 767, 777, McDonald Douglas DC-10, Lockheed L-1011, A-300 Airbus, etc.

jump seat — A compact portable seat positioned slightly behind the pilot's and copilot's seats in an airplane where a flight engineer sits to monitor certain engine-operating instruments and operate some of the auxiliary controls.

jumper — A temporary electrical lead wire used to bypass a circuit for purposes of troubleshooting.

junction — The point at which two conductors or circuits join.

Junction Field Effect Transistor (JFET) — A semiconductor device that uses voltage to control conductivity instead of current as in the normal transistor.

junction box — A metal or fiberglass box for holding the electrical terminal strips to which wire bundles are attached.

junction transistor — A transistor consisting of a single crystal of P- or N-type germanium between two electrodes of opposite types. The center layer is the base and forms junctions with the emitter and collector.

jury strut — A small strut extending from approximately the mid-point of a wing strut to the spar of the wing. Its purpose is to stabilize the main strut against vibrations.

K

K — The abbreviated use for kilo or one thousand. An item that weighs 3 Kg (kilograms) weighs 3,000 grams.

K index — A stability index used to determine the percentage probability of the occurrence of an airmass thunderstorm.

K monel — A high-strength, non-magnetic, heat-treatable, corrosion resistant alloy made up of nickel, copper, and aluminum.

katabatic — Any wind blowing down slope.

katabatic wind — Any wind blowing down slope.

K-band — A radar wavelength.

K-chart — A sheet metal fabrication chart providing the multiplier to use when determining the setback for bends of other than 90°.

keel — A longitudinal member or ridge along the center bottom of a seaplane float or hull.

keeper — A soft iron bar or plate placed across the poles of a magnet when the magnet is not being used in order to keep the magnet from losing any of its magnetism.

Kelvin (K) — The absolute temperature scale with minus 273°C as absolute zero. Used in many engine performance calculations.

Kelvin bridge — A resistance-measuring instrument used for accurate measurements of low resistances.

Kennelly-Heaviside layer — An ionized layer in the upper atmosphere that reflects radio waves to Earth. Also referred to as E-layer or ionosphere.

kerf — The slot or groove left by a cutting tool as it cuts through a material.

kerosene — A thin, colorless, flammable hydrocarbon material distilled from crude oil and used as a base for turbine engine fuel and as a solvent for cleaning parts.

Kett saw — A metal cutting, hand-held, power-operated tool. The head can be turned to any desired angle for cutting sheet stock aluminum. Uses various sizes of blades.

Kevlar® — Trademark of DuPont. A strong, lightweight aramid fiber used as a reinforcement fiber.

keyhole saw — A small U-shaped hand saw with a stiff, narrow blade used to cut a larger hole from a small drilled hole.

keying — The process of modulating a continuous carrier wave (cw) with a key circuit to provide interruptions in the carrier in the form of dots and dashes for code transmission.

keyway — A slot or groove machined into a hole or a shaft onto which a key is fitted.

kHz — Abbreviation for kilohertz (1,000 CPS).

kickback — The reverse rotation of a reciprocating engine due to premature ignition during starting.

Kidde — A manufacturer of fire detection systems.

kilo (K) — A metric term used to indicate one thousand. Kilo is also used as the prefix for kilogram, kilometer, and kilohertz.

kilogram (kg) — One thousand grams. 1 kg = 1,000 grams.

kilohertz — A frequency of 1,000 cycles per second.

kilomega — An outdated term for one thousand million (i.e., a billion or 1,000,000,000). Giga is now used.

kilovolt — 1,000 volts.

kilovolt amperes reactive (KVAR) — A measure of reactive power.

kindling point — That temperature at which combustion can take place for a particular fuel/oxygen combination.

kindling temperature — The temperature required for a material to burn when combined with oxygen.

kinds of operations list — A list found in the pilot's operating handbook (POH) of an aircraft that includes all equipment originally installed in the aircraft at the time of certification. It also includes notations on which equipment must be operational for various types of flight including day and night VFR.

kinematic viscosity — The ratio of absolute viscosity and density, expressed in units of centistokes.

kinesthetic learners — People who prefer to be doing something and primarily absorb information through actual hands-on experience. Kinesthetic learners ascertain more from performing a preflight inspection than from studying a checklist.

kinetic energy — Energy due to motion, defined as one half mass times velocity squared.

kink — A sharp twist or bend in a piece of wire, sheet metal, or a piece of tubing.

Kirchhoff's current law — "The algebraic sum of the currents entering and leaving any junction of conductors is equal to zero." Simply stated, this means that all of the current that arrives at a point must leave that point.

Kirchhoff's voltage law — The law of electrical circuits that states the sum of voltage that drops around a series circuit is equal to the applied voltage.

kirksite — An alloy of aluminum and zinc.

kite — A framework covered with paper, cloth, metal, or other material, intended to be flown at the end of a rope or cable, and having as its only support the force of the wind moving past its surfaces.

Klyston tube — An electron tube in which modulation is accomplished by varying the velocity of the electrons flowing through the tube.

knife edge — A sharp-edged piece of steel or other hard material used as a balance point or fulcrum for precision balance of a control surface or a propeller.

knife edge filter — A piece of metal shaped like a cylinder whose surface is cut with incoming and outgoing sharp cut grooves. When inserted in an oil passageway, contaminants remain in the incoming groove and clean oil squeezes between the knife edge and the casing and through the outgoing groove to the oil jet.

knife edge tip — Thin metal rims on a shrouded-tip turbine blade. These sealing tips establish their own clearance to the shroud ring by contact loading and wear.

knife switch — A switch that uses a blade that fits between two spring-loaded contacts to complete a circuit. When the blade is rotated out of the slot between the contacts, the circuit is opened.

knock — A loud knocking or banging noise made inside a reciprocating engine cylinder during the compression stroke. The knock is an explosion rather than a smooth burning process and is caused by the almost instantaneous release of heat energy from fuel in an aircraft engine caused by the fuel air mixture reaching its critical pressure and temperature.

knot — **1.** One nautical mile per hour, or 1.1508 statute miles per hour. **2.** A round, hard section of a tree branch embedded in a board and that weakens the overall strength of the board.

known traffic — With respect to ATC clearances, aircraft whose altitude, position and intentions are known to ATC.

knuckle pin — The hardened steel pin that holds an articulating rod in the master rod of a radial engine.

knurl — A series of small ridges on the surface of a material to aid in gripping.

Kollsman window — Registered trade name for the altimeter setting window of a Kollsman altimeter. Sometimes erroneously used to describe the altimeter setting window of any altimeter.

Koroseal — Plastic lacing used for support and anti-chafing protection of wires and lines.

Kraft paper — Strong brown paper such as the material of which grocery sacks are made.

Kreuger flap — A leading edge wing flap hinged at the bottom side of an airfoil. When it is actuated, the leading edge bends downward, increasing the overall wing camber and allowing the wing to develop additional lift at lower airspeeds.

L

L/Dmax — The maximum ratio between total lift (L) and total drag (D). This point provides the best glide speed. Any deviation from the best glide speed increases drag and reduces the distance of the glide.

labyrinth air seal — The thin sheet metal rims on turbine blades, either rotating or stationary, that control air leakage from the gas path to the inner portions of the engine. Same as knife edge air seal.

labyrinth oil seal — A main bearing oil seal. It is configured with thread-type grooves that allow gas path air to leak inward to the bearing sump and keep oil mist from escaping. Unlike the carbon seal, which rides on a surface, the labyrinth oil seal has a small clearance between its sealing lands and the rotating shaft.

lacing cord — A strong cotton, linen, or synthetic fiber cord used to rib stitch fabric covering to an aircraft structure. Also referred to as rib-stitching cord.

lack of common experience — In communication, a difficulty that arises because words have different meanings for the source and the receiver of information due to their differing backgrounds.

lacquer — The pigments dissolved in a volatile base (solvents, plasticizers, and thinners) in preparation for spraying as a liquid. Cures to a durable surface when the solvent evaporates.

lag — 1. A delay in time. 2. To fall behind.

lagging current — An occurrence in an AC inductive reactance circuit whereby changes in the voltage occur before changes in the current. Current, therefore, lags the voltage.

lagging material — An insulating material wrapped around aircraft plumbing to prevent the unwanted loss of heat to the outside air.

LAHSO — An acronym for "Land and Hold Short Operation." These operations include landing and holding short of an intersecting runway, a taxiway, a predetermined point, or an approach/departure flightpath.

LAHSO-dry — Land and hold short operations on runways that are dry.

LAHSO-wet — Land and hold short operations on runways that are wet (but not contaminated).

laminar — Arranged in or consisting of thin layers.

laminar flow — The nonturbulent flow of air or viscous fluid over a surface.

laminate — A structure made by bonding together two or more layers of material with resin. It contains no core material.

laminate ply — In composites, one fabric-resin or fiber-resin layer that is bonded to adjacent layers in the curing process.

laminated — Composed of thin layers of material firmly bonded or united together.

laminated core — The core of a coil, transformer, or other electrical device consisting of a stack of thin, soft iron sheets insulated from each other by a film oxide or varnish. Laminated cores are used to minimize eddy currents.

laminated plastic material — A reinforced plastic resin made up of layers of material such as cloth, paper, or wood bonded with plastic resin to form complex shapes or to produce a material with high strength for its weight.

laminated structure — An element or part of an aircraft made up of layers of material bonded together to form complex shapes or to produce a material with high strength for its weight.

laminated wood — Layers of wood bonded together to make a stronger material.

lampblack — The black soot from incompletely burned carbonaceous materials used for making generator brushes. Lampblack is used as a pigment when mixed with rubber for making tires.

land and hold short operations (LAHSO) — Operations that include simultaneous takeoffs and landings and/or simultaneous landings when a landing aircraft is able and is instructed by the controller to hold-short of the intersecting runway/taxiway or designated hold-short point. Pilots are expected to promptly inform the controller if the hold short clearance cannot be accepted.

land breeze — A coastal breeze blowing from land to sea, caused by temperature difference when the sea surface is warmer than the adjacent land. Therefore, it usually blows at night and alternates with a sea breeze, which blows in the opposite direction by day.

land plane — An airplane designed to operate from the surface of land using wheeled landing gear rather than pontoons or skis for landing on water or snow.

lander — A space vehicle designed for landing on a celestial body.

landing area — Any locality either on land, water, or structures, including airports/heliports and intermediate landing fields, used or intended to be used, for the landing and takeoff of aircraft whether or not facilities are provided for the shelter, servicing, or for receiving or discharging passengers or cargo.

landing area [ICAO] — That part of the movement area intended for the landing and takeoff of aircraft.

landing direction indicator — A device that visually indicates the direction in which landings and takeoffs should be made.

landing distance available — The length of runway declared available and suitable for the ground run of an aeroplane landing.

landing flaps — A secondary control surface that increases the overall wing area or changes the camber of the wing. The increased wing area permits a lower landing speed. Increased camber increases the lift and allows lower landing speeds and shorter landing distances.

landing gear — The wheels, floats, skis, and all of the attachments that support an airplane when it is resting on a landing surface.

landing gear door warning system — A group of components that warn of an unsafe landing gear door condition by the use of a horn, indicators, or red warning lights. Used on aircraft with retractable landing gear.

landing gear extended speed — The maximum speed at which an aircraft can be safely flown with the landing gear extended.

landing gear operating speed — The maximum speed at which the landing gear can be safely extended or retracted.

landing gear position indicating system — A group of components that shows the position of the landing gear though the use of lights or instruments. Used on aircraft with retractable landing gear.

landing gear warning system — A group of components incorporated on all retractable gear aircraft to warn of an unsafe landing gear condition. The pilot is warned of an unsafe condition by a warning light and aural device. The horn blows and the light comes on when one or more throttles are retarded and the landing gear is in any position other than down and locked.

landing lights — The high-intensity lights located on the wing or fuselage used to illuminate the runway for landing and taking off at night. These lights also make an aircraft more visible to other aircraft during the day.

landing minimums — The minimum visibility prescribed for landing a civil aircraft while using an instrument approach procedure. The minimum applies with other limitations set forth in FAR 91 with respect to the Minimum Descent Altitude (MDA) or Decision Height (DH) prescribed in the instrument approach procedures as follows:
a. Straight-in landing minimums — A statement of MDA and visibility, or DH and visibility, required for straight-in landing on a specified runway, or
b. Circling minimums — A statement of MDA and visibility required for the circle-to-land maneuver.

Note: Descent below the established MDA or DH is not authorized during an approach unless the aircraft is in a position from which a normal approach to the runway of intended landing can be made and adequate visual reference to required visual cues is maintained.

landing roll — The distance from the point of touchdown to the point where the aircraft can be brought to a stop or exit the runway.

landing sequence — The order in which aircraft are positioned for landing.

landing weight — The weight of the aircraft at touchdown. Often limited by the manufacturer to less than takeoff weight for structural reasons.

landing wires — The streamlined wires attached to the upper wing above the fuselage and extending to near the outboard end of the lower wing in a biplane. These wires brace the wings against the forces that affect the wings during landing. These forces are opposite those encountered during flight.

lap belt — A seat belt or safety belt that crosses a person's lap while seated in an aircraft.

lap joint — A joint in a sheet metal structure where the edge of one sheet overlaps the other. Welding, riveting, or bolting fastens the lap joints together.

lap winding — A method of manufacturing the armature of a DC generator by connecting the ends of each coil wound on the armature to the next adjacent commutator segment with the coils lapping over each other.

lapping — To rub two surfaces together with a very fine abrasive between them in order to produce an extremely close fit.

lapping compound — An abrasive paste used to polish surfaces.

laps — The surface defects in sheet metal caused by the folding over of fins or sharp corners into the surface of the material. Occurs during the rolling process.

lapse rate — The rate of decrease of an atmospheric variable with altitude; commonly refers to a decrease of temperature or pressure with altitude.

lapse rate — The rate of decrease of an atmospheric variable with height; commonly refers to decrease of temperature with height.

large aircraft — Aircraft of more than 12,500 pounds, maximum certificated takeoff weight.

large calorie (Cal.) — A unit of heat energy. It is the amount of heat energy necessary to raise the temperature of 1 kg of water 1 °C.

large scale integration — A method of fabricating integrated circuit (IC) chips to place multiple transistors or logic circuits on one small IC. Large scale integration was an early version that placed hundreds of logic circuits on one chip. Very large scale integration (VLSI) places thousands of logic circuits on a chip and most recently, millions are placed on a single chip.

laser — A device that emits coherent light (light that vibrates in phase) used in many applications, from CD players to powerful metal cutting devices. Acronym for Light Amplification by Stimulated Emission of Radiation.

laser beam welding — The use of laser beam heat for welding engine parts. Currently used to weld titanium fan blades, which could not be welded by conventional methods.

laser memory — A method of storing billions of bits of digital information on a disk such as a CD or DVD.

laser printer — A printer that prints a high quality copy of the information being processed by a computer.

last assigned altitude — The last altitude/flight level assigned by ATC and acknowledged by the pilot.

last chance filter — The final filter located just before the spray nozzle of a turbine lubrication system. Used to prevent foreign matter from clogging the spray nozzle.

latch — A fastening device used to hold a door closed.

latching relay — An electrically operated relay that, once energized, holds the contacts in the energized position by a mechanical latch and can only be released by some mechanical means.

late timing — On a reciprocating engine, the condition where the timing is late (retarded). That is, the ignition occurs very near top dead center (TDC) on the compression stroke or even after TDC. Late timing is an inefficient fuel condition and the causes the engine to develop less than optimum power.

latent heat — The amount of heat required to change the state of a material without changing its temperature (eg., The amount of heat absorbed or released during a change of state from ice to water.)

latent heat of condensation — The amount of heat given off when a substance changes from a vapor to a liquid without changing its temperature.

latent heat of evaporation — The amount of heat absorbed by a substance when it changes from a liquid to a vapor without changing its temperature.

latent heat of fusion — The amount of heat that must be removed from a substance to change it from a liquid to a solid without changing its temperature.

latent heat of vaporization — The amount of heat that must be absorbed by a liquid to change it into a vapor without changing its temperature.

lateral — The span of an airplane from wingtip to wingtip.

lateral axis — An imaginary straight line drawn perpendicularly (laterally) across the fuselage and through the center of gravity. Pitch movement occurs around the lateral axis, and is controlled by the elevator.

lateral navigation (LNAV) — A function of area navigation (RNAV) equipment that calculates, displays, and provides lateral guidance to a profile or path.

lateral separation — The lateral spacing of aircraft at the same altitude by requiring operation on different routes or in different geographical locations.

lateral stability — The stability about the longitudinal axis of an aircraft. Rolling stability or the ability of an airplane to return to level flight due to a disturbance that causes one of the wings to drop.

lateral vibration — An unwanted lateral or side-to-side movement in a helicopter. The cause of lateral vibration is normally an unbalanced main rotor.

lathe — A wood or metal-working tool in which the material is turned about its longitudinal axis and cutting tools are fed into its outer circumference in order to change its shape.

latitude — Measurement north or south of the equator in degrees, minutes, and seconds. Lines of latitude are also referred to as parallels.

launch — The release of an aircraft or rocket for flight.

launching pad — A specially designed platform from which rockets can be fired to launch them.

law of conservation of energy — The law that states that the internal energy of an isolated system remains constant; only forms of energy can be changed.

lay of a control cable — The twist of the strands of a wire cable.

layer — 1. A single thickness. 2. In reference to sky cover, clouds or other obscuring phenomena whose bases are approximately at the same level. The layer can be continuous or composed of detached elements.

layout — A drawing, pattern, or format of a piece of sheet metal in which the locations for all of the bends and drilled holes are indicated.

lay-up — In composites, reinforcing material that is placed in position in the mold.

L-band radar — An airborne radar.

LC circuit — A circuit network containing inductance and capacitance.

LCR electrical circuit — An AC circuit that has inductance (L), capacitance (C), and resistance (R).

The total opposition to the flow of current in an LCR circuit is the vector sum of resistance and the difference between inductive and capacitive reactances.

lead — A heavy, pliant, silvery metallic element with a symbol of Pb and an atomic weight of 82.

lead and lag — In rotorcraft, the fore (lead) and aft (lag) movement of the rotor blade in the plane of rotation.

lead of screw thread — The distance a nut will move forward on a screw if it is turned one revolution.

lead-acid battery — A commonly used secondary cell having lead as its negative plate and lead peroxide as its positive plate. Sulfuric acid and water serve as the electrolyte.

leading current — An AC electrical circuit in which the current that flows in the circuit has more capacitive reactance than it has inductive reactance.

leading edge — The foremost edge of an airfoil section.

leading edge flap — A portion of the leading edge of an airplane wing that folds downward to increase the camber, lift, and drag of the wing. The leading-edge flaps are extended for takeoffs and landings to increase the amount of aerodynamic lift that is produced at any given airspeed.

leading edge mean aerodynamic chord (LEMAC) — The leading edge of the mean aerodynamic chord often used as a location or reference for many aerodynamic measurements in aircraft operations and designs.

lead-lag hinge — A hinge at the root of a helicopter rotor blade with its axis perpendicular to the plane of rotation. Also referred to as the alpha, drag, or hunting hinge.

lead-off question — A question used by an instructor to open up an area for discussion. The purpose is to get the discussion started.

leaf brake — A bending tool used to form straight bends in sheet metal. The material is clamped in the tool, and a heavy leaf folds the metal back over a radius block to form the desired bend. Also referred to as a cornice brake.

leaf spring assembly — A series of flat springs hinged at one end and arched in the center. When a load is applied to the center of the arch it is absorbed as the spring alternately straightens out and returns to its arched shape.

leakage — 1. The breakdown of the dielectric strength of an insulator that allows current to pass through it. 2. An amount of a liquid or gas that leaks into or out of something.

leakage current — The amount of current that flows from the battery terminals to the battery case through any moisture or contamination on top of the battery.

leakage flux — The magnetic flux that does not follow a direct path between the poles of a magnet and which does not provide any useful work.

lean blow out — A condition in jet engine fuel combustion during which the fuel supply is decreased to maintain or reduce engine speed. During this condition the burning can be so slow that the flame is carried out of the combustion chamber and extinguished.

lean flame out — A condition of turbine engine operation in which the fire goes out in the engine because the fuel-air mixture is too lean to support combustion.

lean mixture — A fuel-air mixture in which there is an excess amount of air in the mixture.

learning — A change in behavior as a result of experience.

learning plateau — A learning phenomenon where progress appears to cease or slow down for a time before once again increasing.

learning style — The concept that how a person learns is dependent on that person's background and personality, as well as the instructional methods used.

least significant bit — A bit in a binary number that has the lowest value. It is the bit on the far right of a binary number.

least significant digit — A digit in a decimal number that is the least meaningful for accuracy. It is the digit on the far right of a decimal number.

Leclanche cell battery — A name for a common carbon-zinc cell flashlight battery.

lecture method — An educational presentation usually delivered by an instructor to a group of students with the use of instructional aids and training devices. Lectures are useful for the presentation of new material, summarizing ideas, and showing relationships between theory and practice.

lee side — The downwind side of a mountain range, building, island, etc.

lee wave — Any stationary wave disturbance caused by a barrier in a fluid flow. In the atmosphere when sufficient moisture is present, this wave will be evidenced by lenticular clouds to the lee of mountain barriers. Also referred to as a mountain wave or standing wave.

lee wave region — The upper layer of a two-layer lee-wave system where smooth wave flow dominates and microscale turbulence occasionally occurs.

lee wave system — A system marked by two distinct layers. The upper layer is the lee wave region, which begins just above mountain to level, and the lower layer is the lower turbulent zone.

left brain — A concept that each hemisphere of the brain processes information differently. People with left-brain dominance are more verbal, analytical, and objective.

left-hand rule — **1.** The left-hand generator rule: The fingers of the left hand are arranged in such a way that the thumb, first finger, and second finger point 90° to each other. If the thumb points in the direction of movement of the conductor, the first finger will point in the direction of the lines of flux (north to south), and the second finger will point in the direction of the induced voltage (back voltage, from positive to negative). **2.** The direction of magnetic flux: If the fingers of the left hand encircle a conductor in the direction of the lines of magnetic flux, the thumb will point in the direction of electron flow. **3.** The polarity of an electromagnet: If the coil of an electromagnet is grasped in such a way that the fingers encircle the coil in the same way the electrons are flowing, the thumb will point to the north pole of the electromagnet that is formed by the coil.

left-hand thread — A thread that winds in a counterclockwise direction. All left-hand threads are designated LH.

leg — **1.** One side of a formed angle. **2.** One portion of a trip made with dead reckoning navigation (the distance between turn points.) **3.** The portion of a trip between stops.

legs of a right triangle — The sides of a right triangle joined by the right angle (90°).

LEMAC — The leading edge of the mean aerodynamic chord. Often used as a reference location for aerodynamic measurements in aircraft operations and designs.

lenticular cloud — A cloud shaped like a lens that forms on the downwind side of a mountain. It usually indicates severe air turbulence and should be avoided. They can extend the entire length of the mountain range producing the wave and are called wave clouds or lennies by glider pilots.

Lenz's law — The law of induced current that says the current induced in a conductor will produce a magnetic field that opposes the field producing the original current.

lesson plan — An organized outline for a single instructional period. It is a necessary guide for the instructor in that it tells what to do, in what order to do it, and what procedure to use in teaching the material of a lesson.

letter of authorization (LOA) — A letter from the FAA authorizing use of a Minimum Equipment List (MEL) for a specific airplane. The letter and the MEL together are considered a supplemental type certificate.

level — **1.** The horizontal condition of a body: A flat horizontal surface of an object is level when no part of the object is higher than another. **2.** A spirit or bubble level: An indicating device that has a curved glass tube filled with a slightly colored liquid, except for a small air bubble. The tube is mounted in a housing so when it is place in a parallel position to the ground the bubble will move to the center of the tube indicating a level condition.

leveling lugs — The points on an aircraft on which a level can be placed for leveling the aircraft.

leveling means — A method of checking an aircraft for level flight attitude as specified by the aircraft manufacturer. This can be longitudinal, lateral, and/or both.

leveling scale — A scale built into the aircraft for checking the leveling of the aircraft in conjunction with a plumb bob.

lever — **1.** A device such as a bar used for prying. **2.** A flat bar pivoting on a fulcrum.

Leyden jar — A primitive capacitor. In effect, an apparatus for storing an electric charge on the inside foil lining of a glass jar.

L-filter — An inductor-input filter consisting of an inductor and a capacitor used to smooth the ripple from the output of a rectifier.

licensed empty weight — The empty weight that consists of the airframe, engine(s), unusable fuel, and undrainable oil, plus standard and optional equipment as specified in the equipment list.

life-limited part — A part with a specified number of operating hours or operating cycles before it must be removed for overhaul.

life-support systems — The oxygen and pressurization systems in an aircraft that make it possible for the occupants to function at high altitudes.

lift — One of the four main forces acting on an aircraft. On a fixed-wing aircraft, an upward force created by the effect of airflow as it passes over and under the wing.

lift fan — A turbofan engine with an exhaust duct that can be pointed down to provide upward thrust for vertical or short takeoff. Used in VSTOL (Vertical/Short Takeoff and Land) aircraft.

lift wires — The biplane wing support wires installed between the wings of a biplane to hold the wings in alignment against the forces of lift. Lift wires extend from the inboard end of the lower wing to the interplane struts on the upper wing.

lift-drag ratio — The efficiency of an airfoil section. It is the ratio of the coefficient of lift to the coefficient of drag for any given angle of attack.

lifted index — The common approach to the evaluation of potential instability requirement for thunderstorm formation. It is the difference between the observed 500 mb temperature and the temperature the parcel of air would have if lifted from the boundary layer to the 500 mb level.

lifting body — A wingless aircraft developed by NASA where lift is created by the shape of the craft itself.

light — The electromagnetic radiations of a frequency range visible to the human eye.

light aircraft — An aircraft having a total gross weight of 12,500 lbs. or less. See also light plane.

light emitting diode (LED) — A semiconductor diode that emits light when current flows through it.

light ends — The products of petroleum that boil off first in the process of fractional distillation.

light gun — A handheld directional light signaling device that emits a brilliant narrow beam of white, green, or red light as selected by the tower controller. The color and type of light transmitted can be used to approve or disapprove anticipated pilot actions where radio communication is not available. The light gun is used for controlling traffic operating in the vicinity of the airport and on the airport movement area.

light plane — An aircraft having a total gross weight of 12,500 lbs. or less. Light plane is also used as a colloquial term to mean a small, single-engine airplane.

light year — The distance light travels in one year. Unit of measurement in astronomy equal to 5.88×10^{12} miles.

light-activated silicon control rectifier — In electronics, a semiconductor device that conducts when exposed to light.

lighted airport — An airport where runway and obstruction lighting is available.

lightening hole — A hole cut in a structural part to decrease weight. Strength is often maintained by flanging the area around the hole.

lighter-than-air aircraft — An aircraft that can rise and remain suspended by using contained gas weighing less than the air that is displaced by the gas.

lightning — An immense discharge of static electricity in response to the build-up of electrical potential between cloud and ground, between clouds, within a single cloud, or between a cloud and the surrounding air.

light-up — The point at which combustion occurs in a turbine engine as indicated by an exhaust temperature rise on the cockpit indicator.

lime grease — A grease made up of oil and calcium hydroxide. It does not emulsify in water and is highly resistant to washout in wet environments.

limit load — The maximum load, expressed as multiples of positive and negative G (force of gravity), that an aircraft can sustain before structural damage becomes possible. The load limit varies from aircraft to aircraft.

limit switch — A switch designed to stop an actuator at the limit of its movement.

limiter — A stage in a frequency modulated receiver that limits the amplitude of the signal and thus removes static.

limits — The bounds of travel or size allowed for a component.

limits of size — The applicable maximum and minimum dimensions of a part.

Lindberg fire detection system — A continuous-element-type fire detector consisting of a stainless steel tube containing a discrete element that has been processed to absorb gas in proportion to the operating temperature. As the temperature rises, gas is released, causing the pressure in the stainless steel tube to increase to the operating temperature set point. This closes a switch that actuates the warning light and bell.

line loss — The voltage loss in a conductor. The amount of loss is related to its length.

line maintenance — The inspection and repairs accomplished on the flight line as opposed to shop maintenance.

line of sight radio reception — The clear path between the transmitting and receiving antennas of high-frequency radio signals.

line voltage — The main power line voltage that operates a system.

line voltage regulator — The device used to stabilize the line voltage by sensing and regulating the voltage demands supplied to a piece of electrical or electronic equipment.

linear accelerometer — A device that measures acceleration of a body in a straight line.

linear actuator — An actuator that changes hydraulic or pneumatic pressure into linear motion.

linear amplification — In electronics, an amplifier in which the change in output is proportional to the change in input.

linear control — In electronics, a control device in which the change in output is proportional to the amount the control's adjustment is changed.

linear integrated circuit — In electronics, an integrated circuit (IC) in which the change in output is proportional to the change in input.

linear movement — A movement or progression in which the output or result is directly proportional to the input.

linear operation — The operation in which the output of a device is directly proportional to its input. If the input increases by 10 percent, the output will also be increased by 10 percent.

linear resistance curve — The characteristic illustrated by a load when any increase or decrease in the voltage across the load results in a proportional change in the current through the load.

linen — A fabric made from flax that was a favorite covering material for truss-type airplanes. In the United States, it has been almost totally replaced by grade-A cotton and synthetic fiber.

lines of flux — The lines of magnetic force connecting the poles of a magnet.

link — **1.** A short connecting rod used for transmitting power and/or force. **2.** On the Internet, an identifier indicating connection with another similarly identified element. Links are usually identified by a different color of text, underlining, or button and can be accessed merely by clicking on them with a mouse. The mouse pointer also will usually change from an arrow to a hand when it hovers over a link.

link rod — An articulated rod that connects the pistons in a radial engine to the master rod. There is one less articulated rod than there are cylinders in each row of cylinders in a radial engine since one piston is attached to the master rod.

linseed oil — A solvent used in some aircraft finishes. Also used to coat the inside of steel tubing to prevent rusting. Linseed oil is obtained from flaxseed.

liquid — A fluid that assumes the shape of the container in which it is held.

liquid air — A slightly bluish, transparent liquid that has been changed into a liquid by lowering its temperature to -312°F (-191° C, 81° Kelvin). It is used chiefly as a refrigerant.

liquid crystal — A liquid whose reflectivity varies according to the voltage applied to it.

liquid crystal display — A constantly activated display that consists of segments of a liquid crystal. Uses very low current levels, making it more suitable for displays than light emitting diodes (LED's).

liquid lock — A condition that occurs in reciprocating engines that have the cylinders below the crankcase. Oil leaks past the piston rings and fills the cylinder. The engine cannot be rotated without damage. Also referred to as a hydraulic lock.

liquid nitrogen — Nitrogen that has been changed into its liquid state by lowering its temperature to -195°C (78° Kelvin) or lower.

liquid oxygen (LOX) — Oxygen that has been changed into its liquid state by lowering its temperature to -113°C (160° Kelvin) or lower.

liquid-cooled — A device or machine that is cooled by the use of liquid.

liquid-cooled engine — A reciprocating engine that uses a mixture of water and ethylene glycol to remove excess heat. This mixture flows around the cylinders in jackets and absorbs the heat created by the combustion in the cylinders. This heat is released to the outside air through a radiator.

listening — Hearing your students talk and listening to what they are saying are two different things. Instructors can use a variety of techniques or tools to become better listeners, including do not interrupt, do not judge, think before answering, be close enough to hear, watch nonverbal behavior, be aware of biases, look for underlying feelings, concentrate, avoid rehearsing answers while listening, do not insist on the last word.

liter — A metric unit of volume (1.0567 qt.) used for gaseous or liquid measurement.

lithium — An alkaline-metal element with a symbol of Li and an atomic number of 3. The lightest metallic element known.

lithium cell — One of a family of chemical cell types incorporating lithium in one pole.

lithium grease — A water-resistant, low operating temperature grease made of lithium salts and fatty acids.

lithometeor — In meteorology, dry particles suspended in the atmosphere such as dust, smoke, and haze.

lithosphere — The Earth's most outer area consisting of the Earth's crust and the upper mantle. The lithosphere extends downward toward the center of the Earth approximately 50 to 60 miles.

litmus — A water-soluble powder that turns red in acid solutions and blue in alkaline solutions.

litmus paper — An indicator paper that changes color when it comes in contact with an acid or an alkali. It turns red when wet with an acid, and blue when wet with an alkali.

live center — A lathe component with a sharp-pointed center that fits into the headstock of a lathe and turns with it. Used to locate the exact center of the headstock.

load — An energy-absorbing or energy-using device of any sort connected to a current.

load bank — A heavy-duty resistor used to discharge a storage battery.

load cell — An electronic weighing system component that contains the strain gauges. It is placed between an aircraft jack pad and the jack to measure the weight of the aircraft load.

load chart — A chart used for weight and balance purposes that specifies the location and distribution of weights. It aids the pilot in determining the loaded center of gravity condition.

load factor — The ratio of the load supported by the airplane's wings to the actual weight of the aircraft and its contents. Also referred to as G-loading.

load manifest — An itemized list of weights and moments of a particular load taken on a specific flight. Used by FAR Part 121 and 135 operators.

load, electrical — Any apparatus that uses electrical power to perform a function such as operating a motor.

loading graph — A method of computing the loaded weight and center of gravity of an aircraft.

loading schedule — A document showing where cargo should be stowed and in what weights at specified locations.

loadmeter — A current measuring instrument calibrated in terms of the percentage of the total rated current of the power source.

lobes — The eccentric portions of a cam or camshaft.

LOC mode — The operating position of an automatic pilot when it is receiving its signals from the localizer portion of an instrument landing system.

local action — The formation of tiny chemical cells in one or both of the poles of a chemical cell (battery) due to impurities in the material. Local action can result in the exhausting of the service capacity of a cell or corrosion of the pole pieces.

local airport advisory (LAA) — A service provided by flight service stations or the military at airports not serviced by an operating control tower. This service consists of providing information to arriving and departing aircraft concerning wind direction and speed, favored runway, altimeter setting, pertinent known traffic, pertinent known field conditions, airport taxi routes and traffic patterns, and authorized instrument approach procedures. This information is advisory in nature and does not constitute an ATC clearance.

local Mach number — In aerodynamics, the speed of air flow (Mach number) at a specified location. Because of the aerodynamic shape of an aircraft, the speed of airflow at any point can be greater than the air speed of the aircraft as a whole.

local oscillator — The internal oscillator section of a superhetrodyne circuit.

local traffic — Aircraft operating in the traffic pattern or within sight of the tower, or aircraft known to be departing or arriving from flight in local practice areas, or aircraft executing practice instrument approaches at the airport.

localizer — The component of an ILS that provides course guidance to the runway.

localizer course [ICAO] — The locus of points, in any given horizontal plane, at which the DDM (difference in depth of modulation) is zero.

localizer offset — An angular offset of the localizer from the runway extended centerline in a direction away from the no transgression zone (NTZ) that increases the normal operating zone (NOZ) width. An offset requires a 50 foot increase in DH and is not authorized for CAT II and CAT III approaches.

localizer type directional aid — A NAVAID used for nonprecision instrument approaches with utility and accuracy comparable to a localizer but that is not a part of a complete ILS and is not aligned with the runway.

localizer usable distance — The maximum distance from the localizer transmitter at a specified altitude, as verified by flight inspection, at which reliable course information is continuously received.

locator — An L/MF NDB used as an aid to final approach. Note: A locator usually has an average radius of rated coverage of between 18.5 and 46.3 km (10 and 25 NM)

lock tabs — A washer with tabs that are bent to prevent a nut from loosening.

locked-rotor current — The amount of current flowing through the windings of an electric motor when the motor is prevented from turning. This is the highest current draw of the motor since once the motor starts rotating, counter electromotive force (CEMF) is generated in the windings that opposes voltage and, in turn, current flow through the windings.

lockout debooster — A hydraulic component that decreases the pressure applied to aircraft brakes. Its lockout function shuts off all flow of fluid to the brake in the event of a rupture of the brake line below the debooster.

lockring — A horseshoe-shaped ring that snaps into a groove on a shaft in order to hold the shaft in position.

lockstitch — A modified seine knot used to lock the stitches when hand sewing aircraft fabric. The baseball stitch is used for sewing and it is locked with the seine knot every eight to ten stitches to prevent loosening.

locktab — A mechanical lock used to prevent a nut from coming loose. Its appearance is similar to a washer but with notches cut from the periphery. When a locktab is placed under a nut on the shaft of a bolt, one or more of the external locking tabs are bent up against the flats of the nut to keep it from backing off and becoming loose.

lockwire — A stainless steel, brass, or galvanized steel wire used to exert a pulling motion on the head of screws or bolts to prevent them from loosening. Also referred to as safetywire.

lockwiring — A method of tying two or more screws or bolts together by twisting lockwire between them in such a way that tension is held on the head of each fastener in the direction of tightening.

lodestone — A natural rock having magnetic characteristics.

log — A journal containing a record of activities. Pilots keep a log of their flight time, and ground crews keep logs on the mechanical operating components of the aircraft, airframe, engine, propeller, and rotor to show the amount of time in service, and to record all the maintenance that has been completed on each device.

logarithm — The exponent that indicates the power to which a number is raised to produce a given number. For $5^2 = 25$, the logarithm of 25 to the base 5 is 2.

logarithmic or audio taper potentiometer — A volume control potentiometer whose resistance decreases (volume increases) logarithmically as the control shaft is rotated in a clockwise direction.

logbooks, mechanical — The journals containing records of the total operating time, repairs, alterations or inspections performed, and all AD notes complied with. A mechanical logbook should be kept for the airframe, for each engine, and for each propeller.

logic — Valid reasoning through facts and actuality.

logic circuit — A circuit designed to operate according to the fundamental laws of logic.

logic flowchart — A flowchart that resembles a pert chart, which graphically shows the flow of information through a computer program, and the decisions that must be made at various points. Boxes show information, a diamond shaped figure is a logical decision point, and a parallelogram is a point where data is either put into or taken out of the program.

logic functions — Statements of logic conditions used in digital computers. Usually associated with logic gates such as AND, OR, NOT, NAND, NOR, and exclusive OR. See also AND, OR, NOT, NAND, NOR, and exclusive OR.

logic gate — See logic functions.

logic one — In logic functions, the ON, YES, or TRUE choices that correspond to a binary "1."

logic state — The state or condition (logic one or zero) of a digital electronic conductor.

logic zero — In logic functions, the OFF, NO, or FALSE choices that correspond to a binary "0."

logical one — A YES or a TRUE condition in digital electronics. It is produced by a closed switch or the presence of a voltage.

logical zero — A NO or a FALSE condition in digital electronics. It is produced by an open switch or the absence of a voltage.

long duct turbofan — A design that ducts the cold stream to the rear of the engine and to the atmosphere. The cold and hot streams mix on some engines.

long range navigation (LORAN) — An electronic navigational system by which lines of position are determined by measuring the difference in the time of reception of synchronized pulse signals from fixed transmitters

long waves — In meteorology, the wave-like structure in the contour and westerly wind patterns in the mid- and upper troposphere. Marked by long, (5,000 miles) slow moving wave troughs located frequently along the east coasts of both Asia and North America.

longeron — The main longitudinal strength-carrying member of an aircraft fuselage or engine nacelle.

longitude — Measurement east or west of the Prime Meridian in degrees, minutes, and seconds. The Prime Meridian is 0 degrees longitude and runs through Greenwich, England. Lines of longitude are also referred to as meridians.

longitudinal — Of or pertaining to length.

longitudinal axis — The axis of an airplane that extends through the fuselage from the nose to the tail, passing through the center of gravity. Longitudinal axis is also referred to as the roll axis, which is controlled by the movement of the ailerons.

longitudinal separation — The longitudinal spacing of aircraft at the same altitude by a minimum distance expressed in units of time or miles.

longitudinal stability — Stability about the lateral axis. A desirable characteristic of an airplane whereby it tends to return to its trimmed angle of attack after displacement.

longitudinal wave — A wave in which the particles vibrate in the same direction as the wave as a whole is moving..

long-range communication system (LRCS) — A system that uses satellite relay, data link, high frequency, or another approved communication system that extends beyond line of sight.

long-range navigation (LORAN) — A radio navigation system that utilizes the master and slave stations transmitting timed pulses. The time difference in reception of pulses from several stations establishes a hyperbolic line of position, which can be identified on a loran chart. A fix in position is obtained by utilizing signals from two or more stations.

long-range navigation system (LRNS) — An electronic navigation unit that is approved for use under instrument flight rules as a primary means of navigation, and has at least one source of navigational input, such as inertial navigation system, global positioning system, Omega/very low frequency, or Loran C.

long-term memory — The portion of the brain that stores information which has been determined to be of sufficient value to be retained. In order for it to be retained in long-term memory, it must have been processed or coded in the working memory.

long-wire antenna — A radio energy antenna with a length greater than one-half the wavelength of the frequency of the energy being transmitted or received.

loom — A tubular flexible insulating material used for wire protection.

loop — 1. A control circuit consisting of a sensor, a controller, an actuator, a controller unit, and a follow-up or feedback to the sensor. Also, any closed electronic circuit including a feedback signal that is compared with the reference signal to maintain a desired condition. 2. A flight maneuver executed in such a manner that the airplane follows a closed 360° circle in a vertical plane.

loop antenna — A highly directional sensitive antenna wound in the form of a coil used to find the direction between the loop and the station transmitting the received signal.

loopstick antenna — An antenna with a large number of turns of wire wound on a powdered iron (ferrite) rod to increase the radio signal the coil receives. Loopsticks are particularly useful in small portable radio receivers.

LORAN — An electronic navigational system by which hyperbolic lines of position are determined by measuring the difference in the time of reception of synchronized pulse signals from two fixed transmitters. Loran A operates in the 1750-1950 kHz frequency band. Loran C and D operate in the 100-110 kHz frequency band.

lost communications — Loss of the ability to communicate by radio. Aircraft are sometimes referred to as NORDO (No Radio). Standard pilot procedures are specified in FAR 91. Radar controllers issue procedures for pilots to follow in the event of lost communications during a radar approach when weather reports indicate that an aircraft will likely encounter IFR weather conditions during the approach.

louver — An opening with fixed or movable slanted slats. Also spelled louvre.

low — In meteorology, an area of low barometric pressure, with its attendant system of winds. Also referred to as a cyclone.

low altitude airway structure — The network of airways serving aircraft operations up to but not including 18,000 feet MSL.

low altitude airway structure / federal airways (USA) — The network of airways serving aircraft operations up to but not including 18,000 feet MSL.

low altitude alert system — An automated function of the TPX-42 that alerts the controller when a Mode C transponder-equipped aircraft on an IFR flight plan is below a predetermined minimum safe altitude. If requested by the pilot, LAAS monitoring is also available to VFR Mode C transponder-equipped aircraft.

low altitude alert, check your altitude immediately — A safety alert issued by ATC to aircraft under their control if ATC is aware the aircraft is at an altitude which, in the controller's judgment, places the aircraft in unsafe proximity to terrain/obstructions.

low approach — An approach over an airport or runway following an instrument approach or a VFR approach including the go-around maneuver where the pilot intentionally does not make contact with the runway.

low blower — The lower speed setting of a two-speed internal supercharger.

low bypass turbofan — An engine with a one-to-one bypass ratio. Approximately the same air mass flows across the fan as across the core engine.

low frequency (LF) — The frequency band between 30 and 300 kHz.

low IFR (LIFR) — Weather characterized by ceilings lower than 500 feet AGL and/or visibility less than one statute mile.

low pitch, high RPM setting — The setting of a controllable-pitch propeller that allows the engine to produce its highest RPM with the propeller at its lowest pitch.

low pressure turbine — The turbine rotor that drives the low pressure compressor in a dual or triple spool axial flow gas turbine engine.

lower turbulent zone — The portion of the lee wave system, starting at ground level extending to just above the mountaintop, and marked by turbulence.

low-frequency radio waves — Radio waves of frequency lower than the bottom of the commercial AM broadcast radio band.

low-lead 100-octane aviation gasoline (100 LL) — Gasoline that contains a maximum of 2 ml. of tetraethyl lead per gallon. Normal 100-octane avgas is allowed to contain 4.6 ml. of tetraethyl lead per gallon.

low-level wind shear — Wind shear below 2,000 feet AGL along the final approach path or along the takeoff and initial climb out path.

low-level wind shear alert system (LLWAS) — A system installed at many large airports that continually monitors surface winds at remote sites on the airport. A computer evaluates the wind differences from the remote sites to determine if a wind shear problem exists.

low-pass filter — A filter circuit designed to pass low-frequency signals and attenuate high-frequency signals.

low-pressure compressor — The front section of a dual compressor gas turbine engine Also referred to as the N_1 compressor or low-speed compressor.

low-pressure compressor gas turbine engine — The front section of a dual compressor driven by the last stages of the turbine. Also referred to as the N_1 compressor or low-speed compressor.

low-pressure oxygen system — A gaseous oxygen system formerly used in military aircraft in which the oxygen is stored under pressures of approximately 450 PSI.

low-tension ignition system — A magneto system used for reciprocating engine airplanes that fly at high altitudes. It consists of a rotating magnet, a cam, breaker points, a condenser, a coil with only the primary winding, and a carbon brush-type distributor. The primary current is directed through the distributor to a coil for each individual spark plug. These coils have a primary and a secondary winding that generate the high voltage at the spark plug.

low-tension magneto — An ignition system used for reciprocating engine airplanes that fly at high altitudes. It consists of a rotating magnet, a cam, breaker points, condenser, a coil with only the primary winding, and a carbon brush-type distributor. The primary current is directed through the distributor to a coil for each individual spark plug. These coils have a primary and a secondary winding that generate the high voltage at the spark plug.

low-voltage ignition system — A main ignition system used on turbine engines with a voltage output in the range of approximately 1,000 to 5,000 volts delivered to the igniter plug.

low-wing airplane — An airplane having one main supporting aerodynamic surface flush with the bottom of the fuselage.

lubber line — The reference line on a magnetic compass or directional gyro that represents the heading of a ship or aircraft.

lubricant — A natural or artificial substance used to reduce friction, heat, and wear between moving parts. It also can be used to prevent corrosion on metallic surfaces.

lubricating — The process of applying a lubricant.

Lucite — A transparent, thermoplastic resin used for windshields and side windows of small aircraft. A trademark of the DuPont company.

lug — A projection from a structural member used as an attachment point.

luminance — The intensity of light emitted or scattered from a surface area in a given direction.

luminescence — The emission of light from essentially nonthermal sources such as phosphorescence.

luminous paint — A paint that glows in the dark. Used for marking aircraft instrument dials and pointers.

lye — An alkaline solution consisting of potassium hydroxide or sodium hydroxide.

M

mach — The ratio of the aircraft's true airspeed to the speed of sound.

mach cone — The cone-shaped shock wave produced by an object exceeding the speed of sound.

mach number — The ratio of the speed of an airplane to the speed of sound in the same atmospheric conditions; e.g., MACH.82, MACH 1.6.

mach technique [ICAO] — Describes a control technique used by air traffic control whereby turbojet aircraft operating successively along suitable routes are cleared to maintain appropriate MACH numbers for a relevant portion of the enroute phase of flight. The principle objective is to achieve improved utilization of the airspace and to ensure that separation between successive aircraft does not decrease below the established minima.

machine bolt — The common name for a hex head bolt with uniform threads.

machine language — A language used in a computer system made up of zeros and ones. A special program called a compiler converts a programming language into machine language that can be used by the computer.

machine screw — A screw fastener with uniform threads that can be screwed into a tapped hole or into a nut. The head of a machine screw can be round, flat, truss, oval, or a fillister-type.

machine-sewn fabric seams — Machine-sewn aircraft fabric seams. The most common types of machine-sewn fabric seams include the French fell, folded fell, and plain overlap.

machining — The process of forming the surface by cutting away material by turning, planing, shaping, and milling. Normally accomplished with machine-operated tools such as lathes, milling machines, shapers, and planers.

machinist — A skilled person in the operation of metal-working machine tools such as lathes, shapers, planers, and milling machines.

machmeter — A direct-reading indicator installed in the instrument panel of high-speed aircraft that gives the pilot an indication of his flight Mach number. The internal mechanism of a machmeter includes a bellows for measuring the difference between pitot pressure and static pressure, and includes an aneroid that modifies the output from this differential pressure bellows to correct for the changes in altitude.

mackerel sky — A meteorological condition of clouds that resemble the scales on a mackerel fish. The clouds consist of rows of altocumulus or cirrocumulus clouds.

macroscale — Spatial scales of 1,000 NM or more.

magamp — A contraction of magnetic amplified. An amplifier system using saturable reactors to control an output to obtain amplification. See also magnetic amplifier.

magnesium — A silver-white, malleable, ductile metallic element with a symbol of Mg and an atomic number of 12. Used to produce light alloys for aircraft construction.

Magnesyn system — An AC remote indicating system in which a permanent magnet is used as the rotor. Based upon the synchronous-motor principle, in which the angular position of the rotor of one motor at the measuring source is duplicated by the rotor of the indicator motor. Used in fuel-quantity or fuel-flow measuring systems, position-indicating systems, etc.

magnet — A device or material that has the property of attracting or repelling other magnetic materials. Lines of magnetic flux link its external poles, and a conductor cutting across the flux will have a voltage induced into it.

magnet keeper — A soft iron bar placed across the north and south poles of a U-shaped magnet. The iron bar produces a closed path through which magnetic lines of force pass.

magnet wire — A small-diameter, varnish-insulated copper wire used in coil windings for electromagnets, transformers, motors, and generators.

magnetic amplifier — An electronic control device that uses a saturable reactor. The condition of saturation is controlled by the input signal to modulate the flow of a much larger current in the output circuit. It is essentially a multi-coil transformer that controls the amount of load current allowed to flow in the load winding by a small amount of current in the magnetic core. Changing the amount of DC flowing in the magnetic core changes the permeability of the core. This, in turn, changes the amount of inductive reactance that opposes the AC flowing in the load winding.

magnetic bearing — The magnetic course to go direct to an NDB station.

magnetic brake — 1. A friction brake controlled by an electromagnetic solenoid. The brake can be either actuated or released by electromagnetic action (usually on or off). 2. A brake that uses magnetism to oppose rotation of a disc or drum without any physical contact to create the slowing force. Often used in motor driven devices and in effect reverses the rotational forces on the motor until the motor comes to a stop. If the reverse forces were not removed when the motor stops, the motor would start to turn the opposite direction.

magnetic bubble memory — The memory stored in the form of bubbles or circular areas on a thin film of magnetic media. Used on early digital computers as non-volatile memory, but no longer widely used.

magnetic chuck — A metal machining tool that consists of a special work surface that uses electromagnetism to hold the material being machined.

magnetic circuit — Any complete path of magnetic lines of flux that leaves the north magnet pole of an electrical machine such as a motor or generator and enters the magnetic south pole.

magnetic circuit breaker — A circuit breaker that opens a circuit whenever there is an excess of current flow in the circuit. It works on the principle of electromagnetism. When rated current flow is exceeded, the magnetic field develops enough strength to open a set of contacts and deenergize the circuit.

magnetic compass error — Acceleration error, magnetic dip, and turning error. Acceleration error is inherent in magnetic compasses, caused by the force of acceleration acting on the dip compensating weight when the aircraft accelerates or decelerates on an easterly or westerly heading. In compasses compensated for flight in the Northern Hemisphere, when the aircraft accelerates on an easterly or westerly heading, the compass gives the indication that the aircraft is turning to the North. When the aircraft decelerates on either of these headings, the compass gives the indication that the aircraft is turning to the South. Magnetic dip is an error as the result of the north end of the compass trying to dip toward the magnetic poles in the North and South Hemispheres. This is compensated for by adding weights on the south end of the bar magnet in the Northern Hemisphere, and the north end in the Southern Hemisphere. Turning error is caused by the dip compensating weight. It shows up mostly on turns to or from north or south headings and causes the compass to lead or lag the actual turn.

magnetic course — The path of an airplane as measured from magnetic north.

magnetic deviation — A compass error caused by localized magnetic fields in the airplane attracting the floating magnets in the magnetic compass and deflecting it away from magnetic north.

magnetic drag cup — The aluminum or copper cup surrounding the rotating magnet in a simple mechanical tachometer. Eddy currents are generated in this cup by the rotating magnetic field, resulting in eddy currents and attendant magnetic fields. This resultant magnetic field interacts with those of the rotating magnet and the cup then is displaced axially by the rotation.

magnetic drain plug — Similar to a chip detector, except some types cannot be powered to show contamination on a warning light in the cockpit. The drain plug consists of two small permanent magnets built into it to attract and hold any ferrous metal particles that can be in the lubricating oil system. Ferrous metal chips on the drain plug indicate the possibility of internal engine failure. Usually located in the lower portion of a sump in the scavenge oil subsystem.

magnetic field — The space around a magnet or conductor where magnetic flux is found.

magnetic flux — The invisible lines of magnetic force that exist between the poles of a magnet, and which follow the path of least resistance. Traditionally, they are given the direction from north pole to south pole. When an electrical conductor cuts across the lines of magnetic flux, a voltage is produced in the conductor. One line of magnetic force is called one maxwell.

magnetic flux density — The unit of field intensity is the gauss. An individual line of force, called a maxwell, in an area of one square centimeter produces a field intensity of one gauss.

magnetic heading — The angle between the longitudinal axis of an aircraft and magnetic north.

magnetic hysteresis — The tendency of a material to retain magnetism after the magnetizing influence has been removed. Hysteresis loops are graphs that indicate the magnetic properties of different materials.

magnetic north — True north direction corrected for variation error.

magnetic north pole — The point on the Earth's surface in the Northern Hemisphere where isolines of the Earth's magnetic field converge. Compasses align with the lines of magnetic flux that connect the north and south magnetic poles, with the north-facing pointer directed toward the magnetic north pole. The pole's location is not co-located with the north axis of rotation of the Earth known as true north.

magnetic particle inspection — A nondestructive inspection for ferrous metal parts in which the part is magnetized, producing north and south poles across any discontinuity, either on the surface or subsurface. Iron oxide, sometimes mixed with a fluorescent dye, is attracted and held over the discontinuity. The discontinuity shows up as a line of iron oxide. If using fluorescent dye, an ultraviolet light (a black light) shined on the part shows the discontinuity as an incandescent line.

magnetic pickup RPM system — A newer fan speed indicating system that uses a magnetic pickup in the fan case. Blade motion produces eddy currents that are measured and interpreted electronically and displayed as RPM on a cockpit indicator.

magnetic poles — A suspended magnet swinging freely will align itself with the Earth's magnetic poles. One end is labeled "N," meaning north-seeking. The opposite end of the magnet is labeled "S," meaning south-seeking.

magnetic saturation — A magnet's saturated condition in which all of the magnetic domains are lined up in the same direction, and any increase in the magnetic field is not possible.

magnetic shunt — A piece of soft iron shunted across the air gap of a magnet used in an electrical measuring instrument. The position of the magnetic shunt can be changed to calibrate the instrument by varying the amount of magnetic flux that crosses the air gap.

magnetic variation — The angular difference between the geographic north pole and the magnetic north pole. The exact value of variation depends on the position on the Earth where the measurement is taken.

magnetic wave — The component of a radio wave perpendicular to the antenna.

magnetic yoke — The mechanical support that completes the magnetic circuit between the poles of a generator. Also used to describe any ferrous material within the magnetic field of a magnet. This material serves as a low impedance "path of least resistance" route for magnetic lines of flux.

magnetic-drag tachometer — A simple mechanical tachometer that contains an aluminum or copper drag cup surrounding a rotating magnet. The magnet is connected to the equipment whose RPM is to be measured and the drag cup is attached to the pointer in the tachometer instrument. Eddy currents are generated in this cup by the rotating magnetic field, resulting in eddy currents and attendant magnetic fields. This resultant magnetic field interacts with those of the rotating magnet and the cup is then displaced axially by the rotation.

magnetism — The ability to attract certain materials containing iron and to influence moving electrons.

magneto — A self-contained, permanent-magnet AC generator with a set of current interrupter contacts and a step-up transformer. It is used to supply the high voltage required for ignition in an aircraft reciprocating engine.

magneto safety check — An operational check on an aircraft reciprocating engine in which the magneto switch is placed in the OFF position with the engine idling to ascertain that the switch actually does ground out both magnetos.

magnetometer — An instrument used to measure the intensity of a magnetic field. Also used detect the presence of a metallic object.

magnetomotive force (MMF) — The magnetizing force in a magnetic field. Measured in gilberts or ampere-turns.

magnetosphere — The magnetic field surrounding a celestial object (Earth, Moon, stars, etc.). Charged particles are trapped within the magnetosphere.

magnetron — A vacuum tube that generates power at microwave frequencies. The flow of electrons is controlled by an externally applied magnetic field.

magnitude — A condition of size, quantity, or number.

main bus — A common voltage tie point for electrical circuits.

main fuel system — The fuel distribution system used for all normal engine operating conditions.

main rotor — The rotor that supplies the principal lift to a rotorcraft.

main wheels — The wheels of an aircraft landing gear that support the major part of the aircraft weight.

maintain — **1.** Concerning altitude/flight level, the term means to remain at the altitude/flight level specified. The phrase "climb and" or "descend and" normally precedes "maintain" and the altitude assignment; e.g., "descend and maintain 5,000." **2.** Concerning other ATC instructions, the term is used in its literal sense; e.g., maintain VFR.

maintenance — The inspection, overhaul, repair, preservation, and replacement of parts, excluding preventive maintenance.

maintenance manual — A manual produced by the manufacturer of an aircraft, aircraft engine, or component, which details the approved methods of maintenance.

maintenance planning friction level — The friction level specified in AC 150/5320-12, Measurement, Construction, and Maintenance of Skid Resistant Airport Pavement Surfaces that represents the friction value below which the runway pavement surface remains acceptable for any category or class of aircraft operations but that is beginning to show signs of deterioration. This value will vary depending on the particular friction measurement equipment used.

maintenance release — A return to service approval of an aircraft by an authorized A&P technician or IA. Logged in the appropriate maintenance record.

major alteration — An alteration not listed in the aircraft, aircraft engine, or propeller specification, and one which might appreciably affect weight, balance, structural strength, performance, powerplant operation, flight characteristics, or other qualities affecting airworthiness.

major axis of an ellipse — The longer axis passing through one focus of an ellipse. See also ellipse.

major diameter — The diameter of a bolt or a screw to the tip of the threads.

major overhaul — The complete disassembly, cleaning, inspection, repair, and reassembly of an aircraft, engine, or other aircraft component in accordance with the manufacturer's specifications, and which will return the device to a serviceable condition.

major repair — A repair that, if improperly done, might appreciably affect weight, balance, structural strength, performance, powerplant operation, flight characteristics, or other qualities affecting airworthiness. FAR 43, Appendix A provides guidance on what constitutes major repairs and alterations.

major structural damage — Damage to a structure that requires a major repair in order to return it to proper working or operating condition.

make short approach — Used by ATC to inform a pilot to alter his traffic pattern so as to make a short final approach.

make-and-break ignition — One of the earliest forms of electrical ignitions for internal combustion engines. It produced an arc when two contacts carrying low-voltage current within the cylinder were separated.

male electrical connector — The pin contact that completes a circuit by sliding inside a socket (the female connector).

male fitting — A fitting designed to be placed, screwed, or bolted into another unit.

malfunction — The failure of a part of a component to function; a deviation in the operation of a unit from its intended purpose or design.

malleability — The measure of a material's ability to be stretched or shaped by beating it with a hammer or passing through rollers without breaking.

malleable — The ability to be stretched or shaped by beating with a hammer or passing through rollers without breaking.

mallet — A hammer with a heavy wood, plastic, rubber, or leather head.

mammatus — Bulges or pouches that appear under the anvil of a mature cumulonimbus cloud.

mandatory altitude — An altitude depicted on an instrument Approach Procedure Chart requiring the aircraft to maintain altitude at the depicted value.

mandatory altitude — An altitude depicted on an instrument Approach Procedure Chart requiring the aircraft to maintain altitude at the depicted value.

mandrel — **1.** Lathe: A tapered shaft that fits into a hole used to support and center a device or piece of material so that it can be machined. **2.** Tube bending: A long steel rod with a rounded end inserted into a piece of metal tubing in order to keep the tubing from flattening while it is being bent.

maneuverability — The ability of an aircraft to change directions along a flight path and withstand the stresses imposed upon it.

maneuvering — In aviation, to move through a specific series of changes in direction, speed and position for a specific purpose. Often used in regards to positioning an aircraft in the landing pattern.

manganese — A non-magnetic chemical element of grayish white with a symbol of Mn and an atomic number of 25. Used in the manufacturing of iron, aluminum, and copper alloys.

manganese dioxide — A chemical compound used in carbon-zinc batteries to absorb the hydrogen gas that would otherwise insulate the carbon rod when electrons flow from the zinc can to the carbon.

Manganin — An alloy of copper, manganese, and nickel.

manifest — A list of passengers and cargo carried on any one flight.

manifold — A chamber having several outlets through which a liquid or gas is distributed or gathered.

manifold absolute pressure (MAP) — The absolute pressure, measured in inches of mercury, existing in the intake manifold of an engine. This is the pressure that forces the fuel-air mixture into the cylinders.

manifold pressure — The absolute pressure, measured in inches of mercury, existing in the intake manifold of an engine. This is the pressure that forces the fuel-air mixture into the cylinder.

manometer — An instrument consisting of a glass tube filled with a liquid for measuring the pressure of gases or vapors either above or below atmospheric pressure.

manual depressurization valve — A back-up valve used to control cabin pressurization by manually controlling the outflow of air from the cabin if the automatic system malfunctions.

manufactured rivet head — The preformed head of an aircraft rivet when it is manufactured at the factory. Also referred to as a shop head.

manufacturer — In aviation, a person or a company who manufactures aircraft, aircraft engines, or aircraft components.

manufacturing — The process of taking raw materials and changing them into finished and usable products.

Marconi antenna — A non-directional, quarter-wave antenna utilizing a ground plane that serves as a quarter-wave reflector. Used for transmitting and receiving radio communications in the higher frequency bands.

marginal VFR (MVFR) — Weather characterized by ceilings 1,000 to 3,000 feet AGL and/or visibility three to five statute miles.

marine grommet — A plastic or metal reinforcement ring designed with a special shield used to keep water spray caused by takeoff and landings from entering the structure. It is normally attached to the underside of wings and control surfaces of fabric-covered aircraft and used to reinforce drain holes that are cut into the fabric.

marker beacon (MB) — An electronic navigation facility transmitting a 75 MHz vertical fan or boneshaped radiation pattern. Marker beacons are identified by their modulation frequency and keying code, and when received by compatible airborne equipment, indicate to the pilot, both aurally and visually, that he is passing over the facility.

married needles — A term used regarding an engine-rotor tachometer when the hands are superimposed. One hand indicates engine RPM and the other hand indicates the main rotor RPM.

Marvel balancer — A universal balancer commonly used throughout the helicopter industry.

masking material — Aluminum foil or special paper used during painting to cover areas of an aircraft surface that are not to be sprayed or on which a finish is not to be applied.

masking tape — Paper tape that has a sticky surface on one side and generally comes in rolls of varying widths. Used for masking areas during painting.

Maslow's hierarchy of human needs — A listing by Abraham Maslow of needs from the most basic to the most fulfilling. These range from physical through safety, social, and ego to self-fulfillment.

Masonite — A type of fiberboard.

mass — A measure of the amount of material or matter contained in a body. It is the property of a body that causes the force of gravity to give a body weight.

mass flow rate — The result of a fluid's density and its linear velocity.

mass production — The production of objects in very large quantities in a relatively short time period by the use of complex, and often computerized, equipment.

massive-electrode spark plug — Spark plugs using two, three, or four, large nickel-alloy ground electrodes.

mass-type fuel flowmeter — A fuel-flow measurement system used with turbine engines that indicates the mass flow rather than the volume flow.

mast — In rotorcraft, the component that supports the main rotor.

mast bumping — In rotorcraft, the action of the rotor head striking the mast, occurring on underslung rotors only.

master cylinder — A combination cylinder, piston, and reservoir used in an aircraft brake system. Fluid is stored when the brakes are not applied and is then forced into the system for braking.

master minimum equipment list (MMEL) — A master list developed by the Federal Aviation Administration for an aircraft by make and model delineating the specific equipment allowed to be inoperative during various types of flight operations. The MMEL is the basis for the development of an MEL.

master rod — The only connecting rod in a radial engine attached directly to the crankshaft. All of the other rods connect to the master rod rather than the crankshaft.

master switch — A single switch designed to control electric power to all circuits in a system.

mat — In composites, typically used in the mold making process. Chopped fibers are held together with a binder. When the resin matrix is applied, the binder melts. Typically used with polyester resin systems.

matched gears — Two gears used in a set and only replaced in a set.

matching transformer — An electronic device used to connect a load and a source of electrical power that differ in their impedance. It consists of a transformer with a primary winding that matches the impedance of the source and a secondary winding that matches the load.

mathematics — That branch of science dealing with numbers and their operation.

mating surfaces — Two surfaces that come together to form a seal.

matrix — In composite construction, the material that bonds the fibers together, and distributes the stress to the fibers. Typically in advanced composites, the matrix is a resin.

matter — Any substance that has weight and occupies space.

mature stage — In meteorology, the most intense stage of a thunderstorm. Begins when the precipitation-induced downdraft reaches the ground. Usually lasts about 20 minutes.

maximum allowable zero-fuel weight — The maximum weight authorized for an aircraft excluding its fuel load.

maximum authorized altitude (MAA) — A published altitude representing the maximum usable altitude or flight level for an airspace structure or route segment. It is the highest altitude on a Federal airway, jet route, area navigation low or high route, or other direct route for which an MEA is designated in FAR 95 at which adequate reception of navigation aid signals is assured.

maximum except takeoff power (METO) — The maximum continuous power an engine is allowed to develop without any time restrictions.

maximum landing weight — The maximum authorized weight of the aircraft for landing.

maximum range — The maximum distance an aircraft can travel by flying at the most economical speed and altitude during all stages of flight.

maximum takeoff power — The maximum power an engine is allowed to develop for a limited period of time, usually about one minute.

maximum takeoff weight — The maximum design weight of any aircraft for takeoff.

maximum weight — The maximum allowable weight for an aircraft under any conditions.

maxwell — A unit of magnetic flux. One magnetic line of force.

may — As related to aircraft maintenance, means that such an item is allowed, but not required.

mayday — The international radiotelephony distress signal. When repeated three times, it indicates imminent and grave danger and that immediate assistance is requested.

mean — The average of a number of factors. Often used to indicate the mid-point between two extremes.

mean aerodynamic chord (MAC) — The chord of an imaginary rectangular airfoil having the same aerodynamic characteristics as that of the actual wing.

mean sea level (MSL) — The reference used for measuring altitude above sea level. Mean sea level is the average height of the surface of the sea.

mean solar day — The average time it takes the Earth to rotate about its axis in one day.

measured ceiling — In weather, a classification where the height of the ceiling has been determined (by one of several measures) as opposed to estimated.

measuring circuit — Any combination of resistors, batteries, and meters that make it possible to measure electrical values.

mechanic — In aviation, a person certificated by the FAA as an Airframe and Powerplant (A & P) mechanic. (Mechanics can be certified as either a Powerplant and/or Airframe mechanic.) Aviation Maintenance Technician (AMT) has become the preferred term.

mechanical advantage — The increase in force or speed gained by using such devices as levers, pulleys, gears, or hydraulic cylinders.

mechanical blockage thrust reverser — A thrust reverser usually of the post exit-type (clamshell) used to reverse the hot exhaust stream of a gas turbine engine to help slow the airplane during landings.

mechanical bond — The joining of two or more parts or pieces by mechanical methods such as bolts, rivets, or pins.

mechanical efficiency — The ratio of the brake horsepower delivered to the output shaft of an engine to its indicated horsepower.

mechanical energy — Energy that expresses itself in mechanical movement or the physical production of work.

mechanical linkage — A direct connection between a control and a unit. No remote actuator.

mechanical mixture — A mixture of two or more elements or compounds that can be identified by microscopic examination.

mechanical properties — Those properties that involve a relationship between strain and stress.

mechanical turbulence — The turbulence that results when airflow is slowed by surface friction.

median — In mathematics, the number in an ordered set of values below and above which there is an equal number of values. In the list 1,2,3,7,15,16, and 23, the median is seven since there are three numbers smaller and three numbers greater than seven. Note that this differs from the mathematical mean.

medical certificate — Short for FAA Airman Medical Certificate. Obtained by passing a physical examination administered by a doctor who is an FAA-authorized Aviation Medical Examiner (AME).

medium bypass turbofan — Engines with 2:1 or 3:1 bypass ratios, which is the ratio of the amount of air the fan moves (or bypasses) in relation to the core engine. Also referred to as moderate bypass.

medium frequency (MF) — The band of electromagnetic radiation frequencies that lie between 300 kHz and 3 MHz.

medium-frequency radio transmission — An outgoing signal from a transmitter broadcasting an electromagnetic radiation frequency between 300 kHz and 3 MHz.

medium-scale integration — A method of fabricating integrated circuit (IC) chips to place multiple transistors or logic circuits on one small IC. Medium-scale integration places 10 to 100 circuits on one integrated circuit.

megahertz (MHz) — 1,000,000 cycles per second.

megger — A high-voltage, high-range ohmmeter that has a built-in (often hand-turned) generator for producing the voltage needed to measure insulation resistance and the resistance between a component and electrical ground. Also referred to as a megohmmeter.

megohmmeter — High-resistance measuring instrument incorporating a high-voltage DC generator in the instrument case. Not only does this measure high resistance, it does it with a high enough voltage to cause insulation breakdown if it has been weakened. Also referred to as a megger.

melt — A change in the physical state of a material when it goes from a solid to a liquid as a result of absorbing sufficient heat to produce the change.

melting — The change of state of a solid to a liquid, as ice to water.

melting point — A temperature at which a solid becomes a liquid.

member — Any portion of the aircraft structure essential to the whole.

memory effect — A reduction in the service capacity of nickel-cadmium cells that occurs when cells on standby service are regularly recharged after being discharged to only a small fraction of their full service capacity.

meniscus — The curved upper surface of a column of liquid in a tube. If the liquid wets the tube, the curve will be concave; if the liquid does not wet the tube, the curve will be convex.

mensuration — The act or process of measuring.

Mercator projection — A map projection where parallel meridians and lines of latitude are straight lines at right angles to the meridians. The distance between lines of latitude increases as they move farther away from the equator.

mercerize — See mercerizing.

mercerizing — The process of dipping cotton yarn or fabric into a hot solution of caustic soda. It gives the material greater strength and luster and is stronger and more pliable than untreated fabric.

mercury — A heavy, silver-colored, toxic, liquid, metallic chemical element with a symbol of Hg and an atomic number of 80. Mercury remains in a liquid state under standard conditions of pressure and temperature. Mercury is approximately 13 times as heavy as water.

mercury barometer — A closed glass tube partially filled with mercury, used to determine the pressure exerted by the atmosphere. The standard atmospheric pressure at sea level will hold the mercury in the tube to a height of 760 mm or 29.92".

mercury cell — A primary cell using zinc for the negative electrode, mercuric oxide for the positive electrode, and potassium hydroxide as the electrolyte.

mercury clutch — A centrifugal clutch in which mercury is used to engage the clutch.

mercury oxide cell — A chemical cell using powdered mercuric oxide and powdered zinc as its pole pieces. The electrolyte is a liquid solution of potassium hydroxide.

mercury switch — A switch that makes and breaks the circuit by way of mercury in a glass tube bridging the contacts as the tube is rocked back and forth.

mercury thermometer — A thermometer consisting of a glass tube with an extremely small inside diameter to which is attached a small reservoir containing mercury. A temperature scale marked alongside the tube is used to indicate the temperature when the mercury expands up the tube due to a rise in temperature.

mercury trap — A container in the pick-up tube of a vacuum cleaner used to retrieve spilled mercury. The mercury is sucked up by the cleaner and deposited in the bottle that prevents it from being sprayed out by the discharge of the cleaner.

mercury vapor lamp — A lamp that glows from the excitation of mercury vapor atoms by an electric arc.

mercury-vapor rectifier — A rectifier tube containing mercury, which vaporizes during operation and increases the current-carrying capacity of the tube.

mesh — The engagement of the teeth of gears.

mesh rating — A U.S. sieve number and filtration rating common to fuel filters. Similar to a micron rating, e.g. a 74 micron filter carries an equivalent U.S. sieve number of 200 and has 200 openings per linear inch.

mesopause — The outer extent of the mesosphere, slightly more than 280,000 feet MSL, the boundary between the mesosphere and thermosphere.

mesoscale — Spatial scales from 1 to 1,000 nautical miles.

mesoscale convective complex (MCC) — The nearly circular clusters of thunderstorms 300 NM or more in diameter. MCC develops primarily between the Rockies and the Appalachians during the warmer part of the year.

mesoscale convective systems (MCS) — A large cluster of thunderstorms with horizontal dimensions on the order of 100 miles. MCSs are sometimes organized in a long line of thunderstorms (e.g., a squall line) or as a random grouping of thunderstorms. Individual thunderstorms within the MCS may be severe.

mesosphere — A layer of the atmosphere between the top of the stratosphere or the ionosphere and the exosphere (about 250-600 miles above the Earth).

metal — A chemical possessing most of the following characteristics: usually rather heavy, with a bright and shining surface, malleable, ductile, and a good electrical conductor.

metal chip detector (MCD) — An electrical device for warning the aircrew of ferrous particles in the oil. Consists of two magnets separated by a narrow gap or insulator. The detector is found in the bottom of the oil sump. When ferrous particles from the oil are attracted to the magnets and bridge the gap, a circuit is completed and a warning lamp illuminates in the cockpit.

metal fatigue — A method of work hardening or cold working of a metal that results from flexing or vibration and which increases the brittleness of the material to its breaking point.

metal foil — A very thin sheet of metal such as aluminum foil.

metal-matrix composites (MMC) — In composites, fibers bonded together with a metal as the bonding material.

metal oxide rectifier — An electronic device that enables one-way flow of current through the flow of electrons from the base material to an oxide layer, but not from the oxide layer to the base material.

metal oxide semiconductor capacitor — An electronic device that utilizes metal oxide as a dielectric to create capacitance.

metal oxide semiconductor field effect transistor (MOSFET) — A semiconductor whose gate is insulated from the channel. The MOSFET was originally called the insulated-gate FET (IGFET). The MOSFET has an extremely large input impedance. Because the insulating oxide layer is extremely thin, the MOSFET is susceptible to destruction by electrostatic charges. Special precautions are necessary when handling or transporting MOS devices.

metal sheath — A close-fitting metal cover.

metal spinning — A process of metal forming in which sheet metal is clamped into a lathe along with the male die. A shaping tool is used to force the spinning metal against the die.

metal spraying — A method of covering or repairing a material with a coating of metal. The metal to be used for the coating is melted and sprayed out with hot, high-velocity compressed air.

metal-film resistor — A resistor in which an oxide of metal is deposited as a film onto a base material. The type of metal and the thickness determine the resistance of the device.

metallic — Having the nature of metal or containing metal.

metallic ion concentration cell corrosion — Corrosion that results from a concentration of metallic ions in the electrolyte. The area of high concentration of metallic ions is the cathode.

metallic pigment — Extremely tiny flakes of metal suspended in paint to produce a sheen.

metallic ring test — A test for delaminations in a bonded structure in which a coin or similar object is used to tap on the surface. If the bond is good, a metallic ringing sound will be produced, however, if it is delaminated, a dull thud will be heard.

metallizing — **1.** To replace the fabric covering on an aircraft structure with sheet metal. **2.** A method of metal overlay or metal bonding to repair worn parts

metallurgy — The science and technology dealing with metals and their use.

metalworking tools — Machines and tools used in the construction and repair of sheet metal structures.

metamerism index — A measurement used for scientific color matching. It indicates the way a pigment will look under varying light conditions.

METAR — The international weather reporting code that will be introduced in the U.S., after June 1, 1996.

metastable compound — A chemical compound that has only a slight margin of stability.

meteor — A small particle of matter in the solar system visible only as it burns due to the high temperature caused by friction as the meteor falls through the Earth's atmosphere.

meteorological impact statement — An unscheduled planning forecast describing conditions expected to begin within 4 to 12 hours that can impact the flow of air traffic in a specific center's (ARTCC) area.

meteorological visibility — A measure of horizontal visibility in a given direction near the Earth's surface, based on sighting of objects in the daytime or unfocused lights of moderate intensity at night.

meteorology — The study of weather and atmospheric phenomena.

meter — **1.** A device used to measure, indicate, or record. **2.** The basic unit of length measurement in the metric system equal to approximately 39.37".

meter fix time/slot time — A calculated time to depart the meter fix in order to cross the vertex at the ACLT. This time reflects descent speed adjustment and any applicable time that must be absorbed prior to crossing the meter fix.

meter list display interval — A dynamic parameter that controls the number of minutes prior to the flight plan calculated time of arrival at the meter fix for each aircraft, at which time the TCLT is frozen and becomes an ACLT; i.e., the VTA is updated and consequently the TCLT modified as appropriate until frozen at which

time updating is suspended and an ACLT is assigned. When frozen, the flight entry is inserted into the arrival sector's meter list for display on the sector PVD/MDM. MLDI is used if filed true airspeed is less than or equal to freeze speed parameters (FSPD).

metering — A method of time-regulating arrival traffic flow into a terminal area so as not to exceed a predetermined terminal acceptance rate.

metering airports — Airports adapted for metering and for which optimum flight paths are defined. A maximum of 15 airports can be adapted.

metering device — A device used to measure or control the amount of fluid flow.

metering fix — A fix along an established route from over which aircraft will be metered prior to entering terminal airspace. Normally, this fix should be established at a distance from the airport that will facilitate a profile descent 10,000 feet above airport elevation (AAE) or above.

metering jet — The calibrated orifice in a fluid-flow system used to control the amount of flow for a given pressure drop across the jet.

metering pin — A flow control device such as a tapered pin in an oleo shock absorber used to progressively restrict the passage of fluid from one chamber into the other, cushioning the landing impact. The shape or contour of the metering pin determines the amount of fluid that can flow with the pin in any position other than full in or full out.

metering valve — A valve used to control the flow of a fluid.

meter-kilogram — The amount of work produced when one kilogram of force acts through a distance of one meter.

methanol wood alcohol — A liquid alcohol produced by the distillation of wood pulp.

methyl bromide — A fire extinguishing agent (CH_3Br). More effective than CO_2 from a standpoint of weight, but more toxic than CO_2. It will seriously corrode aluminum alloy, zinc, and magnesium. Methyl bromide cannot be used in areas where harmful concentrations can enter personnel compartments.

methylene chloride — A liquid solvent (CH_2CL_2) used as the active agent in many paint strippers.

methyl-ethyl-ketone (MEK) — A low-cost solvent similar to acetone. Used as a cleaning agent to prepare a surface for painting and as a stripper for certain finishes. Should be used only in well-ventilated areas.

metric horsepower — A measurement of power in the metric system of measurement. One metric horsepower is equal to 1.0139 mechanical horsepower in the system of measurement used in the United States.

metric prefixes — A system of prefixes that indicate multiples and submultiples of ten. Some of the more common prefixes used are:

Tera 10^{12}
Giga 10^{9}
Mega 10^{6}
Kilo 10^{3}
milli 10^{-3}
micro 10^{-6}
nano 10^{-9}
pico 10^{-12}

mho — A unit of electrical conductance; the reciprocal of ohm.

mica — A transparent silicate mineral. It is used as an electrical insulator in capacitors and as an insulator for electric irons and heaters.

Micarta — A phenolic-type thermosetting resin impregnated cloth. It is used as an electrical insulator and for the manufacturing of control pulleys.

mice — Small sheet metal, wedge-shaped tabs inserted into the tail pipe of some older turbine engines to reduce the nozzle opening and increase thrust. Used to "trim" a turbine engine. Also referred to as tail pipe inserts.

micro - One millionth (0.000001) of a unit.

microammeter — An electrical current measuring instrument capable of measuring current flow in millionths of an ampere.

microballoons — Microscopic-size phenolic or glass spheres used to add body with very little weight to a resin when used as a filler or potting compound.

microbarograph — An instrument used in meteorology to measure very small changes in pressure.

microbes — Microscopic forms of animal life. They exist in water and feed on hydrocarbon aircraft fuel. Microbes form a water-entrapping scum on the bottom of jet aircraft fuel tanks.

microbiological corrosion — The deterioration of materials caused directly or indirectly by bacteria, algae, molds, or fungi, either alone or in combination. Microbiological corrosion is significant to aviation when linked to corrosion of airframe components, particularly in fuel tanks and related systems.

microburst — A small downburst with outbursts of damaging winds extending 2.5 miles or less. In spite of its small horizontal scale, an intense microburst could induce wind speeds as high as 150 knots.

microcircuit — An extremely small electronic component that has a large number of circuit elements combined into a single unit.

microelectronics — The branch of electronics that deals with integrated circuits and other small electronic devices.

micro-en route automated radar tracking system (M-EARTS) — An automated radar and radar beacon tracking system capable of employing both short-range (ASR) and long-range (ARSR) radars. This microcomputer driven system provides improved tracking, continuous data recording, and use of full digital radar displays.

microfarad (µF) — One millionth (0.000001) of a farad.

microfiche — Sheets of film that store reduced-size printed and graphic information. One sheet of microfiche can stores between 24 and 288 individual frames or pages of material on a single 4" x 6" sheet of photographic film. The microfiche film is read on a reader that enlarges the image and can often make a printed copy of the desired page(s), if needed.

microfilm — Reproduction of printed material on 35 mm or 16 mm photographic film; used to store vast quantities of written material in a small space. The microfilm is read on a reader that enlarges the image and can make a full size printed copy of the desired page(s), if needed.

microinch — One millionth (0.000001) of an inch.

micrometer — One millionth (0.000001) of a meter. A micrometer is also referred to as a micron.

micrometer caliper — A precision measuring device having a single movable jaw, advanced by a screw. One revolution of the screw advances the jaw 0.025".

micrometer setting torque wrench — A hand-operated torque wrench in which a preset torque is adjusted with a micrometer-type scale. When torque is reached, the handle of the wrench breaks over, indicating the torque to the operator.

micro-microfarad — A unit of capacitance equal to one millionth of a millionth of a farad. Also referred to as picofarad.

micron — **1.** One millionth (0.000001) of a meter, or one thousandth (0.001) of a millimeter (0.000001 meter or 1×10^6 meter). Also referred to as a micrometer. **2.** The pressure measurement in a column of mercury: One micron of pressure is equal to 0.001 millimeter of mercury (1×10^6 meter of mercury) at 0°C. **3.** One micron is normally used to denote the effectiveness of a filter.

micronic filter — A disposable element filter used in hydraulic or pneumatic systems that filters particles as small as one micron.

microorganism — An organism of microscopic size, normally bacteria or fungus.

microphone — A device for converting sound waves to electric signals.

microprocessor — A small central processing unit (CPU) for a microcomputer.

microscale — Spatial scales of 1 n.m. or less.

microscope — An optical instrument used to magnify extremely small objects so they can be seen by the human eye.

microsecond — One millionth (0.000001) of a second.

microshaver — An adjustable metal-cutting tool used for shaving the heads of countersunk rivets so they are flush with the surface.

microshaving — A process in sheet metalwork in which the head of the countersunk rivet is shaved to absolute smoothness with the surface of the skin.

microswitch — An electrical switch used to open or close a circuit with an extremely small movement of the actuator.

microwave landing system — A precision instrument approach system operating in the microwave spectrum that normally consists of the following components:
a. Azimuth Station.
b. Elevation Station.
c. Precision Distance Measuring Equipment.

microwaves — Electromagnetic radiation with a wavelength between infrared and short-wave radio wavelengths (frequency higher than 1 gigahertz and a wavelength shorter than 30 centimeters).

mid RVR — The RVR (runway visual range) readout values obtained from RVR equipment located midfield of the runway.

middle compass locator — A low power, low or medium frequency (L/MF) radio beacon installed at the site of the middle marker of an instrument landing system (ILS).

middle marker — A marker beacon that defines a point along the glide slope of an ILS normally located at or near the point of decision height (ILS Category I). It is keyed to transmit alternate dots and dashes, with the alternate dots and dashes keyed at the rate of 95 dot/dash combinations per minute on a 1300 Hz tone, which is received aurally and visually by compatible airborne equipment.

mid-flap — The middle flap on a triple-slotted segmented flap.

mid-span shrouds — The lugs on fan blades that contact each other to provide a circular support ring. Mid-span shrouds provide strength and reduce vibration.

mid-span weight — A weight placed in the mid-span area of a helicopter rotor blade to add inertia to the blade.

mid-wing airplane — An airplane having its main aerodynamic surface located in the center of the fuselage.

migrate — To move from one place to another. During an inspection, a technician might discover a fastener has moved or migrated from its intended location to another location due to vibration or jostling.

migration — See migrate.

mil — Commonly used to represent one one-thousandth (0.001) of an inch.

mild steel — Steel that contains between 0.05 and 0.25% carbon.

mildew — A gray colored parasitic fungus growth that forms on organic matter.

mildewcide — An additive to dope or sealers used when covering organic materials to inhibit the growth of mildew.

mile (mi) — One statute mile. Equal to 5,280 ft.

mileage break point — On IFR charts, points other than intersections or navaids where the mileage is broken down for reasons such as reception distance. The distances between navaids is in a box. A number without an outlined box indicates mileage between any combination of intersections, navaids, or mileage break points. All distances are in Nautical Miles.

miles-in-trail — A specified distance between aircraft, normally, in the same stratum associated with the same destination or route of flight.

military authority assumes responsibility for separation of aircraft (MARSA) — A condition whereby the military services involved assume responsibility for separation between participating military aircraft in the ATC system. It is used only for required IFR operations that are specified in letters of agreement or other appropriate FAA or military documents.

military operations area (MOA) — Airspace established outside of a Class A airspace area to separate or segregate certain nonhazardous military activities from IFR traffic and to identify for VFR traffic where these activities are conducted.

Military Standards (MS) — The standards used for aircraft hardware in order to maintain a high degree of quality standards in the manufacturing, repair, and maintenance of aircraft. Originated by the U.S. military services.

military training route (MTR) — Airspace of defined vertical and lateral dimensions established for the conduct of military flight training at airspeeds in excess of 250 knots IAS.

mill bit — A tool used with a router to remove metal and honeycomb core for repairs to bonded structure.

mill file — A single-cut file, tapered slightly in thickness and in width for about one-third of its length.

milli — One one-thousandth (0.001) of a unit.

milliameter — An electrical current measuring device calibrated to read in milliamperes. 1000 milliamperes = 1 ampere.

milliampere — One-thousandth (0.001) of an ampere.

millibar — A unit of barometric pressure equal to approximately 0.75 millimeters of mercury. Standard sea level atmospheric pressure is equal to 1,013.2 millibars.

milling machine — A metal-working machine tool with a movable table that feeds the work into a rotating milling cutter.

millivolt — One-thousandth (0.001) of a volt.

MILSPEC — A term used to identify military specifications.

mineral-based hydraulic fluid — A petroleum-based hydraulic fluid consisting essentially of kerosene and additives to inhibit corrosion and minimize foaming. It is dyed red and is identified as MIL-H-5606.

miniature screw — A screw less than 0.06" in diameter, having a slotted head and threaded for assembly with a preformed internal thread.

minicomputer — A small digital computer.

minidisk — A mass storage medium used in digital computers. Most common is the 3-1/2-inch diskette, also known as a floppy. Older computers used 8-inch and 5-1/4-inch minidisks. Most modern personal computers use CD ROM disks as well as 3-1/2-inch diskettes.

minima — Weather condition requirements established for a particular operation or type of operation; e.g., IFR takeoff or landing, alternate airport for IFR flight plans, VFR flight, etc. See minimums.

minimum crossing altitude — The lowest altitude at certain fixes at which an aircraft must cross when proceeding in the direction of a higher minimum enroute IFR altitude (MEA).

minimum descent altitude — The lowest altitude, expressed in feet above mean sea level, to which descent is authorized on final approach or during circle-to-land maneuvering in execution of a standard instrument approach procedure where no electronic glide slope is provided.

minimum enroute altitude (MEA) — Typically the lowest published altitude between radio fixes that guarantees adequate navigation signal reception and obstruction clearance (2,000 feet in mountainous areas and 1,000 feet elsewhere).

minimum enroute IFR altitude — The lowest published altitude between radio fixes that assures acceptable navigational signal coverage and meets obstacle clearance requirements between those fixes. The MEA prescribed for a Federal airway or segment thereof, area navigation low or high route, or other direct route applies to the entire width of the airway, segment, or route between the radio fixes defining the airway, segment, or route.

minimum equipment list (MEL) — A list developed for larger aircraft that outlines equipment that can be inoperative for various types of flight including IFR and icing conditions. This list is based on the MMEL (master minimum equipment list) developed by the FAA and must be approved by the FAA for use. It is specific to an individual aircraft make and model.

minimum friction level — The friction level specified in AC 150/5320-12, Measurement, Construction, and Maintenance of Skid Resistant Airport Pavement Surfaces, that represents the minimum recommended wet pavement surface friction value for any turbojet aircraft engaged in LAHSO. This value will vary with the particular friction measurement equipment used.

minimum fuel — **1.** The minimum fuel specified for weight and balance purposes when computing an adverse loaded center of gravity. It is the quantity of fuel necessary for one half hour of operation at rated maximum continuous power. **2.** Indicates that an aircraft's fuel supply has reached a state where, upon reaching the destination, it can accept little or no delay. This is not an emergency situation but merely indicates an emergency situation is possible should any undue delay occur.

minimum holding altitude — The lowest altitude prescribed for a holding pattern that assures navigational signal coverage, communications, and meets obstacle clearance requirements.

minimum IFR altitudes — Minimum altitudes for IFR operations as prescribed in FAR 91. These altitudes are published on aeronautical charts and prescribed in FAR 95 for airways and routes, and in FAR 97 for standard instrument approach procedures. If no applicable minimum altitude is prescribed in FAR 95 or FAR 97, the following minimum IFR altitude applies:
a. In designated mountainous areas, 2,000 feet above the highest obstacle within a horizontal distance of 4 nautical miles from the course to be flown; or
b. Other than mountainous areas, 1,000 feet above the highest obstacle within a horizontal distance of 4 nautical miles from the course to be flown; or
c. As otherwise authorized by the Administrator or assigned by ATC.

minimum level flight speed — The speed below which a gyroplane, the propeller of which is producing maximum thrust, loses altitude.

minimum navigation performance specification — A set of standards that require aircraft to have a minimum navigation performance capability in order to operate in MNPS designated airspace. In addition, aircraft must be certified by their State of Registry for MNPS operation.

minimum navigation performance specification (MNPS) airspace — Designated airspace in which MNPS procedures are applied between MNPS certified and equipped aircraft. Under certain conditions, non-MNPS aircraft can operate in MNPSA. However, standard oceanic separation minima is provided between the non-MNPS aircraft and other traffic. Currently, the only designated MNPSA is described as follows:
a. Between FL 285 and FL 420;
b. Between latitudes 27° N and the North Pole;
c. In the east, the eastern boundaries of the CTA's Santa Maria Oceanic, Shanwick Oceanic, and Reykjavik;
d. In the west, the western boundaries of CTA's Reykjavik and Gander Oceanic and New York Oceanic excluding the area west of 60° W and south of 38° 30' N.

minimum obstruction clearance altitude (MOCA) — The lowest published altitude in effect between radio fixes on VOR airways, off-airway routes, or route segments that meets obstacle clearance requirements for the entire route segment and that assures acceptable navigational signal coverage only within 25 statute (22 nautical) miles of a VOR.

minimum off-route altitude (MORA) — On Jeppesen Enroute charts, provides clearance from known obstructions within 10 NM of the route centerline.

minimum reception altitude — The lowest altitude at which an intersection can be determined.

minimum safe altitude (MSA) — **1.** The minimum altitude specified in FAR 91 for various aircraft operations. **2.** Altitudes depicted on approach charts that provide at least 1,000 feet of obstacle clearance for emergency use within a specified distance from the navigation facility upon which a procedure is predicated. These altitudes will be identified as Minimum Sector Altitudes or Emergency Safe Altitudes and are established as follows:

Minimum Sector Altitudes — Altitudes depicted on approach charts that provide at least 1,000 feet of obstacle clearance within a 25-mile radius of the navigation facility upon which the procedure is predicated. Sectors depicted on approach charts must be at least 90 degrees in scope. These altitudes are for emergency use only and do not necessarily assure acceptable navigational signal coverage.

Emergency Safe Altitudes — Altitudes depicted on approach charts that provide at least 1,000 feet of obstacle clearance in nonmountainous areas and 2,000 feet of obstacle clearance in designated mountainous areas within a 100-mile radius of the navigation facility upon which the procedure is predicated and normally used only in military procedures. These altitudes are identified on published procedures as "Emergency Safe Altitudes."

minimum safe altitude warning — A function of the ARTS III computer that aids the controller by alerting him when a tracked Mode C equipped aircraft is below or is predicted by the computer to go below a predetermined minimum safe altitude.

minimum sector altitude [ICAO] — The lowest altitude that can be used under emergency conditions to provide a minimum clearance of 300 m (1,000 feet) above all obstacles located in an area contained within a sector of a circle of 46 km (25 NM) radius centered on a radio aid to navigation.

minimum sink airspeed — In gliders, the airspeed, as determined by the performance polar, at which the glider will achieve the lowest sink rate. That is, the glider will lose the least amount of altitude per unit of time at minimum sink airspeed.

minimum vectoring altitude (MVA) — The lowest MSL altitude at which an IFR aircraft will be vectored by a radar controller, except as otherwise authorized for radar approaches, departures, and missed approaches. The altitude meets IFR obstacle clearance criteria. It may be lower than the published MEA along an airway or J-route segment. It may be utilized for radar vectoring only upon the controller's determination that an adequate radar return is being received from the aircraft being controlled. Charts depicting minimum vectoring altitudes are normally available only to the controllers and not to pilots.

minimum-flow stop — Refers to a fuel control design that prevents the power lever from shutting off fuel. A separate shutoff is provided in this case.

minimums — Weather condition requirements established for a particular operation or type of operation; e.g., IFR takeoff or landing, alternate airport for IFR flight plans, VFR flight, etc.

minor alteration — Any alteration not considered to be a major alteration.

minor axis of an ellipse — A straight line that passes through the center of the ellipse and is perpendicular to the major axis.

minor fastener diameter — The diameter of a threaded fastener measured at the thread root.

minor repair — Any repair not considered a major repair or preventive maintenance. FAR 43, Appendix A provides guidance on what constitutes a major repair or alteration and preventive maintenance..

minority carriers — A term used in reference to semiconductor electronic devices. Both electrons and holes are present in a semiconductor. The more abundant charge carriers are called majority carriers; the less abundant are called minority carriers. In N-type semiconductor material, electrons are the majority carriers and holes are the minority carriers. In P-type semiconductor material, the reverse is true. Current leakage in a reverse-bias direction is flow of the minority carriers.

minuend — The number from which the subtrahend is subtracted.

minus — A negative value. Minus values are indicated by using a short dash in front of the value (-4). A minus sign is used in electricity to indicate a negative condition.

minute — **1.** Measurement: An angular measurement equal to $1/60$ of a degree in a 360° circle (21,600 minutes of angle in a circle.) **2.** Time: A unit of time that is equal to $1/60$ of an hour or 60 seconds.

minutes-in-trail — A specified interval between aircraft expressed in time. This method would more likely be utilized regardless of altitude.

mirror image — An object that is an exact duplicate of the original but reversed as if the object were viewed through a mirror.

misalignment — A condition that exists when two mating surfaces do not meet or match as they should.

miscible — The ability of a material to combine or mix with another material.

misfire — The failure of an explosive charge, as in the misfire of a rocket or an engine.

misfiring — The interruption of even firing of a reciprocating engine's cylinders.

missed approach — **1.** A maneuver conducted by a pilot when an instrument approach cannot be completed to a landing. The route of flight and altitude are shown on instrument approach procedure charts. A pilot executing a missed approach prior to the Missed Approach Point (MAP) must continue along the final approach to the MAP. The pilot can climb immediately to the altitude specified in the missed approach procedure. **2.** A term used by the pilot to inform ATC that he is executing the missed approach. **3.** At locations where ATC radar service is provided, the pilot should conform to radar vectors when provided by ATC in lieu of the published missed approach procedure.

missed approach point — A point prescribed in each instrument approach procedure at which a missed approach procedure shall be executed if the required visual reference does not exist.

missed approach procedure [ICAO] — The procedure to be followed if the approach cannot be continued.

missed approach segment — See segments of an instrument approach procedure.

mist — Tiny droplets of water suspended in the air.

mist coat — A very light spray coat of thinner or other volatile solvent with little or no color in it.

miter — A cut to the edges of a board or surface in such a way that they will match or fit together.

miter box — A device used to guide a hand saw at the proper angle to cut wood or metal in order to form a miter joint.

miter square — A small square used for marking the ends of wood or metal for other than right angle cuts.

mixed exhaust — On a turbofan, a design that allows the primary and secondary airstreams to mix prior to leaving the engine. A sound attenuation feature of more modern engines. Same as forced exhaust mixer.

mixed icing — A combination of clear and rime icing. See also clear ice and rime ice.

mixer — 1. A system of bellcranks that prevents the cyclic inputs from changing the collective inputs on a helicopter control system. 2. A circuit in which two frequencies are combined to produce sum and difference frequencies.

mixing ratio — The ratio of the mass of water vapor to the mass of dry air.

mixture — A combination of matter composed of two or more components that retain their own properties.

mixture control — The primary carburetor control for adjusting the fuel-air mixture ratio. It can be either a manual or automatic control, or it can be a combination of both. In the case of the combination, the pilot adjusts for a particular ratio and the automatic control maintains that ratio by compensating for temperature and pressure variations of the atmosphere.

MLS categories — 1. MLS Category I — An MLS approach procedure that provides for an approach to a height above touchdown of not less than 200 feet and a runway visual range of not less than 1,800 feet. 2. MLS Category II — Undefined until data gathering/ analysis completion. 3. MLS Category III — Undefined until data gathering/ analysis completion.

MMM (manufacturers maintenance manual) — A manual developed by the aircraft manufacturer that includes information prepared for the AMT or technician who performs work on units, components, and systems while they are installed on the airplane. It is normally supplied by the manufacturer and approved by the FAA as part of the original process of certification. It will contain the required instructions for continued airworthiness that must accompany each aircraft when it leaves the factory.

mobile charges — Electrons in a semiconductor material that drift within the material from one electrically charged region to another.

mobile test stand — An engine run-in stand that is portable and can be used at multiple locations.

Mobius loop — A continuous loop with one surface and one edge. Made by twisting one end of a long, thin strip one half turn and attaching this end to the other end of the strip. This creates a loop with one edge and one surface.

mock-up — A full-size reproduction of a part or assembly used to determine whether or not all of the components will fit as they are designed. It is also used for instruction when the real object is impractical to use.

mode — 1. The manner of doing some operation. 2. The letter or number assigned to a specific pulse spacing of radio signals transmitted or received by ground interrogator or airborne transponder components of the Air Traffic Control Radar Beacon System (ATCRBS). Mode A (military Mode 3) and Mode C (altitude reporting) are used in air traffic control.

mode (SSR mode) [ICAO] — The letter or number assigned to a specific pulse spacing of the interrogation signals transmitted by an interrogator. There are 4 modes, A, B, C and D corresponding to four different interrogation pulse spacings.

mode C intruder alert — A function of certain air traffic control automated systems designed to alert radar controllers to existing or pending situations between a tracked target (known IFR or VFR aircraft) and an untracked target (unknown IFR or VFR aircraft) that requires immediate attention/action.

model — A copy of a real object, which can be life-size, smaller, or larger than the original.

model number — A manufacturer's designation of a particular piece of equipment.

modem — A modulator-demodulator device used to connect two computers and allow them to communicate.

modification — The change in the design or configuration of an original unit.

modify — 1. To change something such as an alteration or redesign of an original unit. 2. To change a schedule from the original date or time.

modular maintenance — A maintenance procedure that allows replacement of major assemblies, called modules, in a minimum amount of time and expense.

211

The removed module is returned to a repair facility, bench tested, and repaired as needed.

modular structure — Standardized units built up as modules.

modulate — To change. This normally refers to a radio carrier wave being modulated to transmit audio information.

modulated anti-skid system — An anti-skid brake system that senses the rate of deceleration of the wheels. The system maintains just enough pressure in the brakes to hold the tire in a slip condition, yet not allow a skid to develop. It does this by modulating, or continually changing, the pressure in the brake system.

modulated continuous wave — A radio code transmission that consists of a carrier wave modulated with a series of short and long bursts.

modulated light — Light that is modulated by audio frequency AC voltage. Such light can be received by a photo electric cell and fed into an amplifier to recreate the original audio qualities.

modulation — The changing of frequency or amplitude by superimposing an audio frequency on a carrier frequency.

modulator — That portion of a transmitter circuit that modulates the carrier wave.

modulus — The ratio of a stress load applied to the deformation of a material.

moisture — An all-inclusive term denoting water in any or all of its three states.

moisture absorption — The pickup of water vapor from air by a material, in reference to vapor withdrawn from the air only, as distinguished from water absorption, which is the gain in weight due to the absorption of water by immersion.

moisture separator — A device used in a pneumatic system to separate moisture from the air.

moisture-proof — The property of an object that resists absorption of moisture.

mold — The hollow form used to give shape to a laminate part while curing.

mold line — In metal layout, a line used in the development of the pattern used for forming a piece of sheet metal. It is that part of the formed part that remains flat and is formed by the intersection of the flat surfaces of two sides of a sheet metal part.

mold line dimension — The distance from the edge of metal to a mold point or between mold points.

mold release agent — A material applied to the surface of a mold that prevents the molded product from sticking to the mold. Also referred to as a parting agent.

molecule — The smallest particle of an element or compound that retains all the properties of the substance. Composed of one or more atoms.

molybdenum — A metallic element similar to chromium with a symbol of Mo and an atomic number of 42. Used as an alloying agent in most aircraft alloys.

moment — The product of the weight of an object in pounds and the distance from the center of gravity of the object to the datum or fulcrum (the point about which a lever rotates) in inches. Moment is used in weight and balance computations and is expressed in pound-inches. The formula used is: Moment = distance x force.

moment index — The moment divided by a constant such as 200, 1,000, or 10,000. Its use is to simplify weight and balance computations by eliminating large and unwieldy numbers.

momentum — The tendency of a body to continue in motion after being placed in motion.

moment-weight number — An identification number or letter indicating a measurement of both weight and center of gravity and used on rotating airfoils for balancing purposes.

Monel — A nickel-copper alloy that is extremely resistant to corrosion.

monitor — (When used with communication transfer) listen on a specific frequency and stand by for instructions. Under normal circumstances do not establish communications.

monitor alert (MA) — A function of the ETMS that provides traffic management personnel with a tool for predicting potential capacity problems in individual operational sectors. The MA is an indication that traffic management personnel need to analyze a particular sector for actual activity and to determine the required action(s), if any, needed to control the demand.

monitor alert parameter (MAP) — The number designated for use in monitor alert processing by the ETMS (enhanced traffic management system). The MAP is designated for each operational sector for increments of 15 minutes.

monkey wrench — A slang name for an adjustable wrench that has one fixed jaw and one movable jaw.

monocoque — A stressed-skin type of construction in which the stiffness of the skin provides a large measure of the strength of the structure. No truss or substructure is required.

monolithic casting — A casting formed as a single piece.

monomer — A chemical compound that can be polymerized. (Polymerization is a chemical reaction in which two or more molecules are combined to form a larger molecule that contains repeating structural units.)

monoplane — An airplane with one main supporting wing sometimes divided into two parts by the fuselage.

monopropellant — A rocket engine propellant in which the fuel and the oxidizer are both part of a single substance.

monorail — A single rail used to carry cars or objects.

monospar wing — A fundamental wing design that incorporates only one main longitudinal member in its construction.

monostable — The condition of a device that has one stable condition. When disturbed from this, it will attempt to return to its original state.

monostable multivibrator — An electronic circuit that tries to maintain a condition of on or off. When it is disturbed from this position, it will automatically return to its stable condition.

monsoon — A wind that blows in the summer from the sea to a continental interior, bringing copious rain, and in winter blows from the interior to the sea, resulting in sustained dry weather.

Morse code — A system of dots and dashes used for transmitting messages by audible or visual signals. Used in aviation to identify radio navigation facilities.

most significant bit — A bit in a binary number that has the highest value. It is the bit on the far left of a binary number.

most significant digit — A digit in a decimal number that is the most meaningful. It is the digit on the far left of a decimal number.

mothball — To preserve and store surplus airplanes, parts, or equipment for future use.

mothballed — Parts, machinery, or equipment that has been preserved and placed in storage.

mother board — The primary printed circuit board in an electronic device into which all other components are connected.

motion — The act of changing place or position.

motivation — A need or desire that causes a person to act. Motivation can be positive or negative, tangible or intangible, subtle or obvious.

motor bypass — A device in a hydraulic system that prevents a hydraulic motor from receiving excessive fluid. The fluid bypasses the motor.

motor over — The process of rotating the engine with the starter for reasons other than for starting.

motoring — Rotating a turbine engine with the starter for reasons other than starting.

motorization — The adding of an electric motor to equipment normally operated manually or by mechanical means other than electric motors.

mountain breeze — Occurring on a larger scale than downslope winds, it blows down the valley with a return flow, or anti-mountain wind, above the mountaintops.

mountain wave — An atmospheric gravity wave that forms in the lee of a mountain barrier. See lee wave.

mountain wave turbulence (MWT) — Turbulence produced in conjunction with the mountain lee wave.

mounting lug — A lug used to secure an accessory, cylinder, etc.

mounting pad — A provision made on the accessory section of an aircraft engine for attaching such accessories as magnetos, generators or alternators, and fluid pumps.

movement — The moving parts of a device that move in a defined manner, i.e., the inner workings of a watch.

movement area — The runways, taxiways, and other areas of an airport/heliport utilized for taxiing/hover taxiing, air taxiing, takeoff, and landing of aircraft, exclusive of loading ramps and parking areas. At those airports/heliports with a tower, specific approval for entry onto the movement area must be obtained from ATC.

movement area [ICAO] — That part of an aerodrome to be used for the takeoff, landing and taxiing of aircraft, consisting of the maneuvering area and the apron(s).

moving target indicator — An electronic device that permits radar scope presentation only from targets in motion. A partial remedy for ground clutter.

moving-coil meter — A d'Arsonval meter movement. Most commonly used meter movement in DC measuring instruments. A movable coil on which a pointer is mounted rotates in the field of a permanent magnet. The amount of current in the coil determines the strength of the coil's electromagnetic field and the amount the pointer is deflected by the magnetic field of the permanent magnet.

moving-iron meter movement — A d'Arsonval meter movement where the coil is fixed and the magnet (iron) is free to move.

muff — A shroud placed around a section of the exhaust pipe. The shroud is open at the ends to permit air to flow into the space between the exhaust pipe and the wall of the shroud and be heated. This heated air can be used for carburetor heat or cockpit and cabin heat.

mule — An auxiliary hydraulic power supply that can supply fluid under pressure to the aircraft hydraulic system when the engines are not running. The mule is normally used to test the landing gear and flight control systems.

multicell thunderstorm — A group of thunderstorm cells at various stages of development. The proximity of the cells allows interaction that prolongs the lifetime of the group beyond that of a single cell.

multicom — A mobile service not open to public correspondence used to provide communications essential to conduct the activities being performed by or directed from private aircraft.

multi-engine — An aircraft having more than one engine.

multimedia — A combination of more than one instructional medium. This format can include audio, text, graphics, animations, and video. Recently, multimedia implies a computer-based presentation.

multimeter — A piece of electrical test equipment consisting of one meter movement and several shunts, multipliers, and other circuit elements to allow the meter to be used as a voltmeter, ohmmeter, milliammeter, and ammeter. Rectifiers make it usable for AC and well as DC.

multiple runways — The utilization of a dedicated arrival runway(s) for departures and a dedicated departure runway(s) for arrivals when feasible to reduce delays and enhance capacity.

multiple-choice — A test item consisting of a question or statement followed by a list of alternative answers.

multiple-disk brake — An aircraft brake in which a series of discs, keyed to the wheel, mesh and rotate between a series of stationary discs keyed to the axle. The brakes are applied by hydraulically clamping the discs together.

multiple-spar (multi-spar) wing — An airplane wing structure that uses several spanwise structural members to give the wing its strength.

multiplex communications — A method of two-way communication in which two sites can transmit and receive on the same frequency and at the same time.

multiplicand — Any number to be multiplied by another.

multiplier — A number by which another number is multiplied. The factor by which something is multiplied or extended.

multiplier resistor — The resistor in series with a voltmeter movement used to multiply or extend the range of the meter.

multiplier tube — An electron tube designed to amplify or multiply very weak electron currents by means of secondary emission.

multi-spar wing — A fundamental wing design that incorporates more than one spanwise structural member for support.

multivibrator — An oscillator that produces its output by having two transistors or vacuum tubes alternately conduct current. When one conducts, the other is shut off. Conduction alternates between the two.

Mumetal — A nickel alloy of iron, nickel, chromium, molybdenum, and copper. Used in transformer cores and for shielding electronic devices from external magnetic fields.

muriatic acid — Commercial hydrochloric acid (HCl).

mushroom head — A flared head that forms on a pounding tool such as a punch or chisel when it is hammered. This is a dangerous condition, and it must be ground away.

mutual inductance — The inductance of a voltage in one coil due to the magnetic field produced by an adjacent coil. Inductive coupling is accomplished through the mutual inductance of two adjacent coils.

Mylar — A polyester film.

Mylar capacitor — A capacitor that uses Mylar film as a dielectric.

N

N3 — The speed of high pressure compressor in a triple-spool turbine engine.

nacelle — The streamlined enclosure on the wing or fuselage of an aircraft that houses the engine.

NAND gate — A "not and" logic device, which will not have a voltage at its output only when a voltage appears at all of the inputs. Opposite to an "and" gate.

nano — One billionth (0.000000001) of a unit.

nanovoltmeter — A sensitive voltmeter that measures voltages as low as one nanovolt.

nap — The short fiber ends that protrude from the surface of a fabric. When the fabric is doped, these fibers become stiff and must be sanded off.

naphtha — A volatile and flammable hydrocarbon liquid used chiefly as a solvent or cleaning agent.

narrowing grinding — The removable part of the valve seat's top edge in the cylinder of a reciprocating engine.

NAS drawings and specifications — Dimensional and material standards for aircraft fasteners developed by the National Aircraft Standards (NAS) Committee. All drawings and specifications are prefixed by NAS.

NAS stage A — The enroute ATC system's radar, computers and computer programs, controller plan view displays (PVDs/Radar Scopes), input/output devices, and the related communications equipment that are integrated to form the heart of the automated IFR air traffic control system. This equipment performs Flight Data Processing (FDP) and Radar Data Processing (RDP). It interfaces with automated terminal systems and is used in the control of enroute IFR aircraft.

National Airspace System — The common network of United States airspace, air navigation facilities, equipment and services, airports or landing areas, aeronautical charts, information and services, rules, regulations and procedures, technical information, and manpower and material. Included are system components shared jointly with the military.

National Beacon Code Allocation Plan (NBCAP) airspace — Airspace over United States territory located within the North American continent between Canada and Mexico, including adjacent territorial waters outward to about boundaries of oceanic control areas (CTA)/Flight Information Regions (FIR).

National Flight Data Center — A facility in Washington D.C., established by FAA to operate a central aeronautical information service for the collection, validation, and dissemination of aeronautical data in support of the activities of government, industry, and the aviation community. The information is published in the National Flight Data Digest.

National Flight Data Digest — A daily (except weekends and Federal holidays) publication of flight information appropriate to aeronautical charts, aeronautical publications, Notices to Airmen, or other media serving the purpose of providing operational flight data essential to safe and efficient aircraft operations.

national route program (NRP) — The NRP is a set of rules and procedures designed to increase the flexibility of user flightplanning within published guidelines.

national search and rescue plan — An interagency agreement that provides for the effective utilization of all available facilities in all types of search and rescue missions.

natural aging — The aging of solution, heat-treated aluminum alloy material. It is allowed to harden at room temperature following heat treatment.

natural numbers — Positive integers such as 1, 2, 8, and 9. Negative numbers, zero and fractions are NOT natural numbers.

naturally aspirated engine — A reciprocating aircraft engine that is not supercharged, but whose induction air is forced into the cylinders by atmospheric pressure only.

nautical mile — A measure of distance used primarily in navigation. It is equal to 6,076 feet and is one minute of latitude at the equator.

nautical twilight — The periods before sunrise and after sunset when the sun is not more than 12° below the horizon.

NAVAID — Any visual or electronic device, airborne or on the surface, that provides point-to-point guidance information, or position data, to aircraft in flight. NAVAIDs include VORs, ILSs, and DMEs.

NAVAID classes — VOR, VORTAC, and TACAN aids are classed according to their operational use. The three classes of NAVAIDs are:
a. T — Terminal
b. L — Low altitude
c. H — High altitude
Note: The normal service range for T, L, and H class aids is found in the AIM. Certain operational requirements make it necessary to use some of these aids at greater service ranges than specified. Extended range is made possible through flight inspection determinations. Some aids also have lesser service range due to location, terrain, frequency protection, etc. Restrictions to service range are listed in the Airport/Facility Directory.

navigable airspace — Airspace at and above the minimum flight altitudes prescribed in the FARs including airspace needed for safe takeoff and landing.

navigate — To move between sites on the Internet. Often done by means of links or connections between sites.

navigation lights — Lights on an aircraft consisting of a red light on the left wing, a green light on the right wing, and a white light on the tail. FARs require that these lights be displayed in flight during the hours of darkness.

navigational aid (NAVAID) — Any visual or electronic device, airborne or on the surface, that provides point-to-point guidance information, or position data, to aircraft in flight. NAVAIDs include VORs, ILSs, and DMEs.

NBCAP (National Beacon Code Allocation Plan) airspace — Airspace over United States territory located within the North American continent between Canada and Mexico, including adjacent territorial waters outward to about boundaries of oceanic control areas (CTA)/Flight Information Regions (FIR).

N-channel field effect transistor (FET) — A device with a conducting channel of N-type material that is on a P-type substrate. It can conduct in either direction, thus it can pass AC. One end of the channel is called a "drain" and the other a "source." These are brought out to pins for connection into an external circuit. A third connection is to the "gate." The gate can be P-type material. When the gate is made negative in respect to the channel, a depletion area is formed that reduces the channel current and allows control of it. The input impedance of any FET is very high.

neck — **1.** A portion of a fastener's body near the head that performs a definite function such as preventing rotation, etc. **2.** A reduced diameter of a portion of a fastener's shank required for design or manufacturing reasons.

needle and ball indicator — A flight instrument consisting of a rate gyro that indicates the rate of yaw and a curved glass clinometer that indicates the relationship between gravity and centrifugal force. It indicates the relationship between angle of bank and rate of yaw. Also referred to as a turn and slip indicator.

needle bearings — An anti-friction bearing made of hardened steel. The bearing consists of a series of small diameter rollers that ride between two hardened and polished steel races. One race is pressed into the housing, and the other race is pressed onto the rotating shaft.

needle valve — A tapered end fluid control needle valve that fits into a seat or recess to control or restrict a flow of fluid through an orifice.

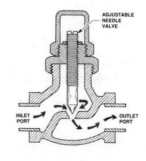

negative — **1.** A condition in which there is an excess of electrons. **2.** An accepted name for the terminal of a battery or power source from which the electrons flow. **3.** During communications, this indicates: "No," "Permission not granted," or "That is not correct."

negative acceleration — Deceleration; slowing down.

negative angle of attack — A flight condition where the angle of attack formed between the relative wind and the chord line of the airfoil is formed with the relative wind being the top leg in relation to the chord line and the so-called top surface of the aircraft. An aircraft in inverted, level flight would have a negative angle of attack.

negative battery terminal — The terminal of a battery from which the electrons leave and enter the circuit.

negative buoyancy — The tendency of an object, when placed in a fluid, to sink because it is heavier than the fluid it displaces.

negative condition — One in which there is an excess of electrons.

negative contact — Used by pilots to inform ATC one of two things. Either that previously issued traffic is not in sight or that they were unable to contact ATC on a particular frequency. The first instance might be followed by the pilot's request for the controller to provide assistance in avoiding the traffic.

negative dihedral — A downward inclination of a wing or other surface. It is the downward angle that is formed between the wings and the lateral axis of the airplane.

negative electrical charge — An unbalanced electrical condition caused by an atom having more electrons than protons.

negative electrical condition — A condition in which there are more negative charges than there are positive charges.

negative electrical resistance — A decrease in current through a device when there is an increase in voltage.

negative feedback — Information or a signal that is fed back into a circuit or device that tends to cause a decrease in the output.

negative ion — An atom that has more electrons than protons spinning around its nucleus.

negative moment — In aircraft weight and balance calculations, a moment resulting from an arm being forward of the datum.

negative pressure — Pressure that is less than atmospheric pressure.

negative pressure relief valve — A valve in an aircraft pressurization system that prevents the outside air pressure from becoming greater than the pressure inside the cabin

negative resistance — A condition in which the normal direct relationship between voltage and current is reversed.

negative stagger — The placement of the wings of a biplane in which the leading edge of the lower wing is ahead of the leading edge of the upper wing.

negative static stability — A condition in which an object disturbed from a condition of rest will tend to move further away from its condition of rest.

negative temperature coefficient — A condition where a conductor or device decreases in resistance as the temperature increases.

negative thrust — The thrust produced when a propeller is moved into the beta range.

negative torque system (NTS) — A system in a turboprop engine that prevents the engine from being overdriven by the propeller. The NTS increases the blade angle when the propellers try to overdrive the engine.

negative transfer of learning — Students interpret new things in terms of what they already know. Some degree of transfer is involved in all learning. Previous learning interferes with students' understanding of the current task.

negative vacuum relief valve — A relief valve used on pressurized aircraft that opens when outside air pressure is greater than cabin pressure.

negative value — A value less than zero.

neon — A gaseous element with a symbol of Ne and an atomic number of 10.

neon bulb — A bulb that glows when neon vapor atoms are excited by an electric arc.

neoprene — An oil-resistant synthetic rubber made by polymerizing chloroprene. Used in items such as seals and locknuts.

net thrust (Fn) — The effective thrust developed by a jet engine during flight, taking into consideration the initial momentum of the air mass prior to entering the engine.

neutral — 1. The condition in which a gear, lever, or other mechanism is not engaged. 2. An electrical condition that is neither positive nor negative.

neutral axis — An imaginary line through the length of a loaded beam where the forces of compression and tension are neutral.

neutral conductor — The conductor of a 3-phase circuit or a single-phase three wire circuit that is of a ground potential. The potential between the neutral and each of the other conductors is equal in magnitude and phase.

neutral flame — A flame used in oxyacetylene welding that is neither carburizing nor oxidizing and that uses the correct ratio of acetylene gas and oxygen.

neutral line — In sheet metal bending, the line near the middle of the sheet that is unaffected by either compression on the inside of the curve, or by stretching on the outside of the curve.

neutral plane — An imaginary line drawn perpendicular to the resultant flux in a generator. For arcless commutation, the neutral plane should lie directly over the plane of the brushes.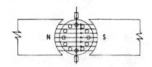

neutral position — The position of the rotating magnet of a magneto between the pole shoes. In the neutral position, no lines of flux flow in the magneto frame.

neutral stability — A system is characterized by neutral stability if, when displaced, it accelerates neither toward nor away from its original position. The atmosphere displays neutral stability when lapse rate is equal to the dry adiabatic lapse rate.

neutral static stability — The condition of an object in which, when once disturbed from a condition of rest, it has neither the tendency to return to a condition of rest nor to depart further from it. It continues in equilibrium in the direction of disturbance.

neutralize — To make balanced or inert by combining equal positive and negative quantities.

neutron — An uncharged particle in the nucleus of an atom. Its mass is essentially equal to that of a proton.

never exceed speed — The speed beyond which an aircraft should never be operated.

newton — The unit of force in the meter-kilogram-second system equal to the force required to impart an acceleration of one meter per second squared to a mass of one kilogram.

Newton's second law — The law of physics that states: "The acceleration produced in a mass by the addition of a given force is directly proportional to the force, and inversely proportional to the mass."

Newton's third law — The law of physics that states: "For every action, there is an equal and opposite reaction."

Newton's first law of motion — The law of physics that describes inertia. "A body at rest tends to remain at rest and a body in motion tends to continue to move at constant speed, along a straight line, unless it is acted upon by an external force."

Newton's second law of motion — "The greater the force acting on an object, the greater the acceleration. And the greater the mass, the less the object will accelerate."

Newton's third law of motion — The law of physics that describes action and reaction. "For every action, there is an equal and opposite reaction."

nibble — To take small bites or quantity.

nibbler — A sheet metal cutting tool that cuts the metal by a series of small nibbles or bites.

nichrome — An alloy of nickel and chromium. It is used for making precision wire-wound resistors.

nick — A sharp-sided gouge or depression with a V-shaped bottom that is generally the result of careless handling.

nickel — A silver-white, hard, malleable, metallic chemical element with a symbol of Ni and an atomic number of 28. Used for plating because of its high resistance to oxidation.

nickel silver — A metal alloy of copper, zinc, and nickel.

nickel-cadmium battery (Nicad) — A battery made up of alkaline secondary cells. The positive plates are nickel hydroxide, the negative plates are cadmium hydroxide, and potassium hydroxide is used as the electrolyte.

night — The time between the end of evening civil twilight and the beginning of morning civil twilight, as published in the American Air Almanac, converted to local time.

night [ICAO] — The hours between the end of evening civil twilight and the beginning of morning civil twilight or such other period between sunset and sunrise as can be specified by the appropriate authority.

Note: Civil twilight ends in the evening when the center of the sun's disk is 6 degrees below the horizon and begins in the morning when the center of the sun's disk is 6 degrees below the horizon.

nimbostratus — A dark gray cloud layer that produces rain or snow.

Nimonic® alloy — A nickel-chromium alloy used for sheet-metal fabrications in gas-turbine engines.

nipple pipe fitting — A short piece of pipe fitting threaded on both ends.

nitrate — A salt or ester of nitric acid (HNO_3). Used in some dopes to protect fabric-covered aircraft.

nitrate dope — A finish for aircraft fabric, consisting of a film base of cellulose fibers dissolved in nitric acid with the necessary plasticizers, solvents, and thinners.

nitric acid — A colorless or yellowish flowing, suffocating, caustic, corrosive, water-soluble liquid (HNO_3) with powerful oxidizing properties.

nitriding — A case hardening in which a steel part is heated in an atmosphere of ammonia (NH_3.) The ammonia breaks down, freeing the nitrogen to combine with aluminum in the steel to form an extremely hard abrasive-resistant aluminum nitride surface. Cylinder walls and crankshaft journals can be nitrided.

nitrite — A salt or ester of nitrous acid (HNO_2).

nitrogen — A colorless, tasteless, odorless, gaseous element forming nearly $4/5$ of the Earth's atmosphere.

no gyro approach — A radar approach/vector provided in case of a malfunctioning gyro-compass or directional gyro. Instead of providing the pilot with headings to be flown, the controller observes the radar track and issues control instructions "turn right/left" or "stop turn" as appropriate.

no gyro vector — See no gyro approach.

no procedure turn (NOPT) — No procedure turn is required nor authorized.

no transgression zone (NTZ) — The NTZ is a 2,000 foot wide zone, located equidistant between parallel runway final approach courses in which flight is not allowed.

noble — Chemically inert or inactive, especially toward oxygen.

noble gas — An inert gas such as neon, argon, krypton, and xenon.

nocturnal inversion — A surface-based stable layer that occurs due to nighttime radiational cooling.

nodal system — A vibration dampening system used by Bell Helicopter to reduce main rotor vibration.

noise — 1. A general term for any loud or unusual sound that is annoying or excessive. 2. Undesired signals within an electronic circuit.

noise suppressor — A device installed in the tailpipe of a turbojet engine to slow the mixing of the exhaust gases with the surrounding air, thus decreasing the intensity of the sound.

no-load current — The electrical current draw of a device when the device is not under load.

Nomex® — Trademark of DuPont. A nylon paper treated material that is made into a honeycomb core material.

nominal rating — The stated value of a quantity or component, which might not be the actual value measured.

nominal resistance of a thermistor — The true resistance of a thermistor at a particular reference temperature. Most manufacturers use 20°C as the reference temperature.

nominal size — The designation used for the purpose of general identification.

nominal value — A stated value that has a tolerance that would allow the actual value to be somewhat different.

nomograph-viscosity index — An ASTM-produced chart used to plot the viscosity change of turbine oils with temperature change.

nonabrasive — Material that will not scratch or scar when rubbed on another surface.

nonabrasive scraper — A scraper that has no abrasive materials attached to it.

non-airworthiness item — An inspection item that if broken or inoperative does not affect the airworthiness of the aircraft.

nonapproach control tower — Authorizes aircraft to land or takeoff at the airport controlled by the tower, or to transit the airport Class D airspace. The primary function of a nonapproach control tower is the sequencing of aircraft in the traffic pattern and on the landing area. Nonapproach control towers also separate aircraft operating under instrument flight rules clearances from approach controls and centers. They provide ground control services to aircraft, vehicles, personnel, and equipment on the airport movement area.

non-aqueous developer — In dye-penetrant inspections, a material that causes a crack to show up. Non-aqueous developer is not water based and is less corrosive to the part.

non-atomizing spray — The application of a material to a surface by a spray gun in which the material is fed in a solid stream rather than in tiny droplets.

non-atomizing spray gun — A spray gun that propels a solid stream from the spray nozzle.

noncommon route/portion — That segment of a North American Route between the inland navigation facility and a designated North American terminal.

noncomposite separation — Separation in accordance with minima other than the composite separation minimum specified for the area concerned.

nondestructive inspection (NDI) — An inspection of aircraft parts, units, components, etc., that doesn't alter or destroy the physical or material properties and/or integrity of the part. It is used to determine the continued serviceability. Also referred to as nondestructive testing.

nondestructive testing (NDT) — See nondestructive inspection (NDI).

nondimensional number — A number that does not have a dimensional value such as a Mach speed.

nondirectional antenna — An antenna that has the ability to receive or transmit equal signals in all directions.

nondirectional beacon — An L/MF or UHF radio beacon transmitting nondirectional signals whereby the pilot of an aircraft equipped with direction finding equipment can determine his bearing to or from the radio beacon and "home" on or track to or from the station. When the radio beacon is installed in conjunction with the Instrument Landing System marker, it is normally called a Compass Locator.

nonferrous metal — A metal that contains no iron.

nonflexible control cable — A grouping of seven or 19 strands of solid wire preformed into a helical or spiral shape. Can be used for straight runs where it does not pass over any pulleys.

noninductive winding — A winding consisting of two parts wound in such a way that the magnetic field from one cancels the other. The result is no inductive load.

noninductive load — An electrical load with no inductance. The entire load is due to resistance and capacitance.

nonlinear output — Any output not directly proportional to the input.

nonlinear scale — The scale of an indicating instrument in which the numbers are spread out at one end and are bunched up at the other.

nonlinear system — Nonuniform in length, width, or output.

nonmagnetic — Metal that does not have the properties of a magnet and/or that cannot be magnetized or attracted by a magnet.

nonmovement areas — Taxiways and apron (ramp) areas not under the control of air traffic.

non-owner liability coverage — An insurance policy against claims arising from bodily injury or damage caused to others or their property while using an aircraft that one does not own.

nonporous — Any material that does not allow a liquid to pass through it.

nonprecision approach (NPA) — An instrument approach based on a navigation system which provides course deviation information, but no glidepath deviation information. VOR, NDB, and LNAV are nonprecision approaches. Vertical Descent Angle (VDA) on some nonprecision approaches may provide a Vertical Descent Angle as an aid in flying a stabilized approach, without requiring its use in order to fly the procedure. This does not make the approach an APV (Approach With Vertical Guidance), since it still must be flown to an MDA and has not been evaluated with a glidepath.

nonprecision approach procedure — A standard instrument approach procedure in which no electronic glideslope is provided; e.g., VOR, TACAN, NDB, LOC, ASR, LDA, or SDF approaches.

nonradar — Precedes other terms and generally means without the use of radar, such as:

a. Nonradar Approach — Used to describe instrument approaches for which course guidance on final approach is not provided by ground based precision or surveillance radar. Radar vectors to the final approach course may or may not be provided by ATC. Examples of nonradar approaches are VOR, NDB, TACAN, and ILS/MLS approaches.

b. Nonradar Approach Control — An ATC facility providing approach control service without the use of radar.

c. Nonradar Arrival — An aircraft arriving at an airport without radar service, or at an airport served by a radar facility and radar contact has not been established or has been terminated due to a lack of radar service to the airport.

d. Nonradar Route — A flight path or route over which the pilot is performing his own navigation. The pilot can be receiving radar separation, radar monitoring, or other ATC services while on a nonradar route.

e. Nonradar Separation — The spacing of aircraft in accordance with established minima without the use of radar; e.g., vertical, lateral, or longitudinal separation.

nonradar separation [ICAO] — The separation used when aircraft position information is derived from sources other than radar.

nonrepairable — Something that cannot be repaired and which, therefore, renders the part nonserviceable.

nonrepairable damage — Damage that requires the aircraft or aircraft component to be replaced.

nonrigid airship — An engine-driven, lighter-than-air aircraft such as a blimp that uses gas pressure to maintain the shape of the craft.

nonscheduled airline — An airline that does not operate according to a regularly published schedule.

nonservo brakes — Brakes that do not use the momentum of the aircraft to assist in the application of the brakes.

nonskid brakes — A feature found in high performance aircraft braking systems that provides wheel antiskid protection. A skid control generator unit measures the wheel rotational speed. As the wheel rotates, the generator develops a voltage and current signal. The signal strength indicates the wheel rotational speed. This signal is fed to the skid control box though the harness. The box interprets the signal and if the wheel is slowing too quickly, it signals a solenoid in the skid control valve to release the brake pressure until the wheel begins to speed up allowing the wheel to continue to rotate without skidding.

nonstandard fastener — A fastener that differs in size, length, material, or finish from established and published standards.

nonstructural — The portion of an aircraft that does not carry any aerodynamic loads.

nontautening dope — A special formulation of aircraft dope used on heat-shrunk polyester fabric. It provides the necessary fill for the fabric, but produces a minimum of shrinkage.

NOR gate — A "not or" logic device that will produce a voltage at its output only when there is no voltage on any input.

NORDO (No Radio). —Loss of the ability to communicate by radio. Standard pilot procedures are specified in FAR 91. Radar controllers issue procedures for pilots to follow in the event of lost communications during a radar approach when weather reports indicate that an aircraft will likely encounter IFR weather conditions during the approach.

norm reference testing (NRT) — System of testing where students are rank ordered in accomplishment of objectives.

normal category airplane — An airplane certificated for nonacrobatic operation.

normal heptane — A liquid hydrocarbon material (C_7H_{17}) having a low critical pressure and temperature and whose detonation characteristics are used in determining the octane rating of aviation gasoline.

normal operating speed — The velocity obtained in level flight at design altitude of the airplane at no more than 70% of normal rated engine power.

normal operating zone (NOZ) — The NOZ is the operating zone within which aircraft flight remains during normal independent simultaneous parallel ILS approaches.

normal rated power — The highest power at which an engine can be operated continuously without damage.

normal refraction — The refraction of a radar beam as it passes through the atmosphere due to changes in air density with height. Usually, because the atmosphere's density decreases rapidly with height, the radar beam will be deflected downward.

normal shock wave — A shock wave formed ahead of an airfoil approaching the speed of sound. It is perpendicular to the path of the airfoil.

normalizing — A heat treatment in which a metal is heated to its critical temperature and allowed to cool slowly in still air. Normalizing relieves stresses in the metal.

normally aspirated engine — An engine that does not compensate for decreases in atmospheric pressure through turbocharging or other means.

normally closed relay — A relay switch consisting of a coil or solenoid, an iron core, and movable contacts controlled by a spring. Normally closed relay contacts are held closed by the spring. When current flows through the solenoid, the contacts are opened by the magnetic pull of the electromagnetic coil.

normally open relay — A relay switch consisting of a coil or solenoid, an iron core, and movable contacts controlled by a spring. Normally open relay contacts are held open by the spring. When current flows through the solenoid, the contacts are closed by the magnetic pull of the electromagnetic coil.

norm-referenced testing (NRT) — System of testing where students are ranked against the performance of other students.

North American route — A numerically coded route preplanned over existing airway and route systems to and from specific coastal fixes serving the North Atlantic. North American Routes consist of the following:
a. Common Route/Portion — That segment of a North American Route between the inland navigation facility and the coastal fix.
b. Non-Common Route/Portion — That segment of a North American Route between the inland navigation facility and a designated North American terminal.
c. Inland Navigation Facility — A navigation aid on a North American Route at which the common route and/or the noncommon route begins or ends.
d. Coastal Fix — A navigation aid or intersection where an aircraft transitions between the domestic route structure and the oceanic route structure.

north mark — A beacon data block sent by the host computer to be displayed by the ARTS on a 360 degree bearing at a locally selected radar azimuth and distance. The North Mark is used to ensure correct range/azimuth orientation during periods of CENRAP.

north pacific — An organized route system between the Alaskan west coast and Japan.

north pole — The north-seeking pole of a magnet.

northerly turning error — In the Northern Hemisphere, a weight is placed on the south-facing end of the compass needle to compensate for dip (See dip). When a turn is made toward the north from a westerly or easterly heading the dip weight causes the compass to overshoot the actual heading. When turning toward the south, the heading is undershot. The effects are reversed in the Southern Hemisphere if the compass is compensated with a weight on the north-facing end of the compass needle.

nose cone — A conical-shaped dome usually attached to the front portion of a fuselage to house the radar antenna or other electronic equipment. A nose cone can also be described as the front cover of propellers, intakes for jet engines, etc., for streamlining and directing the airflow.

nose gear — The forward gear on an aircraft equipped with tricycle landing gear.

nose heavy — A condition that exists on an aircraft in which the center of gravity is ahead of the forward limit.

nose rib — A false, or partial, wing rib extending back from the leading edge only to the main spar. Its purpose is to add smoothness to the leading edge of a wing.

nose section — The forward section of an aircraft.

NOT gate — A logic device having one input and one output. There will be no voltage on the output when a voltage appears at the input.

NOTAM [ICAO] — A notice, containing information concerning the establishment, condition or change in any aeronautical facility, service, procedure or hazard, the timely knowledge of which is essential to personnel concerned with flight operations.
a. I Distribution. Distribution by means of telecommunication.
b. II Distribution. Distribution by mean other than telecommunications.

NOTAM(D) — A (distant) NOTAM that is disseminated for all navigational facilities which are part of the U.S. airspace system, all public use airports, seaplane bases, and heliports listed in the A/FD.

NOTAM(L) — A (local) NOTAM that is distributed locally only and is not attached to an hourly weather report.

notes — Specific instructions on an aircraft drawing.

notice to airmen (NOTAM) — A notice containing information (not known sufficiently in advance to publicize by other means) concerning the establishment, condition, or change in any component (facility, service, or procedure of, or hazard in the

National Airspace System) the timely knowledge of which is essential to personnel concerned with flight operations.
a. NOTAM(D) — A NOTAM given (in addition to local dissemination) distant dissemination beyond the area of responsibility of the Flight Service Station. These NOTAMs will be stored and available until cancelled.
b. NOTAM(L) — A NOTAM given local dissemination by voice and other means, such as telautograph and telephone, to satisfy local user requirements.
c. FDC NOTAM — A NOTAM regulatory in nature, transmitted by USNOF and given system wide dissemination.

Notices To Airmen Publication (NOTAM) — A publication issued every 28 days, designed primarily for the pilot, which contains current NOTAM information considered essential to the safety of flight as well as supplemental data to other aeronautical publications. The contraction NTAP is used in NOTAM text.

nozzle — The tapered end of a duct.

nozzle blades — Any of the blades of a nozzle diaphragm.

nozzle diaphragm — A ring of stationary blades in a turbine engine ahead of the turbine wheel. Used to direct the flow of hot gases into the turbine for maximum efficiency.

nozzle, fuel — The pressure-atomizing unit that receives fuel under high pressure from the fuel manifold and delivers it to the combustor in a highly atomized, precisely patterned spray.

NPN transistor — A three-element semiconductor made up of a sandwich of P-type silicon or germanium between two pieces of N-type material.

N-strut — Struts of a biplane near the wing tips shaped in the form of the letter N.

N-type semiconductor material — A semiconductor material that has been doped (an impurity added) that leaves the outer ring of the valence shells with electrons that are readily given up.

N-type silicon — Silicon that has been doped with an impurity having five valence electrons.

nuclear energy — Energy released as a result of either nuclear fission, nuclear fusion, or radioactive decay. Nuclear fusion is not yet commercially feasible.

nuclear fission — The splitting of an atomic nucleus resulting in the release of energy.

nuclear fusion — The union of atomic nuclei of certain light elements to form heavier nuclei resulting in the release of energy.

nuclei - Plural of nucleus.

nucleonics — The branch of science that deals with atomic nuclei and nuclear energy.

nucleus — 1. The center or core around which other parts are grouped. 2. The center or core of an atom consisting of positively charged protons and uncharged neutrons.

nucleus of an atom — In chemistry, the central portion of an atom.

null — An indicated low or zero point in a radio signal.

null balance — An electrical circuit in which two voltages have cancelled each other out.

null detector — An electrical instrument that determines values by adjusting a known quantity until it is identical with the quantity to be measured. Often utilizes a Wheatstone bridge.

numbers — A symbol used to assign a value of quantity or placement in a sequence.

numerator — The part of a fraction that illustrates a portion of something that makes up a whole. In decimals, the numerator is the number to the right of the decimal point. In a fraction, it is the number above the line and signifies the number of parts of the denominator used.

numerical control — Allows more precise control of operations than analog control. Also referred to as digital control.

numerical weather prediction — Meteorological forecasting using digital computers to solve mathematical equations that describe the physics of the atmosphere; used extensively in weather services throughout the world.

numerous targets vicinity (location) — A traffic advisory issued by ATC to advise pilots that targets on the radar scope are too numerous to issue individually.

nut — An internally threaded collar used to screw onto bolts or screws to form a complete fastening device.

nutation — The wobbling of the axis of a spinning body such as the Earth, or a gyroscope.

nutplate — A nut that can be riveted to the inside of a structure. Bolts and screws can be screwed into the nutplate without a wrench to hold the nut in place.

nylon — A tough, lightweight, elastic polyamide material used especially in fabrics and plastics.

O

objectivity — Characteristic of a measuring instrument when it is free of any personal bias by the person grading the test.

oblique angle — An angle other than a right angle. A view on a mechanical drawing at other than normal front, rear, top, bottom, and side angles.

oblique photography — An aerial photograph taken with a camera pointed at an angle other than straight down.

oblique shock wave — A shock wave attached to the bow and tail of an aircraft flying at a speed greater than the speed of sound. The sides of the oblique shock wave form the Mach cone.

oblique triangle — A three-sided, closed figure that does not contain a right angle.

oblong shape — An object that is longer than it is broad. An elongated circle or square.

obscuration — Denotes sky hidden by surface-based obscuring phenomena and vertical visibility restricted overhead.

observation aircraft — Military aircraft that fly behind enemy lines and observe the movement of troops or supplies.

obsolete — Something no longer in use or in practice.

obstacle — Something that impedes progress or achievement. In avation, an object or terrain at a fixed geographical location (or which is expected to be at a fixed location) within a prescribed area with reference to which vertical clearance is or must be provided during flight operation.

obstacle clearance altitude (height) OCA(H) (ICAO) — The lowest altitude (OCA), or alternatively the lowest height above the elevation of the relevant runway threshold or above the aerodrome elevation as applicable (OCH), used in establishing compliance with the appropriate obstacle clearance criteria.

obstacle free zone (OFZ) — A three dimensional volume of airspace that protects for the transition of aircraft to and from the runway. The OFZ clearing standard precludes taxiing and parked airplanes and object penetrations, except for frangible NAVAID locations that are fixed by function. Additionally, vehicles, equipment, and personnel may be authorized by air traffic control to enter the area using the provisions of FAAO 7110.65, Para 3-1-5, VECHILES/EQUIPMENT/PERSONNEL ON RUNWAYS. The runway OFZ and, when applicable, the inner-approach OFZ, and the inner-transitional OFZ, comprise the OFZ.

a. Runway OFZ — The runway OFZ is a defined volume of airspace centered above the runway. The runway OFZ is the airspace above a surface whose elevation at any point is the same as the elevation of the nearest point on the runway centerline. The runway OFZ extends 200 feet beyond each end of the runway. The width is as follows:

1) For runways serving large airplanes, the greater of:
(a) 400 feet, or
(b) 180 feet, plus the wingspan of the most demanding airplane, plus 20 feet per 1,000 feet of airport elevation.
2). For runways serving only small airplanes:
(a) 300 feet for precision instrument runways.
(b) 250 feet for other runways serving small airplanes with approach speeds of 50 knots, or more.
(c) 120 feet for other runways serving small airplanes with approach speeds of less than 50 knots.

b. Inner-approach OFZ — The inner-approach OFZ is a defined volume of airspace centered on the approach area. The inner-approach OFZ applies only to runways with an approach lighting system. The inner-approach OFZ begins 200 feet from the runway threshold at the same elevation as the runway threshold and extends 200 feet beyond the last light unit in the approach lighting system. The width of the inner-approach OFZ is the same as the runway OFZ and rises at a slope of 50 (horizontal) to 1 (vertical) from the beginning.

c. Inner-transitional OFZ — The inner transitional surface OFZ is a defined volume of airspace along the sides of the runway and inner-approach OFZ and applies only to precision instrument runways. The inner-transitional surface OFZ slopes 3 (horizontal) to 1 (vertical) out from the edges of the runway OFZ and inner-approach OFZ to a height of 150 feet above the established airport elevation.

obstruction — Any object/obstacle exceeding the obstruction standards specified by FAR 77, Subpart C.

obstruction clearance limit (OCL) (ICAO) — The height above aerodrome elevation below which the minimum prescribed vertical clearance cannot be maintained either on approach or in the event of a missed approach.

obstruction light — A light, or one of a group of lights, usually red or white, mounted on a surface structure or natural terrain to warn pilots of the presence of a flight hazard.

obtuse angle — An angle greater than 90°. Also referred to as an open angle.

obtuse triangle — A triangle that contains an angle greater than 90°.

occluded front — The surface front after a cold front overtakes a warm front.

occlusion process — The process by which a cold front overtakes the warm front in a wave cyclone, pushing the warm sector air aloft.

oceanic airspace — Airspace over the oceans of the world, considered international airspace, where oceanic separation and procedures per the International Civil Aviation Organization are applied. Responsibility for the provisions of air traffic control service in this airspace is delegated to various countries, based generally upon geographic proximity and the availability of the required resources.

oceanic display and planning system — An automated digital display system that provides flight data processing, conflict probe, and situation display for oceanic air traffic control.

oceanic navigational error report — A report filed when an aircraft exiting oceanic airspace has been observed by radar to be off course. ONER reporting parameters and procedures are contained in FAAO 7110.82, Monitoring of Navigational Performance In Oceanic Areas.

oceanic published route — A route established in international airspace and charted or described in flight information publications, such as Route Charts, DOD [and Jeppesen] Enroute Charts, Chart Supplements, NOTAMs, and Track Messages.

oceanic transition route — An ATS route established for the purpose of transitioning aircraft to/from an organized track system.

O-condition — A temper designation. The soft or annealed temper condition of a wrought metal product.

octagon — An eight-sided figure with each side having the same length.

octahedron — A solid design or figure with eight plane surfaces.

octal number system — In digital electronics, the number system based on eight units (0-7).

octane rating — The rating system of aviation gasoline with regard to its antidetonating qualities. Fuel with an octane rating of 87 is made up of a mixture of 87% isooctane and 13% heptane.

octave — Musical interval of eight tones or notes. The interval between two frequencies having a ratio of 2 to 1.

odd harmonics — The odd multiples of a frequency.

oersted — A magnetomotive force of 1 gilbert per square centimeter or 79.577 ampere-turns per meter.

off course — A term used to describe a situation where an aircraft has reported a position fix or is observed on radar at a point not on the ATC-approved route of flight.

off-idle mixture — The fuel-air mixture ratio of an aircraft engine in the transition period between idle RPM utilizing the idle jets and higher power settings using the main metering system.

off-route vector — A vector by ATC which takes an aircraft off a previously assigned route. Altitudes assigned by ATC during such vectors provide required obstacle clearance.

offset parallel runways — Staggered runways having parallel centerlines.

offset rivet set — A rivet set used in a hand-held pneumatic riveting gun in which the head is offset from the center line of the shank. Offset rivet sets are used in locations where a straight set cannot be used.

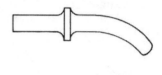

offset screwdriver — A screwdriver in which the blades are oriented at 90° to the shank. An offset screwdriver is used to turn screws when there is not enough clearance in line with the screw to allow a regular screwdriver to be used.

offshore control airspace area — That portion of airspace between the U.S. 12 NM limit and the oceanic CTA/FIR boundary within which air traffic control is exercised. These areas are established to provide air traffic control services. Offshore/Control Airspace Areas can be classsified as either Class A airspace or Class E airspace.

off-the-shelf item — Any standard item, part, or program that can be used in place of a custom part.

ohm (Ω) — The unit of electrical opposition equal to the resistance of a circuit in which an electromotive force of one volt will maintain a current of one ampere.

ohmmeter — An electrical measuring instrument used to measure resistance in a circuit. An ohmmeter measures resistance by calculating the amount of current that flows when a known voltage is applied across an unknown resistance.

Ohm's law — The law that establishes the relationship between current, voltage, and resistance in an electrical circuit. The current in a circuit is directly proportional to the voltage causing it and inversely proportional to the resistance of the circuit.

ohms-per-volt — The measure of a voltmeter's sensitivity. It is found by dividing the number one by the amount of current needed to deflect the meter pointer full scale.

oil canning — A condition of the sheet metal skin of an aircraft that is slightly bulged or stretched between rows of rivets. This bulge will pop back and forth in the same way the bottom of an oil can pops back and forth.

oil circuit breaker — A circuit breaker used in high amperage circuits such as municipal power plants and distribution grids. Oil is used to quench the arcs that form when connecting and disconnecting such circuits.

oil control ring — The piston ring below the compression rings used to control the amount of oil between the piston and the cylinder wall of an aircraft reciprocating engine. It is usually a multi-piece ring and normally fits into a groove with holes to drain part of the oil back to the inside of the piston.

oil cooler — 1. A heat exchanger used to cool the oil. 2. A radiator used to maintain normal operating temperature of lubricating oil. Some coolers utilize air as a cooling agent; others use fuel.

oil dilution — The process of thinning engine oil by adding fuel to it at shutdown to make starting easier in cold weather.

oil film — A light coating of oil sufficient to prevent metal-to-metal contact or to protect metal parts from corrosion.

oil filter — A device for removing impurities and foreign matter from the lubricating oil used in an aircraft engine.

oil hardening — A process of hardening steel using oil as a quenching agent. When red hot steel is immersed in a bath of oil, the steel cools more slowly than by other methods and gives the steel a more uniform hardness.

oil inlet — The fitting on an aircraft engine through which the lubricating oil enters the oil system.

oil jet — A small nozzle opening that directs a stream of oil onto an area to be lubricated such as bearings, gears, etc.

oil outlet — The fitting on an aircraft engine from which the lubricating oil leaves the engine to return to the external reservoir.

oil pan — The removable part of the crankcase where engine lubricating oil is collected by gravity and stored. Oil in the pan is forced through passages in the engine by a pressure pump.

oil passages — Channels or holes in an engine through which lubricating oil flows to lubricate, seal, or cool the engine.

oil pressure indicator — An instrument that indicates the pressure of the oil in the engine lubricating system.

oil scraper ring — A piston ring located at the bottom or skirt end of a piston used to wipe the oil either toward or away from the oil control ring depending on the design of the engine.

oil screen — A fine mesh screen in the engine lubrication system used to stop and hold impurities, preventing their passage through the engine and causing damage.

oil seal — A device used to prevent engine oil from leaking past a moving shaft.

oil separator — A device used to separate oil from the discharge air of a wet-type air pump.

oil slinger — A rotating device used as a centrifugal impeller to direct oil flow, usually away from a bearing sump and toward a scavenge pump.

oil sump — A container built into the lower part of an aircraft engine that holds the supply of lubricating oil.

oil tank pressurizing valve — A check valve used to trap oil vapors in the expansion space above the oil surface to provide a pressurization effect of approximately 3 to 5 PSI within the oil tank.

oil temperature indicator — Indicates oil temperature of the inlet oil to the engine. System usually is of the ratiometer type. Normal temperature is a good signal to the pilot that the engine is warmed up for takeoff.

oil temperature regulator — A control device that maintains the oil temperature within the desired operating range by either passing the oil through the core of the cooler or around the jacket of the cooler.

oil wiper ring — The bottom ring on a piston used to direct oil up between the piston and the cylinder wall for lubrication and sealing.

oil-cooled transformer — Transformers that use oil to cool the coils during the transforming of very high voltages in municipal power grids.

Oilite bushing — A friction bushing made of a bronze material impregnated with oil. Friction of the moving parts generates enough heat to bring the oil to the surface and provide the needed lubrication.

oleo strut — An aircraft landing gear shock strut that absorbs the initial landing impact by the transfer of oil from one chamber to another through a restricting orifice. Taxi shocks are absorbed with compressed air or by a spring.

Omega — An RNAV system designed for long-range navigation based upon ground-based electronic navigational aid signals.

Omega navigation system — An RNAV system designed for long-range navigation based upon ground-based electronic navigational aid signals.

omni — Variable Omni Range (VOR). A navigation system that uses phase comparison of two navigational signals transmitted from ground stations. The signals are in phase when they are received at a location directly magnetic north of the station. An instrument tells the pilot when the needle is off center from the pre-selected radial. If the aircraft gets away from the pre-selected radial, the needle moves out of center to show the pilot which direction to turn in order to bring the aircraft back to the selected radial.

omni bearing selector (OBS) — On a VOR indicator, the course selector. The knob a pilot uses to select the radial from the VOR station along which to fly. The OBS shifts the phase of the course deviation indicator reference signal so the needle will center when the aircraft is on the selected radial.

omni station — The ground station of a VOR.

omnidirectional microphone — A non-directional microphone that picks up sounds equally from all directions.

omnidirectional radio antenna — A non-directional radio antenna that transmits or receives signals equally well in all horizontal directions.

omnirange navigation equipment (VOR) — A phase comparison type of electronic navigation equipment that provides a directional reference (measured from magnetic north) between the airplane and the ground station.

on course — **1.** Used to indicate that an aircraft is established on the route centerline. **2.** Used by ATC to advise a pilot making a radar approach that his aircraft is lined up on the final approach course.

on-condition maintenance — A maintenance concept whereby some components of the engine remain in service as long as they appear airworthy at each inspection. The replace-on-condition concept is opposed to the concept of replacing a component after a "life-limited" time interval. In the case of engines themselves, this means no TBO (time between overhaul) is required.

on-course indication — An indication on an instrument, which provides the pilot a visual means of determining that the aircraft is located on the centerline of a given navigational track, or an indication on a radar scope that an aircraft is on a given track.

one hundred and eighty degree ambiguity — An error inherent in radio direction finding systems, in which the system is unable to determine whether the bearing to the station is as indicated or 180° different than indicated, for example, whether the station is in front of or behind the aircraft.

one-minute weather — The most recent one minute updated weather broadcast received by a pilot from an uncontrolled airport ASOS/AWOS.

one-shot rivet gun — A rivet gun that hits the rivet with a single hard blow for each pull of the trigger. It is used for rivets that are subject to becoming brittle if driven by ordinary rivet guns that deliver a continuous series of blows as long as the trigger is held.

one-to-one vibration — In rotorcraft, a low frequency vibration having one beat per revolution of the rotor. This vibration can be lateral, vertical, or horizontal.

on-speed — The condition in which the actual engine speed is equal to the desired engine speed as set on the propeller control by the pilot.

on-speed condition — A condition of a propeller governor system that maintains the selected RPM by metering to or draining from the propeller piston the exact quantity of oil necessary to maintain the proper blade angle for that RPM.

open angle — The angle through which metal has been bent that is greater than 90°. Produces a "V" with less than 90° between the sides.

open circuit — An incomplete electrical circuit. Does not provide a continuous path for electrons to flow.

open wiring — A wire, wire group, or wire bundle not enclosed in conduit.

open-assembly time — The assigned time between the time when an adhesive is spread on to two surfaces to be joined and the time the surfaces are clamped together.

open-center selector valve — A hydraulic selector valve used in open-center hydraulic systems that allows fluid to flow from the pump to the reservoir when the selector valve is placed in neutral (when none of the actuating cylinders are receiving fluid under pressure).

open-circuit voltage — The measured voltage of a battery or generator when there is no load or flow of electrons in the circuit.

open-end wrench — A solid, nonadjustable wrench with open parallel jaws on one or both ends.

open-tip turbine blades — A blade with no shroud attached at the tip. This blade can withstand higher speed-induced tip loading than the shrouded tip blade. Also referred to as an open perimeter tip.

operant conditioning — Behavior initiated voluntarily by the individual, such as turning on a fuel pump, reading a checklist, or initiating a go-around. Operant behavior is the target of learning and the most important to a flight instructor. It generally consists of three phases: cues are provided to initiate the behavior, the behavior is performed, and consequences associated with that performance are received.

operate — To use, cause to be used, or authorize the use of an aircraft for the purpose of air navigation.

operate — With respect to aircraft, means use, cause to use or authorize to use aircraft, for the purpose (except as provided in FAR 91.13) of air navigation including the piloting of aircraft, with or without the right of legal control (as owner, lessee, or otherwise).

operating center of gravity range — The distance between the forward and rearward center of gravity limits as specified in the Aircraft Specifications or Type Certificate Data Sheets.

operating pressures — The hydraulic or pneumatic pressures to which an object or system is subjected in normal operation.

operating relay time — The measured time from when a relay control switch is closed until the relay contacts are completely closed.

operating time — The time measured from when an engine, component, or unit begins to operate until the operating unit completes its operation or movement.

operating weight — The empty weight of the aircraft along with items carried in the aircraft during flight such as crew, water, food, etc.

operation raincheck — A program designed to familiarize pilots with the ATC system, its functions, responsibilities and benefits. Visiting facilities such as control towers, approach control, or air route traffic control centers can expand an aviation professional's knowledge of ATC services.

operation takeoff — A program that educates pilots on how best to utilize the FSS modernization efforts and services that are available in automated flight service stations (AFSS), as stated in FAA Order 7230.17. One can interact with weather briefers at a local flight service station and learn tips and techniques to get the most out of the services they provide.

operational — A phase of flight wherein an aircraft commander of a State-operated aircraft assumes responsibility to separate his aircraft from all other aircraft.

operational amplifier — A DC amplifier with a high input impedance and a low output impedance. Most basic type of integrated circuits. Also referred to as an op-amp.

operational checks — The inspection of a unit done to determine if it is operating properly within the manufacturer's specifications.

operational control — With respect to a flight, means the exercise of authority over initiating, conducting or terminating a flight.

operational pitfalls — The desire to complete a flight as planned, please passengers, meet schedules, and demonstrate that they have the right stuff can all have an adverse effect on safety by causing pilots to overestimate their piloting skills under stressful conditions. The operational pitfalls are as follows: mind set, scud running, continuing VFR into IMC, operating without adequate fuel reserves, flying outside the envelope, neglect of flight planning, preflight inspections and checklists, getting behind the aircraft, peer pressure, get-there-itis, duck-under syndrome, and descent below minimum enroute altitude.

Operations Limitations Manual — Approved information that must be carried in an airplane, including the speeds, engine operating limits, and any other information vital to the pilot. Also referred to as a flight manual.

opposed-type engine — A reciprocating engine with the cylinders arranged in two horizontal rows, one on either side of the crankshaft. The cylinders are slightly staggered, with the cylinders in one bank slightly ahead of those in the other bank. Staggering the cylinders allows each piston to be connected to a separate throw of the crankshaft. Also referred to as a horizontally-opposed engine.

opposite direction aircraft — Aircraft are operating in opposite directions when:
a. They are following the same track in reciprocal directions; or
b. Their tracks are parallel and the aircraft are flying in reciprocal directions; or
c. Their tracks intersect at an angle of more than 135°.

opposite side — The leg of a right triangle opposite the reference angle.

optical coupler — An optoelectronic device that connects circuits together with no physical electrical connection. Usually consists of a light emitting diode (LED) on the input side connected optically to a light-sensitive electronic device on the output side to reconstruct the electronic input.

optical micrometer — A precision measuring device used to measure the depth of scratches or fissures in the surface of a material by measuring the change in focus of a complex lens. The lens is focused on an undamaged surface of the material that is used as a reference, and then it is focused at the base of the damage. The amount of change in the lens focus is converted into a measure of the depth of the damage.

optical pyrometer — A temperature measuring instrument used to indicate the temperature of molten metal inside a furnace.

option approach — An approach requested and conducted by a pilot that will result in either a touch-and-go, missed approach, low approach, stop-and-go, or full stop landing.

optional equipment — Aircraft equipment approved for installation in an aircraft, but which is not required for airworthiness.

OR gate — A logic device that will have a voltage on its output any time a voltage appears at any one or more of its inputs.

oral quizzing — The most common means of evaluating a student's knowledge. Proper oral quizzing during a lesson promotes active student participation, identifies points that need more emphasis, and reveals the effectiveness of training procedures.

orange peel — A defect in a painted surface that resembles the skin of an orange. It can be caused by errors in paint viscosity, air pressure, spray gun settings, or an improper distance between the spray gun and the surface.

orbital electron — An electron spinning around the nucleus of an atom; different from free electrons, which are able to move from one atom to another.

ordinate — A line parallel to the Y-axis of a graph. A specified distance from the X-axis, used to fix a point.

organic brake linings — Organic material reinforced with brass wool and attached to solid metal backings. Used for single-disk brakes.

organic fabric — A woven material made of natural origin fibers such as cotton or linen. Used in the manufacturing of aircraft fabric covering materials.

organic fibers — Fibers of natural origin such as cotton or linen. Used in the manufacture of aircraft fabric covering materials.

organic lining — The friction material used in spot-type, single disk brakes. It is a composition material in which brass- or copper-wool or particles of brass are embedded to control the coefficient of friction.

organized track system — **1.** A movable system of oceanic tracks that traverses the North Atlantic between Europe and North America the physical position of which is determined twice daily taking the best advantage of the winds aloft. **2.** A series of fixed and charted ATS (Air Traffic Services) routes; i.e., CEP, NOPAC; or flexible and described by NOTAM; i.e., NAT Track Message

orientation — In composites, the alignment of the fibers (0°, 45°, 90°) to the baseline set by the manufacturer for a particular component.

orifice — A small hole of a specific size that meters or controls the flow of a fluid.

orifice check valve — A component in a hydraulic or pneumatic system that allows unrestricted flow in one direction and restricted flow in the opposite direction.

original skin — The skin or metal covering originally used in the manufacturing of an airplane.

O-ring — A sealing device used in a pneumatic or hydraulic system that has a circular cross section and is made in the form of a ring.

ornithopter — An aircraft designed to produce lift by the flapping of its wings.

OROCA (off--route obstruction clearance altitude) — An offroute altitude that provides obstruction clearance with a 1,000 foot buffer in nonmountainous terrain areas and a 2,000 foot buffer in designated mountainous areas within the United States. This altitude may not provide signal coverage from groundbased navigational aids, air traffic control radar, or communications coverage.

orographic — Relating to mountains. In weather, refers to phenomena caused by mountains, i.e., orographic clouds.

orographic lifting — The lifting of an airmass when it encounters a barrier, for example, mountains or a hill.

oronasal oxygen mask — An oxygen mask covering only the mouth and the nose of the wearer.

orphaned airplane — Slang, for an aircraft that is frequently non-airworthy and spends a lot of time in the hangar. In the military or commercial aviation, these aircraft are frequently robbed of parts to keep other aircraft airworthy, thus prolonging the status of the hangar queen. Also referred to as a hangar queen.

orthographic projection — In mechanical drawing, six different views of an object are possible through orthographic projection: front, rear, top, bottom, left side, and right side. Each view is drawn as if the object were placed in a box and only one side of the box opened to reveal one view (either front, rear, top, bottom, left side, or right side) of the object.

oscillate — To swing back and forth with a consistent force or rhythm.

oscillator — **1.** An electronic device that converts DC into AC. **2.** An electronic circuit that produces AC with frequencies determined by the inductance and capacitance in the circuit.

oscillograph — A device for mechanically or photographically producing a graphical representation of an electric signal.

oscilloscope — An electrical measuring instrument with which repeating voltage and current changes can be observed on a cathode-ray tube similar to a small television tube.

Otto cycle — A constant-volume cycle of events used to explain the energy transformation that takes place in a reciprocating engine. In this type of engine, four strokes are required to complete the required series of events or operating cycle of each cylinder. Two complete revolutions of the crankshaft (720°) are required for the four strokes and the spark plug in each cylinder fires once for every two revolutions of the crankshaft.

out — The conversation is ended and no response is expected.

out of ground effect (OGE) hover — In rotorcraft, hovering greater than one rotor diameter distance above the surface. Because induced drag is greater while hovering out of ground effect, it takes more power to achieve a hover out of ground effect.

out of phase — A condition in which two cyclic waves such as voltage and current do not pass through the same point at the same time.

out time — In composites, the time a pre-preg is exposed to ambient temperature, namely, the total amount of time the pre-preg is out of the freezer. This can include shipping time as well as the time it takes to cut off a small piece from the roll.

outer area — Airspace surrounding a Class C airspace area, extending out to a radius of 20 NM ATC provides radar vectoring and sequencing for all IFR aircraft and participating VFR aircraft.

outer area (associated with class C airspace) — Nonregulatory airspace surrounding designated Class C airspace airports wherein ATC provides radar vectoring and sequencing on a full-time basis for all IFR and participating VFR aircraft. The service provided in the outer area is called Class C service and includes: IFR/ IFR-standard IFR separation; IFR/VFR-traffic advisories and conflict resolution; and VFR/VFR-traffic advisories and, as appropriate, safety alerts. The normal radius will be 20 nautical miles, with some variations based on site-specific requirements. The outer area extends outward from the primary Class C airspace airport and extends from the lower limits of radar/radio coverage up to the ceiling of the approach control's delegated airspace excluding the Class C charted area and other airspace as appropriate.

outer compass locator — A low power, low or medium frequency (L/MF) radio beacon installed at the site of the outer marker of an instrument landing system (ILS). It can be used for navigation at distances of approximately 15 miles or as authorized in the approach procedure.

outer fix — 1. A general term used within ATC to describe fixes in the terminal area, other than the final approach fix. Aircraft are normally cleared to these fixes by an Air Route Traffic Control Center or an Approach Control Facility. Aircraft are normally cleared from these fixes to the final approach fix or final approach course. 2. An adapted fix along the converted route of flight, prior to the meter fix, for which crossing times are calculated and displayed in the metering position list.

outer fix time — A calculated time to depart the outer fix in order to cross the vertex at the ACLT. The time reflects descent speed adjustments and any applicable delay time that must be absorbed prior to crossing the meter fix.

outer flame — The enveloping, almost transparent flame that surrounds the bluish-white inner flame or cone in oxyacetylene welding.

outer liner — The annular and can-annular combustion liner outer shell as opposed to its inner shell. Formerly used to refer to the outer case of a can-type combustion chamber.

outer marker — A marker beacon at or near the glide slope intercept altitude of an ILS approach. It is keyed to transmit two dashes per second on a 400 Hz tone, which is received aurally and visually by compatible airborne equipment. The OM is normally located four to seven miles from the runway threshold on the extended centerline of the runway.

outflow boundary — The remnant of a gust front that continues to exist long after the thunderstorms that created it dissipates.

outflow valve — The valve in a pressurized aircraft cabin that maintains the desired pressure level inside the cabin by controlling the amount of air allowed to flow out of the cabin.

outlook briefing — A general overview forecast of the weather for a period 6 to 12 hours in advance.

out-of-rig — A condition of aircraft flight control rigging in which the controls are not properly adjusted, thus preventing the aircraft from being flown without the aid of the pilot touching the controls.

out-of-round — Eccentrically shaped because of damage or wear.

out-of-track — A condition of a helicopter rotor or the propeller of an airplane in which the tips of the blades do not follow the same path in their rotation.

out-of-trim — A condition in an aircraft in which straight-and-level, hands-off flying is impossible due to an aerodynamic load caused by an improperly adjusted trim device.

output — The power or energy a device delivers or produces.

output transformer — A transformer used to match the high impedance of an electronic circuit to the low impedance of the output device, usually a speaker.

outside air temperature (OAT) — The measured or indicated air temperature (IAT) corrected for compression and friction heating. Also referred to as true air temperature.

outside caliper — A measuring device having two movable legs. Used to determine the distance across an object. Once the distance has been established, the actual dimension can be made using a steel scale or a vernier caliper.

outside skin — The outer surface of an aircraft.

over — My transmission is ended; I expect a response.

overall efficiency — In turbine engines, the product of multiplying propulsive efficiency and thermal efficiency.

overbalance — The adding of counterbalancing weight to the extent that the trailing edge of a control surface is above the horizontal position when performing a static balance test. In a balanced condition, the control surface would rest in the neutral position.

overboost — A condition in which the manifold pressure of a reciprocating engine has exceeded the maximum specified by the manufacturer. Can cause damage to internal engine components.

overcompounded motor — A compound-wound motor with more series windings than parallel windings. Tends to try to speed up under increased loads.

overcontrol — Any movement of a control device in excess of that needed for a given condition.

overcurrent protection device — An electronic device that protects a circuit against abnormally high currents.

overhang — In meteorology, the anvil of a thunderstorm, under which hail can occur and a turbulent wake can create severe turbulence.

overhaul — To restore an aircraft, engine, or component to a condition of airworthiness.

overhead cam — The cam of an aircraft reciprocating engine located above the cylinder head. The overhead cam operates the valves directly without the aid of pushrods.

overhead maneuver — A series of predetermined maneuvers prescribed for aircraft (often in formation) for entry into the visual flight rules (VFR) traffic pattern and to proceed to a landing. An overhead maneuver is not an instrument flight rules (IFR) approach procedure. An aircraft executing an overhead maneuver is considered VFR and the IFR flight plan is cancelled when the aircraft reaches the "initial point" on the initial approach portion of the maneuver. The pattern usually specifies the following:
a. The radio contact required of the pilot.
b. The speed to be maintained.
c. An initial approach 3 to 5 miles in length.
d. An elliptical pattern consisting of two 180 degree turns.
e. A break point at which the first 180 degree turn is started.
f. The direction of turns.
g. Altitude (at least 500 feet above the conventional pattern).
h. A "Roll-out" on final approach not less than 1/4 mile from the landing threshold and not less than 300 feet above the ground.

overhead question — In the guided discussion method, a question directed to the entire group to stimulate thought and discussion. An overhead question can be used by an instructor as the lead-off question.

overhead valve — A valve located in the upper part of an aircraft reciprocating engine cylinder head.

overheat warning system — A system that warns of an overheat condition that could lead to a fire.

overinflation valve — A relief valve that opens to relieve excessive air pressure. Used in some of the larger aircraft wheels that mount tubeless tires.

overlapping — To lap over or extend beyond.

overload — To apply a load in excess of that for which a device or structure is designed.

overlying center — The ARTCC facility that is responsible for arrival/departure operations at a specific terminal.

overrunning — In meteorology, when a warm, moist, stable airmass moves over a warm front or a stationary front.

overrunning clutch — Generally, a pawl and ratchet arrangement used on various types of starters. It permits the starter to drive the engine but not allow the engine to drive the starter. The ratchet on the engine side will slip around within the pawls if normal disengagement does not occur.

overseas air commerce — The carriage by aircraft of persons or property for compensation or hire, or the carriage of mail by aircraft, or the operation or navigation of aircraft in the conduct or furtherance of a business or vocation, in commerce between a place in any State of the United States, or the District of Columbia, and any place in a territory or possession of the United States; or between a place in a territory or possession of the United States, and a place in any other territory or possession of the United States.

overseas air transportation — The carriage by aircraft of persons or property as a common carrier for compensation or hire, or the carriage of mail by aircraft, in commerce:
a. Between a place in a State or the District of Columbia and a place in a possession of the United States; or
b. Between a place in a possession of the United States and a place in another possession of the United States; whether that commerce moves wholly by aircraft or partly by aircraft and partly by other forms of transportation.

overshoot — Overshoot is caused by rapid increase in throttle, which causes the controller to overshoot the requirement for the engine boost, resulting in overboost.

overshooting tops — In thunderstorms, very strong updrafts that penetrate the otherwise smooth top of the anvil cloud.

oversize — A question with many possible correct answers.

oversized stud — A stud having a greater diameter than standard for the portion that is pressed or threaded into the stud boss. The external diameter and thread pitch are the same as original standard stud.

overspeed — **1.** A condition in which an engine has produced more RPM than the manufacturer recommends. **2.** The condition in which the actual engine speed is higher than the desired engine speed as set on the propeller control.

overspeed condition — A condition of the propeller operating system in which the propeller is operating above the RPM for which the governor control is set. This causes the propeller blades to be at a lower angle than that required for the desired speed.

overspeed governor — A speed-limiting device. Governors regulate speed through the fuel control.

overspeeding — Exceeding the maximum RPM limits of the engine. Adjusting pitch stops in the prop governor allow RPM limits to be set.

overtemperature — **1.** A condition in which a device has reached a temperature above that approved by the manufacturer. **2.** Any exhaust temperature that exceeds the maximum allowable for a given operating condition or time limit.

overtemperature warning system — A warning system that warns the pilot of an overheat condition. If the temperature rises above a set value in any one section of the overheat sensing circuit, the sensing device turns on a cockpit light indicating the location of the overtemperature.

over-the-top — Above the layer of clouds or other obscuring phenomena forming the ceiling.

overview — A concise presentation of the objective and key ideas, supplemented with appropriate visual aids, to give students a clear picture of what is to come.

overvoltage protector — An electrical circuit protection device used to protect components from damage caused by high voltage surges. If the voltage is excessive, the overvoltage protector opens and protects the component.

oxidation — A chemical action in which a metallic element is combined with oxygen. Electrons are removed from the metal in this process.

oxide — A chemical combination in which oxygen is combined with another element.

oxide film — A layer, coating, or metallic oxide on the surface of a material.

oxidized — **1.** Combined with oxygen. **2.** A substance that causes another to combine with oxygen.

oxidizing flame — An oxyacetylene welding flame in which there is an excess of oxygen passing through the torch. An oxidizing flame can be recognized by its sharp-pointed inner cone and a hissing noise made by the torch.

oxyacetylene — Gas welding that uses oxygen and acetylene.

oxy-gas welding — Gas welding that uses oxygen and a gas such as acetylene.

oxygen — One of the basic elements with a symbol of O and an atomic weight of 8. In the free gas state, it is always O_2 because two atoms of oxygen must combine to form one molecule of oxygen gas. Oxygen is a colorless, odorless, tasteless, gaseous chemical element that makes up about 21% of the Earth's atmosphere. It will not burn but supports combustion and is essential to life processes.

oxygen bottle — Special, high-strength steel cylinder used to store gaseous oxygen under pressure.

oxygen cell corrosion — Corrosion that results from a deficiency of oxygen in the electrolyte.

oxygen concentration cell corrosion — Corrosion that forms between the lap joints of metal where moisture gets trapped.

oxygen manifold — A device for connecting several oxygen masks into one oxygen supply or several oxygen sources into a master manifold.

oxygen mask — A small face mask with special attachments for breathing oxygen.

oxygen plumbing — Tubing and fittings used in the oxygen system to connect the various components.

oxyhydrogen — Gas welding that uses oxygen and hydrogen.

ozone — A variety of oxygen that contains three atoms of oxygen per molecule rather than the usual two. The major portion of ozone in the atmosphere is formed by the interaction of oxygen with the sun's rays near the top of the ozone layer. It is also produced by electrical discharges (lightning storms). Ozone is important to living organisms because it filters out most of the sun's ultraviolet radiation.

ozone hole — The region of the ozone layer that has a lower than normal concentration of O_3.

ozone layer — A layer of O_3 found in the lower stratosphere. Characterized by a relatively high concentration of ozone, this layer is responsible for the increase of temperature in the stratosphere.

ozonosphere — The stratum of the Earth's atmosphere that has a high concentration of ozone and absorbs ultraviolet radiation from the sun. The ozonosphere is approximately 20 to 30 miles above the Earth.

P

P factor — A tendency for an aircraft to yaw to the left due to the descending propeller blade on the right producing more thrust than the ascending blade on the left. This occurs when the aircraft's longitudinal axis is in a climbing attitude in relation to the relative wind. The P factor would be to the right if the aircraft had a counter-clockwise rotating propeller.

P time — Proposed Departure Time. The time a scheduled flight will depart the gate (scheduled operators) or the actual runway off time for nonscheduled operators. For EDCT purposes, the ATCSCC adjusts the "P" time for scheduled operators to reflect the runway off times.

pack carburizing — A heat treatment method for case-hardening steel parts in which the parts are packed in the carburizing compound, placed into a drum, and heated to a temperature of about 1500° F.

package — A complete assembly unit.

packing — The hydraulic seal that prevents fluids from leaking between two surfaces that move in relation to each other.

packing ring — An O-ring used to confine liquids or gases, preventing their passing between a fixed body and a movable shaft.

paint — An aircraft finish that consists of pigments suspended in a solvent-type vehicle. The paint protects and improves the appearance of the aircraft.

paint drier — Any substance added to paint to improve its drying properties.

paint stripper — A chemical material that softens the paint film and loosens its bond to the metal, thus enabling the paint to be easily wiped or washed away.

pal nut — A thin, pressed-steel check nut screwed down over an ordinary nut to prevent it from coming loose or backing off.

palladium — A metallic element with a symbol of Pd and an atomic number of 46.

pan — A hollow depression or shallow metal object used to hold oil.

pancake landing — An aircraft landing procedure in which the aircraft is on an even plane with the runway. As the aircraft reduces speed and lift, it drops to the ground in a flat or prone attitude.

panel — Any separate or distinct portion of an aircraft surface.

pan-head screw — A machine screw or sheet metal screw that has a large, slightly domed head.

pan-pan-pan — The international radio-telephony urgency signal. When repeated three times, indicates uncertainty or alert followed by the nature of the urgency.

pants — Streamlined airplane wheel covers.

paper electrical capacitor — An electrical component that uses two strips of metal foil for its plates and strips of waxed paper as its dielectric.

parabola — A plane curve equally distant from a fixed point and a fixed line.

parabolic light reflector — A light reflector, such as a light bulb, with a curved or parabolic surface.

parabolic microphone — A sensitive, highly directional microphone mounted at the focal point of a parabolic dish capable of picking up sound from long distances.

parachute — A large cloth device shaped like an umbrella used to retard the fall of a body or object through the air.

parachute — A device used or intended to be used to retard the fall of a body or object through the air.

par-al-ketone — A heavy, waxy grease used to protect control cables and hardware fittings from corrosion on seaplanes.

parallax — The apparent change in relationship of two objects when viewed from different locations not in line with the objects.

parallax — Apparent displacement of an object if first viewed from one position and then from another.

parallel — 1. Lines that run in the same direction and will never meet or cross because the distance between them is constant. 2. Having more than one path for electron flow between the two sides of the electron source.

parallel access — In computers, a term used to describe a method of accessing data simultaneously, or all data at the same time.

parallel circuits — Two or more complete circuits connected to the same two power terminals.

parallel ILS approaches — Approaches to parallel runways by IFR aircraft that, when established inbound toward the airport on the adjacent final approach courses, are radar separated by at least 2 miles.

parallel lines — Two or more lines extending in the same direction and at the same distance apart at all points.

parallel MLS approaches — See parallel ILS approaches.

parallel of latitude — Any of the imaginary lines on the surface of the Earth parallel to the equator and representing degrees of latitude on the Earth's surface.

parallel offset route — A parallel track to the left or right of the designated or established airway/route. Normally associated with Area Navigation (RNAV) operations.

parallel operation — In computers, an operation that moves information simultaneously over several lines as opposed to serially, which moves information one bit at a time over a single line. Also used to describe an operation using more than one processor simultaneously.

parallel resonant electrical circuit — A circuit made up of an inductor and a capacitor connected in parallel. If a circuit is resonant, the inductive reactance and the capacitive reactance are equal at a particular frequency. Also referred to as a tank circuit

parallel runways — Two or more runways at the same airport whose centerlines are parallel. In addition to runway number, parallel runways are designated as L (left) and R (right) or, if three parallel runways exist, L (left), C (center), and R (right).

paralleling — Controlling the output of more than one generator in order to share a load equally. This involves matching the voltage outputs, the frequency (of AC generators) and the phase relationship of the AC current (of AC generators).

paralleling generators — An operational procedure in which the output voltages of multi-engine aircraft electrical system generators are adjusted to share the electrical load equally. This involves matching the voltage outputs, the frequency (of AC generators) and the phase relationship of the AC current (of AC generators).

paramagnetic material — A material that becomes magnetic when placed in a strong magnetic field and retains some level of magnetic property as long as the strong magnetic field is present. However, when the strong magnetic field is removed, the material loses its magnetic properties.

parameter — A quantity or constant whose value varies with the circumstances of its application.

parasite drag — Drag caused by the friction of the air flowing over a body. Parasite drag increases as airspeed increases.

parasite drag — That part of total drag created by the form or shape of airplane parts.

parasol wing airplane — An airplane having one main supporting surface mounted above the fuselage on cabane struts.

parcel — A volume of air, small enough to contain uniform distribution of its meteorological properties, but large enough to remain relatively self-contained and respond to all meteorological processes.

Parco lubrizing — Anti-friction coatings that contain solvent-based suspensions of solid lubricants instead of a coloring pigment.

Parkerizing — A method of treating metal parts by immersing them in a solution of phosphoric acid and manganese dioxide. Used to protect the surface from rusting.

Parker-Kalon (PK) screws — Self-tapping sheet metal screws. PK screws are made of hardened steel, and have sharp, coarse threads that combine drilling, tapping and fastening.

parking brake — A mechanical or hydraulic brake system used to prevent an aircraft from moving from its parked position.

part number — An identification number assigned to a particular part or assembly by the manufacturer.

part power trim check — The act of trimming the engine with the power lever against a trim stop or rig pin, then checking the EPR or N_1 speed against a trim curve for ambient conditions. If the correct values are not present, adjustment of the fuel control, referred to as trimming, is required.

partial obscuration — Denotes that 1/8th or more of the sky, but not all of the sky, is hidden by surface-based (excluding precipitation) phenomena in the atmosphere.

partial panel — Controlling the airplane without the benefit of all instrumentation due to failure of one or more instruments in flight. Pilots are required to demonstrate basic attitude instrument flying on partial panel and are also required to fly a partial panel instrument approach during the practical test for the instrument rating.

partial pressure — The gases that make up the atmosphere each exert a partial pressure. When all of the partial pressures are added together, they equal the total atmospheric pressure.

partial-panel flight — Instrument flight without vacuum-powered gyroscopic instruments. Remaining instruments would be the altimeter, airspeed indicator, turn coordinator, vertical speed indicator, and magnetic compass.

particle — A small piece of any substance or matter.

particulates — Very small liquid or solid particles in the atmosphere. When suspended in the atmosphere, they are called aerosols.

parting agent — In composites, a lubricant used to prevent the part from sticking to the mold.

parting film — In composites, a layer of thin plastic to prevent bagging materials from sticking to the part. It can be perforated to vent excess resin. It is removed after cure. Can be used instead of peel ply.

parts manufacturing approval (PMA) — Approval by the Federal Aviation Administration to design and manufacture aircraft replacement parts for sale directly to the public.

pascal — A metric unit of pressure. Equals one newton per square meter.

Pascal's law — A basic law of fluid power that states that pressure in an enclosed container is transmitted equally, undiminished to all points of the container, and acts at right angles to the enclosing walls.

passenger mile — An airline statistic used to track utilization of aircraft. One passenger mile is equal to one passenger traveling one mile. Five passengers traveling 100 miles would equal 500 passenger miles.

passivating — A treatment of corrosion-resistant steels after welding. The purpose is to remove iron from the surface and expose more chromium, allowing a chrome rich oxide film to form and protect the surface from corrosion.

passive electrical circuit — An electrical circuit that does not contain any source of electrical energy such as a battery or generator. Such circuits would include only passive electronic components. One example would be a filter (consisting of inductors and capacitors) in a receiver circuit that is powered only by the received signal.

passive electrical component — An electrical component, such as resistor, capacitor or inductor, that produces no gain in the circuit.

passive satellite — A passive satellite reflects received radio signals back to Earth without amplification.

passive sonar — A device for detecting the presence of an object by the sound it emits in water. Sonar stands for sound navigation ranging.

passive video — Refers to segments of video that are simply watched by the students.

patch — A small piece of material used for strengthening, reinforcing, or covering a hole or weak spot in a structure.

pattern — **1.** A model, guide, or plan used to form or make things. **2.** The flight pattern an aircraft must follow when approaching the airport for landing and when leaving the airport after taking off. Aircraft operating from the airport must follow the same flight pattern in order to reduce the danger of an in-flight collision.

pawl — A pivoted stop in a mechanical device that allows motion one way but prevents it in the opposite direction. It is commonly used in a ratchet mechanism.

payload — That part of the useful load of an aircraft that is over and above the load necessary for the operation of the vehicle. The term used for passengers, baggage, and cargo.

P-channel field effect transistor (FET) — Similar to an N-channel FET but with the types of material reversed. Control is achieved by the application of positive voltage to the gate.

peak alternating current — The greatest amount of current that flows in one alternation of alternating current. The greatest amount of deviation from the zero line of deviation as measured on an oscilloscope.

peak inverse voltage — The maximum voltage that can be applied safely to an electron tube in the direction opposite to normal current flow.

peak value — The maximum value of AC or voltage measured from the zero reference line.

peak wind — The maximum wind speed since the last hourly observation.

peak-to-peak voltage — Absolute value of the difference between the maximum positive and maximum negative values of an AC waveform.

pedestal grinder — A grinder mounted on a pedestal and stands on the floor of the shop.

peel ply — In composites, a layer of fabric used in manufacturing to vent excess resin up into the bleeder material. It prevents bagging materials from sticking to the part, and it leaves a very finely etched surface for painting. It is removed after cure.

peel strength — In composites, the amount of strength it takes a part to resist the stress applied from peeling apart of two plies.

peen — To round over or flatten the end of a shaft or rivet by light hammer blows.

peened surface — A marked surface as from an impact with a blunt instrument. Caused by careless handling or concentrated load sufficient to permanently deform the metal surface.

pencil compass — A drawing instrument that uses a pencil to draw circles or arcs.

pendular action — In rotorcraft, the lateral or longitudinal oscillation of the fuselage due to it being suspended from the rotor system.

pendulum — A body suspended from a fixed point but that is free to swing back and forth or oscillate.

pendulum valves — Gravity-operated air valves over the discharge ports of a pneumatic gyro horizon's rotor housing. When the gyro tilts, the pendulum valves change the airflow from the housing and cause a precessive force that erects the gyro.

pentagrid converter — A five-grid electron tube that serves as a mixer, local oscillator, and first detector in a superheterodyne radio receiver.

pentode — A five-element vacuum tube or electron tube containing five electrodes: cathode, plate, control grid, suppressor grid, and screen grid.

perceptions — The basis of all learning. Perceptions result when a person gives meaning to external stimuli or sensations. Meanings, which are derived from perceptions, are influenced by an individual's experience and many other factors.

percussive welding — A resistance welding process in which an electrical arc and pressure are simultaneously applied across the materials being welded.

perfect dielectric — A dielectric that has no conductivity. A perfect dielectric is an insulator that returns all of the used energy to establish an electric field when the electrical field is removed.

perforate — To create a hole in a material such as paper.

perforated parting film or release film — In composites, a thin layer of plastic film used to prevent bagging materials from sticking to the part. The perforations allow some resin to flow through small holes in the plastic. Used in the same way as peel ply.

performance chart — A chart detailing the aircraft performance that can be expected under specific conditions.

performance envelope — A range of flight conditions and performance for which an aircraft has been designed and tested for. Flying an aircraft outside these parameters is done so at the risk of unknown consequences.

performance number — The anti-detonation rating of a fuel that has a higher critical pressure and temperature than isooctane. Isooctane is used as the reference.

performance-based objectives — A statement of purpose for a lesson or instructional period that includes three elements: a description of the skill or behavior desired of the student, a set of conditions under which the measurement will be taken, and a set of criteria describing the standard used to measure accomplishment of the objective.

period — The time required for one cycle of AC.

period oscillation — The amount of time needed to complete one cycle of an oscillation.

periodic event — Any regularly repeated event.

periodic inspection — Any regularly repeated inspection. An annual or 100-hour inspection of an aircraft is a periodic inspection.

periodic table of chemical elements — A table of chemical elements arranged in the order of their atomic numbers.

periodic vibration — A vibration that has a regularly recurring resonance.

peripheral equipment — Devices that provide support for primary units of equipment. In computers, peripheral equipment would include hard disk drives, floppy drives, modems, and CD ROM drives.

periphery — The outside of a circular or curved figure.

Permalloy — An alloy of iron and nickel used in the manufacture of permanent magnets.

permamold crankcase — An engine crankcase that has been pressure molded in a permanent mold. It is thinner and denser than a sand-cast crankcase.

permanent ballast — A weight that has been permanently installed in an aircraft to bring its center of gravity into allowable limits

permanent echo — Radar signals reflected from fixed objects on the Earth's surface; e.g., buildings, towers, terrain. Permanent echoes are distinguished from "ground clutter" by being definable locations rather than large areas. Under certain conditions they can be used to check radar alignment.

permanent magnet — A ferrous metal or alloy of ferrous metals usually containing nickel and cobalt, in which the magnetic domains are aligned, and tend to remain aligned. Lines of magnetic flux join the poles of the permanent magnet so that an electrical current can be generated when these lines of flux are cut by a conductor.

permanent magnet speaker — A speaker that uses a small permanent magnet to provide the magnetic field for the voice coil.

permanent mold — A mold that can be used repeatedly to produce molded parts. Usually constructed of metal or ceramic materials.

permanent set — A mechanical deformity caused by excessive stress placed on a material.

permeability — The ability of a material to accept and concentrate lines of magnetic flux.

Perminvar — Special alloy used for permanent magnets.

persistence — In cathode-ray tubes (CRTs), the length of time the phosphorescent glow remains after the electron beam has moved on.

persistence forecast — A weather prediction based on the assumption that future weather will be the same as current weather.

person — An individual, firm, partnership, corporation, company, association, joint-stock association, or governmental entity. It includes a trustee, receiver, assignee, or similar representative of any of them.

personal checklists — To help students determine if they are prepared for a particular flight, encourage them to create personal checklists that state their limitations based on such factors as experience, currency, and comfort level in certain flight conditions.

personal computer-based aviation training devices (PCATDs) — A computer based flight simulator that has been authorized by the FAA to be used for a small portion of the instrument time required for instrument ratings. Uses cathode ray tube computer monitor for display.

personality — A set of personal traits and characteristics of an individual.

perspective — The technique of representing the spatial relationship of objects as they appear to the eye.

petroleum — A substance containing chemical energy and used as a fuel for most of engines. It is a natural hydrocarbon product that was at one time plant or animal life, but was buried under billions of tons of Earth. It is obtained as a liquid from deep wells or from oil shale.

petroleum-zinc-dust compound — A material used inside an aluminum terminal lug when swaging the lug onto aluminum wire. The zinc dust abrades the oxides from the aluminum, and the petrolatum prevents its reformation.

pH — A measure of acidity and alkalinity of a solution on a 14 point scale where a value of 7 represents neutrality. 1 indicates maximum acidity and 14 indicates maximum alkalinity.

phantom line — Thin lines made up of alternating long dashes and two short dashes. Used to show an alternate position or a missing part in a mechanical drawing.

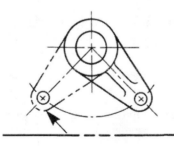

phase — 1. The time difference between an event in the voltage waveform and the equivalent event in the current waveform in an AC circuit. 2. A section or a distinguishable part of a maintenance program or inspection.

phase angle — The number of degrees of generator rotation between the time the voltage passes through zero and the time the current passes through zero in the same direction.

phase change — A change of state.

phase inverter — An electronic device that changes the phase of a signal by 180°.

phase lock — A method of modulating the phase, condition, or state of one electronic oscillating device so it will exactly follow that of another oscillating device.

phase modulation — A method of modulating the carrier wave of a radio signal. The phase angle is changed relative to voltage changes of the input signal.

phase shift — The difference in time between similar points of an output and an input electrical wave form.

phenol-formeldehyde resin — A thermosetting plastic resin reinforced with cloth or paper to make molded plastic objects.

phenolic material — A thermosetting resin or plastic made by condensation of a phenol with an aldehyde and used for molding and insulating.

phenolic plastic — Plastic thermosetting phenolic-formaldehyde resin material, reinforced with cloth or paper.

phenolic resin — A thermosetting resin produced by the condensation of an aromatic alcohol with an aldehyde, particularly of phenol with formaldehyde.

Phillips-head screw — A recessed-head screw designed to be driven with a cross-pointed screwdriver whose point has two distinct tapers and a blunt point. The Reed and Prince is another cross-point driver but it has straight sides, only one taper, and tapers to a point.

phonetic alphabet — Standard words (lexicon) and combinations of words used for each of the letters in the alphabet during radio transmission. Examples of standard phonetic alphabet now in use are: ALFA, BRAVO, CHARLIE, DELTA, ECHO, FOXTROT, GOLF, HOTEL, INDIA, LIMA, NOVEMBER, OSCAR, PAPA, ROMEO, SIERRA, TANGO, UNIFORM, VICTOR, WHISKEY, X-RAY, YANKEE, ZULU.

phosgene — A colorless gas with an unpleasant odor produced when Refrigerant-12 is passed through an open flame. Causes severe respiratory irritation.

phosphate ester-base hydraulic fluid — A synthetic, fire-resistant hydraulic fluid used in high-pressure hydraulic systems of modern jet aircraft. It is identified by the specifications MIL-H-8446.

phosphate film — A dense, insoluble, inorganic film deposited on the surface of a metal treated with a conversion coating.

phosphor coating — A coating for the numerals and pointers on instrument dials. The coating glows when excited with ultraviolet light rays.

phosphorescent paint — A paint that absorbs energy from natural light or from ultraviolet light and which continues to glow after the natural light is removed.

phosphoric acid etchant — That constituent of a conversion coating that microscopically roughens the surface of the metal being treated and deposits a phosphate film.

phosphorus — A chemical element with a sign of P and an atomic number of 15.

photo cell — An electronic device that becomes conductive or produces a voltage when struck by light.

photo reconnaissance — Military activity that requires locating individual photo targets and navigating to the targets at a preplanned angle and altitude. The activity normally requires a lateral route width of 16NM and altitude range of 1,500 feet to 10,000 feet AGL.

photocathodes — An electrode within an electron tube that releases electrons after it has been exposed to light.

photochemistry — That branch of chemistry having to do with the effects of light on chemical reactions.

photoconductive cell — A photoelectric cell that changes its resistance according to the amount of light exposure. Photoconductive cells are used in photographic light meters.

photodiode — A semiconductor diode that can conduct in its reverse direction when light is shown on its junction.

photoelectric characteristics — The changes (positive or negative) produced in the electrical characteristics of a material when exposed to light.

photoelectric material — Any element that emits electrons when exposed to light. Elements such as alkaline metals, cesium, lithium, and rubidium are photoelectric.

photoelectricity — Electricity produced by the action of light on certain photoemissive materials.

photoemissive characteristic — The trait of a material to emit electrons when exposed to light.

photoemissivity — The tendency of a material to emit electrons when exposed to light.

photon — A particle of radiant energy.

photonegative characteristics — The characteristics of a material increasing in resistance when exposed to light.

photosensitive — The property of emitting electrons when struck by light. See also photoemissive.

photothyristor — A semiconductor device that is responsive to visible or infrared radiant energy. May or may not include mounting hardware and/or heat sink.

phototransistor — A transistor that can be forward-biased into conduction by applying light to its emitter-base junction.

photovoltaic cell — A solid-state electrical component that produces a voltage when exposed to light.

phugoid oscillations — Long-period oscillations of an aircraft around its lateral axis. It is a slow change in pitch accompanied by equally slow changes in airspeed. Angle of attack remains constant, and the pilot often corrects for phugoid oscillations without even being aware of them.

physical needs — Needs that encompass the necessities for survival, which include food, rest, exercise, and protection from the elements. Until these requirements are satisfied, students cannot fully concentrate on learning.

physical organism — A perception factor that describes a person's ability to sense the world around them.

physical properties — Those properties of a body that can be determined by methods other than chemical, including weight, strength, and hardness.

physical tables — Tables that list the physical properties and characteristics of materials.

physics — A natural science that deals with matter and energy and their interaction in the various fields of mechanics.

pi (Π) — A mathematical constant representing the ratio between the circumference and diameter of a circle with a value of approximately 3.1415927. Represented by the Greek letter •.

pi filter — A network consisting of two capacitors and an inductor net arranged in the form of the Greek letter pi (π). It is essentially a capacitor-input filter followed by an L-filter.

piano hinge — A continuous metal hinge consisting of hinge bodies attached to both fixed and movable surfaces. A hard steel wire connects the two bodies and serves as the hinge pin.

pickling — **1.** The treatment of a metal surface with acid to remove surface contamination. **2.** Preparing an aircraft engine for longtime storage.

pick-off — That portion of a device or system that removes a signal from a sensor.

pico — One-millionth of a millionth (0.000,000,000,001) of a unit.

picofarad (pf or μμf) — One-millionth (0.000001) of a microfarad.

pictorial diagram — A diagram used by maintenance technicians. It can be either a line drawing, sometimes with shading to emphasize shapes, a picture of a part enlarged to show detail, or a photograph of a piece of equipment, illustrating the overall appearance of a unit, its shape, relative sizes and location of components, interconnecting wires, cables, etc.

pictorial drawing — A drawing consisting of pictures that shows an object as it appears to the eye.

pie chart — A graph drawn in the shape of a circle and divided into pieces like a pie to convey data or proportions.

piezoelectric crystal — A thin crystal that produces a voltage when distorted. Used in strain gauges and vibration detectors. Also produces vibration when a voltage is applied. Used in ultrasonic transducers.

piezoelectric effect — The property of certain crystals that enables them to generate an electrostatic voltage between opposite faces when subjected to mechanical pressure. Conversely, the crystal will expand or contract if subjected to a strong electrical potential.

piezoelectric transducer — An electrical device that enables a mechanical movement to generate an electrical signal.

piezoelectricity — Electricity produced when certain crystalline materials such as quartz are subjected to mechanical pressure.

pig iron — Crude iron, reduced from the iron ore in a blast furnace.

pigment — A powder or paste mixed with a paint finish to give the desired color.

pigtail — A piece of wire that sticks out of a component and allows the component to be installed or tied.

pilot — **1.** A person licensed to operate an airplane, ship, or balloon in flight. **2.** A part that guides another part in its movement.

pilot balloon — In weather, a small balloon that is released and tracked in order to determine wind direction and speed.

pilot balloon observation — In weather, a method of determining wind direction and speed by tracking a pilot balloon. Also referred to as a pibal.

pilot briefing — A service provided by the FSS to assist pilots in flight planning. Briefing items can include weather information, NOTAMs, military activities, flow control information, and other items as requested.

pilot chute — A small parachute attached to the canopy of the main parachute. The pilot chute pulls the main canopy out of the parachute pack so that it can open.

pilot controlled lighting (PCL) — Radio control of lighting is available at selected airports to provide airborne control of lights by keying the aircraft's microphone. The control system consists of a 3-step control responsive to 7, 5, and/or 3 microphone clicks. The 3-step and 2-step lighting facilities can be altered in intensity. All lighting is illuminated for a period of 15 minutes (except for 1-step and 2-step REILs which can be turned off by keying the mike 5 or 3 times, respectively).

Suggested use is to always initially key the mike 7 times; this assures that all controlled lights are turned on to the maximum available intensity. If desired, adjustment can then be made, where the capability is provided, to a lower intensity (or the REIL turned off) by keying the mike 5 and/or three times. Approved lighting systems can be activated by keying the mike as indicated below:

KEY MIKE ( ) times within 5 seconds — FUNCTION:

7 times within 5 seconds — Highest intensity available

5 times within 5 seconds — Medium or lower intensity (Lower REIL or REIL-Off)

3 times within 5 seconds — Lowest intensity available (Lower REIL or REIL-Off)

Due to the close proximity of airports using the same frequency, radio controlled lighting receivers can be set at a low sensitivity requiring the aircraft to be relatively close to activate the system. Consequently, even when lights are on, always key mike as directed when overflying an airport of intended landing or just prior to entering the final segment of an approach. This will assure the aircraft is close enough to activate the system and a full 15 minutes lighting duration is available.

pilot error — An action or decision made by the pilot that was the cause of, or contributing factor, which led to an accident or incident. This definition also includes failure of the pilot to make a decision or take action.

pilot hole — A small hole drilled or punched in sheet metal that is smaller than the bolt or rivet to be used. The pilot hole serves as a guide for final drilling.

pilot in command — The person who: (1) Has final authority and responsibility for the operation and safety of the flight; (2) Has been designated as pilot in command before or during the flight; and (3) Holds the appropriate category, class, and type rating, if appropriate, for the conduct of the flight.

pilot light — Electrical equipment light indicating power is on.

pilot proficiency award program — WINGS is the FAA Pilot Proficiency Award Program designed to encourage general aviation pilots to continue their training. The objective is to provide pilots with the opportunity to establish and participate in a personal recurrent training program. WINGS is an excellent opportunity for pilots to reevaluate their flight proficiency and knowledge. WINGS is open to all pilots holding a recreational certificate or higher with a current medical certificate, when required. After pilots log three hours of dual instruction under the program and attend at least one FAA sanctioned safety seminar, they are eligible to receive and wear a distinctive set of WINGS. They will also receive a certificate of completion. Each twelve-month interval after earning the first set of WINGS, the pilot will be eligible for more WINGS. CFIs can substitute completion of a flight instructor refresher clinic or renewal program for the safety seminar. In addition, they can satisfy the flying portion of the first three phases by providing the instruction for three WINGS candidates — a minimum of nine hours of instruction.

pilot weather report (PIREP) — A report, generated by pilots, concerning meteorological phenomena encountered in flight.

pilot's discretion — When used in conjunction with altitude assignments, pilot discretion means that ATC has offered the pilot the option of starting climb or descent whenever the pilot wishes and conducting the climb or descent at any rate. The pilot can temporarily level off at any intermediate altitude. However, once the pilot has vacated an altitude, the pilot cannot return to that altitude.

pilotage — Navigation by visual reference to landmarks.

pilot-in-command — The person who has final authority and responsibility for the operation and safety of the flight; has been designated as pilot in command before or during the flight; and holds the appropriate category, class, and type rating, if appropriate, for the conduct of the flight.

pilot-induced oscillation (PIO) — Rapid oscillations caused by the pilot's over-controlled motions. PIOs usually occur on takeoff or landings with pitch sensitive gliders and in severe cases can lead to loss of control or damage.

pilot's telephone weather answering service (PATWAS) — A recorded, continuous telephone briefing, forecast for the local area within a 50 nautical mile radius of the station. No longer provided by Flight Service Stations, has been replaced by the Telephone Information Briefing Service (TIBS), an automated service.

pin — A straight cylindrical or tapered fastener designed to perform an attaching or locating function.

pin contacts — Electrical connector contacts, called male contacts, in the form of a set of metal pins in one-half of a connector. These pins fit into sockets, called female contacts, in the other half of a connector.

PIN diode — A junction diode with a region of intrinsic semiconductor between layers of n-silicon and p-silicon.

pin holes — In composites, small holes caused by the mold used.

pin jack — A female receptacle that will accept and hold a small metal pin attached to the end of a wire or a test lead. Pin jacks are used on test equipment such as multimeters.

pin punch — A long punch with straight sides. Used to remove bolts and rivets from tight-fitting holes. See also punch.

pin spanner — A semi-circular wrench with pins that fit into holes around the edge of a circular nut.

pinch-off voltage — In electronics, the reverse bias that must be applied to an FET to reduce source-drain current to a specific value.

pinhole — A tiny defect in a finish caused by a bubble in the paint film.

pinion — A small cogwheel whose teeth fit into a larger gear.

pinion gear — A small gear on a shaft driven by either a sector gear or a toothed rack.

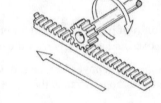

pinked edge — The edge of a fabric material that has been cut into a series of small V's to prevent the material from unraveling.

pinked-edge fabric — A fabric material with an edge that has been cut into a series of small V's to prevent the material from unraveling.

pinked-edge tape — A surface tape whose edges have been cut into small V's.

pinking shears — Scissors that cut fabric in a series of small V's.

pinouts — Diagrams that show the positions of and name the functions of the connections to the pins of an electronic device.

pint — A measure unit of volume equal to $1/2$ quart, $1/8$ gallon, 28.875 cubic inches, or approximately 4.73 x 10^{-4} cubic meter.

pipe threads — The tapered threads on a hollow pipe or a piece of round stock. The taper provides the seal.

piston — 1. A cylindrical member that moves back and forth within a steel cylinder. 2. In a reciprocating engine, a cylindrical member that moves back and forth in the cylinder. Alternately compresses the fuel-air mixture and is pushed by the expanding gases to transfer mechanical power to the crankshaft of the engine.

piston displacement — The total volume swept by the piston of an engine in one stroke (one half revolution) of the crankshaft. The total displacement of an engine would be equal to piston displacement multiplied by the number of pistons.

piston engine — A reciprocating engine.

piston fuel pump — Sometimes used as a main fuel pump in place of the more typical spur gear pump. This pump is capable of delivering fuel at higher pressures than other types and can also vary its output per revolution. A variable displacement-type pump.

piston insulator — Composition insulators between the hydraulic actuating pistons and the pressure plate. Used to prevent heat transferring from the pressure plate into the piston where it would likely damage the seals and the fluid.

piston pin — The hardened steel pin that attaches the small end of a connecting rod into a piston.

piston pin boss — The enlarged area on the interior of a piston that provides additional bearing area for the wrist pin.

piston pump — A pump in a fluid power system used to move fluid and apply pressure to the system.

piston ring grooves — The grooves in the circumference of a piston into which the piston rings fit.

piston rings — Rings made of a special gray cast iron that fit into grooves in the periphery of a piston. Piston rings form a seal between the piston and the cylinder wall.

piston skirt — The lower portion of a piston.

piston-type pump — A hydraulic fluid pressure pump in which fluid is moved by pistons that move up and down in the cylinders of the pump.

pitch — 1. The rotation of an airplane about its lateral axis. 2. The distance between the centers of adjacent rivets in the same row. 3. On threaded fasteners, the distance measured between corresponding points on two adjacent threads.

pitch angle — In rotorcraft, the angle between the chord line of the rotor blade and the reference plane of the main rotor hub or the rotor plane of rotation.

pitch attitude — The angle of the longitudinal axis relative to the horizon. Pitch attitude serves as a visual reference for the pilot to maintain or change airspeed.

pitch axis — The lateral axis of an aircraft or the axis about which an aircraft pitches in a nose-up or nose-down attitude.

pitch of a propeller — Zero-thrust pitch. The distance a propeller would have to advance in one revolution to give no thrust. Also referred to as experimental mean pitch.

pitch of screw thread — The distance from the center of one thread to the center of the next thread.

pitch ratio of propeller — The ratio of the pitch to the diameter.

pitch setting — The propeller blade setting as determined by the blade angle measured in a manner, and at a radius, specified by the instruction manual for the propeller.

pitch, bolt threads — The distance from any point on the thread of a screw to the corresponding point on an adjacent thread.

pitot pressure — Ram or impact pressure used in the measurement of airspeed.

pitot static system — The pressure system for airspeed indicators, altimeters, and vertical speed indicators. It consists of the pitot tube and a static port, along with all of the necessary tubing and moisture traps.

pitot tube — An open-ended tube that faces directly into the relative airstream of an aircraft and picks up the ram, or pitot, pressure to be used in an airspeed indicator.

pitot-static tube — A combination tube with the pick-up for the pitot pressure as well as openings that pick up undisturbed, or static, air pressure.

pitting — The formation of small pockets on the surface of a metal.

pitting corrosion — A metal corrosion in which small, localized pits filled with the corrosive salts form on the surface of the metal.

pivot trunnion — A bearing surface on the top of the landing gear on which the gear rides when folding into the aircraft.

placard — Small statement or pictorial sign permanently fixed in the cockpit and visible to the pilot. Placards are used for operating limitations (e.g., weight or speeds) or to indicate the position of an operating lever (e.g., landing gear retracted or down and locked).

plain bearing — A simple bearing used to support an aircraft engine crankshaft or camshaft and only designed to take loads that are perpendicular to its face.

plain flap — A wing flap in which a portion of the trailing edge of the wing folds down to increase the camber of the wing without increasing the wing area.

plain nut — A simple hex nut that has no provisions for locking.

plain overlap seam — A seam used for machine sewing of aircraft fabric in which the edge of one piece of fabric laps over the edge of the other and one or more rows of stitches hold the pieces together.

plain rib — A rib used to give an airfoil its shape. Also referred to as a former rib.

plain washer — A flat washer used to provide a smooth bearing surface for a nut or to shim between a surface and a nut.

plain weave — A weaving pattern in which the warp and fill fibers alternate; that is, the repeat pattern is warp/fill/warp/fill.

plan position indicator (PPI) scope — A radar indicator scope displaying range and azimuth of targets in polar coordinates.

plan view — The view from the top of an object.

plane of rotation — The plane in which a propeller or a helicopter rotor rotates. It is perpendicular to the crankshaft or the rotor shaft.

plane of symmetry — A vertical plane that passes through the longitudinal axis of an aircraft and divides the aircraft into two symmetrical sides.

planetary gears — A reduction gearing arrangement in which the propeller shaft is attached to an adapter holding several small planetary gears. These gears run between a sun gear and a ring gear, either of which can be driven by the crankshaft, while the other is fixed into the nose section. Planetary gears reduce the propeller speed without reversing the direction of rotation.

planform — The outline of a wing as viewed from above. The shape seen when looking at a top view of a blueprint or plan.

plans display — A display available in URET CCLD (User Request Evaluation Tool Core Capability Limited Deployment) that provides detailed flight plan and predicted conflict information in textual format for requested Current Plans and all Trial Plans.

Plante cell — A secondary cell in which the pole pieces are formed of sheets of lead and lead dioxide. The electrolyte is a dilute solution of sulfuric acid.

plaque — The base for plates of nickel-cadmium (Nicad) batteries. Plaque is covered by nickel, formed under heat and pressure onto a fine mesh nickel screen.

plasma — A gas that is sufficiently ionized so as to affect its electrical properties and behavior. Plasmas are used in low temperature forms for the construction of integrated circuits and for providing light in fluorescent lights. In high temperature applications, they are used for cutting and welding metals and for propulsion for interplanetary probes.

plasma arc welding — A very localized, high temperature form of welding that minimizes distortion of the metal being welded. See also plasma.

plasma coating — The process of applying a thin coating of highly wear-resistant material on the surface of turbine engine parts. The process is accomplished by spraying material under high pressure and high heat onto the surface of the blades.

plasma engine — An engine used in outer space that uses plasma technology for propulsion.

plasma generator — A device that uses electricity, magnetism, and low pressures to generate ionized plasma gases. Also, a device that uses plasma gases to generate electricity.

plasma torch — A device that utilizes hot, high velocity, ionized gases in the cutting and welding of metals.

plastic — Any of the thermoplastic or thermosetting polymers used in modem aircraft construction.

plastic range — The stress range of a material in which, though the material does not fail when subjected to force, the material does not completely return to its original shape but is deformed.

Plastic Wood — A registered trade name for a filler consisting of wood fibers and a resin that sets as plasticizing solvents evaporate.

plasticizer — A chemical used in a lacquer finish to give film its flexibility and resilience.

plate — 1. The electrode in a vacuum tube that serves as the anode receiving the electrons from the cathode. 2. The active element in a storage battery. 3. Metal of a thickness greater than $1/4$". Metal thinner than $1/4$" is referred to as sheet metal.

plate current — In a vacuum tube, the current that flows between the plate and the cathode.

plate glass — A rolled sheet of glass that is ground and polished to provide an undistorted view.

plate power supply — A high-voltage supply of positive direct current for use in vacuum tubes.

plate resistance — In an electron tube, the ratio of a change in plate voltage to a change in plate current given a constant grid voltage.

plate saturation — The condition in an electron tube when the plate will no longer attract electrons as fast as they are emitted by the cathode.

plating — A process in which one metal is used to cover another using a process of electrical deposition. Specifically, chromium and cadmium are useful metals for covering steel.

platinum — A hard, gray metallic element with a symbol of Pt and an atomic number of 78. Platinum has an extremely high melting point and is used for the electrodes of fine-wire spark plugs.

platinum spark plug — A fine-wire electrode spark plug that can operate at very high temperatures.

play — The relative movement between parts. Play is the amount of movement that occurs in a cockpit flight control before the associated flight surface to move.

P-lead — The primary lead of an aircraft magneto. Connected to the ignition switch.

plenum — An enlargement of a duct or an enclosing space in an aircraft engine induction system or air conditioning system. Used to smooth out the pulsations in the flow of the air.

plenum chamber — An enclosed volume of air in which the air is held at a slightly higher pressure than that of the surrounding air.

Plexiglas — A transparent acrylic plastic material used for aircraft windshields and side windows.

pliers — Small, pincher-like hand tools used for holding small objects or for bending and cutting wire.

plies — Sheets of material that are laminated together.

plumb — A weight attached to a line or string and used to indicate vertical direction. Anything that lines up with the plumb line is said to be plumb or in plumb.

plumb bob — A weight attached to a line or string and used to establish a location directly below the point to which the line is attached.

plumb line — The straight line of a string to which a plumb bob is attached and hung. Anything that lines up with the plumb line is said to be plumb or in plumb.

plumbing — Tubing and fittings or connectors used for transmitting fluid within a structure or piece of equipment.

plumbing connection — Threaded connections that join sections of tubing, or which are used to connect the tubing to a component.

plunger — A part of a machine that works with a relatively rapid downward motion.

plutonium — A radioactive metallic element with the symbol Pu and an atomic number of 94.

ply — In composites, one layer of reinforcement in a laminate.

ply rating — A load rating for aircraft tires that relates to the strength of cotton plies. For example, a 20-ply rating nylon tire has the same load rating as a tire with 20 cotton plies.

plywood — Layers of wood glued together so that the grain in each layer is placed 45° or 90° to the others.

PN Junction — In semiconductor devices, a PN junction is fabricated from a single slice of semiconductor with one side doped as P-type and the other side doped as N-type. In operation, the migration of electrons to the P side and the holes to the N side creates a depletion area that resists further migration of ions. The flow of current is then controlled by the types of voltage applied to the ends of the device.

pneudraulic — A combination of air and hydraulic pressure.

pneumatic altimeter — An altimeter that measures height above a given pressure level. Its calibration is based on a specified lapse rate or change in pressure with height.

pneumatic drill motor — An air motor equipped with a chuck to hold twist drills.

pneumatic fire detection system — A system that uses a gas-filled continuous tube. The gas expands when heated and acts on a diaphragm to close an electrical circuit and show a warning light in the cockpit.

pneumatic starter — Starting motor operated by air pressure.

pneumatic system — The power system in an aircraft used for operating landing gear, brakes, wing flaps, etc. with compressed air as the operating fluid.

pneumatic-mechanical fuel control — A fuel control that utilizes pneumatic and mechanical forces to operate its fuel scheduling mechanisms.

pneumatics — The system of fluid power that transmits force by the use of a compressible fluid.

PNP transistor — The three-element semiconductor device made up of a sandwich of N-type silicon or germanium between two pieces of P-type material.

pod — An enclosure housing a complete engine assembly.

pogonip — A dense winter fog containing frozen particles. Forms in deep mountain valleys.

point out — An action taken by a controller to transfer the radar identification of an aircraft to another controller if the aircraft will or can enter the airspace or protected airspace of another controller and radio communications will not be transferred. Also referred to as radar point out.

pointer — A thin strip of movable metal moved over a calibrated scale by an analog instrument mechanism. Also referred to as hands or needles.

point-to-point wiring — An antiquated method of building electronic units. Electronic components were mounted directly on the chassis and interconnected by means of wires that were integral parts of the components' leads or by means of insulated hook up wire.

polar airmass — An airmass with characteristics developed over high latitudes, especially within the sub polar highs. Continental polar air (cP) has cold surface temperatures, low moisture content, and, especially in its source regions, has great stability in the lower layers. It is shallow in comparison with Arctic air. Maritime polar (mP) initially possesses similar properties to those of continental polar air, but in passing over warmer water it becomes unstable with a higher moisture content. See also tropical airmass.

polar easterlies — Surface winds generated by polar highs north of 60° N latitude.

polar front — The semi-permanent, semi-continuous front separating airmasses of tropical and polar origins.

polar front jet stream — One of two jet streams that commonly occur in the westerlies. Associated with the polar front.

polar front model — An idealized representation of events that follow the development of a frontal low. The surface component of the model describes the structure and behavior of fronts and airmasses in the lower atmosphere. The upper air part of the model deals with the associated development of troughs, ridges, and jet streams.

polar track structure — A system of organized routes between Iceland and Alaska that overlie Canadian MNPS Airspace.

polarity — The property of an electrical device having two different types of electrical charges: positive (deficiency of electrons) or negative (excess of electrons).

polarization — A degradation in chemical cell performance, particularly in the case of Leclanche cells, caused by gas formation and the resulting insulation of portions of the pole area.

polarized capacitor — A capacitor that can only be hooked up to direct current of the proper polarity. Electrolytic capacitors are normally polarized, and are labeled for proper polarity connection.

polarized light — Light that has all the light waves in the same plane.

polarized receptacle — A receptacle that has its sockets arranged so that the plug can only be inserted one way. This is done either through the shape of the plug/socket in two prong receptacles or by the arrangement of three or more prongs.

pole — The designation given to the ends of a magnet.

pole shoes — The field assembly part of an electric generator or motor.

poles of a magnet — The north and south poles of a magnet where magnetic lines of flux leave the south pole and reenter at the north pole.

polishing — The process of producing a smooth surface by rubbing it with fine abrasive wheels, belts, or compounds.

polyacrylonitrile (PAN) — The base material used in manufacturing some types of carbon fibers.

polyconic projection — A map projection where the parallels of latitude are arcs of a circle centered on the north or south pole and the meridians are straight lines radiating from the poles as in the spokes of a wheel.

polyester fiber — A synthetic fiber noted for its mechanical strength, chemical stability, and long life. It is used to make woven fabric for covering aircraft structures.

polyester resin — A synthetic resin, usually reinforced with fiberglass cloth or mat, and used to form complex shapes for aircraft structures.

polyethylene — A lightweight, thermoplastic resin material with good chemical resistance. Polyethylene resins are used for making containers for liquids and sheets of protective covering material.

polyethylene plastic material — A lightweight, thermoplastic resin material that has very good chemical- and moisture-resistant characteristics. It is used for plastic sheeting and containers.

Poly-Fiber® — A fabric woven from polyester fibers.

polymer paint — A fast drying, water-based paint that contains vinyl or acrylic resins. When the water in the paint evaporates, it leaves a waterproof film of the plastic resin.

polymerization — The process of joining two or more chemicals with molecules of similar structure, forming a more complex molecule with different physical properties. In this chemical reaction, the material essentially jells.

polymid — A translucent plastic material commonly referred to as nylon.

polyphase alternating current — Three-phase AC electricity produced by more than one set of generator windings.

polyphase electric motor — An induction motor that operates on two-phase or three-phase AC.

polystyrene — A transparent plastic used to make cell cases for some nickel-cadmium batteries.

polyurethane enamel — A two-component, chemically cured enamel finishing system noted for its hard, flexible, high-gloss finish.

polyvinyl chloride (PVC) — **1.** A thermoplastic resin used in the manufacture of transparent tubing for electrical insulation and fluid lines that are not subject to any pressure. **2.** A popular, low-cost, wire insulating material.

pontoon — A float attached to the landing gear of a land airplane to allow it to operate from water.

poor judgment chain — A series of mistakes that can lead to an accident or incident. Two basic principles generally associated with the creation of a poor judgment chain are: (1) one bad decision often leads to another; and (2) as a string of bad decisions grows, it reduces the number of subsequent alternatives for continued safe flight. Aeronautical decision making is intended to break the poor judgment chain before it can cause an accident or incident.

pop-open nozzle — The afterburner nozzle that pops full open at idle for the purpose of efficient engine operation at very low thrust.

poppet valve — A circular-headed, T-shaped valve. Used to seal the combustion chamber of a reciprocating engine, and at the proper time, either admit the fuel-air mixture into the cylinder or conduct the burned exhaust gases out of the cylinder.

porcelain — A hard, smooth surfaced ceramic material.

pores — Small holes or openings on the surface of metals.

porosity — The condition of a material having small pores or small cavities throughout the material.

porous chrome — A plating of hard chromium on bearing surfaces. The surface of the plating consists of tiny cracks in which lubricant can adhere to reduce sliding friction.

porous chrome plating — An electrolytically deposited coating of chromium on the walls of aircraft engine cylinders. The surface contains thousands of tiny cracks that hold oil to provide for cylinder wall lubrication.

porous salt — The residue normally left on the surface of a metal that has been attacked by corrosion.

porpoising — Hunting, or oscillating, around the lateral axis of the aircraft normally caused by an incorrectly functioning automatic pilot.

port side — The left-hand side of an aircraft or ship as one faces the nose of the aircraft or bow.

position error — The error in an airspeed indicator caused by the static source not being exposed to absolutely still air.

position lights — Lights on an aircraft consisting of a red light on the left wing, a green light on the right wing, and a white light on the tail. FARs require that these lights be displayed in flight during the hours of darkness.

position report — A report over a known location as transmitted by an aircraft to ATC.

position symbol — A computer-generated indication shown on a radar display to indicate the mode of tracking

positive — Symbol: +. A condition of electrical pressure caused by a deficiency of electrons.

positive acceleration — An increase in the rate of change of velocity.

positive angle of attack — A flight condition where the angle of attack 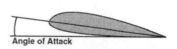 formed between the relative wind and the chord line of the airfoil is formed with the relative wind being the bottom leg in relation to the chord line and the so-called top surface of the aircraft.

positive buoyancy — The tendency of an object, when placed in a fluid, to ascend or float because it is lighter than the fluid it displaces.

positive control — Control of all air traffic, within designated airspace, by air traffic control.

Positive Control Area — An obsolete term that designated what is now referred to as Class A Airspace.

positive electrical charge — An electrical condition caused by a deficiency of electrons.

positive feedback — In electronics, a signal fed from the output section back into the amplifier to reinforce the input signal.

positive ion — An atom that has fewer electrons than protons.

positive logic — The logic used in binary computers. The more positive signal is considered to be equal to a one and the less positive or more negative signal is considered to be equal to a zero.

positive static stability — The condition of stability of an aircraft that causes it, when disturbed from a condition of straight and level flight, to tend to return to straight and level flight.

positive temperature coefficient — A ratio that shows an increase in length, resistance, capacitance, etc. with an increase in temperature.

positive terminal — The terminal of a battery or power source where electrons enter the source after they have passed through the external source.

positive transfer of learning — Since students interpret new things in terms of what they already know, some degree of transfer is involved in all learning. During a learning experience, knowledge or skills they have gained in the past can aid students.

positive vorticity — In weather, the vertical component of vorticity (i.e., a measure of how rapidly the air is spinning) leading to cyclonic flow. It is associated with upward air motions.

positive-displacement pump — A fluid pump that moves a specific amount of fluid each time it rotates. Examples of positive-displacement pumps include gear pumps, gerotor pumps, and vane pumps.

positron — The positive counterpart of an electron, which has the same mass and spin characteristics as an electron, but with a positive electrical charge.

post exit thrust reverser — A thrust reverser used to reverse the hot exhaust stream of a gas turbine engine to help slow the airplane during landings.

postcure — During the curing cycle of a manufactured composite component, the postcure is an additional elevated temperature soak to improve the mechanical properties.

pot life — The usable life of a resin. The time before it begins to thicken after the catalyst and accelerator have been added.

potassium — A metallic element with a symbol of K and an atomic number of 19.

potential — The electrical pressure or voltage caused by dissimilar metals in an acid solution or an electrolyte.

potential barrier — The difference in forward bias potential between the N side and the P side of a semiconductor. This is the amount of forward voltage necessary to begin current flow through a device. The nominal amount of the potential barrier is 0.3 volts for a germanium device and 0.7 volts for a silicon device. OK, but fails to mention the effect of temperature.

potential difference — The difference in voltage that exists between two terminals or two points of differing potential.

potential drop — A drop in voltage in an electrical circuit caused by the resistance of current flow through a resistance.

potential energy — The energy possessed by an object because of its position, configuration, or chemical arrangement of its constituents.

potential instability — A layer of air that is not only potentially unstable, it is conditionally unstable, and has a high moisture content. Potential instability is one of the two basic requirements for the formation of a thunderstorm; the other is initial lift.

potentiometer — **1.** A variable resistor having both ends and its wiper in the circuit. Used as a voltage divider. **2.** An instrument used for measuring differences in electrical potential by balancing the unknown voltage against a known variable voltage.

potentiometer ohmmeter — An ohmmeter circuit in which resistance is measured by placing a known voltage across a standard resistor; then the circuit is opened and the unknown resistor is placed in series. The voltage drop across the standard resistor is read and displaced on the meter as ohms.

pot-life — In composites, the length of time that the resin, mixed with catalyst, will be in a workable state.

potted circuit connector — An electrical circuit connector protected by encapsulating it with an insulating potting compound.

potting compound — **1.** A resin having filler capability, used to fill cells when making minor repairs to damaged honeycomb panels. **2.** A non-hardening, rubber-like material used to moisture proof and protect the wires in certain electrical plugs.

pound — A measure of mass equal to approximately 0.454 kg.

pour point — **1.** The lowest temperature at which a fluid will pour without disturbance. **2.** The lowest temperature at which oil will gravity flow.

powder metallurgy — A development that makes use of powdered metals rather than ingots. A process used to produce superalloys for high heat, high strength turbine components and for bearing material (sintered bearings).

powdered-iron core — A molded, magnetic powdered iron mixed with a binder. Used in magnetic applications when high permeability and low eddy current losses are desired.

power (P) — **1.** The time-rate of doing work. Force times distance, divided by time. Power can be expressed in terms of foot-pounds of work per minute, or in horsepower (HP). One HP is 33,000 ft.-lbs. of work per minute. **2.** The basic unit of electrical power is the watt, and 746 watts of electrical power is equal to one mechanical horsepower. In electrical problems, power is the product of voltage (E) times current (I) (P = E X I). Power in watts delivered to a circuit varies directly with the square of the applied voltage and inversely with the circuit resistance.

power amplifier — An electronic device designed to create an increase in power (voltage times amperage) as opposed to an amplifier designed to create an increase in voltage.

power brake control valve — A pressure regulator between the aircraft hydraulic system and the brake cylinders. The amount of pressure applied to the brakes is directly proportional to the force the pilot puts on the brake pedals.

power control system — A control system in which the normal movement of the controls is assisted by the use of hydraulic or pneumatic actuators to reduce the amount of force the pilot must apply.

power density — In radar, the amount of radar energy per cross-sectional area.

power enrichment system — A carburetor subsystem for a reciprocating engine that increases the fuel mixture when the engine is operating at full power.

power factor — The ratio of the resistance of an electrical circuit to the circuit impedance measured by a wattmeter.

power frequency — Frequency of AC electricity used for heat and light. Commercial power frequency in the U.S. is 60 Hz, and the aircraft power frequency is 400 Hz.

power lever — The cockpit lever that connects to the fuel control unit for scheduling fuel flow to the combustor. Also referred to as power control lever or throttle.

power lever angle (PLA) — A protractor on the fuel control showing movement of the power lever in degrees.

power loading — The ratio of an aircraft's maximum gross weight to the brake horsepower produced by the engines.

power overlap — The time in which two or more cylinders of an engine are simultaneously on the power stroke. The more cylinders an engine can have on the power stroke at one time, the greater the power overlap and the smoother the operation.

power pump — A hydraulic pump driven by the aircraft engine or by an electric motor.

power recovery turbine (PRT) — A power recovery device used on the Wright R-3350 engine. The exhaust gases spin a series of small turbines that are clutched to the crankshaft by fluid-coupling devices.

power section — That portion of a radial engine on which the cylinders are mounted.

power stroke — The movement of the piston of an aircraft reciprocating engine when the piston is forced down by the expanding gases. This is the only time work is accomplished by the engine.

power supply — The part of an electronic circuit that supplies the filament and plate voltages for the operation of the circuit.

power transformer — An electrical power supply transformer that changes voltage to that needed for the operating unit.

power turbine — A turbine rotor connected to an output shaft but not connected to the compressor. Also referred to as free power turbine.

powered-lift — A heavier-than-air aircraft capable of vertical takeoff, vertical landing, and low speed flight that depends principally on engine-driven lift devices or engine thrust for lift during these flight regimes and on non-rotating airfoil(s) for lift during horizontal flight.

powerplant — The complete installation in an aircraft of the engine, propeller, and all of the accessories and controls needed for its proper operation.

powerplant technician — A person who holds a certificate from the FAA authorizing him or her to perform maintenance or inspection on the powerplant, including the propeller, of certificated aircraft.

practical test standards (PTS) — An FAA published list of standards that must be met for the issuance of a particular pilot certificate or rating. FAA inspectors and designated pilot examiners use these standards when conducting pilot practical tests and flight instructors should use the PTS while preparing applicants for practical tests.

practice instrument approach — An instrument approach procedure conducted by a VFR or an IFR aircraft for the purpose of pilot training or proficiency demonstrations.

Pratt truss — A truss structure in which the vertical members carry only compressive loads, and the diagonal members carry only tensile loads. A Pratt truss is used for most fabric-covered wings.

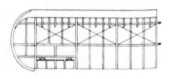

preamplifier — An electronic circuit component that amplifies an extremely weak input signal to a value strong enough to be used in other amplifiers.

prearranged coordination — A standardized procedure that permits an air traffic controller to enter the airspace assigned to another air traffic controller without verbal coordination. The procedures are defined in a facility directive that ensures standard separation between aircraft.

precession — One of the characteristics of a gyroscope that causes an applied force to be felt, not at the point of application, but 90° from that point in the direction of rotation.

precious metal — Highly valued metal because of scarcity. Examples of precious metals are gold, silver, and platinum.

precipitable water — The amount of liquid precipitation that would result if all water vapor were condensed.

precipitate — To condense out of, or to separate from, a mixture.

precipitation — Any or all forms of water particles (rain, sleet, hail, or snow), that fall from the atmosphere and reach the surface.

precipitation attenuation — In radar meteorology, any process that reduces intensity of radar signals from precipitation.

precipitation fog — Fog that develops when rain saturates the air near the ground.

precipitation hardening — Hardening caused by the precipitation of a constituent from a supersaturated solid solution. A process of reheating an alloy and allowing it to cool slowly. Allows crystalline structure to harden uniformly.

precipitation heat treatment — A step in the heat-treating process of aluminum in which the metal, after having been heated to its critical temperature and quenched, is raised to an elevated temperature and held for a period of time. This process artificially ages the metal and increases its strength.

precipitation static — The static heard on communication gear when electricity is discharged from radio antennas or other objects on the aircraft as it flies through clouds containing ice particles.

precipitation-induced downdraft — Downdrafts present inside of a thunderstorm that are induced by rainfall and are much stronger than downdrafts that exist outside of the thunderstorm.

precision approach (PA) — An instrument approach based on a navigation system that provides course and glidepath deviation information meeting the precision standards of ICAO Annex 10. PAR, ILS, and MLS are precision approaches.

precision approach procedure — A standard instrument approach procedure in which an electronic glideslope/glidepath is provided; e.g., ILS, MLS, PAR.

precision approach procedure — A standard instrument approach procedure in which an electronic glide slope/glide path is provided; e.g., ILS/MLS and PAR..

precision approach radar (PAR) — A radar system that uses two separate antenna. One is used to determine the aircraft's horizontal position relative to the extended centerline of the runway, and the other to determine the aircraft's vertical position relative to the glideslope leading to the runway. A ground controller directs the pilot of the aircraft horizontally and vertically to maintain position on centerline and on glideslope.

precision measuring instruments — Instruments capable of making exact measurements.

precision runway monitor (PRM) — Provides air traffic controllers with high precision secondary surveillance data for aircraft on final approach to parallel runways that have extended centerlines separated by less than 4,300 feet. High resolution color monitoring displays (FMA) are required to present surveillance track data to controllers along with detailed maps depicting approaches and no transgression zone.

precision switch — A snap-action switch that actuates at the point when a mechanism reaches a very definite position. These switches typically have a plunger that requires an extremely small movement to trip and close the contacts. When the plunger is released, a spring snaps the contacts open.

predeparture clearance (PDC) — An automated Clearance Delivery system relaying ATC departure clearances from the FAA to the user network computer for subsequent delivery to the cockpit via ACARS (Airline/Aviation VHF data link) where aircraft are appropriately equipped, or to gate printers for pilot pickup.

predrilling — The process of drilling a hole that is slightly smaller than required to enable reaming the hole to the proper size later.

pre-exit thrust reverser — A thrust reverser system installed forward of the exhaust nozzle.

preferential routes — Preferential routes (PDRs, PARs, and PDARs) are adapted in ARTCC computers to accomplish inter/intra-facility controller coordination and to assure that flight data is posted at the proper control positions. Locations having a need for these specific inbound and outbound routes normally publish such routes in local facility bulletins, and their use by pilots minimizes flight plan route amendments. When the workload or traffic situation permits, controllers normally provide radar vectors or assign requested routes to minimize circuitous routing. Preferential routes are usually confined to one ARTCC's area and are referred to by the following names or acronyms:

a. Preferential Departure Route (PDR) — A specific departure route from an airport or terminal area to an enroute point where there is no further need for flow control. It can be included in an Instrument Departure Procedure (DP) or a Preferred IFR Route.

b. Preferential Arrival Route (PAR) — A specific arrival route from an appropriate enroute point to an airport or terminal area. It can be included in a Standard Terminal Arrival (STAR) or a Preferred IFR Route. The abbreviation "PAR" is used primarily within the ARTCC and should not be confused with the abbreviation for Precision Approach Radar.

c. Preferential Departure and Arrival Route (PDAR) — A route between two terminals that are within or immediately adjacent to one ARTCC's area. PDARs are not synonymous with Preferred IFR Routes but can be listed as such as they do accomplish essentially the same purpose.

preferred IFR routes — Routes established between busier airports to increase system efficiency and capacity. They normally extend through one or more ARTCC areas and are designed to achieve balanced traffic flows among high density terminals. IFR clearances are issued on the basis of these routes except when severe weather avoidance procedures or other factors dictate otherwise. Preferred IFR Routes are listed in the Enroute Section. If a flight is planned to or from an area having such routes, but the departure or arrival point is not listed in the Enroute Section, pilots can use that part of a Preferred IFR Route that is appropriate for the departure or arrival point that is listed. Preferred IFR Routes are correlated with DPs and STARs and can be defined by airways, jet routes, direct routes between NAVAIDs, Waypoints, NAVAID radials/DME, or any combinations thereof.

preflight inspection — An aircraft inspection done prior to takeoff to determine that all systems are functioning properly for the intended flight.

pre-flight pilot briefing — A service provided by the FSS to assist pilots in flight planning. Briefing items can include weather information, NOTAMs, military activities, flow control information, and other items as requested. Also referred to as a pilot briefing.

preform — In composites, a preshaped fibrous reinforcement of mat or cloth formed to the desired shape on a mandrel or mock-up before being placed in a mold press.

preformed control cable — Steel aircraft control cable whose individual strands were formed into a spiral before the cable was woven. This relieves the bending stresses within the cable and prevents the strands from spreading out when the cable is cut.

preignition — Ignition occurring in the cylinder before the time of normal ignition. Preignition is often caused by a local hot spot in the combustion chamber igniting the fuel-air mixture.

pre-installation checks — Checks made on a unit before installation.

pre-oiling — A procedure that ensures oil has reached all critical lubrication points before an engine is started. Methods of pre-oiling include accumulators that maintain oil pressure in the system when the engine is not operating and operating the oil pump by itself before starting.

preparation — The first step of the teaching process, consisting of determining the scope of the lesson, the objectives, and the goals to be attained. This portion also includes making certain all necessary supplies are on hand. In using the telling and doing technique of flight instruction, this step is accomplished prior to the flight lesson.

pre-preg — In composites, reinforcing material that is pre-impregnated with resin/catalyst mixture. The resin system is in the B-stage and requires refrigerated storage. When heated, the resins begin to glow and will complete the cure when the temperature is elevated to its cure temperature for the proper amount of time.

prerotation — In a gyroplane, it is the spinning of the rotor to a sufficient r.p.m. prior to flight.

presentation — The second step of the teaching process, consisting of the delivery of the knowledge and skills that make up the lesson. The delivery would be by either the lecture method or demonstration-performance method. In the telling and doing method of flight instruction, this is where the instructor both talks about and performs the procedure.

press brake — A sheet metal bending tool in which the sheet is placed on the bed with the sight line directly under the edge of the clamping bar with the correct bend radius die. The clamping bar is brought down to hold the sheet firmly in place and a bending leaf is raised until it bends the metal to the proper angle.

press fit — A tight interference fit between machine parts. Requires one part to be pressed into the other.

press-to-test light — A light that is tested by pressing the light fixture to complete the circuit to ground. If the light illuminates, the bulb is good.

pressure — Force per unit area.

pressure altimeter — A barometric instrument that indicates altitude. When set to standard sea level pressure of 29.92 inches of mercury, it indicates pressure altitude. When set to the local altimeter setting, it indicates indicated altitude from which true altitude can be calculated.

pressure altitude — The height above the standard pressure level of 29.92 in. Hg. It is obtained by setting 29.92 in the barometric pressure window and reading the altimeter.

pressure capsule — The portion of a structure subjected to pressurization. Usually consists of the cabin and cockpit.

pressure carburetor — A fuel metering system that senses the relationship between impact air pressure and venturi pressure to provide a metering force for the fuel.

pressure casting — A method of casting metal parts by forcing molten metal into permanent molds.

pressure controller — That portion of a turbocharger control system that maintains the desired manifold pressure.

pressure demand oxygen system — A demand oxygen system that supplies 100% oxygen at sufficient pressure above the altitude where normal breathing is adequate. Also referred to as a pressure breathing system.

pressure fed gun — A paint spray gun in which the material is fed to the gun by air pressure on the pot or cup holding the material.

pressure fed spray gun — A paint spray gun that feeds the material to be sprayed into the gun under pressure.

pressure gauge snubber — A unit installed in the pressure gauge line that stabilizes pressures and allows the needle of the pressure gauge to give a steady reading.

pressure gradient — The rate of change of pressure per unit distance at a fixed time.

pressure gradient force — The force that arises because of the pressure of a pressure gradient.

pressure jump — In weather, a sudden change in pressure reading.

pressure line — Tubing that carries hydraulic fluid under pressure from the pump to the selector valve or the control valve.

pressure plate — A heavy, stationary disc in a multiple disc brake. It is provided with a wear surface on one side only, and the pistons press against its backside to force the disk stack over against the back plate.

pressure port — The opening in a device through which pressure is introduced.

pressure pot — A container holding the material to be sprayed. An agitator keeps the material in motion, and a regulator maintains the proper air pressure on the material to feed it to the gun.

pressure ratio — One pressure divided by another, used to describe certain engine functions. See also compressor pressure ratio; engine pressure ratio.

pressure ratio controller — In turbocharged engines, controls the maximum turbocharger compressor discharge pressure (34 + or - .5 in. Hg to critical altitude of 18,000 ft.).

pressure reducing valve — A device that reduces the pressure of a liquid or gas from a high value to a fixed lower value.

pressure regulator — In a hydraulic system, a device that maintains a constant output pressure from a constant displacement pump by bypassing a portion of the fluid back to the inlet side of the pump.

pressure relief valve — In a hydraulic system, a pressure control valve that bypasses fluid back to the return manifold or reservoir in the event the pressure rises above a predetermined level.

pressure seal — A seal installed in a pressure bulkhead to permit a cable to pass through it.

pressure sensing switch — An electrical switch that will open or close when a predetermined pressure is reached in a system.

pressure tank — See accumulator.

pressure transducer — A mechanical-to-electrical device in which an electrical signal is generated proportional to the pressure being sensed. The electrical signal is then transmitted to an instrument on the instrument panel.

pressure transmitter — A mechanical-to-electrical device in which an electrical signal is generated proportional to the pressure being sensed. The electrical signal is then transmitted to an instrument on the instrument panel.

pressure vessel — A pressurized portion of an aircraft that is sealed and pressurized in flight.

pressure, static — The pressure measured in a duct containing air, a gas or a liquid in which no velocity (ram) pressure is allowed to enter the measuring device. Symbol Ps.

pressure, total — Static pressure plus ram pressure. Total pressure can be measured by use of a specially shaped probe that stops a small portion of the gas or liquid flowing in a duct thereby changing velocity (ram) energy to pressure energy. Symbol Pt.

pressure-demand oxygen regulator — An oxygen regulator capable of furnishing 100% oxygen under pressure to force the oxygen into the lungs of the user.

pressurization — A means of increasing the partial pressure of oxygen in the cabin of an airplane flying at high altitude. Accomplished by increasing the air pressure in the cabin to that of an altitude which requires no supplemental oxygen.

pressurization controller — A controller that maintains the pressure in an aircraft cabin at a selected pressure.

pressurized aircraft — Any aircraft in which the cabin area is sealed off and pressurized with air from a cabin supercharger. The cabin can be pressurized to a pressure that compares with a maximum altitude of approximately 8,000 ft.

pressurized ignition system — An ignition system that is pressurized with compressed air (usually from the turbocharger system) to keep high voltage from arcing between electrodes due to low air density at high altitudes.

pressurizing and dump valve — A valve used with a dual line duplex fuel manifold and duplex nozzle system. The pressurizing valve opens at higher fuel flows to deliver fuel to the secondary manifold. At engine shutdown, the dump portion opens to drain fuel overboard from the manifolds. Also referred to as a pressurizing and drain valve.

prestretching — A means of preventing an aircraft control cable from stretching during use by applying a load to the cable equal to 60% of its breaking strength for a specified period of time before installing it in the airplane.

pretest — A test used to determine whether a student has the necessary qualifications to begin a course of study. Also used to determine the level of knowledge a student has in relation to the material that will be presented in the course.

pretrack — A method used by some manufacturers to preset the track of a rotor blade prior to installation. The blade is tested at the factory and marked with appropriate settings so it can be adjusted properly during installation.

prevailing visibility — In the U.S., the greatest horizontal visibility equaled or exceeded throughout half of the horizon circle; it need not be a continuous half.

prevailing westerlies — The dominant west-to-east motion of the atmosphere, centered over middle latitudes of both hemispheres.

prevailing wind — Direction from which the wind blows most frequently.

preventive maintenance — Simple or minor preservative operations and the replacement of small standard parts not involving complex assembly operation as listed in Appendix A of FAR Part 43.

prick punch — A tool used to place reference marks on metal.

primacy — A principle of learning where the first experience of something often creates a strong, almost unshakable impression. The importance to an instructor is that the first time something is demonstrated, it must be shown correctly since that experience is the one most likely to be remembered by the student.

primary air — That portion of the compressor output air used for actual combustion of fuel. Sometimes this term is used to refer to the amount of air flowing through the basic engine portion of a turbofan engine.

primary airstream — The air that passes through the core of the engine.

primary cell — An electrical device that generates electron flow by converting some of its substance into ions that free electrons. Some of the material is destroyed in the process. Primary cells are not rechargeable.

primary circuit — The main circuit in a magneto ignition system. It consists of turns of wire in which the primary current flows.

primary controls — Movable surfaces that cause an aircraft to rotate about its three primary axes. The primary controls of an airplane are the ailerons, elevators, and rudder.

primary current — The alternating or pulsating current that flows in the primary winding of a transformer. Induces a current in the secondary winding.

primary cycle — The most intense portion of a lee wave, located immediately down wind of the mountain.

primary exhaust nozzle — On a turbofan, the hot exhaust nozzle. On an afterburner, the inner exhaust nozzle.

primary fuel — In a duplex fuel nozzle, the fuel that initially flows on starting; usually from the center orifice. Also referred to as pilot fuel.

primary instruments — Those instruments that provide the most essential information during a given flight condition.

primary radar — A radar system using only a transmitter, receiver, antenna, and cathode-ray tube display to indicate radar traffic. No transponder signal is added. Indication on display is referred to as a primary return.

primary structure — The portions of the airplane that would seriously endanger the safety of the airplane if they failed. An aircraft's primary structure includes the wing structure, controls, engine mounts, etc.

primary winding — The winding in a magneto coil through which the current induced by the rotating magnet flows. The breaker points are in series with the primary winding.

primary/support concept — A method for teaching attitude instrument flying, which divides the panel into pitch instruments, bank instruments, and power instruments. For a given maneuver, there are specific instruments used to control the airplane and obtain the desired performance.

prime coats — The first coats of an aircraft finish. Used to bond the topcoats to the base material.

primer — 1. A material applied to a metal before applying the topcoats. Used to bond the topcoats to the base material. 2. A small hand-operated pump used to spray raw gasoline into an engine cylinder to provide fuel for starting.

primer fuel system — A low output system for engine starting. Used primarily where the main fuel system uses vaporizing tube nozzles. Also referred to as a starting fuel system.

primer surfacer — See prime coats.

priming a pump — The act of replacing air in a pump with the liquid to be pumped.

Principal Maintenance Inspector (PMI) — An FAA employee that is the assigned primary contact between the FAA and an airline, repair station, maintenance school, etc.

principal view — The view in an orthographic drawing that shows the most detail of the object.

principles of learning — Concepts that provide insight into effective learning and can provide a foundation for basic instructional techniques. These principles are derived from the work of E. L. Thorndike, who first proposed the principles of effect, exercise, and readiness. Three later principles were added: primacy, recency, and intensity.

print — A copy of a formal engineering drawing.

print tolerance — A notation on an aircraft drawing that describes the tolerance allowed on a particular finished part.

printed circuit board (PCB) — Modern replacement for the electronic component chassis. Consists of a plastic, fiberglass, or other insulating board with bonded copper strips for component interconnection. The components are generally soldered directly on the board.

priority valve — A pressure-actuated hydraulic valve that allows certain actuations before others. One use of this type of valve is to assure that the gear doors will be opened before the landing gear extends.

PRM (Precision Runway Monitoring) — A system that permits simultaneous independent ILS approaches. An instrument landing system (ILS) approach conducted to parallel runways whose extended centerlines are separated by less than 4,300 feet requires the the parallel runways to have a PRM System.

probe — A sensing device that extends into the air stream or gas stream for measuring pressure, velocity, or temperature. In the case of pressure, it is used to measure total pressure

procedure turn (PT) — A maneuver in which a turn is made away from a designated track followed by a turn in the opposite direction to permit the aircraft to intercept and proceed along the reciprocal of the designated track.
NOTE 1: Procedure turns are designated "left" or "right" according to the direction of the initial turn.
NOTE 2: Procedure turns can be designated as being made either in level flight or while descending, according to the circumstances of each individual approach procedure.

procedure turn inbound — That point of a procedure turn maneuver where course reversal has been completed and an aircraft is established inbound on the intermediate approach segment or final approach course. A report of "procedure turn inbound" is normally used by ATC as a position report for separation purposes.

procedures — Step-by-step instructions of how to accomplish something.

process annealing — Heating a ferrous alloy to a temperature close to, but below, the lower limit of the transformation range and then cooling in order to soften the alloy for further cold working.

process control record — In composites, a record of the materials and processes used in making the repair.

Production Certificate — A certificate issued by the FAA to allow the production of a type-certificated aircraft, aircraft engine, or component.

professional — Characterized by or conforming to the technical or ethical standards of a profession. Exhibiting a courteous, conscientious, and generally businesslike manner in the workplace.

profile descent — An uninterrupted descent (except where level flight is required for speed adjustment; e.g., 250 knots at 10,000 feet MSL) from cruising altitude/ level to interception of a glide slope or to a minimum altitude specified for the initial or intermediate approach segment of a nonprecision instrument approach. The profile descent normally terminates at the approach gate or where the glide slope or other appropriate minimum altitude is intercepted.

profile drag — **1.** That portion of an aircraft's drag caused by the air flowing over the surface of the craft. **2.** In rotorcraft, drag incurred from frictional or parasitic resistance of the blades passing through the air. It does not change significantly with the angle of attack of the airfoil section, but it increases moderately as airspeed increases.

prognostic chart (contracted prog) — A chart of expected or forecast conditions.

program — A list of events or procedures. In computers, a program is a series of instructions that tell the computer exactly how it is to receive, store, process, and deliver data for the user.

program flowchart — A chart (such as Pert or Gnatt) that shows the steps to be taken in the execution of a program.

programmable calculator — An electronic calculator that can be programmed by the user. Programming is usually stored in its memory.

programmable indicator data processor — The PIDP is a modification to the AN/TPX-42 interrogator system currently installed in fixed RAPCONs. The PIDP detects, tracks, and predicts secondary radar aircraft targets. These are displayed by means of computer-generated symbols and alphanumeric characters depicting flight identification, aircraft altitude, ground speed, and flight plan data. Although primary radar targets are not tracked, they are displayed coincident with the secondary radar targets as well as with the other symbols and alphanumerics. The system has the capability of interfacing with ARTCCs.

programmable read-only memory (PROM) — An integrated circuit memory device for a digital computer. Programmable read-only memory can be modified once by the user. This is called "burning" the PROM. Thereafter, the PROM acts the same as read-only memory (ROM). A computer with erasable programmable read-only memory (EPROM) or electrically erasable programmable read-only memory (EEPROM) can be modified more than once.

progress report — A report over a known location as transmitted by an aircraft to ATC.

progressive inspection — An inspection identical in scope and detail to an annual inspection, but which allows the workload to be divided into smaller portions and performed in shorter time periods. This allows the aircraft to remain airworthy and in service during the extensive (sometimes up to a year long) inspection period.

progressive taxi — Precise taxi instructions given to a pilot unfamiliar with the airport or issued in stages as the aircraft proceeds along the taxi route.

prohibited area — A prohibited area is airspace designated under part 73 within which no person can operate an aircraft without the permission of the using agency.

prohibited area [ICAO] — An airspace of defined dimensions, above the land areas or territorial waters of a State, within which the flight of aircraft is prohibited.

projection — A defense mechanism used by students to relegate blame for their own shortcomings, mistakes, and transgressions to others, or to attribute their motives, desires, characteristics, and impulses to other people.

prony brake — A device used to measure the usable power output of an engine on a test stand. It consists largely of a hinged collar, or brake, that can be clamped to a drum splined to the propeller shaft. The collar and drum form a friction brake that can be adjusted by a wheel. An arm of known length is attached to the collar and terminates at a point that bears on a scale. As the propeller shaft rotates, the force is measured on the scale. This force multiplied by the lever arm indicates the torque produced by the rotating shaft.

proof load — 1. A load applied to a structure that does not cause permanent deformation. 2. A testing measure to insure the structure will be airworthy.

proof pressure test — A series of tests to show that a pressure capsule will withstand the pressure exerted upon it in service.

prop blast — The colloquial term for the rush of air generated by a propeller.

propeller — A device for propelling an aircraft that, when rotated, produces by its action on the air, a thrust approximately perpendicular to its plane of rotation. It includes the control components normally supplied by its manufacturer.

propeller anti-icer — A system within an airplane that meters a flow of alcohol and glycerine along the leading edge of the propeller blades to prevent the formation of ice on the blades.

propeller blade — The part of a propeller that forms the airfoil and converts the torque of the engine into thrust.

propeller blade angle — The acute angle between the chord of a propeller blade and the plane of rotation.

propeller blade pitch — The distance a propeller will advance if there isn't any slip.

propeller blade tipping — The thin sheet brass or stainless steel covering along the leading edge and around the tip of a wooden propeller that protects the blade from erosion.

propeller boot — A propeller cuff that fits around the base of the propeller blade and causes increased airflow into the engine air inlet for cooling.

propeller boss — The thick, central portion of a fixed-pitch propeller hub.

propeller brake — A friction brake used on turbopropeller engines to prevent the propeller windmilling in flight after it has been feathered.

propeller butt — The blade shank or base of the propeller blade that fits into the propeller hub.

propeller critical range — An operational range where engine speed will cause harmonic vibration in the propeller. Engines are usually placarded against operation in this speed range.

propeller cuff — An airfoil shaped attachment made of thin sheets of metal, plastic, or composite material. Propeller cuffs are mounted on the blade shanks and primarily used to increase the flow of cooling air to the engine nacelle.

propeller diameter — Twice the distance from the center of the propeller hub to the blade tip.

propeller efficiency — Ratio of thrust horsepower to brake horsepower. On the average, thrust horsepower constitutes approximately 80% of the brake horsepower. The other 20% is lost in friction and slippage.

propeller hub — The central portion of a propeller to which the blades are attached and by which the propeller is attached to the engine.

propeller pitch — The acute angle between the chord of a propeller and a plane perpendicular to the axis of its rotation.

propeller protractor — A tool used to measure the blade angle of a propeller. It is made of aluminum alloy with three square sides at 90° angles. A bubble spirit level mounted on one corner or the front of the frame swings out to indicate when the protractor is level. A movable ring is located inside the frame and used to set the zero reference angle for the blade angle measurement.

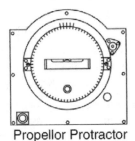

Propellor Protractor

propeller slip — A condition of propeller aerodynamics that equals the difference between the geometric and the effective pitch.

propeller spider — The foundation unit of a controllable pitch propeller. It attaches to the propeller shaft, and the propeller blades ride on bearings on the spider. The spider is enclosed in the propeller hub.

propeller synchronization — In multi engine aircraft, adjusting the propeller controls in order to operate the propellers in unison, eliminating the uncomfortable noise associated with two propellers operating at slightly different rates. This can be done manually or automatically depending on the installed equipment.

propeller thrust — The component of the total air force on the propeller parallel to the direction of flight.

propeller tipping — The thin sheet brass or stainless steel covering along the leading edge and around the tip of a wooden propeller that protects the blade from erosion.

prop-fan — An advanced technology propeller, designed to operate at supersonic tip speeds, Mach 0.8 airspeeds and 20,000 to 35,000 ft. altitude. Referred to as unducted fans, the concept has fallen from favor due to excessive noise.

propjet — An aircraft having a jet engine in which the energy of the jet operates a turbine that drives the propeller.

proportional control — A mechanical control system that creates an output proportional to input, as opposed to a control (such as a switch) that is either off or on.

proposed boundary crossing time (PBCT) — Each center has a PBCT parameter for each internal airport. Proposed internal flight plans are transmitted to the adjacent center if the flight time along the proposed route from the departure airport to the center boundary is less than or equal to the value of PBCT or if airport adaptation specifies transmission regardless of PBCT.

proposed departure time — The time a scheduled flight will depart the gate (scheduled operators) or the actual runway off time for nonscheduled operators. For EDCT (expect departure clearance time) purposes, the ATCSCC (Air Traffic Control System Command Center) adjusts the "P" time for scheduled operators to reflect the runway off times.

proprietary reducers — Thinners or solvents for paints formulated according to and distributed under a trade name of a chemical manufacturer.

propulsive efficiency — In gas turbine engines, external efficiency of an engine expressed as a percentage.

propwash — The force of air blown rearward by the propeller.

protected airspace — The airspace on either side of an oceanic route/track that is equal to one-half the lateral separation minimum except where reduction of protected airspace has been authorized.

proton — The positively charged particles in the nucleus of an atom.

prototype — The first functional unit built as an example of a new design or type.

prototype device - A working model of a design used to test its concept.

protractor — A device for measuring angles in degrees.

protruding head rivet — An aircraft rivet in which the head protrudes above the surface of the metal. Examples include universal-head, round-head, and flat-head rivets.

Prussian blue — A compound used in checking the contact of a valve with a valve seat. A thin coating of Prussian blue is applied to the valve and the valve is pressed against the seat. Prussian blue is transferred to the seat where in order to determine if the contact is uniform.

psychomotor domain — A grouping of levels of learning associated with physical skill levels that range from perception through set, guided response, mechanism, complex overt response, and adaptation to origination.

psychrometer — An instrument for determining the relative humidity of the air by measuring both the wet- and dry-bulb temperatures.

P-type semiconductor material — A semiconductor material that has been doped (an impurity added) so that it leaves the outer ring of the valence shells with holes (absence of electrons) that readily accept electrons.

P-type silicon — Silicon doped by an impurity having three valence electrons.

public aircraft — An aircraft used only in the service of a government or political subdivision. This does not include any government-owned aircraft engaged in carrying persons or property for commercial purposes.

published route — A route for which an IFR altitude has been established and published; i.e., Federal Airways, Jet Routes, Area Navigation Routes, Specified Direct Routes.

puckers — In composites, local areas on pre-preg material where the material has blistered and pulled away from the separator film or release paper.

pucks — The brake linings used on disc brakes.

pull test — A fabric-strength test in which a one-inch sample strip is pulled until it breaks. The strength of the fabric is determined by the force required to break the strip.

pulley — A simple machine in the form of a wheel grooved to accommodate a cable. It is used to guide cables and change direction.

pull-through rivet — A blind mechanically expanded rivet in which the hollow shank is upset by pulling a tapered mandrel through it.

pull-up resistor — A resistor used to limit the current through a two-state device when the device is in its low resistance state and to develop a potential difference when a two-state device is in its high resistance state.

pulsate — To expand and contract rhythmically, yet not change direction.

pulsating direct current — DC that has been chopped by a vibrator or chopper and that changes from zero to maximum and then back to zero. This produces the changing current required for use in a transformer.

pulsation — A beat or rhythmic throb.

pulse — A rhythmic throb in the voltage of an electrical circuit.

pulse amplifier — A wide-band electrical amplifier used to increase the voltage of alternating current.

pulse counter — A device that measures pulses of electrical energy that it receives in a specific interval of time.

pulse generator — An electronic circuit designed to produce sharp pulses of voltage.

pulse-echo — An ultrasonic non-destructive inspection used to detect the presence of internal damage or faults in a construction material.

pulse-echo method of ultrasonic inspection — A method of detecting metal thickness or internal damage by introducing a pulse of ultrasonic energy into a part and timing its travel through the material and back to the point of injection.

pultrusion — In composites, a manufacturing process that pulls the resin impregnated fibers through a shaping die to form a shape. The curing process also is done while it is in the die.

pumice — An extremely fine natural abrasive powder used for polishing metal surfaces.

pump — A mechanical device used to move a fluid. A pump is not a pressure producing machine as pressure can be produced only when a flow of fluid is restricted.

pump cavitation — The formation of partial vacuums in a liquid caused by a moving pump rotor. Pump cavitation creates turbulence in the pump cavity, reducing pump efficiency.

punch — 1. A short, tapered steel rod used for driving pins, bolts, or rivets from holes. 2. A device used to cut holes in paper, thin metal, or gasket material by shearing the material between close-fitting male and female dies.

punch test — A test of the strength of aircraft fabric while it is on the airplane. A pointed, spring-loaded plunger is pushed into the fabric, and the amount of force required to penetrate the fabric indicates its strength.

puncture — A hole pierced in a material.

purge — To cleanse a system by flushing.

push fit — An interference fit in which the parts can be assembled by hand-pushing them together rather than having to drive or press them.

push rod — The component in a reciprocating engine that transmits the movement of the cam to the rocker arm to open the valves.

push to-test light — A light fixture for an indicator light that can be pressed to complete a circuit. Illumination of the light indicates the bulb is in operating condition.

push-button electrical switch — An electrical switch actuated on or off by a push button. Each time the button is pressed, it opens or closes the circuit.

pusher propeller — A propeller that fits onto an engine whose shaft points to the rear of the aircraft. The thrust pushes the aircraft through the air rather than pulling it.

push-pull amplifier — An electronic amplifier that has two output circuits whose output voltages are equal but 180° out of phase with each other. Also referred to as a balanced amplifier.

push-pull rod — A rigid control rod used to move a component by alternately pushing it and pulling it.

puzzle — A question with so many parts that it is difficult to figure out what is being asked, let alone the answer.

pylon — The structure that holds a turbine engine pod to the wing or fuselage of the aircraft.

pyrometer — A temperature measuring instrument used to indicate temperatures that are higher than can be measured with a mercury thermometer.

Q

Q factor — The performance measurement of an inductance coil. It is the ratio of the inductive reactance to the resistance of a coil.

Q factor of a coil — The performance measurement of an inductance coil. It is the ratio of the inductive reactance to the resistance of a coil.

Q-band radar — Radar operating in a frequency range of 36 to 46 gigaherz (wave length of approximately 7 millimeters). Used primarily in Very Large Array (VLA) intergalactic research.

QEC unit — A completed assembly of a basic engine and the necessary components for a particular airframe installation. A quick engine change (QEC) unit consists of an engine with all of the accessories and propeller already installed. Minimizes the time it takes to replace an engine. Also referred to as a quick engine change assembly (QECA) or a QEC kit.

QFE — Height above airport elevation (or runway threshold elevation) based on local station pressure.

QNE — Altimeter setting 29.92 inches of mercury, 1013.2 hectopascals or 1013.2 millibars.

QNH — The barometric pressure as reported by a particular station.

Q-springs — A system that provides artificially produced feedback of the control surface movement to the pilot. Necessary because large, high-speed aircraft normally utilize hydraulically actuated controls that provide no natural feedback to the pilot.

quad-clamp — A quick-attach-detach clamp used to attach accessories to their gearbox mounting pads.

quadrant — **1.** The location of an aircraft's pivoting engine control levers. Some aircraft use push-pull knobs in the instrument panel. **2.** A quarter part of a circle, centered on a NAVAID, oriented clockwise from magnetic north as follows: NE quadrant 000-089, SE quadrant 090-179, SW quadrant 180-269, NW quadrant 270-359. **3.** In mathematics, one of four equal segments defined by the intersection of the two main axes of the rectangular coordinate system.

quadrantal error — In navigation, the angular error in a measured bearing due to the presence of metal structures and engines in the vicinity of the direction-finding antenna.

quality control (QC) — A management and inspection function that controls quality, standards, and performance in the manufacturing and repair of aircraft, aircraft engines, and components.

quantity — An amount or portion. The total amount of a particular thing.

quantum — A quantity or amount.

quantum theory — A branch of physical theory involving the transference or transformation of energy in an atomic or molecular scale based on the concept of the subdivision of radiant energy into finite quanta.

quart — A liquid measurement of volume equal to $1/4$ gallon, 2 pints, or 57.75 cubic inches.

quarter-sawed wood — Lumber cut at 45° across the annual rings.

quarter-turn cowl fastener — A quick-release cowling fastener that requires only $1/4$ turn to either fasten or release. Examples: Dzus or Camloc fasteners.

quarter-wave antenna — The length of a radio antenna that is one quarter of the wavelength of the frequency for which the antenna is used.

quartz — A mineral found in nature in the form of a six-sided crystal.

quartz crystal — A thin slice of quartz used in electronic devices to control frequency. A quartz crystal vibrates when an electrical pulse is applied. The frequency is dependent on the size, shape, and thickness of the crystal.

quartz glass — Glass constructed of pure quartz.

quartz lamp — A lamp constructed with quartz glass. Quartz glass does not absorb ultraviolet rays as regular glass does.

quartz oscillator — An electronic frequency generator that uses a quartz crystal to control frequency.

quartz-iodine lamp — Lamps with a tungsten filament in a nitrogen-argon gaseous medium to which bromine has been added, enclosed in an envelope of quartz. These lamps have an operating life two to three times that of standard incandescent lamps and do not blacken with age. Now referred to as tungsten-halogen "quartz" lamps.

quasi-stationary front — A stationary or nearly stationary front. Conventionally, a front that is moving at a speed of less than 5 knots is generally considered to be quasi-stationary. Commonly referred to as a stationary front.

quench hardening — A ferrous alloy hardened by austenitizing and then cooling rapidly enough so that some or all of the austenite transforms to martensite.

quenching — The rapid cooling of a metal as part of the heat-treating process. The metal is removed from the furnace and submerged in a liquid such as water, oil, or brine.

queuing — The placement, integration, and segregation of departure aircraft in designated movement areas of an airport by departure fix, EDCT, and/or restriction.

quick engine change assembly (QECA) — An assembly made up of an engine, propeller, all accessories, along with all of the necessary cowling and engine mounts. Used to minimize downtime when an engine change is required. Also referred to as a QEC unit or QEC kit.

quick look — A feature of NAS Stage A and ARTS that provides the controller the capability to display full data blocks of tracked aircraft from other control positions.

quick-break fuse — A fuse that contains a spring that pulls the fuse link apart earlier and more quickly than a normal fuse.

quick-break switch — A switch that has spring loaded contacts that snap open and closed. This helps prevent arcing during the opening and closing of the circuit.

quick-disconnect coupling — A fluid coupling that incorporates a check valve to prevent fluid loss upon connecting or disconnecting.

quicksilver — Mercury. A heavy, silver-colored, toxic, liquid, metallic chemical element with a symbol of Hg and an atomic number of 80. Mercury remains in a liquid state under standard conditions of pressure and temperature. Mercury is approximately 13 times as heavy as water.

quill shaft — A hardened steel shaft with a hollow cross section and splines on each end. Torsional flexing of the shaft is used to absorb torsional vibrations.

quota flow control — A flow control procedure by which the Central Flow Control Function (CFCF) restricts traffic to the ARTC Center area having an impacted airport, thereby avoiding sector/area saturation.

quotient — The number resulting from the division of one number by another.

R

rabbet — A groove cut into the edge or face of a board so that another board or panel can be fitted into it. Allows a T- or L-shaped assembly to be constructed.

rabbet plane — A woodworking tool used to create rabbets in a surface.

race — A grooved, hardened, and polished steel surface on which a bearing is supported.

racetrack procedure (ICAO) — A procedure designed to enable the aircraft to reduce altitude during the initial approach segment and/or establish the aircraft inbound when the entry into a reversal procedure is not practical.

rack — A straight piece of metal that has teeth cut into one side.

rack and pinion — A set of gears arranged in such a way that a rotary motion of the pinion gear is changed into linear motion of the rack.

rack and pinion actuator — A rotary actuator in which the fluid acts on the rack of gear teeth cut on a piston. As the piston moves it rotates a mating pinion gear.

radar — A device that, by measuring the time interval between transmission and reception of radio pulses and correlating the angular orientation of the radiated antenna beam or beams in azimuth and/or elevation, provides information on range, azimuth, and/or elevation of objects in the path of the transmitted pulses.
a. Primary Radar — A radar system in which a minute portion of a radio pulse transmitted from a site is reflected by an object and then received back at that site for processing and display at an air traffic control facility.
b. Secondary Radar/Radar Beacon (Air Traffic Control Radar Beacon System (ATCRBS)) — A radar system in which the object to be detected is fitted with cooperative equipment in the form of a radio receiver/transmitter (transponder). Radar pulses transmitted from the searching transmitter/receiver (interrogator) site are received in the cooperative equipment and used to trigger a distinctive transmission from the transponder. This reply transmission, rather than a reflected signal, is then received back at the transmitter/receiver site for processing and display at an air traffic control facility.

radar advisory — Information or advice provided to pilots based on radar observations.

radar altimeter — Aircraft equipment that makes use of the reflection of radio waves from the ground to determine the height of the aircraft above the surface.

radar altitude — Distance from the ground determined by radar altimeter, which makes use of the reflection of radio waves from the ground.

radar approach — An instrument approach procedure that utilizes Precision Approach Radar (PAR) or Airport Surveillance Radar (ASR).

radar approach control facility — A terminal ATC facility that uses radar and nonradar capabilities to provide approach control services to aircraft arriving, departing, or transiting airspace controlled by the facility.
a. Provides radar ATC services to aircraft operating in the vicinity of one or more civil and/or military airports in a terminal area. The facility can provide services of a ground controlled approach (GCA); i.e., ASR and PAR approaches. A radar approach control facility can be operated by FAA, USAF, US Army, USN, USMC, or jointly by FAA and a military service.
b. Specific facility nomenclatures are used for administrative purposes only and are related to the physical location of the facility and the operating service generally as follows:
1) Army Radar Approach Control (ARAC) (Army).
2) Radar Air Traffic Control Facility (RATCF) (Navy/FAA).
3) Radar Approach Control (RAPCON) (Air Force/FAA).
4) Terminal Radar Approach Control (TRACON) (FAA).
5) Air Traffic Control Tower (ATCT) (FAA). (Only those towers delegated approach control authority.)

radar beacon transponder — An electronic device that receives an interrogation or transmission from ground radar and responds with a coded transmission that appears on the traffic controller's radar scope.

radar beam — A narrow beam of electromagnetic radiation used in radar units to determine distance, height, size, and composition of desired target items.

radar contact — Term used by ATC to advise a pilot that the aircraft is identified on radar.

radar echo — The returning radar signal that is reflected from an object.

radar identified aircraft — An aircraft, the position of which has been correlated with an observed target or symbol on the radar display.

radar monitoring — The radar flight-following of aircraft, whose primary navigation is being performed by the pilot, to observe and note deviations from its authorized flight path, airway, or route. When being applied specifically to radar monitoring of instrument approaches; i.e., with precision approach radar (PAR) or radar monitoring of simultaneous ILS/MLS approaches, it includes advice and instructions whenever an aircraft nears or exceeds the prescribed PAR safety limit or simultaneous ILS/ MLS no transgression zone.

radar navigational guidance — Vectoring aircraft to provide course guidance.

radar point out — An action taken by a controller to transfer the radar identification of an aircraft to another controller if the aircraft will or may enter the airspace or protected airspace of another controller and radio communications will not be transferred.

radar required — A term displayed on charts and approach plates and included in FDC Notams to alert pilots that segments of either an instrument approach procedure or a route are not navigable because of either the absence or unusability of a NAVAID. The pilot can expect to be provided radar navigational guidance while transiting segments labeled with this term.

radar route — A flight path or route over which an aircraft is vectored. Navigational guidance and altitude assignments are provided by ATC.

radar separation — Radar spacing of aircraft in accordance with established minima.

radar service — A term that encompasses one or more of the following services based on the use of radar that can be provided by a controller to a pilot of a radar identified aircraft.
a. Radar Monitoring — The radar flight-following of aircraft, whose primary navigation is being performed by the pilot, to observe and note deviations from its authorized flight path, airway, or route. When being applied specifically to radar monitoring of instrument approaches; i.e., with precision approach radar (PAR) or radar monitoring of simultaneous ILS/MLS approaches, it includes advice and instructions whenever an aircraft nears or exceeds the prescribed PAR safety limit or simultaneous ILS/ MLS no transgression zone.
b. Radar Navigational Guidance — Vectoring aircraft to provide course guidance.
c. Radar Separation — Radar spacing of aircraft in accordance with established minima.

radar service [ICAO] — Term used to indicate a service provided directly by means of radar.
a. Monitoring — The use of radar for the purpose of providing aircraft with information and advice relative to significant deviations from nominal flight path.
b. Separation — The separation used when aircraft position information is derived from radar sources.

radar service terminated — Used by ATC to inform a pilot that he will no longer be provided any of the services that could be received while in radar contact. Radar service is automatically terminated, and the pilot is not advised in the following cases:

a. An aircraft cancels its IFR flight plan, except within Class B airspace, Class C airspace, a TRSA, or where Basic Radar service is provided.
b. An aircraft conducting an instrument, visual, or contact approach has landed or has been instructed to change to advisory frequency.
c. An arriving VFR aircraft, receiving radar service to a tower-controlled airport within Class B airspace, Class C airspace, a TRSA, or where sequencing service is provided, has landed; or to all other airports, is instructed to change to tower or advisory frequency.
d. An aircraft completes a radar approach.

radar summary chart — A weather product derived from the national radar network that graphically displays a summary of radar weather reports.

radar surveillance — The radar observation of a given geographical area for the purpose of performing some radar function.

radar traffic advisories — Advisories issued to alert pilots to known or observed radar traffic that can affect the intended route of flight of their aircraft.

radar traffic information service — Advisories issued to alert pilots to observation of radar identified and nonidentified aircraft targets on an ATC radar display that may be in such proximity to the position or intended route of flight of their aircraft as to warrant their attention.

radar vector — A heading issued by a radar controller to the pilot of an aircraft to provide navigational guidance.

radar vectoring [ICAO] — Provision of navigational guidance to aircraft in the form of specific headings based on the use of radar.

radar weather echo intensity levels — Existing radar systems cannot detect turbulence. However, there is a direct correlation between the degree of turbulence and other weather features associated with thunderstorms and the radar weather echo intensity. The National Weather Service has categorized radar weather echo intensity for precipitation into six levels. These levels are sometimes expressed during communications as "VIP LEVEL" 1 through 6 (derived from the component of the radar that produces the information - Video Integrator and Processor). The following list gives the "VIP LEVELS" in relation to the precipitation intensity within a thunderstorm:
Level 1. WEAK
Level 2. MODERATE
Level 3. STRONG
Level 4. VERY STRONG
Level 5. INTENSE
Level 6. EXTREME

radarsonde observation — An upper air observation used to determine winds, and other meteorological data, by tracking the range, elevation, and azimuth of a radar target carried aloft, usually by a balloon.

radial — A navigational signal generated by a VOR or VORTAC. . Each VOR station has 360 radials. The radials are assigned numbers that pertain to their position around the magnetic compass card and are considered being drawn away from the VOR station in a particular magnetic direction.

radial engine — A reciprocating aircraft engine in which all of the cylinders are arranged radially, or spoke-like, around a small crankcase.

radial inflow turbine — A turbine wheel that receives its gases at the blade tips and guides the air inward and outward to the exhaust duct. It looks similar to a radial outflow compressor and is used extensively in APUs.

radial lead — An electrical component lead that protrudes from the outside of a component.

radial outflow compressor -An impeller-shaped device that receives air at its center and slings it outward at high velocity into a diffuser for increased pressure.

radial ply tire — A tire in which layers of metal or synthetic casing plies are constructed at right angles to the tread. The plies are folded around the wire beads and back against the tire sidewall, completely encompassing the tire body.

radial velocity — Motion toward or away from Doppler RADAR.

radian — A unit of angular measurement equal to the angle between two radii, separated by an arc equal to the length of the radius. Radians are used in the measurement of angular velocity.

radiant energy — Energy due to any form of electromagnetic radiation, for instance, from the sun.

radiant heat — The transfer of heat energy from a heat source to the surrounding air. Rooms that have heated water tubing embedded in the floor use radiant heat.

radiation — The emission of energy by a medium and transferred, either through free space or another medium, in the form of electromagnetic waves.

radiation — Energy radiated in the form of electromagnetic waves.

radiation fog — Fog characteristically resulting when radiational cooling of the Earth's surface lowers the air temperature near the ground to or below its initial dewpoint on calm, clear nights.

radiation sensing detector — A fire detection system that utilizes heat-sensitive units to complete an electrical circuit when the temperature rises to a preset value.

radiation shield — An insulation blanket made with layers of aluminum foil used around turbine engines to prevent heat radiating from the engine into the structure.

radical sign — The mathematical symbol "√ "placed before a quantity to show its root is to be extracted.

radio — 1. A device used for communication. 2. Used to refer to a flight service station; e.g., "Seattle Radio" is used to call Seattle FSS.

radio altimeter (RA) — A device that measures the height of an aircraft above the terrain by means of the transmissions of a continuous wave, constant amplitude, frequency modulated signal. The difference in the frequency of the reflected signals at any time is read on the RA indicator in feet above the ground.

radio beacon — An L/MF or UHF radio beacon transmitting nondirectional signals whereby the pilot of an aircraft equipped with direction finding equipment can determine his bearing to or from the radio beacon and "home" on or track to or from the station. When the radio beacon is installed in conjunction with the Instrument Landing System marker, it is normally referred to as a compass locator.

radio control — Operating a unit or device by means of radio transmissions.

radio detection and ranging (Radar) — A device that, by measuring the time interval between transmission and reception of radio pulses and correlating the angular orientation of the radiated antenna beam or beams in azimuth and/or elevation, provides information on range, azimuth, and/or elevation of objects in the path of the transmitted pulses.

radio direction finding — A method of determining the direction from which signals are received. When an aircraft is within reception range of a radio station, a directional antenna provides a means of fixing a bearing from the station to the aircraft. The bearings from two or more receivers are plotted on a chart to determine the position of the aircraft with reasonable accuracy.

radio frequency — The electromagnetic waves that radiate from the antenna of a radio transmitter. The frequency is higher than the audible frequency range and below the frequency range of heat and light.

radio frequency energy — Electromagnetic energy used for communication. Generally in the frequency range of 3 kilohertz to 300 gigahertz.

Radio Magnetic Indicator (RMI) — A movable-card instrument on which the card automatically rotates to reflect the aircraft's magnetic heading. Two needles display the relative bearing to ADF and VOR stations.

radio marker beacon — A low-powered, single-frequency radio transmitter used to designate specific navigational points. Marker beacon receivers receive the signal only when flying directly over the radio marker beacon.

radio transmitter — An electronic device that transmits electromagnetic waves for the purpose of communicating with associated receiver(s).

radioactivity — The property possessed by some elements or isotopes of emitting electrons or alpha particles as their atomic nuclei disintegrate.

radio-frequency cable — A transmission line in which the center conductor is surrounded by an insulator and a braided outer conductor. All of this is enclosed in a weatherproof outer insulator.

radio-frequency interference (RFI) — Undesired electromagnetic radiation that interferes with radio communication. Often the result of improperly filtered radio transmitters and other electronic equipment.

radio-frequency spectrum — The range of frequencies used for radio communication. Commonly used bands are:

BAND	FREQUENCY RANGE
Very Low Frequency (VLF)	3-30 kHz
Low Frequency (LF)	30-300 kHz
Medium Frequency (MF)	300-3,000 kHz
High Frequency (HF)	3-30 MHz
Very High Frequency (VHF)	30-300 MHz
Ultra High Frequency (UHF)	300-3,000 MHz
Super High Frequency	3-30 GHz
Extremely High Frequency (EHF)	30-300 GHz

radiography — The system of nondestructive inspection using X-rays or gamma rays to determine the condition of a part without disassembly.

radiosonde — A balloon-borne instrument for measuring pressure, temperature, and humidity aloft.

radiosonde observation — a sounding made by the instrument. If the balloon is tracked, winds can also be determined.

radium — A radioactive element with a symbol of Ra and an atomic number of 88.

radium poisoning — A poisoning associated with the handling of luminous materials used for marking the dials on some older aircraft instruments.

radius — 1. The distance from the center of a circle or sphere to its outer circumference; one half the diameter of a circle. 2. Also, the radial part of an object.

radius bar — The part of the cornice brake top leaf that has an accurately ground radius at its edge for bending sheet metal. It is used to obtain a bend radius appropriate to the alloy and thickness of a material.

radius block — A metal block around which sheet metal is bent to obtain a specific bend radius.

radius dimpling — A method of preparing thin sheet metal in which a cone-shaped male die is forced into the recess of a female die with either a hammer or a pneumatic rivet gun. The material can be riveted with flush rivets after the dimpling process is complete.

radius gauge — A precision gauge with an accurately cut inside and outside radii, used to measure the radius of a bend.

radius of turn — The amount of horizontal distance an aircraft uses to complete a turn.

radome — The protective covering of a radar antenna. A radome is constructed of lightweight honeycomb core material electrically transparent so it will not interfere with the transmitted and received pulses of electrical energy.

rag wing — A common slang term for a fabric covered airplane.

rain — Precipitation with water drops larger than drizzle and that fall in relatively straight, although not necessarily vertical, paths as compared to drizzle that falls in irregular paths.

rain bands — Streaks of rain that spiral into a storm, lines of convergence associated with CB's (cumulonimbus) clouds and shower activity.

rain gauge — An instrument used to measure the amount of rain that has fallen in a given locale.

rain shadow — The drier downwind side of a mountain.

rain shower — Precipitation in the form of rain from a cumuliform cloud; characterized by sudden onset and cessation, rapid change of intensity, and usually by rapid change in the appearance of the sky.

ram air pressure — The air pressure caused by the forward motion of an aircraft. Ram air pressure is slightly above ambient air pressure.

ram air temperature rise — The increase in temperature caused by the ram compression of the air as an aircraft passes through the air at a high rate of speed. The rate of temperature increase is proportional to the square of the speed of the aircraft.

ram pressure rise — The pressure rise in the inlet due to the forward speed of the aircraft.

ram ratio — The ratio of ram pressure to ambient pressure in a jet engine.

ram temperature rise — The inlet temperature rise due to inlet ram pressure rise.

ramp — The apron or paved surface around a hangar used for parking aircraft.

ramp and soak — In composites, a curing process in which the temperature is slowly raised at a given rate to the final cure temperature and held for a specific amount of time. After that time, the temperature is slowly lowered to room temperature. For example, if the final cure temperature for a part is 250°F, the temperature would be ramped up to 250° at a rate of 8° per minute. Once it reaches 250°, the temperature is held there for 1 hour and 30 minutes, and then the temperature is lowered to 80° at a rate of 5° per minute. This process is typically done by using a temperature controller found on hot patch bonding equipment.

ramp weight — The total weight of the aircraft while on the ramp. It differs from takeoff weight by the weight of the fuel that will be consumed in taxiing to the point of takeoff.

random altitude — An altitude inappropriate for direction of flight and/or not in accordance with FAA Order 7110.65, paragraph 4-5-1.

random route — Any route not established or charted/published or not otherwise available to all users.

range-height indicator (RHI) scope — A radar display with height as the vertical axis and range as the horizontal axis. A vertical cross section of the cloud masses in a particular azimuth can be determined.

Rankine Temperature — The absolute temperature scale using degrees Fahrenheit with minus 460°F as absolute zero. It is used in many engine performance calculations.

rapid decompression — The almost instantaneous loss of cabin pressure in aircraft with a pressurized cockpit or cabin.

rapid exit taxiway (ICAO) — A taxiway connected to a runway at an acute angle and designed to allow landing airplanes to turn off at higher speeds than are achieved on other exit taxiways thereby minimizing runway occupancy times.

rarefied air — Of or pertaining to less air. Often used to describe the Earth's upper atmosphere.

rasp — A coarse file with raised cutting points used for scraping soft materials such as wood or plastic.

ratchet — A mechanism that consists of a toothed wheel and a bar, or pawl, which allows the wheel to rotate in one direction but prevents its backward motion.

ratchet coupling — A toothed wheel into which a pawl drops so that motion can be imparted to the wheel in only one direction.

ratchet handle — A handle with a ratchet coupling used to turn socket-type wrenches. A ratcheting mechanism in the handle allows the socket to be turned in one direction while the handle is moved with a back-and-forth movement. By reversing the catch in the ratchet body, the same back-and-forth movement of the handle will rotate the socket in the opposite direction.

rate gyro — A device used to measure the rate of an aircraft's rotation about its vertical axis through 360°. A rate gyro is not affected by the roll or pitch of the aircraft but aligns itself with the fore and aft line.

rate of burning — The time required for a specific amount of fuel-air mixture to burn or to release its heat energy.

rate of climb — The rate, measured in feet per minute, at which vertical motion occurs.

rate of turn — The change of heading per unit of time. For example, 180° per minute, or 3° per second.

rate of yaw — The rate, in degrees per second, at which an aircraft rotates about its vertical axis.

rate signal — A signal proportional to a rate of change.

rated 2 1/2-minute OEI (one engine inoperative) power — With respect to rotorcraft turbine engines, means the approved brake horsepower developed under static conditions at specified altitudes and temperatures within the operating limitations established for the engine under FAR Part 33, and limited in use to a period of not more than 2 ½ minutes after the failure of one engine of a multiengine rotorcraft.

rated 2-minute OEI (one engine inoperative) power — With respect to rotorcraft turbine engines, means the approved brake horsepower developed under static conditions at specified altitudes and temperatures within the operating limitations established for the engine under FAR part 33, for continued one-flight operation after the failure of one engine in multiengine rotorcraft, limited to three periods of use no longer than 2 minutes each in any one flight, and followed by mandatory inspection and prescribed maintenance action.

rated 30-minute OEI (one engine inoperative) power — With respect to rotorcraft turbine engines, means the approved brake horsepower developed under static conditions at specified altitudes and temperatures within the operating limitations established for the engine under FAR Part 33, and limited in use to a period of not more than 30 minutes after the failure of one engine of a multiengine rotorcraft.

rated 30-second OEI (one engine inoperative) power — With respect to rotorcraft turbine engines, means the approved brake horsepower developed under static conditions at specified altitudes and temperatures within the operating limitations established for the engine under FAR part 33, for continued one-flight operation after the failure of one engine in multiengine rotorcraft, limited to three periods of use no longer than 30 seconds each in any one flight, and followed by mandatory inspection and prescribed maintenance action.

rated continuous OEI (one engine inoperative) power — With respect to rotorcraft turbine engines, means the approved brake horsepower developed under static conditions at specified altitudes and temperatures within the operating limitations established for the engine under FAR Part 33, and limited in use to the time required to complete the flight after the failure of one engine of a multiengine rotorcraft.

rated horsepower — The maximum horsepower an engine is approved to produce under a given set of circumstances.

rated maximum continuous power — With respect to reciprocating, turbopropeller, and turboshaft engines, means the approved brake horsepower that is developed statically or in flight, in standard atmosphere at a specified altitude, within the engine operating limitations established under Part 33, and approved for unrestricted periods of use.

rated maximum continuous thrust — With respect to turbojet engine type certification, means the approved jet thrust that is developed statically or in flight, in standard atmosphere at a specified altitude, without fluid injection and without the burning of fuel in a separate combustion chamber, within the engine operating limitations established under Part 33 of this chapter, and approved for unrestricted periods of use.

rated maximum continuous augmented thrust — With respect to turbojet engine type certification, means the approved jet thrust that is developed statically or in flight, in standard atmosphere at a specified altitude, with fluid injection or with the burning of fuel in a separate combustion chamber, within the engine operating limitations established under Part 33 of this chapter, and approved for unrestricted periods of use.

rated takeoff augmented thrust — With respect to turbojet engine type certification, means the approved jet thrust that is developed statically under standard sea level conditions, with fluid injection or with the burning of fuel in a separate combustion chamber, within the engine operating limitations established under Part 33 of this chapter, and limited in use to periods of not over 5 minutes for takeoff operation.

rated takeoff power — With respect to reciprocating, turbopropeller, and turboshaft engine type certification, means the approved brake horsepower that is developed statically under standard sea level conditions, within the engine operating limitations established under Part 33, and limited in use to periods of not over 5 minutes for takeoff operation.

rated takeoff thrust — With respect to turbojet engine type certification, means the approved jet thrust that is developed statically under standard sea level conditions, without fluid injection and without the burning of fuel in a separate combustion chamber, within the engine operating limitations established under Part 33 of this chapter, and limited in use to periods of not over 5 minutes for takeoff operation.

rated thrust — A manufacturer's guaranteed thrust as specified on the Type Certificate.

rate-of-climb indicator — A rate of air pressure change device that indicates the rate at which an airplane is climbing or descending in feet per minute. Also referred to as a vertical speed indicator.

rate-of-temperature rise indicator — A thermocouple fire detection system that depends on a rapid rate of temperature rise in order to operate. It will not indicate an overheat condition that has developed slowly.

rating — A statement that, as a part of a certificate, sets forth special conditions, privileges, or limitations.

ratio — The relationship between one number and another number expressed as a fraction. A proportion.

ratiometer indicator — A DC remote-indicating system whose pointer movement is determined by the ratio of current-flow between two resistors or portions of a special variable resistor.

rationalization — A defense mechanism that students employ when they cannot accept the real reason for their behavior. This permits them to substitute excuses for reasons; moreover, they can make those excuses plausible and acceptable to themselves. A subconscious technique for justifying actions that otherwise would be unacceptable.

rattail file — A steel tool that has a circular cross section with hardened ridged surfaces, used in smoothing, grinding, or boring. Also referred to as a round file.

rawhide mallet — A mallet made of rawhide wound into a tight cylinder used to form sheet metal without scratching it.

RC circuit — A circuit containing both resistance and capacitance.

RC time constant — The time required to charge a capacitor to 63.2% of its full-charge state through a given resistance.

reach — The length of the shell thread of a spark plug. For 18 mm spark plugs, long-reach plugs are threaded for $^{13}/_{16}$" and short-reach plugs for ½".

reactance — The opposition to the flow of AC made by an induction coil or a capacitor. Reactance is expressed in ohms.

reaction — A response to a stimulus or force. An event or sequence of events caused by another action.

reaction engine — A jet engine that receives thrust only from the reaction caused by hot gases being expelled.

reaction formation — A defense mechanism where students protect themselves from dangerous desires by not only repressing them, but by actually developing conscious attitudes and behavior patterns that are just the opposite. A student can develop a who-cares-what-other-people-think attitude to cover up feelings of loneliness and a hunger for acceptance.

reaction turbine — A stator vane and rotor blade arrangement, whereby the vanes form straight ducts and the blades form convergent ducts. Gases leaving the trailing edge of the blades turn the rotor. The design is common to turbine-type accessories such as air-starters.

reactive current — A current in an AC circuit not in phase with the voltage.

reactive metal — A metal such as aluminum or magnesium, which reacts with oxygen to form corrosion.

reactive power — The power in a reactive AC circuit that subtracts from the power that would be present in a purely resistive circuit. This represents energy that is alternately stored and released by the action of inductors and capacitors.

reactor — 1. A device in an AC circuit used to add reactance to it. 2. A small coil installed in an alternating current circuit to furnish inductive reactance and cancel some of the capacitive reactance in the circuit.

read back — Repeat my message back to me.

reader — A machine that magnifies the information on microfilm or microfiche so it can be read. Some readers can make printed paper copies.

readiness — A principle of learning that the eagerness and single-mindedness of a person toward learning affect the outcome of the learning experience.

real number — A number that is rational or irrational, not imaginary.

real power — The power in an AC circuit that is the product of the voltage and the current in phase with that voltage. Real power equals the voltage multiplied by the current multiplied by the power factor.

ream — To enlarge and smooth a drilled hole with a precision cutting tool called a reamer. A hole is reamed in preparation for work with close-tolerance parts.

reamer — A sharp edged cutting tool used for enlarging or tapering drilled holes in preparation for work with close-tolerance parts.

rebreather bag — A bag connected to an oxygen mask that allows expired air to be mixed with a fixed rate of oxygen. It allows users to rebreath a portion of each expired breath.

rebreather oxygen mask — An oxygen mask with a rebreather bag attached.

receiver — In communication, the listener, reader, or student who takes in a message containing information from a source, processes it, reacts with understanding, and changes behavior in accordance with the message.

receiver autonomous integrity monitoring (RAIM) — A technique whereby a civil GNSS (Global Navigation Satellite System) receiver/processor determines the integrity of the GNSS navigation signals without reference to sensors or non-DoD integrity systems other than the receiver itself. This determination is achieved by a consistency check among redundant pseudorange measurements.

receiving controller — A controller/facility receiving control of an aircraft from another controller/ facility.

receiving facility — See receiving controller.

recency — A principle of learning that things learned today are remembered better than things that were learned some time ago. The longer time that passes, the less that will be remembered. Instructors use this principle when summarizing the important points at the end of a lecture in order that the students will better remember them.

reciprocal — A reverse bearing, opposite in direction by 180°.

reciprocating engine — An engine that converts the heat energy from burning fuel into the reciprocating movement of the pistons. This movement is converted into a rotary motion by the connecting rods and crankshaft.

reciprocating motion — An alternating back-and-forth motion.

reciprocating saw — A power-driven saw that cuts material by driving a blade back and forth.

recirculating fan — A fan in an aircraft cabin comfort system that circulates the air in the cabin without taking in any outside air.

reclaimed oil — Lubricating oil that has already been used and is restored to a useful state by removing impurities.

reconditioning a cell — Procedure of servicing a nicad battery cell by completely discharging it and then bringing it and all other cells in the battery back up to a full state of charge. Reconditioning equalizes the cells for optimum performance.

reconformance — The automated process of bringing an aircraft's Current Plan Trajectory into conformance with its track.

rectangle — A closed plane figure with four sides and four right angles. The opposite sides of the figure are parallel and equal in length.

rectifier — A device that converts alternating current into direct current. In effect, it is an electron check valve.

rectifier bridge — A rectifier using four diodes arranged in a bridge circuit.

rectify — In electricity, to change alternating current into direct current.

recurring — Happening again and again at regular or frequent intervals.

recurring Airworthiness Directive — An Airworthiness Directive that requires compliance at regular hourly or calendar time periods.

red brass — A copper-zinc alloy containing approximately 85% copper and 15% zinc. It is somewhat stronger than commercial bronze and is hardened by cold working.

red line — A red mark on an aircraft instrument that indicates a maximum allowable operating condition.

red rust — A nonmagnetic iron oxide.

red-line condition — The maximum safe condition at which a unit can operate. Most aircraft instruments are marked with a red line indicating the boundaries of their operating limits.

reduce speed to (speed) — An ATC procedure used to request pilots to adjust aircraft speed to a specific value for the purpose of providing desired spacing. Pilots are expected to maintain a speed of plus or minus 10 knots or 0.02 mach number of the specified speed.

reduction factor — A constant (usually 100 or 1,000) by which a moment is divided to obtain a moment index. A moment index is used to simplify weight and balance computations.

reduction gear train — The gear arrangement of an operating unit in which the output shaft turns slower than the input shaft. An example of this is a gear arrangement in an aircraft engine that allows the engine to turn at a faster speed than the propeller.

reduction gears — The gear arrangement in an aircraft engine that allows the engine to turn at a faster speed than the propeller.

Reed and Prince screw — A recessed-head, cross-point screw driven by a special cross-point screwdriver whose tip has a single taper.

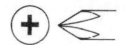

reed valve — A thin, leaf-type valve located in the valve plate of a reciprocating-type air conditioning compressor to control the inlet and outlet of the refrigerant.

reface — To resurface an object in order to remove imperfections or wear marks.

reference datum — An imaginary vertical plane at or near the nose of an aircraft from which all horizontal distances are measured for weight and balance calculations.

reference datum — An imaginary vertical plane from which all horizontal distances are measured for balance purposes.

reference designator — A combination of letters and numbers used to identify a component or assembly. Reference designators are stackable, meaning a full reference designator identifies the major assembly, subassemblies, and the actual component.

reference dimensions — The dimensions on an installation drawing required to illustrate the relationship between two parts.

reference junction — One of the two junctions in a thermocouple system. The reference junction is held at a constant or stable temperature to serve as a reference for the measuring junction.

reference pressure — The pressure in the outflow valve control established by cabin air pressure flowing through the cabin air filter and orifice. It is metered by the reference pressure metering valve.

refining — A petroleum extraction process in which crude oil is broken down into all of its different parts.

reflective — A reflective student considers all possibilities or alternatives before making a decision.

reflector — A surface that reflects light.

refrigerant — A fluid used in an air conditioning system to absorb heat from the cabin and carry it to a radiator where it can be transferred to the outside air.

refrigerant 12 — Dichlorodifluoromethane, a chemical compound used in many aircraft air conditioning systems. Also referred to as R-12.

regeneration — A process in an electrical circuit in which part of the output is fed back to the input to causes amplification.

regenerative braking — A method of slowing a motor by switching the wiring effectively to make it into a generator. This applies a load to the rotating mass and slows the motor.

regional checkout — A checkout that goes beyond learning about a particular airplane and encompasses learning how to fly in a specific region. Before flying or giving flight instruction in any unfamiliar environment, obtain a regional checkout from a qualified CFI who is experienced in that geographical area. For example, mountain flying offers some breathtaking scenery and wonderful experiences, but it also has some unique challenges and can be extremely dangerous to inexperienced pilots.

registration certificate — The document in an airplane that contains the name and address of the person to whom the airplane is registered.

regulated power supply — An electrical device that converts AC voltage to DC output voltage that remains constant under changing load conditions.

Reid vapor pressure — The amount of pressure acting on a liquid to hold the vapors in the liquid at a given temperature.

reinforce — To strengthen by adding extra support or material.

reinforced shell — The outer skin of an aircraft reinforced by a complete framework of structural members.

reinforcement — In composites, material used to strengthen the matrix. Fiber reinforced plastic is an example. Fibers are used to reinforce the plastic material.

reinforcing tape — A narrow, woven cotton or polyester tape used as reinforcement at stitching attachments.

rejuvenation — The restoring of resilience to a dope film by opening up the film with potent solvents and allowing the plasticizers in the rejuvenator to replace those that have migrated from the dope.

rejuvenator — A finishing material consisting of potent solvents and plasticizers used to restore resilience to weathered and cracked dope film.

relative bearing — An angular relationship between two objects measured in degrees clockwise from the twelve o'clock position of the first object.

relative humidity — The ratio of the existing amount of water vapor in the air at a given temperature to the maximum amount that could exist at that temperature; usually expressed in percent.

relative motion — The motion of one object with relation to another.

relative movement — The movement of one object with relation to another.

relative vorticity — Rotation due entirely to the flow of air at the point being measured. Absolute vorticity is relative vorticity plus rotation imparted by the turning of the Earth (planetary vorticity).

relative wind — The airflow caused by the motion of the aircraft through the air. Relative wind, also called relative airflow is opposite and parallel to the direction of flight.

relaxation oscillator — An electronic oscillator that produces pulses at an interval dependent on one or more RC time constants.

relay — An electrically operated remote switch with contacts that are closed or opened by an electromagnetic field.

relay question — Used in response to a student's question. The question is redirected to the group in order to stimulate discussion.

release film — In composites, a layer of plastic material used in the vacuum bagging process that does not allow resin to bleed through it. It will not bond to the part when the resins cure. Perforated release film will allow some resin to bleed through.

release time — A departure time restriction issued to a pilot by ATC (either directly or through an authorized relay) when necessary to separate a departing aircraft from other traffic.

release time [ICAO] — Time prior to which an aircraft should be given further clearance or prior to which it should not proceed in case of radio failure.

reliability — Producing consistent results on multiple tests.

relief — The amount one plane surface is set below or above another plane, usually for clearance or for economy in machining.

relief hole — A hole drilled in a flat sheet metal part to allow intersecting bends to be made. The relief hole prevents the metal from buckling.

relief map — A map that shows land forms with contour lines, shading, or color.

relief tube — A urinal on an aircraft that drains overboard. The discharge area around these tubes is an area highly susceptible to corrosion.

relief valve — A valve that limits the pressure in a system by releasing unwanted pressure at a preset value.

reluctance — 1. The relative difficulty with which magnetic domains can be aligned. 2. The opposition to magnetic flux. The opposite of permeability.

remainder — The result of subtracting one number from another or the amount left when one number does not divide evenly into another.

remanufactured engine — An engine assembled by the manufacturer or his authorized agent using used parts that are held to the new parts' dimensional limits. A remanufactured engine is given zero time records and usually the same warranty and guarantee as a new engine.

remote (slave) scope — A cathode-ray tube that duplicates the display of a primary radar scope.

remote communications air/ground (RCAG) facility — An unmanned VHF/UHF transmitter/receiver facility used to expand ARTCC air/ground communications coverage and to facilitate direct contact between pilots and controllers. RCAG facilities are sometimes not equipped with emergency frequencies 121.5 MHz and 243.0 MHz.

remote communications outlet(RCO) — An unmanned communications facility remotely controlled by air traffic personnel. RCOs serve FSSs. RTRs (Remote Transmitter/Receivers) serve terminal ATC facilities. An RCO or RTR can be UHF or VHF and will extend the communication range of the air traffic facility. There are several classes of RCOs and RTRs. The class is determined by the number of transmitters or receivers. Classes A through G are used primarily for air/ground purposes. RCO and RTR class O facilities are nonprotected outlets subject to undetected and prolonged outages. RCO (O's) and RTR (O's) were established for the express purpose of providing ground-to-ground communications between air traffic control specialists and pilots located at a satellite airport for delivering enroute clearances, issuing departure authorizations, and acknowledging instrument flight rules cancellations or departure/landing times. As a secondary function, they can be used for advisory purposes whenever the aircraft is below the coverage of the primary air/ground frequency

remote control — The control of an object with radio signals from a remote location.

remote transmitter/receiver — See remote communications outlet.

Rene metal — A nickel chromium alloy used in the manufacturing of gas turbine engines.

repair — To restore an item to a condition of practical operation or to its original condition.

repairman license — A certificate issued by the FAA to a person employed by a repair station, a certified commercial operator, or a certified air carrier required to provide a continuous airworthiness maintenance program. The repairman license is valid only as long as the person holding the license is employed by the company for which the certificate was issued.

repeater indicator — An instrument that repeats the information via a master indicator.

repeating decimal — A fraction that cannot be expressed as a definite number. The fraction $^1/_3$ is a repeating decimal (0.333333333).

report — Used to instruct pilots to advise ATC of specified information; e.g., "Report passing Hamilton VOR."

reporting point — A geographical location in relation to the reported position of an aircraft.

repression — A process in which a person subconsciously forgets or excludes unpleasant or anxiety producing information.

repression — Theory of forgetting where a person is more likely to forget information that is unpleasant or produces anxiety.

repulsion — The force that tends to cause objects to move away from each other.

repulsion motor — The repulsion motor consists of stator (field) windings that are powered by input power and rotor windings that have current induced by the magnetic field of the stator windings. The rotor windings are connected to the commutator and the brushes are connected to each other completing a circuit through windings within the rotor. The closed-loop circuits in the rotor are effectively the short-circuited secondaries of a transformer, where the motor's field windings are the primary coil. The currents induced in the rotor create a magnetic field that repels that of the field winding (Lenz's law). This repulsion causes the rotor to turn because the brushes are offset from the field poles, so that the repulsive forces are pushing on the rotor tangential to the rotation axis of the rotor. The brushes can be shifted to control the speed and direction of rotation.

request full route clearance — Used by pilots to request that the entire route of flight be read verbatim in an ATC clearance. Such request should be made to preclude receiving an ATC clearance based on the original filed flight plan when a filed IFR flight plan has been revised by the pilot, company, or operations prior to departure.

required equipment — Equipment determined by the FAA to be necessary for an aircraft to be considered airworthy.

required inspection item (RII) — A designation of items of maintenance and alteration that could result in a failure, malfunction, or defect endangering the safe operation of the aircraft if an inspection is not properly performed or if improper parts or materials are used.

required navigation performance (RNP) — A statement of the navigational performance necessary for operation within a defined airspace. The following terms are commonly associated with RNP:
a. Required Navigation Performance Level or Type (RNP-X). A value, in nautical miles (NM), from the intended horizontal position within which an aircraft would be at least 95-percent of the total flying time.

b. Required Navigation Performance (RNP) Airspace. A generic term designating airspace, route(s), leg(s), operation(s), or procedure(s) where minimum required navigational performance (RNP) have been established.
c. Actual Navigation Performance (ANP). A measure of the current estimated navigational performance. Also referred to as Estimated Position Error (EPE).
d. Estimated Position Error (EPE). A measure of the current estimated navigational performance. Also referred to as Actual Navigation Performance (ANP).
e. Lateral Navigation (LNAV). A function of area navigation (RNAV) equipment that calculates, displays, and provides lateral guidance to a profile or path.
f. Vertical Navigation (VNAV). A function of area navigation (RNAV) equipment that calculates, displays, and provides vertical guidance to a profile or path.

rescue coordination center (RCC) — A search and rescue (SAR) facility equipped and manned to coordinate and control SAR operations in an area designated by the SAR plan. The U.S. Coast Guard and the U.S. Air Force have responsibility for the operation of RCCs.

rescue co-ordination centre [ICAO] — A unit responsible for promoting efficient organization of search and rescue service and for co-ordinating the conduct of search and rescue operations within a search and rescue region.

reserve — To keep something back or set it apart for use in an emergency or for a special purpose.

reservoir — A place where an extra supply of something is kept, for example, a tank in which fluid is stored.

residual — The remainder or anything that is left over.

residual charge — The remaining electrical charge left on capacitor plates after they have been discharged.

residual fuel or oil — The fuel or oil that is trapped in the lines and, therefore, not usable. In weight and balance computations, residual fuel and oil are considered to be part of an aircraft's empty weight.

residual magnetic flux — The magnetism that remains in the core of an electromagnet after the magnetizing current no longer flows in the coil. Also referred to as residual magnetism.

residual magnetism — The magnetism that remains in the core of an electromagnet after the magnetizing current no longer flows in the coil. Also referred to as residual magnetic flux.

residual magnetism testing inspection — An inspection for magnetism that might be left in a ferrous part following magnetic particle inspection. The test is conducted with a magnetic indicator that discloses any remaining magnetic field. It is important that all residual magnetism be removed so that ferrous particles are not attracted to it.

resignation — A negative self-concept is the factor that contributes most to a student's failure to a remain receptive to new experience, and which creates a tendency to reject additional training.

resin — A family of natural or synthetic fluids or semi-solid substances that can be changed into a solid such as plastic through the addition of appropriate catalyzers.

resin rich — In composites, an area that has an excess amount of matrix. A resin rich laminate usually is more brittle and weighs more than laminates with the proper amount of resin.

resin ridge — In composites, a ridge of excess resin that contains only resin.

resin starved — In composites, an area deficient in resin. A resin-starved part will not exhibit the structural strength that a part made with the proper amount of resin.

resin system — In composites, a mixture of resin and ingredients required for the intended processing method and final product.

resin transfer molding (RTM) — In composites, a manufacturing process in which the resin/catalyst mixture is pumped into a two-sided mold in which a fabric reinforcement has been placed. The part is then heated and cured.

resistance (R) — The opposition to the flow of electrons offered by a device or material. Opposition by resistance causes a loss of power.

resistance decade box — An electrical test device that provides the capability to insert selected values of resistance into a circuit.

resistance furnace — A high-temperature furnace heated by passing current through various high resistance materials.

resistance thermometer — A temperature-measuring device based on the linear change in resistance of a material as temperature varies. This device is connected to a Wheatstone bridge to determine resistance and thus temperature.

resistance welding — The fusion of metals by clamping them together and passing a high amperage electrical current through the joint. The resulting heat melts the metal and the pressure causes the two pieces to fuse together. Spot and seam welding are forms of electrical resistance welding.

resistance wire — A wire made from a high resistance material, used in wire-wound resistors and heating elements.

resistive circuit — In electronics, a circuit constructed entirely of resistive elements (no capacitive or inductive components). In such a circuit, the voltage and current are in phase.

resistive current — The current in a circuit where current is in phase with voltage.

resistivity — The ability of a material to resist the flow of electrons. It is the opposite of conductivity.

resistor — An electrical circuit element used to provide a voltage drop by dissipating some of the electrical energy in the form of heat.

resistor color code — A color code marking system used to identify the resistance value of carbon resistors. Measured in ohms. The resistor is marked by either three or four 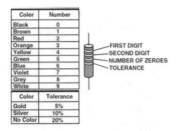 colored bands. The first colored band (nearest the end of the resistor) indicates the first digit. The second colored band indicates the second digit. The third colored band indicates the number of zeros to be added to the two digits derived from the first and second bands. If there is a fourth colored band, it is used as an indication of percentage of tolerance.

resistor power dissipation rating — The amount of power a resistor can safely dissipate in the form of heat under controlled conditions.

resistor spark plug — A shielded spark plug with a composition resistor installed in the barrel. The resistor limits the current stored in the capacitive effect of the shielding. This minimizes electrode erosion and increases the life of the spark plug.

resolution — The ability to distinguish between indications on a display. One example would be the ability to distinguish between two radar returns displayed on a cathode-ray tube.

resolution advisory — A display indication given to the pilot by the traffic alert and collision avoidance systems (TCAS II) recommending a maneuver to increase vertical separation relative to an intruding aircraft. Positive, negative, and vertical speed limit (VSL) advisories constitute the resolution advisories. A resolution advisory is also classified as corrective or preventive.

resonance — A frequency in any given RLC (Resistance, Inductance, Capacitance) AC circuit at which the capacitive reactance is equal to the inductive reactance.

resonance method of ultrasonic inspection — A method of detecting material thickness or indications of internal damage by injecting variable frequency ultrasonic energy into a material. A specific frequency of energy produces the greatest return in a given thickness of material. When the equipment is calibrated for a specific thickness and this thickness changes, an aural or visual indication is given.

resonant circuit — An electrical circuit in which the inductive reactance is equal to the capacitive reactance. They will be equal at a particular frequency called the resonant frequency. The circuit can be made resonant by either varying the frequency until resonance occurs, or holding the frequency constant and varying inductance or capacitance to achieve resonance.

resonant frequency — The frequency of a source of vibration that is exactly the same as the natural vibration frequency of a structure.

resonate — To vibrate at a certain frequency. A mechanical system is said to resonate when its natural vibration frequency is the same as that of the force applied. The level of vibration of an object increases immensely as that frequency is reached and will be less on either side of that frequency.

resource use — An essential part of ADM training. Since useful tools and sources of information may not always be readily apparent, it is important to teach students how to recognize appropriate resources. Resources must not only be identified, but students must develop the skills to evaluate whether they have the time to use a particular resource and the impact its use will have upon the safety of flight.

respirator — A device worn over the mouth and nose to prevent the harmful inhaling of dangerous substances.

responses — Possible answers to a multiple-choice test item. The correct response is often called the keyed response, and incorrect responses are called distractors.

restart — The act of starting an engine after it has been operating and then shut down.

restricted area — Airspace designated under FAR Part 73, within which the flight of aircraft, while not wholly prohibited, is subject to restriction. Most restricted areas are designated joint use and IFR/VFR operations in the area can be authorized by the controlling ATC facility when it is not being utilized by the using agency. Restricted areas are depicted on enroute charts. Where joint use is authorized, the name of the ATC controlling facility is also shown.

restricted area [ICAO] — An airspace of defined dimensions, above the land areas or territorial waters of a State, within which the flight of aircraft is restricted in accordance with certain specified conditions.

restrictor — An orifice for reducing or restricting the flow of a fluid.

resultant flux — The flux in a magnetic circuit of an aircraft magneto. Resultant of the flux of the rotating permanent magnet and the flux that surrounds the primary windings when primary current is flowing.

resultant lift — The vector sum of the magnitude and direction of all of the lift forces produced by an airfoil.

resultant relative wind — In rotorcraft, airflow from rotation that is modified by induced flow.

resume normal speed — Used by ATC to advise a pilot that previously issued speed control restrictions are deleted. An instruction to "resume normal speed" does not delete speed restrictions that are applicable to published procedures of upcoming segments of flight, unless specifically stated by ATC. This does not relieve the pilot of those speed restrictions that are applicable to FAR 91.117.

resume own navigation — Used by ATC to advise a pilot to resume his own navigational responsibility. It is issued after completion of a radar vector or when radar contact is lost while the aircraft is being radar vectored.

retard — To slow or delay the progress of something.

retard breaker points — An auxiliary set of breaker points in a magneto equipped with the Shower of Sparks starting system. These points are operative only during the starting cycle and open later than the run, or normal, points. This provides a late or retarded spark. Also referred to as retard points.

retard points — An auxiliary set of breaker points in a magneto equipped with the Shower of Sparks starting system. These points are operative only during the starting cycle and open later than the run, or normal, points. This provides a late or retarded spark. Also referred to as retard breaker points.

retarder — A slow-drying solvent used to prevent blushing or to provide a more glossy finish by allowing the finish a longer flow-out time.

retention — There are five principles that promote deep learning and enhance retention of course material: praise stimulates remembering, recall is promoted by association, favorable attitudes aid retention, learning with all the senses is most effective, and meaningful repetition aids recall.

retentivity — The ability of a material to retain its magnetic properties.

retina — The photosensitive portion of the eye, made up of cells called rods and cones, that is connected to the optic nerve.

retirement schedule — A list of limited life parts and the times when they must be replaced. This list records the part, serial number, time installed, and removal time.

retort — A laboratory vessel with a long tube used to distill substances by heat.

retract — To pull in or draw back.

retractable gear — This is a pilot controllable landing gear system, whereby the gear can be stowed alongside or inside the structure of the airplane during flight.

retractable landing gear — Landing gear that folds into the aircraft structure to reduce parasite drag.

retraction test — The portion of an aircraft inspection in which the airplane is put on jacks and the landing gear cycled through its retraction and extension sequences.

retread — In tire recapping, a tire that has been renewed to serviceable condition. Tires that meet injury limitations can be recapped. Retreading or recapping means reconditioning of a tire by renewing the tread, or renewing the tread plus one or both sidewalls.

retreating blade — In rotorcraft, any blade, located in a semicircular part of the rotor disc, where the blade direction is opposite to the direction of flight.

retreating blade stall — In rotorcraft, a stall that begins at or near the tip of a blade in a helicopter because of the high angles of attack required to compensate for dissymmetry of lift. In a gyroplane the stall occurs at 20 to 40 percent outboard from the hub.

retrofit — To furnish something with new or modified parts or equipment not available or considered necessary at the time the part was manufactured.

return flow — The upper branch of a thermal circulation.

return stroke — Lightning that is visible to the eye and marks the path of the positive charge of the step leader back into the clouds.

return to service — The completion of all applicable maintenance records and forms after maintenance has been performed on an aircraft that will allow the aircraft to be legally flown.

revalidate — To confirm, sanction or make something valid again.

reverse bias — The polarity relationship between a power supply and a semiconductor that does not allow conduction of current.

reverse idle — A power lever position where the thrust reversers are deployed, but engine power is idle.

reverse pitch — An angle to which the propeller blade can be turned in order to provide reverse thrust from the propeller.

reverse polarity welding — In electric arc welding, reverse polarity occurs when the electrode is positive and the work is negative. The opposite polarity is called straight polarity. Straight polarity puts more heat into the work piece, while reverse polarity minimizes the heat in the work piece.

reverse question — A response to a question. Rather than give a direct answer to the student's query, the instructor can redirect the question to another student to provide the answer.

reverse riveting — A process of driving aircraft rivets in which the manufactured head is bucked by holding it in a rivet set supported in a bucking bar and upsetting the shank with a flush rivet set.

reverse-current relay — A relay incorporated into a generator circuit to disconnect the generator from the battery when the battery voltage is greater than generator voltage.

reverse-flow annular combustor — A combustor design that forms an S-shaped path in which the gases flow from the diffuser to the exhaust. This design shortens the entire engine length because the liner is coaxial to the turbines rather than in front of them as in a conventional annular combustor.

reversible-pitch propeller — A propeller system with a pitch change mechanism that includes full reversing capability. When the pilot moves the throttle controls to reverse, the blade angle changes to a pitch angle and produces a reverse thrust, which slows the airplane down during a landing.

reversing mechanism — A linkage that reverses the direction of movement between two parts.

review and evaluation — The fourth and last step in the teaching process, which consists of a review of all material and an evaluation of the students. In the telling-and-doing technique of flight instruction, this step consists of the instructor evaluating the student's performance while the student performs the required procedure.

revision — In mechanical drawings, a change in dimensions, design, or materials.

revision block — That portion of an aircraft drawing that contains a record of all of the revisions. Symbols are used to indicate changes and their locations.

revolutions per minute (RPM) — The number of complete revolutions a body makes in one minute.

revolved section — A detail on an aircraft drawing in which the external view shows the shape of a part's cross section as though it were cut out and revolved.

Reynolds Number — A dimensionless ratio that relates how smoothly a fluid flows. At low velocities fluid flow is smooth or laminar. As the fluid flows more rapidly, it reaches a velocity, known as the critical velocity, at which the motion changes from laminar to turbulent, with the formation of eddy currents and vortices that disturb the flow.

RF alternating current — Radio frequency (RF) is a term that refers to alternating current (AC) that, when fed into an antenna, generates an electromagnetic field suitable for communications. These frequencies range from nine kilohertz (9 kHz) to thousands of gigahertz (GHz).

rheostat — A variable resistor having only two terminals. It is normally used in a circuit to drop voltage by dissipating some of the energy as heat.

rhetorical question — A question asked to stimulate group thought. Normally answered by the instructor, it is more commonly used in lecturing rather than in guided discussions.

rhomboid — A parallelogram in which the angles are oblique and the adjacent sides are unequal.

rhumb line — In navigation, a rhumb line crosses all lines of longitude at the same angle.

rib — The structural member of an airfoil that gives it the desired aerodynamic shape.

rib cap — A thin, narrow strip of material usually glued to wooden ribs or riveted to metal ribs to enlarge the contact surface for the attachment of skin to the wing or flight control surface.

rib lacing — The attachment of fabric to an aircraft structure with rib stitching cord. A series of loops around the structure and through the fabric are secured with seine knots. Also referred to as rib stitching.

rib stitching — The attachment of fabric to an aircraft structure with rib stitching cord. A series of loops around the structure and through the fabric are secured with seine knots. Also referred to as rib lacing.

rib stitching cord — A strong cotton, linen, or polyester fiber cord used to stitch or lace fabric to an aircraft structure.

ribbon direction — In composites, on a honeycomb core, the way the honeycomb can be separated. The direction of one continuous ribbon.

ribbon parachute — A parachute that consists of strips or ribbons of material instead of a solid fabric. This type of chute is used in instances where it is desirable to have less opening shock, since air escapes between the ribbons.

rich flameout — A condition of turbine engine operation in which the fire goes out in the engine because the fuel-air mixture is too rich to support combustion.

rich solvent — A slow-drying solvent.

ridge/ridge line — In meteorology, an elongated area of relatively high atmospheric pressure; usually associated with and most clearly identified as an area of maximum anticyclonic curvature of the wind flow (isobars, contours, or streamlines).

rigging — The final adjustment and alignment of an aircraft and its flight control system that provide the proper aerodynamic characteristics.

rigging fixture — A template designed to measure control surface travel.

rigging pins — The pins that can be inserted into control system components to hold the controls in their neutral position for rigging the control cables and rods.

right angle — A 90° angle. Formed by the perpendicular intersection of two straight lines.

right brain — A concept that each hemisphere of the brain processes information differently. Students with right-brain dominance are characterized as being spatially oriented, creative, intuitive, and emotional. They can be very good with art or music and can easily put together the big picture.

right-hand rule for the direction of magnetic flux — A method of determining the relation of the flow of magnetic flux to the direction of current flow in a conductor. The right hand is wrapped around the conductor with the thumb pointing in the direction of current flow (positive to negative), and the fingers will point in the direction of the flow of magnetic flux from the north pole to the south pole.

right-hand rule of electric generators — A method of determining the relative directions of current flow, magnetic flux, and motion of a conductor when the conductor is passing through a magnetic field. The right hand should be held with the thumb pointing vertically, the first finger pointing away, and the second finger pointing to the left (the thumb and two fingers will be at right angles to each other). The thumb points in the direction the conductor is moving. The first finger points in the direction of the lines of magnetic flux (north pole to south pole), and the second finger points in the direction of current flow (positive to negative) within the conductor.

right-hand thread — A thread that, when viewed axially, winds in a clockwise and receding direction. All threads are right-hand threads unless otherwise designated.

rigid airship — A dirigible that has a rigid framework covered with cloth fabric and a cabin suspended underneath the frame that houses the crew and engines for driving it through the air.

rigid rotor — In rotorcraft, a rotor system permitting blades to feather but not flap or hunt.

rigid tubing — Fluid lines made of thin wall aluminum alloy, copper, or stainless steel used in an aircraft when there is no relative movement between the ends of the tube.

rigidity — The state of being rigid.

rigidity in space — The principle that a wheel with a heavily weighted rim spun rapidly will remain in a fixed position in the plane in which it is spinning.

rime ice — The rough textured ice that forms on a surface, such as the leading edge of an airplanes wing, when flown through supercooled fog or certain types of clouds.

rime icing — The formation of a white or milky and opaque granular deposit of ice formed by the rapid freezing of supercooled water droplets as they impinge upon an exposed aircraft.

ring and tube assembly — The ring of outer combustion chambers on a multiple-can-type combustor engine.

ring cowl — A streamlined covering over the cylinders of a radial engine.

ring gear — One of the gears in a turboprop negative torque signal and prop reduction gear system. When a predetermined negative torque is applied to the reduction gearbox, the ring gear moves forward against the spring force due to a torque reaction generated by helical splines. In moving forward, the ring gear pushes two operating rods through the reduction gear nose.

ring grooves — The grooves in the circumference of a piston into which the piston rings fit.

ringworms — A circular pattern of cracks in a brittle dope finish that results from a blunt object pressing against the fabric.

rip panel — A free floating balloon's air dump panel that can be opened by pulling a rip-cord. Air is dumped for landings or in the event of an emergency.

ripple — A small periodic variation in the voltage level of a direct current power supply.

ripple filter — A filter consisting of capacitors or inductors, or both, that attenuates the AC component in the output of a rectifier circuit. The ripple is the amount of AC that remains on the DC.

ripple frequency — The frequency of the ripple in DC voltage when rectified from an AC current. A full-wave rectifier produces a ripple frequency of twice the input AC frequency while a half-wave rectifier produces a ripple frequency that is the same as the input frequency.

ripsaw — A saw with course teeth used for cutting lumber along the direction of its grain.

risk elements — How to assess risk is a skill that students need to learn to make effective decisions. When students are faced with making a decision regarding a flight, ask them to evaluate the status of the four risk elements: the pilot in command, the aircraft, the environment, and the operation.

risk elements in ADM — In aeronautical decision making the four elements considered are the pilot, aircraft, environment, and type of operation.

risk management — Part of the decision making process. Situational awareness, problem recognition, and good judgment are used to reduce risks.

rivet — A small metal pin with a specially formed head on one end used to fasten sheet metal parts together by upsetting the shank to form a clamping head.

rivet cutter — A tool used to cut rivets to the required length.

rivet gauge — The transverse pitch or distance between rows of rivets.

rivet gun — A hand-held pneumatic riveting hammer used to vibrate aircraft rivets against a heavy bucking bar to form the upset head.

rivet pitch — The distance between the center of rivet holes in adjacent rows. Also referred to as rivet spacing.

rivet set — A tool that fits into a rivet gun used to hammer against the manufactured head of a rivet so the bucking bar can form the upset head on the opposite side of the skin. Also referred to as a rivet snap.

rivet spacing — The distance between the center of rivet holes in adjacent rows. Also referred to as rivet pitch.

rivet squeezer — A heavy, tong-like clamping machine used to squeeze the ends of a rivet to form an upset head.

riveting burr — A small plain washer placed on a rivet before the rivet is upset. Provides a larger area of contact on the part than would have been provided with the rivet alone.

Rivnut — A patented, hollow, blind rivet manufactured by the B.F Goodrich Company, in which the inside of the shank is threaded. The upset rivet can be used as a blind nut.

RNAV approach — An instrument approach procedure that relies on aircraft area navigation equipment for navigational guidance.

RNAV way point (W/P) — A predetermined geographical position used for route or instrument approach definition or progress reporting purposes that is defined relative to a VORTAC station position.

road reconnaissance — Military activity requiring navigation along roads, railroads, and rivers. Reconnaissance route/route segments are seldom along a straight line and normally require a lateral route width of 10NM to 30NM and an altitude range of 500 feet to 10,000 feet AGL.

robot — An automated mechanism built to do functions ordinarily assigned to humans.

robotics — The research and development of technology that deals with the design, application, construction, and maintenance of automated mechanisms.

rock wool — A mineral wool made by blowing a jet of steam through molten rock and used chiefly for heat and sound insulation

rocker arm — A pivoted arm, as in an aircraft reciprocating engine, used to transfer pushrod motion to a valve stem.

rocker arm boss — That portion of the cylinder head of an aircraft engine that provides support for the rocker arm shaft.

rocket — A device propelled by the high-velocity ejection of gases produced by internal ignition of solid or liquid fuels.

rocket assisted takeoff — An auxiliary means of assisting an aircraft when taking off. The system consists of small rockets fastened to the aft structure. When a heavily loaded aircraft rotates for takeoff, the jets are fired. The added boost provides extra thrust needed for takeoff. Once airborne the rockets are jettisoned. Often abbreviated as JATO.

rocket fuel — Any fuel specifically developed for rocket engines.

rocket ship — Any spacecraft that uses rocket engines for propulsion.

rocketry — The study, theory, and development of rockets.

rocking shaft — A shaft or rod in an instrument that oscillates or rocks upon its bearings, but does not revolve.

Rockwell hardness tester — A machine used to determine the hardness of a material by using a calibrated weight to press either a diamond pyramid or a hardened steel ball into the material. A dial indicator measures the depth of penetration and specifies it with a Rockwell number.

rods — The cells concentrated on the outside of the foveal area of the eye that are sensitive to low light and not to color.

roger — During communications, this indicates: I have received all of your last transmission. "Roger" should not be used to answer a question requiring a yes or no answer.

role model — A person whose behavior in a particular role is imitated by others.

roll — The motion of an aircraft about its longitudinal axis. This motion is controlled by the ailerons.

roll cloud — The dense and horizontal cloud band occasionally found parallel to gust fronts. Also used to describe the rotor clouds associated with mountain lee waves.

roll pin — A pressed-fit pin made of a roll of spring steel. The spring force tending to unroll the pin holds it tight in the hole.

roll threading — Applying a thread to a bolt or screw by rolling the piece between two grooved die plates, one of which is in motion, or between rotating grooved circular rolls.

roller bearing — An antifriction bearing with hardened steel rollers between two hardened steel races.

rollout RVR — The RVR (Runway Visual Range) readout values obtained from RVR equipment located nearest the rollout end of the runway.

root — The supporting base or structure, as in the wing root, connected to the fuselage.

root mean square (RMS) — The value of direct current equivalent to an alternating current sine wave. The value is 0.707 times the peak value of one alternation.

rosette weld — A weld made through a small hole in a piece of steel tubing to weld an inner tube to the outer tube to prevent relative movement.

rosin — A light yellow-colored resin. Rosin remains after oil of turpentine has been distilled from the oleoresin of pine trees.

rosin core solder — A soft solder made chiefly of tin and lead alloys used primarily for bonding copper, brass, and coated iron. Rosin core solder is a hollow wire filled with rosin. During the soldering process the rosin serves to clean corrosion from the components being soldered in order to make a sound bond.

rosin joint — A soldered electrical connection in which the rosin, not the solder, holds the connection. Rosin joints are not considered to be airworthy connections.

rotary breather — A rotating set of vanes or a centrifuge device through which oil laden air from the vent subsystem passes. Deaeration takes place and the oil is returned to the sump while the air exits back to the atmosphere.

rotary pick-off — A device attached to a rotating object that generates a signal proportional to the amount of rotation.

rotary radial engine — An aircraft engine popular in World War I, in which the propeller was attached to the crankcase and the pistons were attached to a crankshaft mounted on the airframe. When the engine ran, the cylinders, crankcase, and propeller all rotated and the crankshaft remained stationary.

rotary solenoid — An electromagnet whose movable core is rotated by current through the coil.

rotary switch — A switch that consists of one or more circular wafers, each with multiple contacts. A central shaft rotates a wiper from contact to contact, switching connection from one circuit to another. When multiple wafers are stacked on the shaft, multiple circuits can be selected simultaneously.

rotating wing — The rotors of a helicopter.

rotation — The act of turning about an axis.

rotational velocity — In rotorcraft, the component of relative wind produced by the rotation of the rotor blades.

rote learning — A basic level of learning where the student has the ability to repeat back something learned, with no understanding or ability to apply what was learned.

rotor — 1. The rotating element in an alternator. It is excited by direct current, and the interlacing fingers on the faces of the rotor form the alternating north and south poles.  2. The rotating blades of a helicopter. 3. The portion of a turbine compressor that spins. 4. Either compressor or turbine. A rotating disk or drum to which a series of blades is attached. 5. In meteorology, a turbulent circulation under mountain-wave crests, to the lee and parallel to the mountains creating the wave. Glider pilots use the term rotor to describe any low-level turbulent flow associated with mountain waves.

rotor brake — A device used to stop the rotor blades of a rotorcraft on shutdown. This can be either a hydraulic or mechanical mechanism.

rotor disc — The area within the tip path plane of a helicopter's rotor. Also referred to as the disc area.

rotor disc area — In rotorcraft, the area swept by the blades of the rotor. It is a circle with its center at the hub and has a radius of one blade length.

rotor force — The force produced by the rotor in a gyroplane. It is comprised of rotor lift and rotor drag.

rotor streaming — In meteorology, a phenomenon that occurs when the air flow at mountain levels can be sufficient for wave formation, but begins to decrease with altitude above the mountain. In this case, the air downstream of the mountain breaks up and becomes turbulent, similar to rotor, with no lee waves above.

rotorcraft — A heavier-than-air aircraft that depends principally for its support in flight on the lift generated by one or more rotors.

rotorcraft-load combination — The combination of a rotorcraft and an external-load, including the external-load attaching means.

rotorcraft-load combinations — Class A, Class B, Class C, and Class D rotorcraft-load combinations, as follows:

Class A— one in which the external load cannot move freely, cannot be jettisoned, and does not extend below the landing gear.

Class B— one in which the external load is jettisonable and is lifted free of land or water during the rotorcraft operation.

Class C— one in which the external load is jettisonable and remains in contact with land or water during the rotorcraft operation.

Class D— one in which the external-load is other than a Class A, B, or C and has been specifically approved by the Administrator for that operation.

round file — A steel tool that has a circular cross section with hardened ridged surfaces, used for smoothing, grinding, or boring. Also referred to as a rattail file.

round head — A fastener with a semi-elliptical top surface and a flat bearing surface.

round off — 1. To make an edge round as by filing. 2. In mathematics, to change a fraction to the closest whole number.

round-nose pliers — Pliers with round jaws. Used to form loops in wire and thin strips of sheet metal.

route — A defined path, consisting of one or more courses in a horizontal plane, which aircraft traverse over the surface of the Earth.

route minimum off-route altitude (route MORA) — This is an altitude derived by Jeppesen. The Route MORA altitude provides reference point clearance within 10 NM of the route centerline (regardless of the route width) and end fixes. Route MORA values clear all reference points by 1000 feet in areas where the highest reference points are 5000 feet MSL or lower. Route MORA values clear all reference points by 2000 feet in areas where the highest reference points are 5001 feet MSL or higher. When a Route MORA is shown along a route as "unknown" it is due to incomplete or insufficient information.

route segment — 1. A part of a route. Each end of that part is identified by: a. A continental or insular geographical location; or b. A point at which a definite radio fix can be established. 2. As used in Air Traffic Control, a part of a route that can be defined by two navigational fixes, two NAVAIDs, or a fix and a NAVAID.

route segment [ICAO] — A portion of a route to be flown, as defined by two consecutive significant points specified in a flight plan.

router — An electric machine that uses a high-speed, rotary cutting tool. In aircraft maintenance, routers are often used to remove damaged honeycomb core from a bonded aircraft structure.

routine service items — The items in a progressive inspection that consists of a thorough check of the aircraft, engines, appliances, components, and systems normally without disassembly.

roving — A bundle of filaments that are twisted together for weaving into fabric.

rubber cement — An adhesive made from rubber that has not gone through the vulcanization process but has been dissolved in solvent.

rudder — The vertically hinged control surface used to effect horizontal changes in course or rotate an aircraft about its vertical axis. The pilot operates the rudder by the movement of the foot pedals in the cockpit.

rudder pedals — The foot-operated controls in an airplane that move the rudder.

ruddervator — A pair of control surfaces on the tail of an aircraft arranged in the form of a "V." These surfaces, when moved in the same direction vertically by the control wheel, serve as elevators. When moved in the same direction horizontally by the rudder pedals, they serve as a rudder.

rule — A straightedge usually with etched, calibrated, dimensional markings.

rumble — A combustor noise caused by choking and unchoking of the turbine nozzle. Caused by improper fuel scheduling.

run — 1. A defect in a painted surface caused by too much finish being applied. The material gathers and attempts to flow off the surface. 2. To set something in motion, such as running an engine.

run-in — The period of time an aircraft engine is operated to seat the moving parts after an overhaul.

running fit — A fit between moving parts that allows them to move freely.

runout — A term frequently used interchangeable with eccentricity but which normally refers to the amount which the outside surface of one component moves with respect to the outside surface of another component. As such, it includes eccentricity, angularity and bow. The amount of runout is usually expressed in terms of Total Indicator Reading (TIR).

runout check — A dial indicator check for measuring the plane of rotation of a rotor shaft or disc.

runway — A defined rectangular area on a land airport prepared for the landing and takeoff run of aircraft along its length. Runways are normally numbered in relation to their magnetic direction rounded off to the nearest 10 degrees; e.g., Runway 1 (010°), and Runway 25 (250°).

runway centerline lighting — Flush centerline lights spaced at 50-foot intervals beginning 75 feet from the landing threshold and extending to within 75 feet of the opposite end of the runway.

runway condition reading — Numerical decelerometer readings relayed by air traffic controllers at USAF and certain civil bases for use by the pilot in determining runway braking action. These readings are routinely relayed only to USAF and Air National Guard Aircraft.

runway edge lights (ICAO) — Are provided for a runway intended for use at night or for a precision approach runway intended for use by day or night. Runway edge lights shall be fixed lights showing variable white, except that:
a. in the case of a displaced threshold, the lights between the beginning of the runway and the displaced threshold shall show red in the approach direction; and
b. a section of the lights 600m or one-third of the runway length, whichever is the less, at the remote end of the runway from the end at which the take-off run is started, may show yellow.

runway edge lights (USA) — Lights used to outline the edges of runways during periods of darkness or restricted visibility conditions. The light systems are classified according to the intensity or brightness they are capable of producing: they are the High Intensity Runway Lights (HIRL), Medium Intensity Runway Lights (MIRL), and the Low Intensity Runway Lights (RL). The HIRL and MIRL systems have variable intensity controls, where the RLs normally have one intensity setting.
a. The runway edge lights are white, except on instrument runways amber replaces white on the last 2,000 feet or half of the runway length, whichever is less, to form a caution zone for landings.
b. The lights marking the ends of the runway emit red light toward the runway to indicate the end of runway to a departing aircraft and emit green outward from the runway end to indicate the threshold to landing aircraft.

runway end identifier lights — Two synchronized flashing lights, one on each side of the runway threshold, which provide rapid and positive identification of the approach end of a particular runway.

runway gradient — The average slope, measured in percent, between two ends or points on a runway.

runway heading — The magnetic direction that corresponds with the extended runway centerline, not the painted runway number.

runway heading — The magnetic direction that corresponds with the runway centerline extended, not the painted runway number. When cleared to "fly or maintain runway heading," pilots are expected to fly or maintain the heading that corresponds with the extended centerline of the departure runway. Drift correction shall not be applied; e.g., if the actual magnetic heading of the runway centerline of Runway 4 is 044°, a heading of 044° should be flown..

runway in use/active runway/duty runway — Any runway or runways currently being used for takeoff or landing. When multiple runways are used, they are all considered active runways. In the metering sense, a selectable adapted item that specifies the landing runway configuration or direction of traffic flow. The adapted optimum flight plan from each transition fix to the vertex is determined by the runway configuration for arrival metering processing purposes.

runway incursion — Any occurrence at an airport involving an aircraft, vehicle, person, or object on the ground that creates a collision hazard or results in loss of separation with an aircraft taking off, intending to takeoff, landing, or intending to land.

runway lights — Lights having a prescribed angle of emission used to define the lateral limits of a runway. Runway lights are uniformly spaced at intervals of approximately 200 feet, and the intensity can be controlled or preset.

runway markings — **1.** Basic marking — Markings on runways used for operations under visual flight rules consisting of centerline markings and runway direction numbers and, if required, letters. **2.** Instrument marking — Markings on runways served by nonvisual navigation aids and intended for landings under instrument weather conditions, consisting of basic marking plus threshold markings. **3.** All-weather (precision instrument) marking — Marking on runways served by nonvisual precision approach aids and on runways having special operational requirements, consisting of instrument markings plus landing zone markings and side strips.

runway overrun — In military aviation exclusively, a stabilized or paved area beyond the end of a runway, of the same width as the runway plus shoulders, centered on the extended runway centerline.

runway profile descent — An instrument flight rules (IFR) air traffic control arrival procedure to a runway published for pilot use in graphic and/or textual form and may be associated with a STAR. Runway Profile Descents provide routing and may depict crossing altitudes, speed restrictions, and headings to be flown from the enroute structure to the point where the pilot will receive clearance for and execute an instrument approach procedure. A Runway Profile Descent can apply to more than one runway if so stated on the chart.

runway safety area — A defined surface surrounding the runway prepared, or suitable, for reducing the risk of damage to airplanes in the event of an undershoot, overshoot, or excursion from the runway. The dimensions of the RSA vary and can be determined by using the criteria contained within Advisory Circular 150/5300-13, Chapter 3. The design standards dictate that the RSA shall be:

a. Cleared, graded, and have no potentially hazardous ruts, humps, depressions, or other surface variations;
b. Drained by grading or storm sewers to prevent water accumulation;
c. Capable, under dry conditions, of supporting snow removal equipment, aircraft rescue and fire fighting equipment, and the occasional passage of aircraft without causing structural damage to the aircraft; and,
d. Free of objects, except for objects that need to be located in the runway safety area because of their function. These objects shall be constructed on low impact resistant supports (frangible mounted structures) to the lowest practical height with the frangible point no higher than 3 inches above grade.

runway temperature — The air temperature immediately above a runway at approximately wing level.

runway use program — A noise abatement runway selection plan designed to enhance noise abatement efforts with regard to airport communities for arriving and departing aircraft. These plans are developed into runway use programs and apply to all turbojet aircraft 12,500 pounds or heavier; turbojet aircraft less than 12,500 pounds are included only if the airport proprietor determines that the aircraft creates a noise problem. Runway use programs are coordinated with FAA offices, and safety criteria used in these programs are developed by the Office of Flight Operations. Runway use programs are administered by the Air Traffic Service as "Formal" or "Informal" programs.

a. Formal Runway Use Program — An approved noise abatement program defined and acknowledged in a Letter of Understanding between Flight Operations, Air Traffic Service, the airport proprietor, and the users. Once established, participation in the program is mandatory for aircraft operators and pilots as provided for in FAR 91.129.
b. Informal Runway Use Program — An approved noise abatement program that does not require a Letter of Understanding, and participation in the program is voluntary for aircraft operators/pilots.

runway visibility value (RVV) — The visibility determined for a particular runway by a transmissometer. A meter provides a continuous indication of the visibility (reported in miles or fractions of miles) for the runway. RVV is used in lieu of prevailing visibility in determining minimums for a particular runway.

runway visual range (RVR) — An instrumentally derived value, based on standard calibrations, that represents the horizontal distance a pilot will see down the runway from the approach end. It is based on the sighting of either high intensity runway lights or on the visual contrast of other targets whichever yields the greater visual range. RVR, in contrast to prevailing or runway visibility, is based on what a pilot in a moving aircraft should see looking down the runway. RVR is horizontal visual range, not slant visual range. It is based on the measurement of a transmissometer made near the touchdown point of the instrument runway and is reported in hundreds of feet. RVR is used in lieu of RVV and/or prevailing visibility in determining minimums for a particular runway.

a. Touchdown RVR — The RVR visibility readout values obtained from RVR equipment serving the runway touchdown zone.
b. Mid RVR — The RVR readout values obtained from RVR equipment located midfield of the runway.
c. Rollout RVR — The RVR readout values obtained from RVR equipment located nearest the rollout end of the runway.

runway visual range (RVR) — An instrumentally derived horizontal distance a pilot should see down the runway from the approach end; based on the sighting of high intensity runway lights.

rust — A reddish brown, crusty coating of hydrated ferric oxide that forms on iron and iron-containing materials when the material is exposed to moist air.

S

S.A.E. — Society of Automotive Engineers.

saber saw — A hand-held, electrically operated jigsaw that uses a short, stiff blade. It operates using reciprocating motion.

sacrificial corrosion — A method of corrosion protection in which a surface is plated with a metal less noble than itself. Any corrosion will attack the plating rather than the base metal.

saddle-mount oil tank — An externally mounted sheet metal tank with a contour, which allows the tank to mount around the curvature of a turbine engine compressor case.

safety alert — A safety alert issued by ATC to aircraft under their control if ATC is aware the aircraft is at an altitude that, in the controller's judgment, places the aircraft in unsafe proximity to terrain, obstructions, or other aircraft. The controller can discontinue the issuance of further alerts if the pilot advises he is taking action to correct the situation or has the other aircraft in sight.
  a. Terrain/Obstruction Alert — A safety alert issued by ATC to aircraft under their control if ATC is aware the aircraft is at an altitude that, in the controller's judgment, places the aircraft in unsafe proximity to terrain/obstructions; e.g., "Low Altitude Alert, check your altitude immediately."
  b. Aircraft Conflict Alert — A safety alert issued by ATC to aircraft under their control if ATC is aware of an aircraft that is not under their control at an altitude that, in the controller's judgment, places both aircraft in unsafe proximity to each other. With the alert, ATC will offer the pilot an alternate course of action when feasible; e.g., "Traffic Alert, advise you turn right heading zero niner zero or climb to eight thousand immediately."
The issuance of a safety alert is contingent upon the capability of the controller to have an awareness of an unsafe condition. The course of action provided will be predicated on other traffic under ATC control. Once the alert is issued, it is solely the pilot's prerogative to determine what course of action, if any, he will take.

safety belt — A belt designed specifically to fasten a person to an object such as an aircraft seat to prevent injury or falling.

safety factor — A structural design feature. Safety factor is the ratio of the maximum load a structural member is designed to support compared to the maximum probable load to which it will be subjected.

safety gap — A specifically designed space in a high-tension magneto that allows a spark to jump without damaging the magneto's internal parts. This can occur in the event that the spark plug lead is disconnected from the spark plug.

safety glass — A glass that does not leave potentially dangerous jagged fragments when it is broken. Some safety glass is laminated with a tough plastic membrane between layers of glass, while others consist of tempered glass.

safety needs — A level of Maslow's Hierarchy of human needs, which includes protection against danger, threats, and deprivation.

safety program manager — An FAA employee who designs, implements, and evaluates the Aviation Safety Program within the FAA Flight Standards District Office (FSDO) area of responsibility.

safety valve — A relief valve that opens to relieve pressure in a container if the pressure rises above a predetermined value.

safety wire — Soft wire made of galvanized low-carbon steel, annealed stainless steel, or brass, used to prevent nuts and bolts from vibrating loose.

safety wiring — A method of fastening bolts or screw-heads together with soft wire to help prevent loosening.

safetying — Installing a device in order to help prevent an attachment from loosening.

sail back — A maneuver during high wind conditions (usually with power off) where float plane movement is controlled by water rudders, opening and closing cabin doors, etc.

sailplane — A high performance glider.

sal ammoniac — Ammonium chloride.

salient pole — The field pole of a motor or generator that extends out from the field frame toward the armature.

salt — The result of the combination of an alkali with an acid. Salts are generally porous and powdery in appearance and are the visible evidence of corrosion in a metal.

samarium — A metallic element with a symbol of Sm and an atomic number of 62.

same direction aircraft — Aircraft are operating in the same direction when:
  a. They are following the same track in the same direction; or
  b. Their tracks are parallel and the aircraft are flying in the same direction; or
  c. Their tracks intersect at an angle of less than 45 degrees.

sand casting — A casting in which the mold is made by shaping special casting sand around a pattern. The pattern is removed, and molten metal is poured into the mold. When the molten metal has hardened, the mold is broken away.

sandbag — A heavy canvas or leather bag filled with coarse sand. It is used as a surface to form sheet metal by the bumping process. A bag filled with lead shot is referred to as a shot bag.

sanding coat — A coat of surfacer or heavy bodied material that is applied to a material and then sanded off to fill small surface imperfections resulting in a smoother surface for subsequent coats.

sandpaper — An abrasive paper made by bonding grains of sand to the surface.

sandwich construction — A bonded structure in which a core of material is inserted between two face sheets of metal or fiberglass cloth. Core materials include metallic or plastic honeycomb and end-grain balsa wood. Sandwich construction is used when high strength, light weight materials are required.

sandwich structure — In composites, a thick, low density, core (Usually foam or honeycomb) between thin faces of high strength material.

satellite — A man-made object that rotates around the Earth. A small object that rotates around a larger object.

saturate — A material that is thoroughly soaked, or penetrated, to the extent that it cannot absorb any more liquid.

saturated adiabatic lapse rate (SALR) — The rate of decrease of temperature with height as saturated air is lifted with no gain or loss of heat from outside sources. Unlike the dry adiabatic lapse rate (DALR), the SALR is not a constant numerical value but varies with temperature, being greatest at low temperatures.

saturated adiabatic process — The rate at which saturated air cools as it ascends. It is less than the dry adiabatic lapse rate because adiabatic cooling is offset partially by the release of latent heat. The difference is larger at higher temperatures.

saturated air — Air containing the maximum amount of water vapor it can hold at a given temperature (100% relative humidity).

saturated vapor — The vapor state above a liquid in which no further vaporization can take place without an increase in temperature.

saturated vapor pressure — The partial pressure of water vapor at saturation.

saturation — A state of equilibrium where the same amount of molecules are leaving a water surface as are returning and the vapor pressures are balanced.

saturation current — In electronics, the condition where the maximum amount of current able to flow in a transistor or vacuum tube has been reached. The application of additional current to the base or gate has no effect on the output.

sawtooth wave — The waveform produced by a relaxation oscillator in which the voltage rises slowly and drops rapidly.

say again — Used to request a repeat of the last transmission. Usually specifies transmission or portion thereof not understood or received; e.g., "Say again all after ABRAM VOR."

say altitude — Used by ATC to ascertain an aircraft's specific altitude/flight level. When the aircraft is climbing or descending, the pilot should state the indicated altitude rounded to the nearest 100 feet.

say heading — Used by ATC to request an aircraft heading. The pilot should state the actual heading of the aircraft.

Saybolt Seconds Universal (SSU) Viscosimeter — A device that measures the viscosity of lubricating oils by giving the time in seconds it takes for 60 cu. cm of oil to flow through its calibrating orifice. Aviation 80 engine oil has an SSU viscosity of 79.2, and Aviation 100 oil has an SSU viscosity of 103.0.

S-band radar — Radar frequency range between 1,550 and 5,200 MHz.

scalar — A variable that only has magnitude such as temperature and pressure compared to vector.

scale — 1. A graduated measure. 2. An oxide of iron sometimes formed on the surfaces of hot forged fasteners.

scale effect — The change in any force coefficient such as the drag coefficient due to a change in the value of Reynold's number.

scale model — A smaller version of the original model made in the same proportions.

scalene triangle — A three-sided form in which all sides and angles are unequal.

scales of circulations — The typical horizontal dimension size and lifetime of an individual circulation. See also macroscale, mesoscale, and microscale.

scarf joint — A joint used for the construction or repair of a wooden aircraft or component. The two parts to be joined are cut with a shallow taper and glued together.

scarf patch — A flush repair to plywood skins where the slope must be shallower than 1 in 12, or about 12 times the thickness of the plywood.

scavenge — To remove an undesirable material from an area.

scavenger pump — A constant displacement pump in an engine that picks up oil after it has passed through the engine and returns it to the oil reservoir.

scheduled maintenance — The inspection and replacement of components that are planned in advance on a day, month, or operating hour basis.

schematic diagram — A graphical presentation used to explain the operation of a system without showing its mechanical details or physical layout. Also referred to as a schematic drawing.

schematic drawing — A graphical presentation or diagram used to explain the operation of a system without showing its mechanical details or physical layout. Also referred to as a schematic diagram.

scientific notation — In mathematics, a form of shorthand. Numbers are displayed by using powers of ten to indicate very large or very small numbers. For example, 4,000,000 would be written as 4×10^6 and 0.000004 would be written as 4×10^{-6}.

Scintilla magneto — A Swiss-designed and built magneto. It is the forerunner of the current Bendix magneto.

scoop — An air inlet that projects beyond the immediate surface of an aircraft structure.

scope — 1. A contraction of oscilloscope. 2. Extent and detail of an inspection or repair.

score — A deep scratch mark or line made across a piece of material making it possible to break the material along the line.

scoring — Deep scratches on the surface of a material caused by foreign particles between moving parts.

scraper ring — The bottom ring on a piston whose function is to scrape the lubricating oil away from the cylinder wall. This prevents oil from getting into the combustion chamber of the cylinder.

scratching surface — Narrow, sharp, shallow markings or lines resulting from the movement of a particle or sharp pointed object across a surface. The most common cause is carelessness in handling and any part can be so affected. Note: To distinguish it from scoring, scratching is not considered to be caused by engine operation.

screech liner — A perforated liner within an afterburner, designed to combat destructive vibrations that cause metal fatigue and noise emissions.

screeching — A shrill, high-pitch noise that comes from a gas turbine engine caused by the instability of combustion in the engine.

screeding tool — A tool used to smooth out or level plastic resins used in bonded structure manufacturing or repair.

screen — A frame covered with a netting or mesh material used to cover an opening to prevent the entry of foreign objects.

screen grid — The electrode in a four-element vacuum tube used to minimize the interelectrode capacitance between the plate and the control grid.

screen-type filter — A fluid filter with an element that consists of a wire screen.

screw dowel — A dowel pin provided with a straight or tapered thread for threading into a material.

screw pitch gauge — A gauge with a series of V-notches cut along one edge. Used to check the number of threads, or pitch, of a screw or bolt.

screwdriver — A hand tool used for turning screws.

scriber — A hardened steel or carbide-tipped, pointed tool used to scratch lines on metal for cutting.

scribing — The process of marking a line on metal with the use of a scriber.

scroll combustor — Used widely on auxiliary power units in conjunction with a radial inflow turbine, a scroll combustor fits around the turbine nozzle that has vane openings at its inner perimeter. The vanes direct the gases inward onto the turbine.

scroll shear — A floor- or bench-mounted sheet metal cutting tool used to cut irregular lines in a piece of sheet metal without having to cut to the edge of the sheet.

scud — Small patches of low clouds that usually form below heavier overcast clouds.

scuffing surface — A dulling or moderate wear of a surface resulting from a slight amount of rubbing. Usually caused by improper clearance and insufficient lubrication. Parts affected are rollers, rings, and steel parts bolted together.

scupper — A recess around the filler neck of a fuel tank. It collects any fuel spilled during the fueling operation and drains it overboard rather than allowing it to enter the aircraft structure.

sea anchor — An apparatus towed in the water to retard a ship's drift and to keep the vessel facing into the wind.

sea breeze — A coastal breeze blowing from sea to land. It occurs in the daytime when the land surface is warmer than the sea surface.

sea breeze front — The boundary between the cool, inflowing marine air in the sea breeze and the warmer air over land.

sea lane — A designated portion of water outlined by visual surface markers for and intended to be used by aircraft designed to operate on water.

sea level — A reference height used to determine standard atmospheric conditions and altitude measurements.

sea level engine — A reciprocating aircraft engine having a rated takeoff power that is producible only at sea level.

sea level pressure — The atmospheric pressure at mean sea level.

sea smoke — Steam fog, evaporation fog.

seal — A component or material used to prevent fluid leakage between two surfaces.

sealant — A material used to form a seal between surfaces. Sealants differ from gaskets in that they are usually liquid or semi-solid.

sealed compartments — The compartments in an aircraft structure that are sealed off and used as fuel tanks.

seam welding — A method of electrical resistance welding that forms a continuous line of weld instead of individual spots. See also spot welding.

seaplane — An airplane designed to land and take off from the surface of water.

search and rescue — A service that seeks missing aircraft and assists those found to be in need of assistance. It is a cooperative effort using the facilities and services of available Federal, state and local agencies. The U.S. Coast Guard is responsible for coordination of search and rescue for the Maritime Region, and the U.S. Air Force is responsible for search and rescue for the Inland Region. Information pertinent to search and rescue should be passed through any air traffic facility or be transmitted directly to the Rescue Coordination Center by telephone.

search and rescue facility — A facility responsible for maintaining and operating a search and rescue (SAR) service to render aid to persons and property in distress. It is any SAR unit, station, NET, or other operational activity that can be usefully employed during an SAR Mission; e.g., a Civil Air Patrol Wing, or a Coast Guard Station.

search engine — Software that reviews a database to find and display information. Used to find information on the Internet.

seasoned lumber — Lumber that has been dried or had its moisture content reduced to a specified amount.

seat — A place or space on which something sits.

seat belt — A belt designed specifically to fasten a person to an object such as an aircraft seat to prevent injury or falling.

seated — A condition in which moving parts have worn together until they are fitted together and a minimum of leakage exists between them.

second in command — A pilot who is designated to be second in command of an aircraft during flight time.

secondary — Something at a second level of importance, rank or value.

secondary air — That portion of compressor output air used for cooling engine parts and combustion gases.

secondary airstream — The air that passes through the fan portion of a turbofan engine.

secondary cell — A storage cell or an electrical device in which electrical energy is converted into chemical energy and stored until needed then converted back into electrical energy. Secondary cells do not produce electricity; they merely store it.

secondary coil — The secondary winding of a transformer.

secondary control surfaces — Control surfaces, such as trim tabs, servo tabs, and spring tabs, that reduce the force required to actuate the primary controls.

secondary current — The current that flows in the secondary winding of a transformer. Secondary current is induced in the secondary winding of a coil by the collapse of the primary coil circuit current.

secondary emission — In electron tubes, the emission of electrons from a surface when struck by high-velocity electrons from the cathode.

secondary exhaust nozzle — On a turbofan, the cold exhaust fan nozzle. On an afterburner, the aft or outer exhaust nozzle. In this instance, it is made up of moveable flaps that change the geometry of the nozzle in different modes of engine operation.

secondary fuel — Refers to the duplex fuel nozzle and the fuel that flows at higher power settings from the secondary orifice. Also referred to as the main fuel.

secondary stall — A demonstration that a flight instructor shows a student pilot. It is normally caused by poor stall recovery technique, such as attempting to climb prior to attaining sufficient flying speed. A secondary stall can occur as a result of increasing angle of attack beyond the critical angle during recovery from a preceding stall.

secondary structure — In aircraft and aerospace applications, a structure that is not critical to flight safety.

secondary voltage — The voltage in a circuit produced across the secondary winding of a transformer.

secondary winding — Output winding of a transformer.

second-class lever — A lever commonly used to help in overcoming larger resistance with relatively small effort. The second-class lever has the fulcrum at one end while the effort is applied at the other end. The resistance is somewhere between these points. A wheelbarrow is an example of a second-class lever.

section line — The crosshatching used in a cutaway section of an aircraft drawing to identify a type of material.

Sectional Aeronautical Chart — Chart designed for visual navigation of slow or medium speed aircraft. Topographic information on these charts features the portrayal of relief, and a judicious selection of visual check points for VFR flight. Aeronautical information includes visual and radio aids to navigation, airports, controlled airspace, restricted areas, obstructions, and related data. Each chart covers 6° to 8° of longitude and approximately 4° of latitude. With a scale of 1:500,000 the resolution of terrain features is very good. Also refereed to as a sectional chart.

sectional view — Drawing details obtained by cutting away part of an object to show the shape and construction of the object at the cutting line.

sectioning — Marking a drawing by suitable crosshatching or other symbols to indicate the material in the cutaway view of a part.

sector — A part of a circle bounded by any two radii and the arc included between the two radii.

sector gear — A portion of a gear that appears to have been purposely cut from a whole gear wheel. It consists of the hub as a pivot point and a portion of the rim with the teeth. Sector gears generally drive a smaller pinion gear for small angular movements.

sector list drop interval — A parameter number of minutes after the meter fix time when arrival aircraft will be deleted from the arrival sector list.

sector visibility — Meteorological visibility within a specified sector of the horizon circle.

securing strap — A strap used to secure an oil or fuel tank to the airframe or engine.

sediment — Any matter that settles to the bottom of a liquid or container.

see and avoid — When weather conditions permit, pilots operating IFR or VFR are required to observe and maneuver to avoid other aircraft. Right-of-way rules are contained in FAR 91.

seesaw rotor — A term used for a semi-rigid rotor.

seesaw rotor system — A rotor system using a semi-rigid rotor

segment — A division or a section. A part of a figure.

segmented circle — An on-ground visual indicator that provides traffic pattern information. It consists of "L" shaped indicators

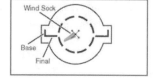

around the periphery of a circle that indicate pattern direction on various runways. Often includes a tetrahedron or wind sock in the center to indicate wind direction.

segmented rotor brake — A heavy-duty multiple disc brake used on large, high speed aircraft. Stators are keyed to the axle and contain high-temperature lining material. The rotors, keyed into the wheel, are made in segments to allow for cooling and the large amounts of expansion encountered.

segments of an instrument approach procedure — An instrument approach procedure can have as many as four separate segments depending on how the approach procedure is structured.

ICAO
a. Initial Approach – That segment of an instrument approach procedure between the initial approach fix and the intermediate approach fix or, where applicable, the final approach fix or point.
b. Intermediate Approach – That segment of an instrument approach procedure between either the intermediate approach fix and the final approach fix or point, or between the end of a reversal, race track or dead reckoning track procedure and the final approach fix or point, as appropriate.
c. Final Approach – That segment of an instrument approach procedure in which alignment and descent for landing are accomplished.
d. Missed Approach Procedure – The procedure to be followed if the approach cannot be continued.

USA
a. Initial Approach – The segment between the initial approach fix and the intermediate fix or the point where the aircraft is established on the intermediate course or final course.
b. Intermediate Approach – The segment between the intermediate fix or point and the final approach fix.
c. Final Approach – The segment between the final approach fix or point and the runway, airport or missed approach point.
d. Missed Approach – The segment between the missed approach point, or point of arrival at decision height, and the missed approach fix at the prescribed altitude.

seize — Equipment failure in which moving parts fuse or bind because of friction, pressure, or excessive temperature.

SELCAL — A contraction of selective calling referring to an automatic signaling system used in aircraft to notify a particular aircraft of an incoming call.

selected ground delays — A traffic management procedure whereby selected flights are issued ground delays to better regulate traffic flows over a particular fix or area.

selection-type test items — Questions where the student chooses from two or more alternatives provided. True-false, matching, and multiple-choice type questions are examples of selection-type test items.

selective call system (SELCAL) — A system that permits the selective calling of individual aircraft over radiotelephone channels linking a ground station with the aircraft.

selective plating — A process of electroplating only a section of a metal part.

selectivity — The ability to select or choose from among several choices.

selector switch — A multi-pole switch that takes the place of several switches. It is used to connect a single conductor to one of several other conductors.

selector valve — A hydraulic flow control valve that directs hydraulic pressure to one side of an actuator and connects the other side to the system return line.

selenium — A chemical element in the sulfur family with a symbol of Se and an atomic number of 34. Used in photoelectric devices because it changes its conductivity with a change in temperature.

selenium rectifier — A rectifier using laminated plates of metal, usually iron, that have been coated with selenium on one side. It is designed to develop a unidirectional current-carrying characteristic. The flow of electrons from the conductive metal to the selenium occurs more readily than the flow in the opposite direction.

self concept — A perception factor that ties together how people feel about themselves with how well they will receive further experiences.

self-accelerating — The ability of a turbine engine to produce enough power to accelerate.

self-aligning bearing — A rod end bearing consisting of a ball fitted into a socket that maintains alignment between the operating control and the unit being controlled.

self-centering chuck — A drill motor chuck with interconnecting jaws. They move at the same time to hold a drill bit when loosening or tightening.

self-concept — A perception factor that ties together how people feel about themselves with how well they will receive further experiences.

self-demagnetization — The process in which a magnet loses its magnetism if allowed to sit for long periods without a keeper bar.

self-excited generator — A generator whose field excitation is taken directly from the armature rather than from an outside source. A self-excited DC generator depends on residual magnetism. Without residual magnetism, it would be impossible to start a self-excited generator once stopped.

self-extinguishing — The ability of a material to automatically stop burning as soon as the outside source of the flame is removed.

self-fulfillment needs — Occupy the highest level of Maslow's pyramid. They include realizing one's own potential for continued development, and for being creative in the broadest sense. Maslow included various cognitive and aesthetic goals in this echelon.

self-healing capacitor — In electronics, an electrolytic capacitor that restores the dielectric film if it is damaged by over-voltage.

self-induction — The generation of back voltage in a conductor. This occurs when lines of flux created by the AC current in the conductor alternately expand and contract across the conductor. The moving lines of flux create a back voltage opposite in direction to the original voltage.

self-launch glider — A glider equipped with an engine, allowing it to be launched under its own power. When the engine is shut down, a self-launch glider displays the same characteristics as a non-powered glider.

self-locking nut — A nut designed with a built-in locking device that grips the threads of a bolt when the nut is tightened. This prevents the nut from loosening because of vibration. Locking devices can be fiber or non-fiber depending on their intended purpose.

self-tapping screw — A wood or metal screw that cuts its own threads as it is screwed into sheet metal or wood.

Selsyn system — A DC-type of synchro remote indicator system.

selvage edge — A manufactured woven edge on fabric that runs the length of the fabric or in the warp direction. It is removed for all fabrication and repair work.

semi — A prefix meaning a half or part of something.

semiautomatic operation -A device that is partially automatic and partially manual in its operation.

semi-cantilever — An externally braced wing.

semicircular — In the form of a half circle.

semiconductor — An insulating material treated with certain impurities that add free electrons to act as current carriers, carrying the flow of current in one direction and blocking it in the other.

semiconductor diodes — A semiconductor type of electron check valve. A device that allows the flow of electrons in one direction but not the opposite.

semi-monocoque — A stressed-skin structure in which the skin is supported by a lightweight framework to

provide extra rigidity. Most of the larger modern aircraft are of semi-monocoque construction.

semirigid rotor — In rotorcraft, a rotor system in which the blades are fixed to the hub but are free to flap and feather.

semispherical — In the shape of half a sphere. Dome-shaped.

sender — In a fuel quantity system, a measuring device located in a reservoir or fuel tank. The sender consists of a float mounted on an arm that rides on the top of the liquid. The arm is free to float and is connected to a variable resistor. Any change in the fluid level sends a signal to an indicator on the pilot's instrument panel showing the amount of fluid in the tank.

sense antenna — A non-directional radio antenna that picks up a signal with equal strength from all directions.

sensible heat — Heat added to a substance that causes a change in the temperature without changing the substance's physical condition. Can be felt and measured. Opposite of latent heat

sensitive altimeter — A pneumatic altimeter in which a pointer makes a complete revolution every thousand feet. It also has an adjustable barometric scale allowing it to be adjusted to the existing barometric pressure.

sensitive relays — A relay that operates electromagnetically with very small current. When the relay closes it controls a larger current to operate other devices.

sensitivity — A measure of the ability of something to be sensitive or responsive to very small changes in external conditions.

sensitized paper — Chemically treated paper used for making photographic prints of drawings.

sensor — A device used to actuate signal-producing devices in response to changes in physical conditions.

sensory register — That portion of the brain that receives input from the five senses. The individual's preconceived concept of what is important will determine how much priority the register will give in passing the information on to the rest of the brain for action.

separation — **1.** Phenomenon occurring when a moving stream of fluid flowing past a body separates from the surface of that body. **2.** Spacing between aircraft, levels or tracks.

separation minima — The minimum longitudinal, lateral, or vertical distances by which aircraft are spaced through the application of air traffic control procedures.

separator — **1.** In composites, a permeable layer that also acts as a release film. This could be in the form of a peel ply or a porous Teflon®-coated fiberglass. Often placed between lay-up and bleeder to facilitate the excess resin wicking into the bleeder. It is removed from the laminate after cure. **2.** In batteries, a porous material that keeps the positive and negative plates apart and prevents them from shorting.

sequence valve — A mechanically actuated hydraulic valve causing a sequential action of certain actuators. Wheel-well doors, for example, must open and contact a sequence valve before the landing gear can extend.

sequential logic devices — A digital memory device whose output depends not only on current input but on past inputs as well.

serial number — A number indicating placement within a series and used as a means of identification.

serial operation — In computers, operations are conducted one at a time rather than several operations occurring simultaneously (parallel operation).

serialists — A learning style that starts with the components and pieces them together to understand the whole. Serialists prefer to start at the beginning and examine the material in order.

series circuit — A circuit allowing only one path for electron flow from the source through the load back to the source.

series ohmmeter — An ohmmeter circuit where resistance is measured by determining the amount of current flow through the unknown resistor by placing a known voltage in series with the meter and the unknown resistor.

series resonant circuit — An AC circuit that has a capacitor and an inductor connected in series. The capacitive reactance and the inductive reactance are equal at the specific frequency of the circuit.

series RLC circuit — An AC circuit where resistance, capacitance, and inductance are arranged so that all current must flow through all three elements.

series wound generator — A generator where the field and armature are connected in series.

series-parallel circuit — A circuit consisting of groups of parallel components connected in series with other components.

series-wound motor — An electric motor with electromagnetic field coils connected in series with the armature. Series-wound motors are used as starter motors because of their high starting torque.

serrations — A formation resembling the teeth along the cutting edge of a saw.

service — A generic term that designates functions or assistance available from or rendered by air traffic control. For example, Class C service would denote the ATC services provided within a Class C airspace area.

service bulletin — Information issued by the manufacturer of an aircraft, aircraft engine, or component that detail maintenance procedures to enhance safety or improve performance of the product.

service capacity — A measurement of the amount of electrical energy that can be obtained from a chemical cell. Measured under controlled conditions and given in ampere-hours.

service ceiling — The height above standard sea level beyond which an airplane can no longer climb more than 100 feet per minute.

service life — The expected length of time a unit, part, component, or piece of equipment is expected to operate satisfactorily.

service manual — A manual issued by the manufacturer of an aircraft, aircraft engine, or component and approved by the FAA. It describes the approved methods of servicing and repairing the component.

serviceable — Equipment or parts in a condition that allows them to be returned to operational status on an aircraft.

servicing diagram — Information furnished by the manufacturer of an aircraft showing the proper access to all of the items or components requiring servicing.

servo — A motor or other form of actuator, that after receiving a small signal from the control device, exerts a large force to accomplish the desired work.

servo altimeter — An altimeter where the aneroid mechanism moves a rotary pick-off. Its signal is amplified to drive a servomotor moving the drums and pointers.

servo brake — A self-energizing drum brake that increases braking action above that which would be applied without any assistance.

servo fuel — An intermediate metered fuel in the Bendix RS fuel injection system controlling the opening of the flow control valve based on the airflow into the engine.

servo loop — An automatic control system that sends a signal to a servomotor to move a control device. The loop signal to a servomotor stops the servomotor when the control is moved the appropriate amount.

servo system — An automatic control system that senses changes in movement such as lowering the flaps. It sends a feedback signal to the control motor to stop moving the flaps when the correct position is obtained.

servo tab — An adjustable tab attached to the trailing edge of a control surface. The tab moves opposite the direction of the control and aids the pilot in moving the control.

servomechanism — Automatic device controlling large amounts of power using small inputs. A feedback system allows it to produce only the required amount of control.

servomotor — A motor that receives a signal due to the action of the control system causing a mechanical movement of a primary control. Servomotors have the ability to move in either direction when the current of the correct polarity is sent to the servomotor.

servo-type carburetor — A carburetor using pressure drops across a servo metering jet to control the amount of metered fuel. It is proportional to the amount of air allowed to flow to the cylinders.

sesquiplane — A biplane where the area of one of the wings is less than one half the area of the other. The smaller wing is usually the lower wing.

set screw — A small headless screw used to secure a wheel, pulley, or knob onto a shaft.

setback — The distance between the mold line and the bend tangent line on a sheet metal layout. For 90° bends, setback is equal to the inside radius of the bend plus the thickness of the metal being bent.

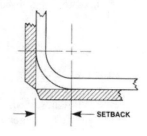

settling with power — In helicopters, a transient condition of downward flight (descending through air that has just been accelerated downward by the rotor) during which an appreciable portion of the main rotor system is being forced to operate at angles of attack above maximum. Blade stall starts near the hub and progresses outward as the rate of descent increases.

severe thunderstorm — A thunderstorm having a much greater intensity, larger size, and longer lifetime than an airmass thunderstorm. Associated weather includes wind gusts of 50 knots or more, and/or hail three-quarters of an inch diameter or larger and/or strong tornadoes.

severe weather avoidance plan — An approved plan to minimize the effect of severe weather on traffic flows in impacted terminal and/or ARTCC areas. SWAP is normally implemented to provide the least disruption to the ATC system when flight through portions of airspace is difficult or impossible due to severe weather.

severe weather forecast alerts — Preliminary messages issued in order to alert users that a Severe Weather Watch Bulletin (WW) is being issued. These messages define areas of possible severe thunderstorms or tornado activity. The messages are unscheduled and issued as required by the National Severe Storm Forecast Center at Kansas City, Missouri.

sewed seam — A seam in aircraft fabric made with a series of stitches joining two or more pieces of material.

sewed-in panel repair — A repair to fabric aircraft covering with a panel extending from the leading edge to the trailing edge and from rib to rib sewn in place. All of the seams are suitably reinforced with surface tape.

sewed-patch repair — A repair to aircraft fabric covering where a patch is sewed into place and the seams are covered with surface tape.

sextant — An instrument used to measure the angular distance between the horizon and a navigational star to determine geographic position or location.

Seyboth fabric tester — A patented, hand-operated precision device for testing the relative strength of installed aircraft fabric. It measures the amount of force required to punch a hole in the fabric with a specially shaped punch.

S-glass — In composites, the S stands for structural fiberglass. This type of fiberglass is for much of the structural use in advanced composite structures. A magnesium aluminosilicate composition that is especially designed to provide very high tensile strength glass filaments.

shaded-pole motor — A low-torque AC induction motor. The rotating field is provided by the inductive action of shaded poles on diametrically opposed pole pieces.

shaft horsepower (SHP) — Turboshaft engines are rated in shaft horsepower and calculated by use of a dynamometer device. Shaft horsepower is exhaust thrust converted to a rotating shaft.

shaft runout — An inspection performed on a rotating shaft or component to determine the straightness of the item.

shaft turbine — A turbine engine used to drive an output shaft commonly used in helicopters.

shall — As related to aircraft maintennace, means that the item is required, by the FAA, to be accomplished.

shank — A straight, narrow, essential part of a rotating body. An example would  be the portion of a twist drill beyond the flutes. (portion held by the drill chuck.)

shank of a drill — The part of the body of the twist drill that is round and smooth, not including the tip and flutes. The portion secured in the chuck of the drill motor.

shaping — A primary method of teaching that involves the use of carefully designed stimuli and the correct reinforcers for appropriate behavior. An instructor should decide what behavior is desired from the students. Each time they demonstrate the correct behavior, provide positive reinforcement to help shape or develop this behavior.

shaving — A cutting operation in which thin layers of material are removed from the outer surfaces of the product.

shear — A stress exerted on a material that tends to slide it apart.

shear failure — The failure of a riveted or bolted joint caused by the rivets shearing rather than the sheet tearing.

shear nut — A thin nut used on clevis bolts to prevent the bolt from falling out. Shear nuts are only suitable for use in shear applications and must never be used in tensile applications.

shear pin — A specially designed pin used in the drive shaft of engine driven pumps to protect the accessory drive train if the pump should seize.

shear point — An intentionally weakened point on a shaft such as in a dual element fuel pump. The shear point is designed to break away if one element becomes jammed and leaves the other element still functioning.

shear section — A narrow portion of a drive coupling designed to shear in case of pump seizure and prevent damage to either the pump or to the engine.

shear strength — The amount of force required to shear a pin, bolt, or rivet.

shear stress — Stress exerted on a material that tends to slide it apart.

shear wave — A wave in which the particles vibrate at right angles to the direction of the wave. Also referred to as a transverse wave.

shearing gravity waves — Short atmospheric gravity wave disturbances that develop on the edges of stable layers in the presence of vertical shears. Wave amplitudes can grow and overturn causing turbulence.

shears — A cutting tool similar to scissors used to cut sheet metal.

sheave — A wheel with a grooved center used as a pulley.

sheet metal — Metal of any thickness up to $1/4$". Metal of a greater thickness is referred to as metal plate.

sheet metal drawing — A forming process where sheet metal is pressed between dies to form the desired compound-curved shape.

sheet metal layout — The pattern of a sheet metal part before forming, cutting, or drilling. Patterns allow a great degree of accuracy in laying out the finished part. Sheet metal layout consists of flat layout, duplication of patterns, or projection through a set of points.

shelf cloud — A cloud that indicates the rising air over the gust front. Associated with the updraft of a multicell thunderstorm it is located just above the gust front at low levels.

shelf life — The period of time a material can be stored and remain suitable for use. In composites, the time span that a product will remain useful. This should be listed on the label. Temperature during storage will affect the shelf life.

shell — The outer structure of an atom formed by the rotating electrons around the nucleus.

shell-type transformer — A transformer encased in steel containing the magnetic lines of flux.

shielded cable — An electrical conductor encased inside a braided metal shielding. The shielding intercepts radiated electrical energy and conducts it to ground rather than allowing it to cause radio interference.

shielded ignition cable — An electrical cable enclosed in a metal braid. Used to carry high voltage from the distributor of the magneto to the spark plug. Its purpose is to prevent radio interference caused by electromagnetic radiation.

shielded spark plug — A spark plug completely encased in a steel shell. The radiated energy from the spark is conducted to ground through the shielding and helps prevent radio interference.

shielded-arc welding — A method of gas welding in which an inert gas such as argon, helium, or carbon-dioxide is used as a covering shield around the arc. Preventing the atmosphere from contaminating the weld results in a stronger, more ductile, and more corrosion resistant weld.

shielding — Metal covers placed around electric and electronic devices to prevent the intrusion of external electrostatic and electromagnetic fields.

shim — A thin piece of metal used to fill in a space between two objects in order to adjust a preload or the clearance between bearing parts.

shimmy — A rapid and violent oscillation of a nose wheel or tail wheel of an airplane often caused by excessive wear in the support bearings.

shimmy damper — A hydraulic snubbing cylinder installed between the nose wheel fork and the landing gear structure. It is used to minimize shimmying of the nose wheel during takeoffs and landings.

shock absorber — A device built into the landing gear of an aircraft to absorb the energy of the landing impact.

shock loading — Stress loading for an extremely short duration.

shock mounted — Any device attached to the airframe with shock mounts to minimize the transmission of vibration or shock from one unit into another.

shock mounts — A shock absorbing attachment used to mount an engine or instrument panel to an airframe to minimize vibration.

shock stall — Turbulent airflow on an airfoil that occurs when the speed of sound is reached. The shock wave distorts aerodynamic airflow causing loss of lift and stall.

shock strut — An aircraft landing gear shock strut that absorbs the initial landing impact by the transfer of oil from one chamber to another through a restricting orifice. Taxi shocks are absorbed with compressed air or by a spring.

shock wave — A compression wave formed when a body moves through the air at a speed greater than the speed of sound.

shop head — The upset head of an aircraft rivet.

shop head rivet — The head formed on a rivet when it is driven.

Shore scleroscope — A hardness tester used for metal, plastic, and rubber.

short circuit — A path for electrons to flow from one electrical potential to another without completing a useful circuit.

short range clearance — A clearance issued to a departing IFR flight that authorizes IFR flight to a specific fix short of the destination while air traffic control facilities are coordinating and obtaining the complete clearance.

short stack — An exhaust system for aircraft reciprocating engines consisting of a short exhaust pipe attached to the exhaust port of the cylinder. Short stacks use no collector system.

short take off and vertical landing (STOVL) — A fixed-wing aircraft capable landing vertically and of clearing a 15-meter (50-foot) obstacle within 450 meters (1500 feet) of commencing takeoff run.

short takeoff and landing (STOL) — The ability of an aircraft to take off and land in a distance of 1,000 feet (sometimes 1,500 ft.) or less while clearing 50-foot obstacles. These obstacles are located just after takeoff or just prior to landing.

short takeoff and landing aircraft — An aircraft that, at some weight within its approved operating weight, is capable of operating from a STOL runway in compliance with the applicable STOL characteristics, airworthiness, operations, noise, and pollution standards.

short wave trough — Troughs in the mid- and upper troposphere and lower stratosphere that correspond to developing frontal lows. Short wave troughs are smaller in scale than long waves. They move toward the east, averaging 600 nautical miles per day.

shorting switch — A multi-pole switch where one circuit is completed before another circuit is opened.

shot peening — A process used to strengthen metal parts by blasting its surface with steel shot.

shoulder bolts — A bolt where the threaded portion is smaller than the shank. It is often used for the installation of plastic materials to prevent over tightening.

shoulder-wing airplane — An airplane having one main supporting wing surface mounted near the top but not directly on top of the fuselage. Also referred to as the mid wing.

show — Unless the context otherwise requires, means to show to the satisfaction of the Administrator.

shower — Precipitation from a cumuliform cloud; characterized by sudden onset and cessation, rapid change of intensity, and usually by rapid change in the appearance of the sky; showery precipitation can be in the form of rain, ice pellets, or snow.

shower of sparks — The induction-vibrator-type starting system used in some Bendix magnetos. A vibrator directs pulsating direct current into the primary circuit of one of the magnetos when the points are closed. When the retard points open, pulsating current flows to ground through the primary coil and induces a high voltage into the secondary winding. As long as both sets of points remain open, a shower of sparks occurs at the spark plug.

show-type finish — A glass-like aircraft finish achieved by applying many coats of dope and repeated sanding and polishing.

shrink fit — Interference fit between parts when the female part is heated and the male part is chilled before they are assembled. When they reach the same temperature, they are essentially locked together.

shrinking — The act of compressing a material into a smaller volume or area.

shrinking block — A sheet metal forming tool that clamps metal to prevent it from buckling while hammering its edge to shrink it.

shrink-wrap — A method used to protect products from dirt and dust while they are held in storage. The part to be shrink-wrapped is covered with a film of transparent thermoplastic material. When heat is applied to the film, the film shrinks to encase the part.

shroud — A cover or housing used to aid in confining an air or gas flow to a desired path.

shroud ring — A stationary sealing ring positioned just outside the tip plane of rotating airfoils. Sometimes it is the inner part of an outer casing.

shrouded-tip turbine blade — A blade with tip platforms that fit one to another to form a circular support ring. Often, the shrouds have thin abradable rims attached at their outer edge to act as air seals.

shunt — An accurately calibrated resistor placed in parallel with a meter movement for measuring current. Current flows through the shunt and produces a voltage drop proportional to the amount of current. The ammeter movement measures this voltage drop and displays it in amperes.

shunt circuit — A circuit that has several paths for electrons to flow.

shunt ohmmeter — An ohmmeter circuit used for measuring low resistances. The unknown resistor is placed in parallel (shunt) with the meter and resistance is measured by the amount of current the unknown resistor takes from the meter.

shunt-wound generator — A generator in which the field and armature are connected in parallel.

shunt-wound motor — A motor in which electromagnetic field windings are connected in parallel with the armature.

shut-off valve — A flow-control valve used to shut off or stop a flow of fluid.

shuttle valve — A valve, mounted on critical components, that directs system pressure into the actuator for normal operation but switches to emergency fluid when the emergency system is actuated.

side bands — The frequency bands on each side of a carrier frequency produced by modulation.

sideslip — An uncoordinated flight condition in which the aircraft moves downward and toward the inside of the turn. Sideslip can also result from the use of rudder to maintain heading in an engine-out situation.

sidestep maneuver — A visual maneuver accomplished by a pilot at the completion of an instrument approach. It permits a straight-in landing on a parallel runway, which must not be more than 1200 feet to either side of the approach runway.

3sight gauge — A glass tube or window attached to a reservoir or tank that shows the quantity of fluid in the container.

sight glass — A liquid level indicator located on the outside of a reservoir that provides a visual indication of the level of liquid in the reservoir.

sight line — A mark on a flat sheet of metal set even with the nose of the radius bar of a cornice or leaf brake. This placement puts the bend tangent line at the beginning of the bend.

SIGMET (WS) — This advisory describes conditions of higher intensity, which pose hazards to all aircraft, including:
a. severe icing not associated with thunderstorms,
b. severe or extreme turbulence or clear air turbulence not associated with thunderstorms,
c. dust storms, sand storms, or volcanic ash lowering surface visibilities to below three miles, and
d. volcanic eruptions.

sigmet information [ICAO] — Information issued by a meteorological watch office concerning the occurrence or expected occurrence of specified enroute weather phenomena that can affect the safety of aircraft operations.

signal — The intelligence or directive portion of a radio wave.

signal generator — A test unit designed to produce reference electric signals that can be applied to electronic circuits for testing purposes.

signal-strength meter — In electronics, an instrument that indicates the signal strength being received.

significant digits — Significant digits are those that are statistically significant. In measurements, the last significant digit is the first one estimated (interpolated between two measured markings). For computations, a set of rules determines the number of significant digits. All non-zero digits are counted as significant. Zeroes that have non-zero digits to the left of them are significant zeroes. Examples: 0.005140 has four significant digits. 31.000 has five significant digits.

significant meteorological information — See SIGMET

significant point — A point, whether a named intersection, a NAVAID, a fix derived from a NAVAID(s), or geographical coordinate expressed in degrees of latitude and longitude. Established for the purpose of providing separation, as a reporting point, or to delineate a route of flight.

silencer — A device used in cabin air distribution systems designed to minimize the noise created by pulsations in the air delivered from the cabin supercharger.

silica gel — A desiccant used as a drying or moisture absorbing agent. Silica gel is often used to package products that can be damaged by excess moisture.

silicon — A natural element with a symbol of Si and an atomic number of 14. Used to produce N- and P-type semiconductor devices having excellent thermal characteristics. **1.** N-Type: Silicon doped with an impurity having five valence electrons. **2.** P-Type: Silicon doped with an impurity having three valence electrons.

silicon carbide — An abrasive used in the manufacture of grinding stones and abrasive papers.

silicon controlled rectifier (SCR) — A gated rectifier that allows current to flow only during that portion of the cycle when the gate has been triggered by a positive pulse.

silicon glaze — A shiny, brown, glass-like deposit on the nose insulator of a spark plug resulting from use in sandy or dusty conditions. This glaze is non-conductive at low temperatures, but at high temperatures it becomes conductive and causes the plug to short.

silicon solar cell — A semiconductor device that develops a voltage when struck by light.

silicon steel — A steel alloy that contains silicon.

silicone rubber — An elastic material made from silicone elastomers. Silicone rubber is used with fluids that attack other natural or synthetic rubbers.

silver — A white, precious, metallic chemical element with a symbol of Ag and an atomic number of 47. Very malleable and a good conductor of electricity.

silver brazing — Brazing (connecting two close fitting parts by the use of molten metal being drawn into the joint by capillary action) using a silver alloy.

silver solder — An alloy of silver, copper, and nickel used for hard soldering. It produces a joint that is stronger than soft solder, but not as strong as some forms of brazing.

silvered-mica capacitor — A capacitor made up of a mica dielectric coated with silver that makes up the plates of the capacitor.

simple flaps — Wing flaps lowered by pivoting them about a point near their leading edge. They change the airfoil section of the wing but do not affect the wing area.

simple fraction — A fraction in which both the numerator and the denominator are whole numbers.

simple machine — A device that transforms energy or changes the direction of a force. The six simple machines include the lever, pulley, wheel and axle, inclined plane, screw, and gear.

simple motion — Newton's law of motion, which states that objects in motion tend to stay in motion.

simplex communications — A method of communication where only one transmitter location can transmit at a time while the other receives.

simplex fuel nozzle — A nozzle with one spray orifice and one spray pattern.

simplex nozzle — A fuel discharge nozzle for turbine engines fed from a single fuel manifold.

simplified directional facility (SDF) — A NAVAID used for nonprecision instrument approaches. The final approach course is similar to that of an ILS localizer except that the SDF course can be offset from the runway, generally not more than 3 degrees, and the course can be wider than the localizer, resulting in a lower degree of accuracy.

simulate — To have the characteristics or appearance of something that is real.

simulated flameout — A practice approach by a jet aircraft (normally military) at idle thrust to a runway. The approach can start at a runway (high key) and can continue on a relatively high and wide downwind leg with a high rate of descent and a continuous turn to final. It terminates in a landing or low approach. The purpose of this approach is to simulate a flameout.

simulator — An enclosed housing that duplicates all of the controls, instruments, furnishings, and environment of an actual airplane cockpit. The simulated environment reproduces the same sensations and indications found in actual flight.

simultaneous ILS approaches — An approach system permitting simultaneous ILS/MLS approaches to airports having parallel runways separated by at least 4,300 feet between centerlines. Integral parts of a total system are ILS/MLS, radar, communications, ATC procedures, and appropriate airborne equipment.

simultaneous MLS approaches — See simultaneous ILS approaches.

sine — A trigonometric function found in a 90° triangle. It is defined as the ratio of the length of the side opposite an angle to the length of the hypotenuse.

sine curve — **1.** A graphic representation of the relationship between an angle and its sine. **2.** The curve showing the relationship between the voltage or current and the angle through which the related rotary generator has turned.

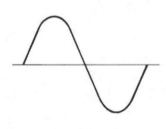

sine wave — The wave form of alternating current produced by a rotary generator. Its amplitude at any time is proportional to the sine of the angle through which the generator has turned.

single direction routes — Preferred IFR Routes that are sometimes depicted on high altitude enroute charts and which are normally flown in one direction only.

single flare — A flare used for aircraft rigid tubing with a flange at the end of the tube. When a single flare is folded back over itself, it is referred to as a double flare.

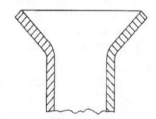

single frequency approach — A service provided under a letter of agreement to military single-piloted turbojet aircraft that permits use of a single UHF frequency during approach for landing. Pilots will not normally be required to change frequency from the beginning of the approach to touchdown except that pilots conducting an enroute descent are required to change frequency when control is transferred from the air route traffic control center to the terminal facility.

single sideband — Radio communications in which one of the two sidebands used in amplitude modulation is suppressed.

single spread — A method of applying adhesive to only one surface of a bonded joint.

single-acting actuator — A linear hydraulic or pneumatic actuator that uses fluid power for movement in one direction and a spring force for its return.

single-axis autopilot — An automatic flight control device that controls the airplane only around the roll axis.

single-crystal turbine blade — A high temperature strength blade with no grain boundaries. It is manufactured by an advanced casting process that produces the blade from a single crystal of metal.

single-cut file — A hand file with a single row of teeth extending across the piece at an angle.

single-engine absolute ceiling — In multi engine aircraft, the density altitude an airplane is capable of reaching and maintaining with the critical engine feathered and the other at maximum power. This assumes the airplane is at maximum weight and in the clean configuration, flying in smooth air. This is also the density altitude at which V_{XSE} and V_{YSE} are the same airspeed. If flying above this altitude and the engine fails, the plane will inevitably descend until it reaches the single-engine absolute ceiling.

single-engine service ceiling — The maximum density altitude at which the single-engine best rate-of-climb airspeed (V_{YSE}) produces a 50 f.p.m. rate of climb. The ability to climb 50 f.p.m. in calm air is necessary simply to maintain level flight for long periods in turbulent air. This ceiling assumes the airplane is at maximum gross weight in the clean configuration, the critical engine (if appropriate) is inoperative, and the propeller is feathered. In comparison, the multi-engine service ceiling is the density altitude at which the best rate-of-climb airspeed (V_Y) will produce a 100 f.p.m. rate of climb at maximum gross weight in the clean configuration.

single-face repair — A repair to a bonded structure in which the damage extends through one face sheet and into the core material.

single-loop rib-stitching — A method of attaching fabric covering to the ribs of an aircraft using only one loop of ribstitch cord per stitch.

single-piloted aircraft — A military turbojet aircraft possessing one set of flight controls, tandem cockpits, or two sets of flight controls but operated by one pilot. C-onsidered single-piloted by ATC when determining the appropriate air traffic service to be applied.

single-point fueling — A method of fueling the aircraft from a single point. It consists of a pressure fueling hose and a panel of controls and gauges that permit one person to load or unload any of the fuel tanks of an aircraft. The panel has valves connecting the various tanks to the main fueling manifold. Fuel flows to each tank until the tank has reached the desired level. Also referred to as pressure fueling.

single-point grounding — A method of electrically grounding a circuit by connecting all of the ground wires to a single point.

single-servo brakes — Brakes that use the momentum of the aircraft to wedge the lining against the drum and assist in braking when the aircraft is rolling forward.

single-sideband — A radio transmission in which only one of the sidebands in a signal is used.

single-spool compressor — An axial flow compressor with a single rotating mass.

single-spool turbine engine — A jet engine that has only one turbine section connected to the compressor section. Twin-spool turbines have two sections of each, operating at different speeds.

single-throw switch — In electricity, a switch that has two positions, ON and OFF. Can have multiple poles to turn ON and OFF multiple circuits simultaneously.

sintered material — A heated material that has become a coherent mass without melting. Used as an alternative to casting for small parts.

sintered metal — A porous material made by fusing powdered metal under heat and pressure.

sinusoidal curve — In mathematics, a curve with angle degrees along the horizontal axis and sines of the angles along the vertical axis. This curve generally illustrates the varying output voltage values of an AC generator as it turns through 360°.

siphon — A device that moves liquid from one container to a lower point. It consists of a flex tube in the liquid and a suction device to start the flow of liquid. Gravity and atmospheric pressure on the surface of the liquid carry the liquid from the container to a point below the container.

siphon tube — A tube installed in a CO_2 fire extinguisher cylinder assuring that CO_2 directed to the discharge nozzle remains in its liquid state.

sites — Internet addresses that provide information and often are linked to other similar sites.

Sitka spruce — A tall spruce tree that has needle-shaped leaves, drooping cones, and berry-like fruit. In the selection of wood for aircraft repairs, spruce is considered the standard to which other woods are compared.

situational awareness — The accurate perception and understanding of all the factors and conditions within the four fundamental risk elements that affect safety before, during, and after the flight.

sizing — Material used in the manufacture of some fabrics to stiffen the yarn for ease of weaving. Aircraft fabric should not contain sizing.

sizing up a situation — A colloquial term used to denote the process of analyzing a situation and deciding what should be done.

sketch — A simple drawing made without the use of drafting instruments. It usually shows a minimum of detail and lacks precision.

ski plane — An airplane whose wheels have been replaced with skis for operation in snow or ice.

skid — In a skid, the rate of turn is too great for the angle of bank, and the ball moves to the outside of the turn.

skid fin — A longitudinal vertical airfoil usually placed above the upper wing of a biplane to increase its lateral stability.

skid shoes — Plates attached to the bottom of a helicopter skid-type landing gear to protect the skid.

skills and procedures — The procedural, psychomotor, and perceptual skills used to control a specific aircraft or its systems. They are the stick and rudder or airmanship abilities that are gained through conventional training, are perfected, and become almost automatic through experience.

skin — The smooth outer cover of an aircraft.

skin antenna — An antenna that is flush with the aircraft skin.

skin effect — The tendency of high frequency alternating currents to flow in the outer portion of a conductor. Skin effect can be reduced by using stranded rather than solid wire.

skin friction — Aerodynamic drag caused by air flowing over the surface of the aircraft.

skip distance — The distance from a transmitter to the point where a reflected sky wave first reaches the Earth.

skip welding — A welding technique used to prevent warping of the material. Skip welding is a series of short welded beads evenly spaced to anchor the entire length of the weld. The remaining gaps are filled in after the skip welds are tacked in.

skull-cap spinner — A round nosed cover for the hub of a fixed-pitch propeller. Normally the same diameter as the hub, it is usually attached to a bracket secured by propeller mounting bolts.

skunk works — Slang. A loosely structured corporate research and development unit formed to foster innovation. Most famous was Lockheed Aircraft's Skunk Works®, headed up for years by Kelly Johnson.

sky wave — The portion of a radio wave reflected from the ionosphere.

Skydrol hydraulic fluid — A synthetic, nonflammable, ester-base hydraulic fluid used in modern high-temperature hydraulic systems.

slab — An unfinished bar of metal.

slag — A completely fused and brittle by-product caused by the flux used in welding. Slag separates from the metal in the process of welding.

slam acceleration — Improper operation of a turbine engine when the power control lever is moved forward too rapidly. A rich flameout is possible because the fuel is metered before the airflow has increased sufficiently. More sophisticated fuel controls in recent turbine engines have minimized this problem.

slant range — In radar, the line-of-sight distance from radar to target as opposed to ground distance (from radar to the position on the ground below the target). Also the line-of-sight distance from an aircraft to a navigation aid on the ground as opposed to the ground distance (from a point under the airplane to the navigation aid).

slant visibility — In weather, the distance from an airborne observer to the farthest point visible on the ground.

slant-line distance — See slant range.

slash — A radar beacon reply displayed as an elongated target.

slat — A movable auxiliary airfoil on the leading edge of a wing. It is closed in normal flight but extends at high angles of attack. This allows air to continue flowing over the top of the wing and delays airflow separation.

slave relay — A relay that operates in response to the actions of a relay in a primary circuit (activates an associated circuit).

slaved gyro — Directional gyro slaved to the output of a flux valve. Gives a directional gyro synchronous attributes.

sleet — Small, transparent or translucent, round or irregularly shaped pellets of ice. They can be (1) hard grains that rebound on striking a hard surface or (2) pellets of snow encased in ice.

sleeve — A tube or tube-like part that fits over or around another part.

slide rule — A ruler-like instrument with a center slide. Both the inner slide and fixed outer sections are marked with logarithmic scales. Slide rules are used for mathematical calculations; however electronic calculators or computers are more widely used.

slide switch — In electricity, a switch that consists of a slider that moves across contacts to make and break one or more circuits.

slide valve — A mechanical valve that opens and closes by sliding a gate across a port.

slide-wire rheostat or potentiometer — In electricity, a variable resistor that changes resistance by moving a slider along a length of resistance wire. The amount of resistance is proportional to the effective length of the wire.

sliding support — A duct support attached to flexible bellows. The sliding support allows movement of the bellows while the duct is under pressure.

sling psychrometer — In weather, a device used to measure the relative humidity. It consists of two thermometers, one of which has wetted cloth wrapped around its bulb. Both thermometers are swung on the end of a cord. Due to evaporation, the thermometer with the wetted cloth will read a lower temperature than the dry thermometer. A chart of the differences is consulted to determine the relative humidity.

slinger ring — A tubular ring mounted around the hub of a propeller where deicing fluid is directed and slung out onto the blades.

slip — **1.** Propeller: The difference between geometric and effective pitch of a propeller. Slip can be expressed as a percentage of the mean geometric pitch or as a linear dimension. **2.** Aircraft flight maneuver: A sideways aircraft maneuver used to descend at a steep angle or compensate for crosswind. **3.** In a slip the rate of turn is too slow for the angle of bank and the ball moves to the inside of the turn.

slip clutch mechanism — A typical installation on electric starter drives designed to prevent sudden high torque to the engine. The clutch plates slip until torque on the engine side matches the torque on the clutch side.

slip joint — A connection or joint in an induction system that remains airtight even as the cylinders expand and contract with the temperature changes.

slip ring — A smooth circular ring used to put field current into the armature of a DC alternator.

slip stick — A slang term to describe a slide rule.

slippage mark — A painted mark between a tire and a wheel that indicates slippage. If the mark of the tire and wheel are not lined up, the tire has slipped on the wheel.

slip-roll former — A metal working machine used to shape sheet metal into cylindrical and curved shapes.

slipstream — A stream of air pushed back by a revolving aircraft propeller.

sloshing sealing compound — A synthetic rubber sealant poured inside a metal fuel tank to seal the tank and prevent fuel leakage. It is poured into the tank and then sloshed around until all surfaces are covered.

slot — A fixed, nozzle-shaped opening near the leading edge of a wing that ducts air onto the top surface of the wing. Its purpose is to increase lift at higher angles of attack.

slot time — See meter fix time/slot time.

slot-headed screw — A screw with a single, straight groove cut across its head to fit the blade of a slotted screwdriver.

slotted flap — A trailing edge wing flap that forms a duct when the flap is lowered. Air forced through the duct is held down on the upper surface of the flap allowing more flap extension before airflow separation.

slotted nut — Similar to a castle nut. A hexagon nut that has grooves cut across its top to fit a cotter pin or safety wire passing through a hole in the shank of a bolt.

slow taxi — To taxi a float plane at low power or low RPM.

slow-blow fuse — A fuse that can take momentary overloads but opens the circuit when excess current flows are sustained. Often used in motor circuits where current is high until the motor begins to turn.

sludge — A heavy, slimy deposit in aircraft lubricating oil resulting from oxidation of the oil and contamination by water.

sludge chamber — Tubes or thin sheet metal chambers installed in the throws of an aircraft engine crankshaft. Sludge is forced into these chambers by centrifugal force and held there until engine overhaul.

slug — A gravitational unit of mass. A force of one pound creates an acceleration of one foot per second.

slugging — A malfunction in a vapor cycle air-conditioning system where liquid refrigerant enters the compressor.

small aircraft — Aircraft having a certificated gross weight of less than 12,500 lbs.

small-scale integration — In electronics, the most basic of integrated circuits, consisting of up to nine gates or transistors.

smaze — A mixture of fog vapor, smoke, dust (haze), and industrial smoke, with a lower moisture content than smog.

smile — The curved dimple around the edge of a rivet head caused by a rivet set not being held squarely against the head while bucking the rivet.

smog — A mixture of smoke and fog.

smoke — In weather, small particles of carbon, the result of combustion, which cause restrictions to visibility. Not to be confused with smog, which is a combination of smoke and haze or fog.

smoke detector — A system that can detect the presence of smoke in an unoccupied aircraft compartment before high temperatures actuate the fire warning system.

smolder — To burn without a flame.

smoothing filter — A filter consisting of capacitors or inductors, or both, that attenuates the AC component in the output of a rectifier circuit.

snake drill — A long, flexible driving mechanism, one end of which is designed to be put in a drill chuck and the other designed to hold a twist drill.

snap ring — A small, spring-loaded, ring-type fastening device that fits into a groove either outside of a shaft or inside of a hole. Spring tension holds the ring in place.

snap roll — An abrupt airplane maneuver where the airplane rotates rapidly about its longitudinal axis. The maneuver begins by entering an accelerated stall with abrupt elevator movement, then adding full aileron and rudder control in the direction of desired snap. Neutralizing all three controls terminates the maneuver.

snap-action electrical switch — An electrical switch that uses spring snaps to open and close the contacts when the switch is moved. Positive action minimizes arcing

problems caused by slow opening or closing of the circuit.

snips — Hand shears used for cutting sheet metal.

snow — Water vapor that changes directly into crystals of ice flakes when frozen in the upper air.

snow depth — The depth of the snow actually on the ground.

snow grains — Precipitation of very small, white opaque grains of ice, similar in structure to snow crystals. The grains are fairly flat or elongated, with diameters generally less than 0.04 inch (1 mm.).

snow pellets — Precipitation consisting of white, opaque approximately round (sometimes conical) ice particles having a snow-like structure, and about 0.08 to 0.2 inch in diameter; crisp and easily crushed, differing in this respect from snow grains; rebound from a hard surface and often break up.

snubber — The portion of a hydraulic actuator that arrests the motion of a piston at the end of its stroke. A snubber helps cushion the stopping action.

soaking — Holding a metal at a specified temperature and time for the purpose of heat treating or annealing.

soap — Material mixed with water and used for cleaning. Soap is produced by mixing alkali and potash with fat or oil.

soap bubble test — A method of testing for leaks in gas systems under pressure. A special non-flammable soap solution is brushed over the suspected fittings. If there is a leak escaping gas causes the soap to bubble.

soapstone — A soft stone having a soapy feeling and composed essentially of talc. Used to mark steel parts prior to welding.

soaring — Flying without the use of an engine. The pilot uses rising air to remain aloft and reduce the rate of aircraft descent.

social needs — A level of Maslow's hierarchy of needs. After physical and safety needs are met, it becomes possible for students to satisfy their social needs. The need to belong and to associate, as well as to give and receive friendship and love.

socket wrench — A small cylindrical shaped wrench internally broached to fit the nut. It is equipped with a square hole in its top that fits a square drive on the wrench handle.

socket-head screw — A screw with a hex-shaped head.

sodium — A silver-white, metallic, alkaline chemical element with a symbol of Na and an atomic number of 11.

sodium bicarbonate — A white powdery crystalline compound. When mixed with water sodium bicarbonate neutralizes spilled battery acid. Also referred to as baking soda.

sodium-vapor lamp — A bulb that glows from the excitation of sodium vapor atoms by an electric arc.

soft magnetic material — A metal such as iron that readily accepts lines of magnetic flux and is easily demagnetized.

soft solder — A physical alloy of lead and tin used to join non-structural metal parts or to increase the electrical conductivity of a twisted wire joint.

soft-faced hammer — A hammer with a wood, plastic, or rubber face on the head.

software — The programs, procedures, and documentation associated with a computer system.

solar cell — A silicon semiconductor device that converts solar energy into electricity.

solar declination — The latitude where the sun is directly overhead.

solar elevation angle — The angle of the sun above the horizon measured in degrees.

solar-radiation — Electromagnetic radiation from the sun.

soldered splice — A splice in electrical wiring made by twisting the wires together and then flowing soft solder over the joint. This type of splice is not recommended for aircraft use because it tends to be brittle.

soldering iron — An electrically heated hand-held tool used to melt solder.

solderless connection — A terminal attached to an electrical conductor by crimping it onto the wire.

solderless splice — A tubular fastener used to connect two or more wires. The fastener is crimped onto the wires.

solenoid — A remotely activated electrical device. A coil of wire with a movable core is activated by completing an electrical circuit. The movable core can be used to open or close another electrical circuit or to achieve some other mechanical action.

solid — 1. A geometric figure that has three dimensions: length, width, and height. 2. One of the three states of matter having definite volume and shape and being relatively firm or hard.

solid conductor — A wire made up of a single strand of metal covered with insulating material.

solid fuel — Fuel such as wood or coal or one of the molded solid propellant materials used in rocket engines.

solid solution — A mixture where two or more elements or compounds combine with each other at an elevated temperature and remain in a combined solid state when rapidly cooled.

solidity ratio — In rotorcraft, the ratio of the total rotor blade area to total rotor disc area.

solid-state device — An electronic device that controls the flow of electrons without any mechanical operations. Also referred to as a semiconductor.

solo flight — Refers to one person piloting an aircraft in flight with no passengers.

solstice — Noon the first day of summer and the first day of winter, when the sun has reached its highest and lowest latitudes, respectively.

soluble — Any substance that can be combined with another substance. An example of this is mixing water and crystalline sugar to form sugar water.

solution — 1. A state in which base metal and alloying agents are united to form a single, solid metal. 2. A combination of a solid, liquid, or gaseous substance homogeneously mixed with a liquid, gas, or solid.

solution heat treatment — A heat treatment of aluminum alloy where metal is raised to its heat treating temperature, held until uniform throughout, and then quenched. This process covers alloying agents held in solid solution, which increases the strength of the metal.

solvent — A liquid used for dissolving and cleaning materials.

sonar — A system that detects objects in the water by sending out sound waves and receiving the echoes. Operates similar to radar. The acronym for SOund NAvigation Ranging.

sonic — Relating to the speed of sound. Subsonic is speed below the speed of sound; transonic is speed bridging the speed of sound; and supersonic refers to speed greater than the speed of sound. The speed of sound is approximately 761 miles per hour at sea level, 59°F.

sonic boom — A vibrational disturbance, heard as a loud noise, caused by an airplane moving faster than the speed of sound.

sonic cleaning — A method of cleaning parts using high intensity sound waves in a cleaning fluid.

sonic frequencies — High frequency vibrational disturbances that the human ear can detect, normally considered to be the frequencies between 20 and 20,000 Hz.

sonic soldering — A method of soldering certain metals such as aluminum that are difficult to solder because they quickly build up surface oxides. The tip of a sonic soldering iron heats the surface by vibrating it at a sonic rate while simultaneously penetrating any surface oxide..

sonic speed — The speed sound travels through a medium.

sonic vibration — A high-frequency vibration caused by sound energy.

soot — A black residue created during oxyacetylene welding when the burning acetylene gas does not have enough oxygen to be completely consumed.

sound — A mechanical radiant energy transmitted by longitudinal pressure waves in the material medium. Its frequency is perceived by the human ear.

sound suppressor — Same as noise suppressor in reference to engines. Also, a device on a test cell that reduces the sound of exhaust stack noises.

sound waves — Longitudinal wave motion through matter. Longitudinal waves vibrate back and forth longitudinally in the direction of wave propagation.

sounding — 1. In meteorology, an upper-air observation; a radiosonde or rawinsonde observation. 2. The electrode of a field-effect transistor similar to the emitter of an ordinary transistor. 3. In communication, the sender, speaker, transmitter, or instructor who composes and transmits a message made up of symbols, which are meaningful to listeners and readers.

south geographic pole — The pole located at the 90th degree of south latitude. The south geographic pole of the Earth is located at the southern end of the Earth's axis.

south pole of a magnet — The pole of a magnet where lines of flux enter the magnet. Lines of flux are considered to leave the magnet at the north pole.

space charge — The electric charge carried by a cloud of electrons in the space between electrodes of an electron tube.

space shuttle — A reusable aerospace vehicle designed for carrying passengers and cargo into Earth's orbit and subsequently returning for landing.

spacers — Devices or components used to take up space between two objects.

spaghetti — Insulating tubing slipped over wires.

spalling — A bearing defect in which chips of the hardened bearing surface are broken out.

span — 1. Length dimension of a beam. 2. In rotorcraft, the dimension of a rotor blade or airfoil from root to tip.

span loading — The ratio between the weight of an airplane and the span of its wings.

span of an airfoil — The length of an airfoil measured from tip to tip.

spanner — A wrench consisting of a hook-shaped arm with a pin in its hooked end. Used for turning a ring-shaped nut.

spanner nut — A shaft-retaining nut tightened using the notches in its face rather than from its outer surfaces as is done with a conventional hex-nut.

spanwise — From wing tip to wing tip.

spar — The main, or principle, spanwise structural member of a wing or other airfoil.

spar varnish — A phenolic modified oil that cures by oxidation rather than evaporation of its solvents. Produces a tough, highly water-resistant film.

spark — Very brief discharge of electrical energy between two conductors separated by air or other gas.

spark coil — A step-up transformer that produces high voltage for spark plugs.

spark plug — A component in an aircraft engine that converts high voltage electricity from the magneto into a high intensity spark for igniting fuel in the combustion chamber of a reciprocating engine.

spark plug bushing — A bronze or steel insert in the cast-aluminum cylinder head of a reciprocating engine into which the spark plug. is screwed.

spark plug resistor — A composition resistor installed in the barrel of most shielded spark plugs. The resistor limits the current stored by the capacitive effect of the shielding, minimizing the erosion of the spark plug electrodes.

spark suppressor — A device inside a magneto, such as a capacitor, that is placed across a set of contacts to keep the spark from jumping across the contact points as they open.

spark test — A common means of identifying various ferrous metals. In this test, the piece of iron or steel is held against a revolving grinding stone and the metal is identified by the sparks thrown off. Each ferrous metal has its own peculiar spark characteristics. The spark streams vary from a few tiny shafts to a shower of sparks several feet in length.

spark-ignition — A method of providing ignition of the fuel-air mixture inside the cylinder of a reciprocating engine by an electric spark.

spatial disorientation — A feeling of balance instability caused by a conflict between the information relayed by central vision and peripheral vision, which has virtually no references with which to establish orientation.

spatula — A broad flat instrument used for spreading soft materials.

speak slower — Used in verbal communications as a request to reduce speech rate.

special activity airspace (SAA) — Any airspace with defined dimensions within the National Airspace System wherein limitations can be imposed upon aircraft operations. This airspace can be restricted areas, prohibited areas, military operations areas, air ATC assigned airspace, and any other designated airspace areas. The dimensions of this airspace are programmed into URET CCLD (User Request Evaluation Tool Core Capability Limited Deployment) and can be designated as either active or inactive by screen entry. Aircraft trajectories are constantly tested against the dimensions of active areas and alerts issued to the applicable sectors when violations are predicted.

special emergency — A condition of air piracy, or other hostile act by a person(s) aboard an aircraft, which threatens the safety of the aircraft or its passengers.

special fastener — A fastener that differs in any respect from recognized standards.

special flight permit — A flight permit issued to an aircraft that does not meet airworthiness requirements but is capable of safe flight. A special flight permit can be issued to move an aircraft for the purposes of maintenance or repair, buyer delivery, manufacturer flight tests, evacuation from danger, or customer demonstration. Also referred to as a ferry permit.

special instrument approach procedure — A procedure approved by the FAA for individual operators, but not published in FAR 97 for public use.

special use airspace — Airspace of defined dimensions identified by an area on the surface of the Earth wherein activities must be confined because of their nature and/or wherein limitations can be imposed upon aircraft operations that are not a part of those activities.

Types of special use airspace are:
a. Alert Area — Airspace that can contain a high volume of pilot training activities or an unusual type of aerial activity, neither of which is hazardous to aircraft. Alert Areas are depicted on aeronautical charts for the information of nonparticipating pilots. All activities within an Alert Area are conducted in accordance with Federal Aviation Regulations, and pilots of participating aircraft as well as pilots transiting the area are equally responsible for collision avoidance.
b. Controlled Firing Area — Airspace wherein activities are conducted under conditions so controlled as to eliminate hazards to nonparticipating aircraft and to ensure the safety of persons and property on the ground.
c. Military Operations Area (MOA) — A MOA is airspace established outside of Class A airspace area to separate or segregate certain nonhazardous military activities from IFR traffic and to identify for VFR traffic where these activities are conducted.

d. Prohibited Area — Airspace designated under Part 73 within which no person may operate an aircraft without the permission of the using agency.
a. e.Restricted Area — Airspace designated under FAR Part 73, within which the flight of aircraft, while not wholly prohibited, is subject to restriction. Most restricted areas are designated joint use and IFR/VFR operations in the area can be authorized by the controlling ATC facility when it is not being utilized by the using agency. Restricted areas are depicted on enroute charts. Where joint use is authorized, the name of the ATC controlling facility is also shown.
e. Warning Area — A warning area is airspace of defined dimensions extending from 3 nautical miles outward from the coast of the United States, that contains activity that can be hazardous to nonparticipating aircraft. The purpose of such warning areas is to warn nonparticipating pilots of the potential danger. A warning area can be located over domestic or international waters or both.

special VFR (SVFR) conditions — Weather conditions that are less than basic VFR weather conditions, but permit flight under VFR within the lateral boundaries of the surface areas of Class B, C, D, or E airspace designated for an airport. ATC authorization must be received.

special VFR flight [ICAO] — A VFR flight cleared by air traffic control to operate within Class B, C, D, and E surface areas in meteorological conditions below VMC.

special VFR operations — Aircraft operating in accordance with clearances within Class B, C, D, and E surface areas in weather conditions less than the basic VFR weather minima. Such operations must be requested by the pilot and approved by ATC.

specialized instructional services — Ground or flight instruction that is oriented to building knowledge and skills in areas other than training pilots for specific certificates or ratings. Some examples include flight reviews, instrument proficiency checks, and transition training. Others involve aircraft checkouts for tailwheel, complex, high-performance, and high-altitude airplanes. In addition, a flight instructor can help experienced military pilots transition to general aviation aircraft or have an opportunity to provide instruction in homebuilt airplanes. Regional or local checkouts are additional areas where specialized instruction can benefit pilots and increase safety.

specific fuel consumption — Number of pounds of fuel consumed in one hour to produce 1 HP.

specific gravity — The ratio of the weight of a given volume of a material to the same volume of pure water.

specific gravity adjustment — A fuel control adjustment that changes the fuel scheduling for use of fuels with differing specific gravities.

specific heat — 1. The amount of heat required to raise the temperature of one gram of a substance one degree Celsius. 2. The ratio of the amount of heat required to raise the temperature of a body 1°, compared with the amount of heat required to raise the temperature of an equal mass of water 1°.

specific humidity — The percentage, by weight, of water vapor in an air sample.

specific thrust — A ratio of mass airflow and net thrust. One means of comparison between engines.

specific weight — Density expressed in pounds per cubic feet (lb./cu. ft.).

specifications — Data concerning dimensions, weights, performance, locations, etc.

spectrometric oil analysis — A system of oil analysis in which a sample of oil is burned and the resulting light is examined for its wavelengths. This test can determine the amount of metals in the oil and can give warning of an impending engine failure.

spectro-photometer — A special device used to determine the way a surface reflects light waves of all frequencies. It is used to analyze paint pigments.

speed — The act of moving swiftly, or the rate of movement.

speed adjustment — An ATC procedure used to request pilots to adjust aircraft speed to a specific value for the purpose of providing desired spacing. Pilots are expected to maintain a speed of plus or minus 10 knots or 0.02 mach number of the specified speed.

Examples of speed adjustments are:
a. "Increase/reduce speed to mach point (number)."
b. "Increase/reduce speed to (speed in knots)" or "Increase/reduce speed (number of knots) knots."

speed brakes — A control system that extends from the airplane structure into the slipstream to produce drag and slow the airplane.

speed handle — A crank-shaped handle used to turn socket wrenches more quickly.

speed of light — The speed at which light travels in a vacuum. 299,792.5 km/sec or 186,282 miles/sec.

speed of sound — The speed at which sound waves travel. At sea level, under standard atmospheric conditions, sound travels 760 mph, 340 m/sec, or 1,116 ft./sec. Referred to as Mach-one. Symbol M.

speed segments — Portions of the arrival route between the transition point and the vertex along the optimum flight path for which speeds and altitudes are specified. There is one set of arrival speed segments adapted from each transition point to each vertex. Each set can contain up to six segments.

speed sensitive switch — An automatic, flyweight-operated sequencing switch driven by the engine gearbox. Used for completing electrical circuits for starting, ignition, fuel, etc.

speed to fly — In gliders, the optimum speed through the (sinking or rising) air mass to achieve either the furthest glide or fastest average cross-country speed depending on the objectives during a flight.

speeder spring — The control spring used in a centrifugal governor to establish a reference force opposed by the centrifugal force of the spinning flyweights.

speed-rated engine — A gas turbine with a rated thrust guaranteed to occur at a certain speed.

sphere — A geometric shape enclosed by a surface on which all points are an equal distance from an enclosed point called the center.

spherical — Having the form of a sphere.

spider — The portion of a propeller assembly used to support propeller blades.

spike — A transient condition of increased voltage/current in an electrical circuit when the circuit is first closed.

spillage — The movement of air from the bottom of the wing to the top, outward and upward over the wing tip. It is the cause of wing tip vortices.

spin — An aggravated stall that results in an airplane descending in a helical, or corkscrew, path.

spin certification — To be eligible for the flight instructor-airplane or flight instructor-glider practical test, an applicant must present a logbook endorsement from an appropriately certificated and rated flight instructor certifying spin competency.

spin training — When applying for an initial flight instructor certificate, CFI students need to accomplish specific spin training in accordance with FAR Part 61.183.

spindle — The threaded part of a micrometer turned by the thimble that moves in and out of the frame.

spinner — The streamlined, bullet-shaped fairing that encloses a propeller hub assembly. A spinner streamlines the propeller installation and contributes to engine cooling.

spiral — A maneuver in which an airplane descends in a helix of small pitch and large radius, with the angle of attack within the normal range of flight angles.

spiral flutes — Twisted grooves that run from one end of an object to the other. The grooves on a twist drill are spiral flutes.

spirit level — A measuring tool used to determine the relationship between a body and the horizon. The measuring element is a curved glass tube filled with liquid, but having a single bubble. The position of the bubble in the tube is used to indicate the relationship.

spirit varnish — A wood finishing material made of resin dissolved in solvent. The varnish forms a hard, resin film on the wood when it dries.

splayed patch — A flush repair to a wood surface where the edges of the patch are tapered, but the slope is steeper than allowed in scarfing operations.

splice — A process where two ends of a material are joined together.

splice connectors — Devices, such as insulated solderless connectors, used for permanently connecting two ends of electrical wiring.

splice knot — A knot used for joining two pieces of waxed rib-stitch cord. Unlike a square knot, a splice knot will not slip.

spline — Any of a series of uniformly spaced ridges on a shaft, parallel to its axis and fitting inside corresponding grooves in the hub of a part.

splined shaft — A shaft with a series of grooves that meshes with a drive unit 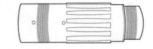 similarly configured so that the two units rotate as one, but are able to move laterally without interrupting the rotation.

split flaps — Wing flaps in which a portion of either the underside or the trailing edge of the wing splits and folds downward to increase lift and drag.

split lock washer — A heavy spring, steel lock washer split at an angle across its face and twisted. Used with machine screws or bolts where the self-locking or castellated-type nut is not appropriate. The spring action of the washer provides enough friction to prevent loosening of the nut from vibration.

split needled — The position of the two hands on a helicopter's engine/rotor tachometer. The two hands are not superimposed and the engine RPM and rotor RPM are unequal.

split needles — A helicopter tachometer having two needles: one shows engine speed, the other shows rotor system speed. When the clutch is fully engaged and the rotor is coupled to the engine, both needles show as one. When the needles are split, it indicates that the rotor clutch is not fully engaged.

split spool compressor/turbine — Multiple-spool compressor/turbine. A turbine engine with two or more separate spools (paired compressor and turbine) that allows the engine to be more efficient and responsive to varying power requirements.

split steel lock washer — A heavy steel washer split and twisted to provide enough tension between the nut and the surface of the material to prevent the nut from loosening.

split-lock keys — Split, tapered, cylindrical wedges used to lock the valve spring retainers to the stem of a poppet valve in an aircraft reciprocating engine.

split-phase induction motor — An electric motor with an auxiliary winding that is out of phase with the main winding. The auxiliary winding helps to start rotate the motor and is disconnected when the motor reaches operating speed.

split-phase motor — An AC motor that utilizes an inductor or capacitor to shift the phase of the current in one of two field windings. This causes the resultant field to have a rotational effect.

spoiler — Any device used to spoil lift by disrupting the airflow over an aerodynamic surface.

spoilers — Devices on the tops of wings to disturb (spoil) part of the airflow over the wing. The resulting decrease in lift creates a higher sink rate and allows for a steeper approach.

spokeshave — A small woodworking tool with a blade mounted between two handles. The blade is drawn toward the operator and is used to round the edges of sawn lumber.

spongy brakes — A brake malfunction caused by air in the hydraulic fluid. Since air is compressible, the braking action will not have a positive feel. It will feel as though there were a sponge or spring between the brake and the brake pedal.

sponson — 1. A flange, or stub, projecting from the side of a flying boat hull to increase the beam of the hull and improve the lateral stability of the aircraft on the water. 2. A projection from the side of the fuselage to support auxiliary equipment such as weapons or cameras.

spontaneous combustion — A condition that exists when the temperature of a substance reaches its kindling temperature and self-ignites. No spark is required to start the fire.

spontaneous ignition — See spontaneous combustion.

spool — In an axial flow compressor, the spool shaped drum on which several stages of compressor blades are mounted.

spot check — The random selection and inspection of manufactured products. The parts checked represent the quality of all the parts in that particular batch.

spot facing — The process of using a rotary tool to remove a small amount of surface material around a hole.

spot welding — An electrical resistance welding where current is passed through sheets of metal stacked together. When metal between the electrodes melts, it forms a button of metal, joining the sheets.

spotlight — A strong beam of brilliant light used to illuminate a particular area.

spot-type fire detection system — Bi-metallic thermoswitches that close to initiate a fire warning signal any time the temperature in an area reaches a predetermined value.

sprag clutch — A clutch joining two rotating shafts. The clutch will ratchet and disengage when the driven shaft turns faster than the driving shaft.

sprag mount — An adjustable bracing system used on the Bell 47 series helicopters.

spray bar — An afterburner fuel nozzle that protrudes into the exhaust stream.

spray paint gun — An atomizing gun and reservoir device that sprays liquid paint or finishing material onto a surface being painted.

spray painting — A method of applying finish to a surface using an atomizing gun and a reservoir filled with properly thinned paint.

spray strip — Metal strips mounted on the side of a flying boat hull used to divert water away from the aircraft.

spreader bar — A horizontal bar separating the floats of a twin-float seaplane.

spring coupling — A spring loaded device in a gear drive train that protects a system from excessive shock loads.

spring steel — Steel containing carbon in percentages ranging from 0.50 to 1.05%. In the fully heat-treated condition, spring steel is very hard, withstands high shear and wear, and resists deformation. Used for making flat springs and wire for coil springs.

spring tab — An auxiliary airfoil set into a control surface that, under conditions of high control forces, acts as a servo-tab providing an aerodynamic assist for the pilot.

springback — The amount that metal springs back after it has been bent to a specific angle. This must be allowed for when making bends in sheet metal.

spring-loaded — A condition where a spring holds one part in a specific relationship with another. Spring loading usually allows for some movement but returns the parts to their original relationship.

springwood — The light portion of wood that represents fast-growing wood from the spring of the year. Slower growing wood of the summer is darker. The two colors create the rings seen in the cross section of a tree.

sprue — The filler hole in a mold. The resulting extension (the shape of the filler hole) of the part is also referred to as the sprue.

spur and pinion reduction gear system — A gear system used in reduction gearing. In a prop reduction gearing system, the spur and pinion gears consist of a large driving gear, or sun gear, splined to the shaft. There is also a large stationary gear, called a bell gear, and a set of small spur planetary pinion gears mounted on a carrier ring. When the engine is operating, the sun gear rotates. Because the planetary gears are meshed with this ring, they rotate around the sun gear. The ring in which the planetary gears are mounted rotates the prop shaft in the same direction as the crankshaft but at a reduced speed.

spur gear — An external toothed gear.

squall — A sudden increase in wind speed by at least 15 knots to a peak of 20 knots or more and lasting for at least one minute. Essential difference between a gust and a squall is the duration of the peak speed.

squall line — A line of thunderstorms often located along or ahead of a vigorous cold front. Squall lines can contain severe thunderstorms. The term is also used to describe a line of heavy precipitation with an abrupt wind shift but no thunderstorms, as sometimes occurs in association with fronts.

square — 1. A plane geometric shape having four equal sides with all four angles being right angles. 2. The mathematical process of multiplying a number by itself.

square engine — An engine whose bore and stroke are equal.

square file — A double-cut file with a square cross section, tapered lengthwise. Used for filing slots for keyways.

square knot — A knot made up of opposite loops, each one enclosing the parallel sides of the other.

square mil — An area equivalent to a square having sides 1 mil (0.001 in.) in length.

square root — The factor of a number that, when squared, will give the number. For example, the square root of 25 is ± 5.

square wave — The waveform of a multi-vibrator oscillator where the leading edge and the trailing edge of the wave are both vertical.

squaring shears — A large, floor-mounted sheet metal tool used to make square cuts across sheets of metal.

squat switch — An electrical switch mounted on one of the landing gear struts. It is used to sense when the weight of the aircraft is on the wheels.

squawk (mode, code, function) — Activate specific modes/codes/functions on the aircraft transponder; e.g., "Squawk three/alpha, two one zero five, low." See also transponder.

squealing brakes — A noise made by glazed brakes chattering at such a high frequency that the sound resembles a squeal rather than a hammering.

squeegee — A long-handled rubber scraper blade used to remove liquid from a surface.

squeeler tip — A tip of reduced thickness at the outer end of rotor blades. This section is designed to wear away rather than damage the shroud ring if tip loading forces cause contact.

squeeze bottle — A soft plastic bottle that, when squeezed, forces the bottle contents out through its top.

squeeze riveter — A pneumatic or hydraulic riveting gun in which sets for both the manufactured head and the upset head are mounted in the jaws of the large clamp. When the squeeze-gun is actuated, the jaws come together just enough to form the proper size upset head. Also referred to as a rivet squeezer.

squelch — A circuit in a communications receiver holding the output volume down until a signal is received.

squelch circuit — See squelch.

squib — A small electric device used to ignite a charge. Found on aircraft where an action needs to be controlled remotely, such as a fire extinguisher bottle in an engine nacelle.

squirrel-cage induction motor — An induction motor whose rotor resembles a squirrel cage.

SRM (structural repair manual) — A manual developed by the manufacturer that covers all items not listed as minor maintenance, including instructions for structural repair, major component removal, installation, and adjustment, setup, etc. Contains manufacturer-approved data for major repairs and replacement.

St. Elmo's Fire — A luminous brush discharge of electricity from protruding objects, such as masts and yardarms of ships, aircraft, lightning rods, steeples, etc., occurring in stormy weather.

stabilator — A single-piece, horizontal tail component that combines the functions of a stabilizer and an elevator.

stability — 1. The property of a body that causes it, when disturbed, to return to its original condition. 2. In meteorology, a state of the atmosphere in which the vertical distribution of temperature is such that a parcel will resist displacement from its initial level.

stabilizer — The fixed horizontal and vertical tail surfaces having the elevators and rudder hinged to the trailing edges.

stabilizer bar — A dynamic component used on some Bell helicopters to insure rotor stability.

stable operation — An operating condition where there is no appreciable fluctuation in any of the operational variables.

stable oscillation — An oscillation whose amplitude does not increase.

stage — In turbine engine construction, a single turbine wheel having a number of turbine blades.

stagger — The longitudinal relationship of the wings of a biplane. If the upper wing is forward of the lower wing, the airplane is said to have positive stagger.

stagger angle — Refers to blade twist design in an impulse-reaction turbine blade.

staggered ignition — Dual ignition timed so that two firing impulses do not occur at the same time.

staggered rings — A method of orienting the compression and oil control rings on a piston so that the gaps are not aligned with each other. This improves the overall compression of the engine.

staggered timing — A reciprocating engine ignition timing method using a dual ignition system to provide spark plug firing at different points in the combustion cycle.

staggerwires — The wire between the cabane struts of a biplane. These wires are used to adjust the stagger.

staging/queuing — The placement, integration, and segregation of departure aircraft in designated movement areas of an airport by departure fix, EDCT (expect departure clearance time), and/or restriction.

stagnation point — The point on the leading edge of an airfoil where the airflow separates and results in some going over the surface and some below.

stain — An agent applied to wood to change the color and/or emphasize the grain. A protective finish such as varnish is usually applied over the stain.

stainless steel — Steel, containing appreciable quantities of chromium and nickel and used for applications where resistance to corrosion is important.

stake — Small, bench-mounted, anvil-type sheet metal tool.

staking — A term used to denote the swaging of terminals onto an electrical conductor.

stall — Condition that occurs when the critical angle of attack is reached and exceeded. Airflow begins to separate from the top of the wing, leading to a loss of lift. A stall can occur at any pitch attitude or airspeed.

stall strip — A spoiler attached to the inboard leading edge of some wings causing the center section of the wing to stall before the tips. This assures aileron control throughout the stall.

stall warning transmitter — A device that produces a signal to warn the pilot of an impending stall.

stall/spin awareness — Stall/spin awareness training is designed to instill a trigger or early warning in the mind of a pilot, which causes the proper reaction to potential stall/spin situations in an immediate, positive way. This training is required by the FAR's. Many stall/spin accidents occur as a result of pilot distraction while maneuvering close to the ground.

stalled torque — The amount of rotative force provided by a motor when properly powered with the rotor held stationary.

stalling angle — The angle of attack at which point the smooth flow of air over the wing ceases.

stall-warning system — A system that warns the operator when the aircraft is approaching the critical stall angle.

stand by — Means the controller or pilot must pause for a few seconds, usually to attend to other duties of a higher priority. Also means to wait as in "stand by for clearance." The caller should reestablish contact if a delay is lengthy. "Stand by" is not an approval or denial.

standard — The test of quality required for a particular purpose.

standard altimeter setting — Altimeter set to the standard pressure of 29.92 in. Hg, or 1013.2 Mb.

standard atmosphere — At sea level, the standard atmosphere consists of a barometric pressure of 29.92 inches of mercury (in. Hg.) or 1013.2 millibars, and a temperature of 15°C (59°F). Pressure and temperature normally decrease as altitude increases. The standard lapse rate in the lower atmosphere for each 1,000 feet of altitude is approximately 1 in. Hg. and 2°C (3.5°F). For example, the standard pressure and temperature at 3,000 feet mean sea level (MSL) is 26.92 in. Hg. (29.92 - 3) and 9°C (15°C - 6°C).

standard atmospheric conditions — See standard atmosphere.

standard barometric pressure — The weight of gases in the atmosphere sufficient to hold up a column of mercury 760 mm high (approximately 30") at sea level (14.7 psi). This pressure decreases with altitude.

standard briefing — The most complete weather picture, tailored to your specific flight. Usually the briefing includes adverse conditions, a weather synopsis, current weather, forecast weather, forecast winds and temperatures aloft, alternate routes, NOTAMs, ATC delays, and request for PIREPs.

standard cell — A cadmium-mercury cell made in a specially shaped glass container. The two electrodes are covered with an electrolyte of cadmium sulfate. Voltage produced by a standard cell is 1.018636 volts at 20°C. Also referred to as a Weston standard or a Weston normal cell.

standard day — Sea level, dry air, 59° F (15° C), no wind, 40° North or South latitude.

standard fastener — A fastener that conforms in all respects to recognized standards.

standard instrument approach procedure — See instrument approach procedure.

standard instrument departure (SID) — A preplanned instrument flight rule (IFR) air traffic control departure procedure printed for pilot use in graphic and/or textual from. SIDs provide transition from the terminal to the appropriate enroute structure.

standard lapse rate — For 1,000 feet of altitude in the lower atmosphere (below 36,000 feet), the standard pressure lapse rate is 1.00 in. Hg., and the standard temperature lapse rate is 2°C (3.5°F).

standard rate turn — A turn of three degrees per second.

standard sea level pressure — A standard value of pressure used as a reference for making aerodynamic computations. It is 14.7 lb./sq. in., 29.92 inches of mercury, or 1013.2 millibars.

standard sea level temperature — A surface temperature of 59<\#161>F or 15<\#161>C. See also standard atmosphere.

standard temperature — 15° C (59° F).

standard terminal arrival (STAR) charts — Charts designed to expedite air traffic control arrival procedures and to facilitate transition between enroute and instrument approach operations. Each STAR procedure is presented as a separate chart and can serve a single airport or more than one airport in a given geographical location.

standard terminal arrival — A preplanned instrument flight rule (IFR) air traffic control arrival procedure published for pilot use in graphic and/or textual form. STARs provide transition from the enroute structure to an outer fix or an instrument approach fix/arrival waypoint in the terminal area.

standard terminal arrival route (STAR) — A preplanned instrument flight rule (IFR) air traffic control arrival procedure published for pilot use in graphic and/or textual form. STARs provide transition from the enroute structure to an outer fix or an instrument approach fix/arrival waypoint in the terminal area.

standard-frequency signal — The National Institute of Standards and Technology (NIST) broadcasts time and frequency information via WWV in Fort Collins, Colorado. The information is broadcast on frequencies of 2.5, 5, 10, 15, and 20 MHz.

standard-rate turn — A rate of change in aircraft direction of 3° per second, completing a 360° turn in two minutes. Higher speed jet aircraft use a half-standard-rate turn of 1½ ° per second, completing a 360° turn in four minutes.

standing waves — Stationary waves occurring on an antenna or transmission line as a result of two waves, identical in amplitude and frequency, traveling in opposite directions along the conductor.

standpipe — A vertical standing pipe in a tank or reservoir. It allows a space for a reserve of fluid between the top of the standpipe and the bottom of the tank. Reserve fluid is drawn from the bottom of the tank.

staple — **1.** Wire: A u-shaped piece of fine wire with sharp-pointed ends, driven into a surface to hold or fasten material to it. **2.** Textile material: The average length of textile material. Cotton fibers. **3.** A chief item: Any chief item, part, or element of a raw material.

stapler — A machine or tool used for driving staples through a piece of wood or stack of paper for the purpose of binding them together.

starboard side — The right-hand side of an aircraft or ship as viewed from aboard the craft and facing the front of the craft.

start winding — An auxiliary winding that is out of phase with the main winding. The auxiliary winding helps to start rotate the motor and is disconnected when the motor reaches operating speed.

starter — A unit that uses electrical, pneumatic, or hydraulic energy to rotate an engine for starting.

starter solenoid — An electrically operated switch that uses a small current controlled from the cockpit to close the high current-carrying contacts in the starter circuit.

starter-generator — A combined unit used on turbine engines The device acts as a starter for rotating the engine, and after running, internal circuits are shifted to convert the device into a generator.

starting torque — The amount of rotating force a motor develops during start. Series-wound motors develop high starting torque while shunt- (parallel) wound motors develop very low starting torque.

starved area — In composites, an area in a plastic part that has an insufficient amount of resin to wet out the reinforcement completely.

starved joint — In composites, an adhesive joint that has been deprived of the proper film thickness of adhesive due to insufficient adhesive spreading or to the application of excessive pressure during the lamination process.

statcoulomb — The amount of charge on each of two bodies 1 cm apart that causes them to exert a force of 1 dyne on each other. The statcoulomb is the charge resulting from the addition of approximately 2×10^9 electrons to a body.

state aircraft — Aircraft used in military, customs and police service, in the exclusive service of any government, or of any political subdivision, thereof including the government of any state, territory, or possession of the United States or the District of Columbia, but not including any government-owned aircraft engaged in carrying persons or property for commercial purposes.

state of charge — A measurement of the percent of charge condition of a battery. For lead-acid batteries, the state of charge is measured by determining the specific gravity of the electrolyte. The state of charge of a nickel-cadmium battery can only be determined by a measured discharge.

statement of demonstrated ability (SODA) — The official term for a waiver. This is a form that can be issued in conjunction with a student's medical exam. SODAs can only be issued by the FAA's federal air surgeon and are granted for a condition normally requiring a denial that is not necessarily a safety factor.

static — **1.** Still. Not moving. A condition of rest. **2.** The noise produced in a radio or television receiver by atmospheric or man-made electrical disturbances.

static balance — **1.** A condition of balance that does not involve any dynamic forces. **2.** When a body stands in any position as the result of counterbalancing and/or reducing heavy portions, it is said to be in standing or static balance.

static charge — The electrical charge that builds up on a nonconductive surface by friction. Friction between the airframe and the air creates a static charge.

static discharger — A device used to dissipate static electricity from a control surface before it builds up to a highly charged state.

static electricity — An electrical charge that can be built up on a nonconductive surface by friction.

static flux — Concentration of lines of flux in the frame of a magneto due to the rotation of the magnet. At full register, lines of flux are at maximum, while at neutral positions, lines of flux are at minimum.

static friction — The friction on an object when an attempt is made to slide the object along a surface. Once in motion the object slides more easily.

static instability — The characteristic of an aircraft that, when disturbed from a condition of rest, tends to move it further from its original condition.

static interference — The noise in a radio caused by static electricity moving between two structures having no common ground.

static port — A small hole, flush with the side of the aircraft, through which static pressure is taken to operate the airspeed indicator, altimeter, and vertical speed indicator.

static pressure — Atmospheric pressure measured at a point where there is no external disturbance and the flow of air over the surface is smooth.

static pressure pickup — A part of the static instrument system. The location on the surface of an aircraft where static air pressure is picked up from port holes or static ports. This data is supplied to the altimeter, airspeed indicator, and vertical speed indicator.

static radial engine — An engine with cylinders radiating out from a small central crankcase. A single-throw crankshaft is used for each row of cylinders. Most single-row radial engines have an odd number of cylinders, but two or more rows can be used if more power is required.

static restrictions — Those restrictions that are usually not subject to change, fixed, in place, and/or published.

static RPM — The maximum RPM a reciprocating engine can produce when the aircraft is not moving through the air. The static RPM is lower than the RPM the engine develops while airborne because forward movement rams additional air pressure into the carburetor inlet.

static stability — The initial tendency to return to a state of equilibrium when disturbed from that state.

static stop — In rotorcraft, a device used to limit the blade flap, or rotor flap, at low r.p.m. or when the rotor is stopped.

static system — Plumbing that connects the altimeter, airspeed indicator, and vertical speed indicator to the outside static air source of the airplane. An alternate source is usually included in this system.

static temperature — A temperature measurement of air not in motion.

static test — A method of testing the structural integrity of an airplane to determine its ability to withstand loads that could possibly be encountered in flight.

static thrust — The thrust produced by a turbine engine not moving through the air.

static tube — A cylindrical tube with a closed end and a number of small openings normal to the axis, pointed upstream, and used to measure static pressure.

static wick — A small device made of metal braid or graphite-impregnated cotton attached to the trailing edge of a control surface to dissipate accumulated electrical charges into the air.

station — The location of a point within an aircraft identified by distance in inches from the datum.

station "0"(zero) — The reference point from which fuselage stations are measured. On some aircraft the point might be the firewall or the leading edge of the wing. Stations in front of this would have negative station numbers and those aft would have positive station numbers. Station zero is sometimes placed at a point in space in front of the aircraft so that all station numbers are positive.

station declination — The orientation with respect to true north of VHF transmitted signals. The orientation is originally made to agree with the magnetic variation (an uncontrollable global phenomenon) at the site. Hence station declination (fixed by man) may differ from changed magnetic variation until the station is reoriented.

station pressure — The actual atmosphere pressure at the observing station.

station web — A built-up section located at some point of applied force, such as attachments for wings, stabilizer, etc.

stationary front — Same as quasi-stationary front.

stationary reservations — Altitude reservations that encompass activities in a fixed area. Stationary reservations can include activities, such as special tests of weapons systems or equipment, certain U.S. Navy carrier, fleet, and anti-submarine operations, rocket, missile and drone operations, and certain aerial refueling or similar operations.

stator — 1. The stationary part of an electrical machine such as a motor or alternator. 2. The stationary portion of an axial flow turbojet compressor. 3. The discs in a multiple-disc brake that are keyed to the axle and do not rotate.

stator case — The outer engine casing that houses either compressor or turbine stator vanes.

stator vane — Stationary vane, either compressor or turbine.

statute mile — A measure of land distance equal to 5280 ft. or 1.609 km.

stay — A cable or wire, loaded in tension only, used as a structural member in a truss.

steady-state condition — The condition of any system that exists when all of the measured values are stable.

steady-state flight — A condition when a rotorcraft is in straight-and-level, unaccelerated flight, and all forces are in balance.

steam — An invisible vapor created when water is heated to the boiling point

steam fog — Sea smoke, evaporation fog.

steatite — The massive form of talc. Used for making ceramic insulating material for high-voltage systems.

steel — A hard and tough carbon-iron alloy.

steel wool — An abrasive material made of long, fine, steel shavings and used for scouring steel parts.

steering damper — A hydraulically actuated device used to absorb shimmy vibrations from the nose wheel of an aircraft.

stellite — An extremely hard, wear-resistant metal used for valve faces and stem tips. Contains cobalt, tungsten, chromium, and molybdenum.

stem — The part of a multiple-choice test item consisting of the question, statement, or problem.

step — A break in the form of the bottom of a float or hull. Designed to diminish resistance, lessen the suction effects, and improve longitudinal control.

step leader — The first of a series of events that make up lightning. Nearly invisible to the eye, it is the path that carries electrons from the base of the clouds to the ground, creating an ionized channel for the subsequent discharge.

step taxi — To taxi a float plane at full power or high RPM in order to obtain a speed that causes the plane to rise up on the planing portion of the floats.

step turn — A maneuver used to put a float plane in a planing configuration prior to entering an active sea lane for takeoff. The STEP TURN maneuver should only be used upon pilot request.

step up coil — A transformer where a secondary winding has more turns than the primary. The voltage in the secondary winding will be stepped up.

stepdown fix — A fix permitting additional descent within a segment of an instrument approach procedure by identifying a point at which a controlling obstacle has been safely overflown.

step-down transformer — A device that steps down voltages and is made with an iron core, a primary winding, and a secondary winding. The step-down transformer has more turns of wire in the primary winding than the secondary winding. This difference determines the stepped down secondary voltage.

stepped solvents — Solvents in a finish that have different rates of evaporation. Some evaporate almost instantly, while others evaporate slowly. This variance provides the desired film.

stepped stud — A stud replacement for one that has been stripped out. The hole is drilled and tapped for a larger stepped stud.

stepping relay — A rotary electrical switch that switches from one set of contacts to another whenever its solenoid receives a pulse.

step-up transformer — A device that steps up voltages and is made with an iron core, a primary winding, and a secondary winding. The step-up transformer has less turns of wire in the primary winding than the secondary winding. This difference, depending on the turns-ratio, determines the step-up of the secondary voltage.

stereo route — A routinely used route of flight established by users and ARTCCs identified by a coded name; e.g., ALPHA 2. These routes minimize flight plan handling and communications.

sterile cockpit — A crew resource management concept that specifically prohibits crewmember performance of nonessential duties or activities while the aircraft is involved in taxi, takeoff, landing, and all other flight operations conducted below 10,000 feet MSL, except for cruise flight.

stiffener — A structural member attached to an aircraft skin for the purpose of making it stiffer. It is quite often an extruded angle or a formed hat-shaped section.

stiffness — The relationship of load and deformation. The ratio between the applied stress and resulting strain.

stimulants — Drugs that excite the central nervous system and produce an increase in alertness and activity

Stoddard solvent — A petroleum product similar to naphtha used as a solvent or cleaning agent.

stoichiometric — A chemical relationship in which all of the constituents are used in the reaction. In the case of a stoichiometric mixture, all of the oxygen and hydrocarbon fuel are used.

STOL aircraft — An aircraft that can perform short takeoff and landing procedures.

stop — A device used to limit the throw or travel of a control.

stop altitude squawk — Used by ATC to inform an aircraft to turn-off the automatic altitude reporting feature of its transponder. It is issued when the verbally reported altitude varies 300 feet or more from the automatic altitude report.

stop and go — A procedure wherein an aircraft will land, make a complete stop on the runway, and then commence a takeoff from that point.

stop burst — Used by ATC to request a pilot to suspend electronic countermeasure activity.

stop buzzer — Used by ATC to request a pilot to suspend electronic countermeasure activity.

stop countersink — A countersink with a collar that doesn't allow the cutter to cut too deeply into the metal skin.

stop drill — A hole drilled in the end of a crack in aircraft structural material to distribute stresses and stop the crack from proceeding further.

stop nut — A nut used to prevent another nut from backing off due to vibrations. A stop nut is either a self-locking nut or a nut that can be torqued against the primary nut while the primary nut is being held in place.

stop squawk (mode or code) — Used by ATC to tell the pilot to turn specified functions of the aircraft transponder off.

stop stream — Used by ATC to request a pilot to suspend electronic countermeasure activity.

stopover flight plan — A flight plan format that permits in a single submission the filing of a sequence of flight plans through interim full-stop destinations to a final destination.

stopway — An area beyond the takeoff runway, no less wide than the runway and centered upon the extended centerline of the runway, able to support the airplane during an aborted takeoff, without causing structural damage to the airplane, and designated by the airport authorities for use in decelerating the airplane during an aborted takeoff.

storage battery — A secondary cell. An electrical device in which electrical energy is converted into chemical energy and stored. When needed it is converted back into electrical energy.

storage life — In composites, the period of time during which a liquid resin, packaged adhesive, or pre-preg can be stored under specified temperature conditions and remain suitable for use. The storage life should be printed on the label. Also referred to as shelf life.

storm detection radar — A radar optimized to show returns of precipitation rather than clouds as in a normal weather radar.

straight mineral oil — Oil, such as petroleum, derived from a mineral source as opposed to oils derived from plants and animals. Straight mineral oil would have no additives. Often used during break-in of an engine.

straight peen hammer — A metal beading hammer with one flat face and one face with a vertical edge.

straight roller bearings — Roller bearings used where the bearing is subjected to radial loads only.

straight shank drill — A twist drill with a straight shank, distinguishable from a twist drill with a tapered shank.

straightedge — Wood, metal, or plastic having a perfectly straight edge used in drawing straight lines, or to check for the straightness of a piece of material.

straight-in approach-IFR — An instrument approach wherein final approach is begun without first having executed a procedure turn, not necessarily completed with a straight-in landing or made to straight-in landing minimums.

straight-in approach-VFR — Entry into the traffic pattern by interception of the extended runway centerline (final approach course) without executing any other portion of the traffic pattern.

straight-in landing — A landing made on a runway aligned within 30⌋ of the final approach course following completion of an instrument approach.

straight-in landing minimums — A statement of MDA and visibility, or DH and visibility, required for straight-in landing on a specified runway.

straight-in minimums — See straight-in landing minimums.

straight-polarity arc welding — Electric arc welding where the electrode is connected to the negative terminal of the power supply.

straight-run gasoline — Gasoline that is refined from crude oil using the fractional distillation process to produce straight-run gasoline. In this method, the crude oil is heated at atmospheric pressure in a heating container. The various hydrocarbon liquids in the crude oil vaporize first, followed by those of higher boiling points. Straight-run gasoline has an octane rating of approximately 70.

strain — 1. Deformation in a material caused by stress. 2. The process of exerting a force beyond the normal physical capacity of the material.

strain gauge — An extremely tiny conductor bonded to the surface of a component on which strain is to be measured. When the surface stretches the cross section of the strain gauge becomes smaller and its resistance increases. The strain gauge is extremely sensitive and measures the change.

strain hardening — The increase in strength and hardness of a metal by work hardening or cold working. Strain hardening is normally done after a piece of material has been heat treated. If an aluminum alloy is not heat-treatable, strain hardening is the only way it can be hardened.

strainer — A very fine mesh screen located in the fuel system and used to remove impurities.

strand — Normally, an untwisted bundle or assembly of continuous filaments used as a unit. Sometimes a single fiber or filament is referred to as a strand.

stranded conductor — An electrical conductor made up of many strands of wire covered with an insulating material.

stranded wire — Electrical wire made up of many smaller wire strands.

strap pack — A tension-torsion system using sheet steel lamination to carry the loads of the rotor blades to the head. Used by Hughes Helicopters.

strategic planning — Planning whereby solutions are sought to resolve potential conflicts.

stratification — Formed in layers.

stratiform — Descriptive of clouds of extensive horizontal development, as contrasted to vertically developed cumuliform clouds; characteristic of stable air.

stratiform clouds — Clouds formed in layers.

stratocumulus — Stratified cumulus consisting of low gray clouds, which are formed in layers, and that often cover the whole sky especially in winter.

stratopause — The top stratosphere, approximately 160,000 feet MSL.

stratosphere — The first layer above the troposphere extending to a height of approximately 160,000 feet, with a composition much like the troposphere.

stratus — Low gray uniform clouds that generally extend over a large area at altitudes of 2000 to 7000 feet (600 to 2100 meters).

streamline flow — A fluid flow without turbulence. All lines of flow are in straight lines.

streamlined — Having a shape or contour that presents a minimum resistance to the air with a minimum of turbulence.

strength — The ability of a material to withstand forces that attempt to deform it. The ability of a material to resist stress without breaking.

strength-to-weight ratio — The ratio of a material's strength to its weight.

stress — The internal resistance or change in shape or size expressed in force per unit area. A stress concentration is an area where the level of an applied stress causes a notch, void, hole, or inclusion.

stress analysis — A mathematical determination of the loads experienced by a structure under specific circumstances.

stress corrosion — Intergranular corrosion that forms within metals subject to tensile stresses and in a corrosive environment. Exposure to such an environment alone would not have caused corrosion. Tends to separate the grain boundaries.

stress crack- In composites, external or internal cracks in a plastic caused by tensile stresses less than that of its short-time mechanical strength. The stresses that cause cracking can be present internally or externally or can be combinations of these stresses.

stress management — Personal analysis of stress being experienced and the application of appropriate coping mechanisms.

stress relieve — A general term describing the process of relieving internal stresses within metals by controlled heating and cooling. Annealing and normalizing are more specific terms for this process.

stress riser — A location on a part that due to shape, or due to a defect, is more prone to failure as the result of stress than the rest of the part. This could be due to a drastic change in cross-section or where the part has been gouged or scratched. Stresses become concentrated at such locations.

stressed-skin structure — Aircraft skin designed to carry the tension and compression stresses of structural loads. Stressed-skin structured aircraft have few internal structural members.

stretching — A sheet metal forming operation in which the material is mechanically stretched over dies to form compound curves.

stringer — A thin metal or wood strip running the length of the fuselage to fill in the shape of the formers.

stroboscope — A device that can be adjusted to flash at specific rates. It can be used to "freeze" the apparent motion of propellers, pulleys, belts, rotors, etc., in order to inspect the action as if it is stopped.

stroboscope tachometer — A variation of a stroboscope. The stroboscope tachometer is shown on a rotating unit and adjusted until apparent rotation is stopped. The sequenced flashing rate of the scope is then read to determine the RPM of the unit being observed.

stroke — In a reciprocating engine, the distance a piston travels from bottom dead center (BDC) to top dead center (TDC). Stroke is two times the crankshaft throw.

structural adhesive — In composites, adhesive used for transferring required loads between two cured parts. An adhesive can also be used to bond metal to a composite structure.

structural bond — A bond that joins basic load-bearing parts of an assembly. The load can be either static or dynamic.

structural failure — When a structure fails to withstand the stresses imposed upon it.

structural icing — The formation of ice on the exterior or structure of an aircraft.

structural machine screws — Machine screws having an unthreaded portion of the shank and made of high-strength alloy steel. Used in place of an aircraft bolt to carry shear loads and some tensile loads.

structural member — Any part of an aircraft structure designed to carry loads or stress.

structural steel — An alloy steel used for parts of an aircraft subjected to high structural loads.

strut — **1.** A compression member in a truss. **2.** The external bracing on a non-cantilever airplane. **3.** The stub wing assembly through which thrust loads are transmitted from a pod-mounted turbine engine into the fuselage.

strux — In composites, a foam like material used to form structural sections for stiffening.

stub antenna — A short, UHF, quarter-wavelength antenna normally used for radar beacon transponders or distance measuring equipment.

stud — A headless bolt that has threads on each end. One end often has coarse  threads for screwing into a casting, while the other end has fine threads to accept a nut.

stuffing box — A box through which a rotating shaft passes. It is packed with a material that inhibits leakage around the shaft.

styrene — A liquid hydrocarbon used in the manufacture of certain synthetic resins to improve their workability. Also, any of various synthetic plastics made from styrene by polymerization or copolymerization.

Styrofoam — A rigid polymer of styrene plastic material.

subassembly — An assembly that is a component of a larger assembly.

subfreezing temperature — Below freezing. Any temperature below the freezing point of water, which is 32° F or 0° C.

subject matter knowledge codes — The subject matter knowledge codes establish the specific reference for the knowledge standards on FAA knowledge examinations. These codes are associated with each FAA knowledge question and are referred to in the results returned to the examinee after an examination.

sublimation — Process by which a solid is changed to a gas without going through the liquid state. Such as, from ice to water vapor.

submerged-arc welding — A method of electric arc welding in which a bare rod, covered with granulated flux, is used as an electrode. The granulated flux melts in the arc and flows ahead of the weld to prevent the formation of oxides in the bead.

subrogate — A legal term, which means to pursue action against a third party determined to be responsible for an accident, and attempt to recover damages from them over the amount of the deductible.

subsidence — A slow descending motion of air in the atmosphere over a rather broad area; usually associated with divergence and stable air.

subsonic flight — Flight when the air flowing over the aircraft structure is moving slower than the speed of sound.

subsonic inlet — A divergent-shaped duct that acts as a subsonic diffuser.

subsonic speed — Speed below the speed of sound.

subsonic-diffuser — A divergent diffuser where the airstream spreads out to increase pressure as axial velocity decreases.

substandard — Unacceptable quality in a manufactured object.

substitute — The replacement of an object with the same or better quality material and which meets all of the specifications of the original.

substitute route — A route assigned to pilots when any part of an airway or route is unusable because of NAVAID status. These routes consist of:
a. Substitute routes shown on U.S. Government charts [and published in the Jeppesen Chart NOTAMs (Enroute)].
b. Routes defined by ATC as specific NAVAID radials or courses.
c. Routes defined by ATC as direct to or between NAVAIDs.

substitutions — Users are permitted to exchange CTA's (controlled time of arrival). Normally, the airline dispatcher will contact the ATCSCC with this request. The ATCSCC shall forward approved substitutions to the TMU's who will notify the appropriate terminals. Permissible swapping must not change the traffic load for any given hour of an EQF Program.

substrate — The supporting material on which an integrated circuit chip is built.

subsystem — An operating unit or assembly that is a component of a larger system.

subtropical jet stream — One of two jet streams commonly associated with the westerlies. Located near 25 to 30° latitude, it reaches its greatest strength in the wintertime and is nonexistent in the summer.

suction — The act of producing negative pressure.

suction cup gun — A paint gun in which material held in a cup is attached to the gun and drawn into the air by suction.

suction gauge — An aircraft instrument used to measure negative pressure or suction in an aircraft vacuum system.

suction relief valve — A control valve in an instrument pneumatic system that provides a constant negative pressure. It opens the system to the outside air when the vacuum rises above the preset value.

suction vortex — A small vortex, about thirty feet in diameter, embedded in a tornado funnel.

sudden stoppage — A condition in which the aircraft engine has come to a complete stop in less than one revolution, usually caused by the propeller hitting an immovable object. Sudden stoppage requires a special inspection to determine internal engine damage.

sulfate radical — A combination of chemical elements that acts as though it were only one atom. In the case of a sulfate radical, the SO_4 behaves in the chemical action of battery charging or discharging as though it were only one element.

sulfated — The condition of plates in a discharged lead-acid battery. The lead has turned to lead sulfate. If allowed to remain for a long period of time, the sulfate becomes impossible to remove by normal charging action.

sulfur — A pale yellow, non-metallic chemical crystal element with a symbol of S and an atomic number of 16.

sum — **1.** The result obtained by adding two or more numbers together. **2.** Total.

sump — **1.** A low area in a fuel tank where water will normally collect. **2.** The component of an aircraft engine used to hold the lubricating oil.

sump jar — A small jar in the vent line of a battery box containing a pad wet with a chemical such as bicarbonate of soda or boric acid. Fumes given off by the battery while it charges are neutralized by this material.

sun gear — The center gear in a planetary gear system around which the planetary gears rotate.

sunset and sunrise — The mean solar times of sunset and sunrise as published in the Nautical Almanac, converted to local standard time for the locality concerned. Within Alaska, the end of evening civil twilight and the beginning of morning civil twilight, as defined for each locality.

super high frequency — The frequency band between 3 and 30 gigahertz (gHz). The elevation and azimuth stations of the microwave landing system operate from 5031 MHz to 5091 MHz in this spectrum.

super high radio frequency — Frequencies between 3.0 and 30.0 gigahertz having wavelengths between 100 and 10 mm.

superadiabatic lapse rate — A lapse rate greater than the dry adiabatic lapse rate.

supercell thunderstorm — A severe thunderstorm that almost always produces one or more of the extremes of convective weather: Very strong horizontal wind gusts, large hail, and/or tornadoes. The supercell can occur anywhere in the mid-latitudes, but by far the favored area is the southern Great Plains of the United States. The supercell is so named because it requires extreme instability and a special combination of boundary layer and high-level wind conditions.

supercharger — An engine or exhaust driven air compressor used to provide additional pressure to the induction air so the engine can produce additional power.

supercharger control system — The system of controlling the supercharger to maintain a constant manifold pressure as the altitude changes.

supercharging — Increasing the cylinder pressure of a reciprocating aircraft engine by introducing compressed air into the cylinder on the intake stroke.

superconduction — The effect experienced at very low temperatures (approaching absolute zero) when atomic vibration ceases. Conduction electrons are free to drift through conductors without opposition or loss of energy.

superconductivity — A reaction of certain chemical elements when they are cooled and held at or near absolute zero degrees. At absolute zero degrees, these elements lose almost all of their electrical resistance and become strongly diamagnetic.

supercooled water — Water that has been cooled below the freezing point, but is still in a liquid state. Supercooled water forms solid ice as soon as it is disturbed.

supercritical wing — An aerodynamic wing design that enhances the range, cruising speed, and fuel efficiency of jet aircraft by producing weaker shock waves that create less drag and permits high efficiency.

superheat — The heat energy added to a gas after evaporation has been completed.

superheated vapor — Vapor that has been heated above its boiling point.

superheated water — The process of heating water to a temperature above 212°F when it normally changes from a liquid state to a vapor. Water can be superheated by heating it in a pressurized container. This is because more heat energy has to be added before the molecules move fast enough to become steam.

superheterodyne — A radio receiver circuit that mixes the radio frequency signal received with a frequency produced in a local oscillator to create an intermediate frequency.

super-refraction — A condition where radar beams are bent more than normal due to anomalies in temperature or humidity.

supersaturated — Any solution (such as sugar and water) in which a solid is dissolved until no more of the solid can be added. This amount varies according to the temperature of the liquid. As the temperature decreases more of the solid recrystallizes and drops out of the solution.

supersede — To make obsolete by replacement. When a manufacturer issues new maintenance manuals, the new manuals supersede the old ones.

supersonic — A speed greater than the speed of sound.

supersonic aerodynamics — The branch of aerodynamics that deals with the theory of flight at speeds faster than the speed of sound. Aero means pertaining to air; dynamics is that branch of physics that considers bodies in motion and the forces that produce changes of bodies in motion.

supersonic diffuser — A converging diffuser where the supersonic airstream pressure is raised as velocity decreases.

supersonic nozzle — A divergent-shaped duct designed to allow gases to expand outward faster than they accelerate rearward.

supersonic speed — Mach 1.0 to Mach 5.0.

supersonic transport — The British-French Concorde and the Russian TU-144. The only two types of supersonic commercial aircraft presently in existence.

superstructure — Framework attached to an aircraft truss structure to provide the desired aerodynamic shape. It is usually covered with lightweight sheet metal or aircraft fabric.

supplemental type certificate (STC) — A certificate authorizing an alteration to an airframe, engine, or component, which has been granted an Approved Type Certificate.

supplemental weather service location — Airport facilities staffed with contract personnel who take weather observations and provide current local weather

supplementary angles — Two angles that add up to 180°.

supply-type test items — Questions where the student supplies answers as opposed to selecting from choices provided. Essay or fill-in-the-blank type questions are examples of supply-type test items.

support clamp — A clamp used to support various fluid lines or wire bundles connected to the aircraft structure.

suppressor grid — The electrode in a pentode vacuum tube used to suppress secondary emissions from the plate.

SUPPS — Refers to ICAO Document 7030 Regional Supplementary Procedures. SUPPS contain procedures for each ICAO Region which are unique to that Region and are not covered in the worldwide provisions identified in the ICAO Air Navigation Plan. Procedures contained in chapter 8 are based in part on those published in SUPPS.

surface air temperature — In meteorology, the temperature of the air measured at 1.5 meters (about 5 feet) above the ground.

surface area — The airspace contained by the lateral boundary of the Class B, C, D, or E airspace designated for an airport that begins at the surface and extends upward.

surface corrosion — Oxidation across the surface of a metal. Primary protection against this corrosion is achieved by surface treatment such as painting, anodizing, or alodining.

surface friction — The resistive force that arises from the combination of skin friction and turbulence near the Earth's surface.

surface movement guidance and control system (SMGCS) — Provisions for guidance and control or regulation for facilities, information, and advice necessary for pilots of aircraft and drivers of ground vehicles to find their way on the airport during low visibility operations and to keep the aircraft or vehicles on the surfaces or within the areas intended for their use. Low visibility operations for this system means reported conditions of RVR 1200 or less.

surface tape — Strips of fabric made of the same material used to cover an aircraft structure. It is applied over the seams, rib-stitching, and edges to give the surface a smooth, finished appearance. Sometimes referred to as finishing tape.

surface tension — A cohesive condition that exists on the surface of a liquid because of molecular attraction.

surface treatment — **1.** Any treatment that changes the chemical, physical, or mechanical properties of a surface. **2.** In composites, a material (size or finish) applied to fibrous material during the forming operation or in subsequent processes. The process is used to enhance bonding capability of fiber to resin.

surface visibility — Visibility observed from eye-level above the ground.

surface-based inversion — An inversion with its base at the surface, often caused by cooling of the air near the surface as a result of terrestrial radiation, especially at night.

surficant — Wetting agent. In lubricating systems surficant adheres to contaminants causing them to drop out of the fuel and settle to the bottom of the tank as sludge.

surge — The abrupt loss of the efficiency of the axial flow compressor in a turbine engine when the angle of attack of the compressor blades becomes excessive. Also referred to as a compressor stall.

surging — A change in engine RPM or power in an oscillatory manner. It is usually caused by a malfunction in the fuel control system.

SURPIC — A description of surface vessels in the area of a Search and Rescue incident including their predicted positions and their characteristics.

surveillance approach (ASR) — An instrument approach wherein the air traffic controller issues instructions, for pilot compliance, based on aircraft position in relation to the final approach course (azimuth), and the distance (range) from the end of the runway as displayed on the controller's radar scope. The controller will provide recommend-ed altitudes on final approach if requested by the pilot.

sustained speed — The average wind speed over a one- or two-minute period.

sustaining speed — The speed of the engine compressor and turbine at which a turbine engine can keep itself running without having to depend on power from the starter.

swage — To squeeze together.

swaged terminals — Solderless terminals fastened to an electrical conductor by the swaging process.

sweat solder — A method of soldering two pieces of metal together. Both pieces are tinned with solder and then heated to form a joint without the use of additional solder.

Sweeny tool — A gear reduction type torque wrench used to remove the mast nut from helicopter rotors.

sweepback — A wing design in which the wings do not form right angles with the longitudinal axis but instead are angled backward from the wing root to the wingtip.

sweptback wing — A wing planform in which the tips of the wing are farther back than the wing root.

swing a compass — The process of aligning the aircraft on a series of known magnetic headings and adjusting the compensating magnets to bring the compass heading as near the magnetic heading as possible.

swirl frame — The inlet case on some turboshaft engines that act as an inlet particle separator.

swirl vanes — Air circulation vanes that surround fuel nozzles creating a small vapor retaining vortex. Fuel vapor trapped in this way is recirculated and utilized more efficiently.

Swiss pattern files — A set of precision files used for delicate metal work.

switching diode — A semiconductor device that behaves like an open switch at low voltage levels, but acts like a closed switch when voltage rises to a trigger level.

syllabus — A step-by-step, building block progression of learning with provisions for regular review and evaluations at prescribed stages of learning. The syllabus defines the unit of training, states by objective what the student is expected to accomplish during the unit of training, shows an organized plan for instruction, and dictates the evaluation process for either the unit or stages of learning.

symbol — Graphic representation used to represent shape, size, or material on a mechanical drawing.

symbols — In communication, words, gestures, and facial expressions are formed into sentences and paragraphs that mean something to the receiver of the information.

symmetrical — A condition in which both halves of an object are the same.

symmetrical airfoil — An airfoil with the same shape on both sides of its center line. The location of the center of pressure of a symmetrical airfoil changes very little as its angle of attack varies.

symmetrical laminate — In composites, a laminate in which the stacking sequence of plies below its midline is a mirror image of the stacking sequence above the midline.

symmetry check — A rigging check of an aircraft where measurements are made to determine that points on both sides of the airplane are equidistant from the center line.

synchro — A synchronous device in which a movable element is slaved to a similar element in a master unit or transmitter.

synchro system — See synchro.

synchronize — To cause two events to occur at exactly the same time.

synchronous motor — An AC motor in which the rotor is an electromagnet and the stator has a pulsating magnetic field from the AC flowing within it. The rotor must have a starting device, but once it is running, it will operate at a constant speed, which is determined by the frequency of the power source.

synchronous speed — The speed at which a synchronous motor rotates. It is dependent on the frequency of the alternating current that excites the field and the number of poles in the stator.

synchrophasing — In multi engine aircraft, a form of propeller synchronization in which not only the r.p.m. of the engines are held constant, but also the position of the propellers with relation to each other.

synchroscope — An instrument showing the relationship of the engine speeds on a multi-engine aircraft.

synoptic chart — In weather, a map or chart that depicts meteorological conditions over a large area at a given time.

synthetic fibers — Man-made products such as fiberglass, polyester, and polyamide fibers used in the production of aircraft covering fabric.

synthetic oil — A lubricating oil with a synthetic rather than a petroleum base. It tends to be less likely to oxidize and form sludge than petroleum oils. Synthetics are used extensively in turbine engines and are gaining wide acceptance in reciprocating engines.

synthetic rubber — Any of several types of man-made products that have characteristics similar to natural rubber.

system — A group of parts or components that work together in order to accomplish a common goal.

system discharge indicator — A yellow disc or blow-out plug on the side of an aircraft. When blown out, it indicates that the fire extinguishing system has been discharged normally rather than by an overheat condition.

system pressure regulator — The hydraulic component that controls hydraulic system pressure. It unloads the pump when a pre-selected pressure is reached. It also brings the pump back online when the pressure drops to the desired kick-in pressure.

system strategic navigation — Military activity accomplished by navigating along a preplanned route using internal aircraft systems to maintain a desired track. This activity normally requires a lateral route width of 10 NM and altitude range of 1,000 feet to 6,000 feet AGL with some route segments that permit terrain following.

T

tab — A small auxiliary control surface hinged to an aircraft primary control surface. Tabs can be used to assist in the movement of a primary control surface or as a means of trimming an aircraft.

TACAN — An acronym for Tactical Air Navigation System. An electronic navigation system that provides an indication of the distance the aircraft is from a ground station. TACAN operates in the UHF range of the radio frequency spectrum.

TACAN-only aircraft — An aircraft, normally military, possessing TACAN with DME but no VOR navigational system capability. Clearances must specify TACAN or VORTAC fixes and approaches.

tachometer — An instrument that measures the rotating speed of an engine. For reciprocating engines, the tachometer reads in RPMs. In most turbine engines, the tachometer reads in percent of the maximum RPM.

tachometer cable — The flexible cable used to drive a mechanical tachometer from the engine. It is made of two layers of steel wire, spiraled in opposite directions about a central core.

tachometer generator — A small electrical generator supplying current at frequencies proportional to the speed of the unit on which it is mounted.

tachometer generator and indicator — The generator has three-phase AC output of approximately 20V. It is a gearbox-driven accessory that produces power for a motor-driven indicator. The indicator provides a RPM indication in the cockpit.

tack — In composites, stickiness of the adhesive of a pre-preg material.

tack coat — In painting, a very light coat of material sprayed on a surface and allowed to stay until the solvents evaporate. It is then covered with the full wet coat of material.

tack rag — A rag, slightly damp with thinner, used to wipe a surface after it has been sanded to prepare it for the application of the next coat of finish.

tack weld — Small temporary welds along a welded seam made for the purpose of holding the pieces of metal in position until the weld is completed.

tacking — Hand-sewn temporary stitches removed prior to machine sewing.

tackle — A pulley in the form of a wheel mounted on a faced axis and supported by a frame. The wheel, or disk, is normally grooved to accommodate a rope. The frame that supports the wheel is called a block. Block and tackle consists of a pair of blocks. Each block contains one or more pulleys and a rope connecting the pulley(s) of each block.

tacky — A gluey or sticky finish.

tactical air navigation — An ultra-high frequency electronic rho-theta air navigation aid that provides suitably equipped aircraft a continuous indication of bearing and distance to the TACAN station.

tag wire — Thin diameter wire used to tie identification tags to objects.

tail boom — A spar or outrigger connecting the tail surfaces to a pod-type fuselage.

tail cone — **1.** The conical-shaped portion of a turbine engine exhaust system used to produce the proper area increase for the gases as they leave the engine. **2.** The rearmost part of an aircraft fuselage.

tail load — A downward aerodynamic force produced by the tail of an airplane to maintain dynamic longitudinal stability.

tail pipe — That portion of the exhaust system of an aircraft engine through which the gases leave the aircraft.

tail rotor — In helicopters, a rotor turning in a plane perpendicular to that of the main rotor and parallel to the longitudinal axis of the fuselage. It is used to counteract the torque of the main rotor and to provide movement about the yaw axis of the helicopter.

tail section — The rear portion, or empennage, of an airplane.

tail skid — **1.** On modem jet aircraft, a portion of the structure designed to absorb shock in case the tail strikes the runway during rotation for takeoff. **2.** A skid that supports the tail of an airplane on the ground. Tail skids were used before the advent of tail wheels.

tail surface — A stabilizing or control surface in the tail of an aircraft.

tail wheel — A small wheel located at the rear of the fuselage of an airplane having a conventional landing gear. It is used as a support for the tail when the airplane is on the ground.

tail wind — A wind blowing in the same direction the airplane is flying.

tail-heavy — A condition of balance in an aircraft in which the center of gravity is behind the aft limit.

tailpipe inserts — Small, sheet metal, wedge-shaped tabs that are inserted into the tailpipe of some older engines to reduce the nozzle opening and increase thrust. Adjustment of thrust is now done at the fuel control.

tailwheel checkout — To act as PIC of a tailwheel airplane, FAR 61.31(g) requires a demonstration of competency in normal and crosswind takeoffs and landings, wheel landings (unless the manufacturer has recommended against such landings), and go-around procedures. Seek a comprehensive tailwheel checkout from a qualified instructor for each make and model of tailwheel airplane that will be used for instruction. If one has logged PIC time in a tailwheel airplane prior to April 15, 1991, then a tailwheel endorsement is not required in one's logbook.

tailwind — Any wind more than 90 degrees from the flight path of the aircraft. For takeoff, any wind more than 90° from the magnetic heading of the runway.

tailwind component — The portion of the wind that acts directly on the tail of the airplane.

takeoff — The beginning of flight in which an airplane is accelerated from a state of rest to that of normal flight. The final breaking of contact with the land or water.

takeoff area — Any locality either on land, water, or structures, including airports/heliports and intermediate landing fields, used or intended to be used, for the takeoff of aircraft whether or not facilities are provided for the shelter, servicing, or for receiving or discharging passengers or cargo.

takeoff briefing — A tool that pilots can use for takeoff planning where they verbally rehearse the entire takeoff and departure prior to taking the active runway. By conducting a takeoff briefing, multi-engine operations can be performed safer. The briefing should include both normal and emergency procedures just prior to taxiing onto the runway. Among other things, this enables pilots to be better prepared to handle engine failures during the various phases of the takeoff profile. This takeoff briefing should review appropriate actions for an engine failure prior to VMC, before the landing gear is retracted, and after the airplane is climbing in the clean configuration.

take-off distance available (TODA) — The length of the take-off run available plus the length of the clearway, if provided.

takeoff power — The brake horsepower approved by the engine manufacturer for takeoff. This can be limited to a given amount of time such as one minute or five minutes.

takeoff power — 1. With respect to reciprocating engines, the brake horsepower developed under standard sea level conditions, and under the maximum conditions of crankshaft rotational speed and engine manifold pressure approved for the normal takeoff, and limited in continuous use to the time shown in the approved engine specification. 2. With respect to turbine engines, the brake horsepower developed under static conditions at a specified altitude and atmospheric temperature, and under the maximum conditions of rotor shaft rotational speed and gas temperature approved for the normal takeoff, and limited in continuous use to the time shown in the approved engine specification.

take-off run available (TORA) — The length of runway declared available and suitable for the ground run of an airplane taking off.

takeoff safety speed — A referenced airspeed obtained after lift-off at which the required one-engine-inoperative climb performance can be achieved.

takeoff thrust — With respect to turbine engines, means the jet thrust that is developed under static conditions at a specific altitude and atmospheric temperature under the maximum conditions of rotorshaft rotational speed and gas temperature approved for the normal takeoff, and limited in continuous use to the period of time shown in the approved engine specification.

takeoff weight — The weight of an aircraft at liftoff. Also referred to as maximum takeoff weight.

tandem — One behind the other.

tandem bearings — The placement of two ball bearings so the thrust load is shared by both bearings.

tandem wing — A configuration having two wings of similar span, mounted in tandem.

tang — The portion of a knife blade or file that fits into the handle.

tangent — 1. A line that contacts the circumference of a circle without penetrating to the inside of the circle. 2. A trigonometric function that is the ratio of the lengths of the side opposite and the side adjacent to the angle in a right triangle.

tangent point (TP) — The point on the VOR/DME RNAV route centerline from which a line perpendicular to the route centerline would pass through the reference facility.

tank — A container or reservoir used to hold liquids.

tank circuit — A parallel resonant circuit including an inductance and a capacitance.

tank selector valve — A selector valve controlled by the pilot with which they can select the fuel tank from which they desire to operate the engine.

tantalum carbide — A rare, corrosion-resistant, metallic chemical element mixed with carbon. Used in cutting tools and instruments.

tap — A tool used to cut threads on the inside of a hole in metal, fiber, or other material.

tap drill — A twist drill used to drill a hole before it is tapped. Charts have been developed to determine the correct size tap drill to use to obtain the correct dimensions for a tapped hole.

tap extractor — A tool used to extract taps that have broken off in the hole. It is equipped with projecting fingers that enter the flutes of the tap. The tap is then backed out of the hole by turning the extractor with a wrench.

tape — In composites, a term used for thin unidirectional material that is usually no wider than 12 inches. The material may or may not be a prepreg.

tape laying — In composites, a manufacturing process where prepreg tapes are laid across or overlapped to build up a shape. The parts are sometimes vacuum bagged and cured. This process can be automated by the use of tape laying equipment

tape measure — A narrow strip (usually made of cloth or steel tape) with graduations of centimeters, inches, feet, etc. used for measuring.

taper — A gradual decrease in width or thickness from one end of an object to the other.

taper in plan only — A gradual change (usually a decrease) in the chord length along the wing span from the root to the tip with the wing sections remaining geometrically similar.

taper in thickness ratio only — A gradual change in the thickness ratio along a wingspan with the chord remaining constant.

taper pin — A device used for fastening concentric shafts together to prevent relative motion between them. The tapered pin is pressed into a tapered hole.

taper reamer — A reamer used to smooth and "true" tapered holes and recesses.

taper tap — A hand-operated thread cutting tap used to start the tapping process in a drilled hole. The tap tapers for the first six or seven threads.

tapered crankshaft — The crankshaft to which a propeller is mounted by fitting over a tapered end.

tapered propeller shaft — See tapered crankshaft.

tapered punch — A hand punch tapered in length and used to drive pins, bolts, or rivets from their holes.

tapered roller bearings — An anti-friction bearing made of hardened steel cylinders rolling between two cone-shaped, hardened steel races. Tapered roller bearings are designed to carry both thrust and radial loads.

tapered-shank drill — A twist drill that has a tapered shank and is held in the chuck by friction

tapped hole — A hole in a casting or other material with threads cut on the inside.

tapped resistor — A wire-wound resistor that has taps along the length of the wire in order to provide a choice of multiple fixed resistances.

tapped stud hole — A hole in which threads have been cut for installation of a stud.

tappet — The component in an aircraft reciprocating engine that rides on the face of the cam and transmits a reciprocating motion to the push rods to open the poppet valves in the engine cylinders. The hydraulic valve lifters normally fit inside the tappets.

tare weight — The weight of all items such as blocks or chocks used to hold an airplane on the scales when it is being weighed. Tare weight must be subtracted from the scale reading to determine the weight of the aircraft.

target — In radar: Generally, any discrete object that reflects or retransmits energy back to the radar equipment. Specifically, an object of radar search or surveillance.

target blade — The identification of one blade of a helicopter during electronic balancing. It is the blade with the double interrupter.

target symbol — A computer-generated indication shown on a radar display resulting from a primary radar return or a radar beacon reply.

tarmac — A hard surfaced area of an airport used for aircraft parking, tie-down, and servicing.

tarnish — A stain, blemish, or dull surface.

tarpaulin — A cover made of a large piece of heavy, waterproof material fitted along its edges with eyelets so ropes can be used for tying.

tasks — Knowledge areas, flight procedures, or maneuvers within an area of operation of a practical test standard. Each task includes a list of the type of aircraft category or class to which it applies and a reference to the applicable regulation or publication.

tautening dope — Aircraft dope consisting of nitrocellulose and a plasticizer. The dope is applied to the fabric surface to produce tautness, increase strength, protect the fabric, waterproof the fabric, and make it airtight.

taxi — The movement of an airplane under its own power on the surface of an airport (FAR 135.100-Note). Also, it describes the surface movement of helicopters equipped with wheels.

taxi into position and hold — Used by ATC to inform a pilot to taxi onto the departure runway in takeoff position and hold. It is not authorization for takeoff. It is used when takeoff clearance cannot immediately be issued because of traffic or other reasons.

taxi lights — Lights similar to landing lights on an aircraft, but specifically aimed to illuminate the runway or taxiway when the airplane is taxiing.

taxi patterns — Patterns established to illustrate the desired flow of ground traffic for the different runways or airport areas available for use.

taxi weight — The maximum weight allowed for ground maneuvering. Also referred to as ramp weight.

taxiway — Airport pavement that allows aircraft to taxi from the terminal or parking area to the runway.

taxonomy of educational objectives — A system of sorting learning outcomes into the three domains of cognitive, affective and psychomotor and rank ordering learning levels from least to most complex within each domain.

TCAS I — A TCAS (traffic collision avoidance system) that utilizes interrogations of, and replies from, airborne radar beacon transponders and provides traffic advisories to the pilot.

TCAS II — A TCAS (traffic collision avoidance system) that utilizes interrogations of, and replies from airborne radar beacon transponders and provides traffic advisories and resolution advisories in the vertical plane.

TCAS III — A TCAS (traffic collision avoidance system) that utilizes interrogation of, and replies from, airborne radar beacon transponders and provides traffic advisories and resolution advisories inthe vertical and horizontal planes to the pilot.

teaching — The systematic and deliberate creation of practical instructional events (experiences) that are conducive to learning.

teaching lecture — An oral presentation that is directed toward desired learning outcomes. Some student participation is allowed.

teardown area — The area in an overhaul shop where equipment is received, inventoried, cleaned, and disassembled for overhaul.

technician — Aviation Maintenance Technician (AMT). A person skilled in repairing aircraft who has been issued a certificate from the FAA authorizing the holder to repair aircraft.

Tedlar® — In composites, a material used on the surface as a waterproof barrier.

tee fitting — A plumbing connector in the shape of a T.

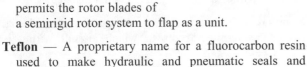

teetering hinge — In rotorcraft, a hinge that permits the rotor blades of a semirigid rotor system to flap as a unit.

Teflon — A proprietary name for a fluorocarbon resin used to make hydraulic and pneumatic seals and backup rings.

teleflex cable — A cable used to connect engine accessories to cockpit controls. Also used as a feedback cable.

telegraphing — In composites, dimpling of the fabric into the honeycomb core.

telemetering — A system of sending data over great distances by radio.

telephone information briefing service (TIBS) — Telephone recording of weather and/or aeronautical information.

telescope — To extend or collapse a series of interconnecting sections in order to make a longer or shorter linear assembly.

telescoping gauge — A precision measuring device that consists of a spring-loaded rod telescoping inside a tube. The gauge is adjusted to the width of a bore or a hole and locked. It is then removed from the hole and the length is measured with a micrometer.

telling and doing technique — A technique of flight instruction that consists of the instructor first telling about and demonstrating the new procedure. This is followed by the student telling and the instructor doing. Third, the student talks about the new procedure while doing it. Last, the instructor evaluates while the student accomplishes the procedure.

temper — The condition of hardness or softness of metal.

temperature — In general, the degree of hotness or coldness as measured on some definite temperature scale by means of any of various types of thermometers. Also, a measure of the direction heat will flow proportional to the mean kinetic energy of the molecules.

temperature amplifier — An electronic device used to amplify the exhaust temperature signal before being passed on to the fuel control. Used to assist in scheduling fuel flow.

temperature bulb — A temperature sensor installed at points such as the air intake or exhaust manifold. Output is transmitted to temperature gauges in the aircraft cockpit or to the fuel control.

temperature coefficient of resistance — The rate of change in resistance per degree centigrade of temperature change.

temperature datum system — An electronic circuit in an electrohydromechanical fuel control. Used to assist in scheduling fuel flow.

temperature gradient — The change of temperature divided by the distance over which the change occurs.

temperature inversion — In meteorology, an increase in temperature with height — a reversal of the normal decrease with height in the troposphere.

temperature ratio — A ratio of two engine temperatures used in certain performance calculations.

temperature scales — 1. Celsius, formerly known as centigrade, is based on the freezing point of water as 0° and its boiling point as 100°. Absolute zero is equal to -273° on the Celsius scale. 2. Fahrenheit is based on the freezing point of water as 32° and its boiling point as 212°. Absolute zero is equal to -460° on the Fahrenheit scale. 3. Kelvin is based on the freezing point of water as 273°K and its boiling point as 373°. Rankine is based on the freezing point of water as 492° and its boiling point as 672°.

temperature-dewpoint spread — The difference between the air temperature and the dewpoint.

tempering — A heat-treatment process in which some of the hardness is removed from a metal to increase its toughness and decrease its brittleness.

template — A pattern made of any suitable material to permit the layout of parts with a minimum expenditure of time and effort. It should be rigid and accurate and have pertinent data marked on it.

tensile load — An external force that tends to lengthen or stretch a body.

tensile strength — The ability of an object to resist forces tending to stretch or lengthen it.

tensile stress — The forces that attempt to pull an object apart.

tensiometer — A measuring instrument used to determine the installed tension of aircraft control cables.

tension — The stress produced in a body by forces acting along the same line but in opposite directions.

tension adjusters — Devices installed in an aircraft control system that maintain a constant cable tension, regardless of the temperature-caused dimensional changes in the airplane.

tension regulators — See tension adjusters.

tension torsion bar — A strap made of layers of sheet steel used to absorb tension of centrifugal loads between the rotor blades and the hub. The tension torsion bar also absorbs the tension in the torque that results from blade pitch changes.

tension torsion strap — A strap made of wire used to serve the same purpose on a helicopter rotor head as the tension torsion bar.

tentative calculated landing time (TCLT) — A projected time calculated for adapted vertex for each arrival aircraft based upon runway configuration, airport acceptance rate, airport arrival delay period, and other metered arrival aircraft. This time is either the VTA (vertex time of arrival) of the aircraft or the TCLT/ACLT (actual landing time calculated) of the previous aircraft plus the AAI (arrival aircraft interval), whichever is later. This time will be updated in response to an aircraft's progress and its current relationship to other arrivals.

tera — Trillion. A metric prefix equal to 1×10^{12}.

terminal — 1. A connecting fitting in the form of a ring that attaches to the end of a wire. Used for connection to a battery, terminal strip, or other component. 2. A keyboard and video monitor that allows a user to interface with the computer.

terminal area — A general term used to describe airspace in which approach control service or airport traffic control service is provided.

terminal area facility — A facility providing air traffic control service for arriving and departing IFR, VFR, Special VFR, and on occasion enroute aircraft.

terminal control area (TCA) — A control area normally located where air traffic routes converge in the vicinity of one or more airfields. No longer valid for U.S. – still valid for foreign airspace.

terminal control area [ICAO] — A control area normally established at the confluence of ATS routes in the vicinity of one or more major aerodromes.

terminal doppler weather radar (TDWR) — Installed at many U.S. airports vulnerable to thunderstorms and microbursts. The use of Doppler radar provides a narrower radar beam and with greater power, a more comprehensive wind shear picture is available for wind shear prediction.

terminal forecast (FT) — Provides weather conditions expected to occur within a five nautical mile radius of the runway complex at an airport.

terminal radar service area — Airspace surrounding designated airports wherein ATC provides radar vectoring, sequencing, and separation on a full-time basis for all IFR and participating VFR aircraft. The AIM [and Jeppesen ATC Section] contains an explanation of TRSA. TRSA's are depicted on VFR aeronautical charts. Pilot participation is urged but is not mandatory.

terminal radar service area (TRSA) (USA) — Airspace surrounding designated airports wherein ATC provides radar vectoring, sequencing and separation on a full-time basis for all IFR and participating VFR aircraft. Service provided in a TRSA is called Stage III Service. Pilots' participation is urged but is not mandatory.

terminal strip — A strip of insulating material that contains terminal posts to which aircraft wiring is attached.

terminal velocity — The speed of an aircraft at which the drag has reached such an amount that the airplane will no longer accelerate.

terminal very high frequency omni-directional range station (TVOR) — A very high frequency terminal omni-range station located on or near an airport and used as an approach aid.

terminal VFR radar service — A national program instituted to extend the terminal radar services provided instrument flight rules (IFR) aircraft to visual flight rules (VFR) aircraft. The program is divided into four types of service referred to as basic radar service, terminal radar service area (TRSA) service, Class B service and Class C service. The type of service provided at a particular location is contained in the Airport/Facility Directory.
a. Basic Radar Service: These services are provided for VFR aircraft by all commissioned terminal radar facilities. Basic radar service includes safety alerts, traffic advisories, limited radar vectoring when requested by the pilot, and sequencing at locations where procedures have been established for this purpose and/or when covered by a letter of agreement. The purpose of this service is to adjust the flow of arriving IFR and VFR aircraft into the traffic pattern in a safe and orderly manner and to provide traffic advisories to departing VFR aircraft.
b. TRSA Service: This service provides, in addition to basic radar service, sequencing of all IFR and participating VFR aircraft to the primary airport and separation between all participating VFR aircraft. The purpose of this service is to provide separation between all participating VFR aircraft and all IFR aircraft operating within the area defined as a TRSA.
c. Class C Service: This service provides, in addition to basic radar service, approved separation between IFR and VFR aircraft, sequencing of VFR aircraft, and sequencing of VFR arrivals to the primary airport.
d. Class B Service: This service provides, in addition to basic radar service, approved separation of aircraft based on IFR, VFR, and/or weight, and sequencing of VFR arrivals to the primary airport(s).

terminal voltage — The potential difference or voltage across the terminals of a power supply such as a battery of cells. As soon as the power source begins to supply current to a circuit its terminal voltage falls because some electric potential energy is lost in driving current against the supply's own internal resistance.

terminating decimal — A decimal fraction that ends with a whole number.

terneplate — Lead-coated thin sheets of steel used in some older aircraft for the construction of fuel tanks and for the tipping on their wooden propellers.

terrain following — The flight of a military aircraft maintaining a constant AGL altitude above the terrain or the highest obstruction. The altitude of the aircraft will constantly change with the varying terrain and/or obstruction.

terrestrial radiation — The radiation emitted by the Earth and its atmosphere.

tertiary — Something at a third level of importance, rank, or value.

test — 1. To submit a unit, component, etc. to conditions that will show its quality, strength, etc. 2. A set of questions, problems or exercises for determining whether a person has a particular knowledge or skill.

test club — A wide-blade, short-diameter propeller used for applying a load to a freshly overhauled aircraft engine for its initial run-in.

test item — A question, problem, or exercise that measures a single objective and calls for a single response.

test stand — A stationary structure where engines, units, components, etc., can be mounted for testing.

test switch — A switch used to test a system in order to determine its operational condition.

tetraethyl lead — A heavy, oily, poisonous liquid ($Pb(C_2H_5)_4$) mixed into aviation gasoline to increase its octane rating.

tetrahedron — A device normally located on uncontrolled airports and used as a landing direction indicator. The small end of a tetrahedron points in the direction of landing. At controlled airports, the tetrahedron, if installed, should be disregarded because tower instructions supersede the indicator.

tetrode — An electron tube having four active electrodes.

T-handle — A T-shaped handle used for turning sockets.

that is correct — During communications, this indicates: The understanding you have is right.

theodolite — An optical instrument used to measure vertical and horizontal angles. Theodolites are used for surveying and weather observations.

theoretical pitch — An assumed pitch of the propeller blades. Also referred to as as geometric pitch.

therapeutic adapter — An adapter for a continuous-flow oxygen mask that allows oxygen to flow approximately three times the normal rate. It is used for passengers that have heart or respiratory problems.

thermal — Rising air that lifts because it is warmer than surrounding air.

thermal anti-icing system — A heated leading edge of the wing and tail surfaces to prevent the formation of ice.

thermal circuit breaker — A circuit breaker that opens a circuit when an excessive amount of current flows through it. Made up of a bimetallic set of contacts that open when heated.

317

thermal circulation — The movement of air resulting from differential heating.

thermal coefficient of resistance — The amount a material's resistance changes with a change in temperature.

thermal conduction — The transfer of heat energy from one object to another.

thermal conductor — A material, such as metal, that can easily transfer heat energy.

thermal cutout switch — A circuit breaker, or switch, that breaks the circuit at a predetermined temperature.

thermal decomposition — A chemical action that decomposes a material into simpler substances by the action of heat.

thermal efficiency — The ratio of the amount of heat energy converted into useful work to the amount of heat energy in the fuel used.

thermal expansion — The expansion of a substance due to heat.

thermal fatigue — A condition in turbine metals caused by heating and cooling each time a power setting is changed.

thermal index (TI) — In soaring, an index used to indicate the probability of thermals forming. It is the temperature, for any given level, of the air parcel having risen at the dry adiabatic lapse rate (DALR) subtracted from the ambient temperature. Experience has shown that a TI should be -2 for thermals to form and be sufficiently strong for soaring flight.

thermal insulator — Materials such as paper, wood, etc. that are poor conductors of heat energy.

thermal output — Amount of heat being discharged.

thermal relief valve — A pressure relief valve installed in a static portion of a hydraulic system to relieve pressure built up due to heat induced expansion.

thermal runaway — A condition existing in a nickel-cadmium battery when the cell resistances become unbalanced because of temperature. The resistance of some cells decreases and allows the cells to take more current, which lowers their resistance even further and creates more current. This action continues until the battery is seriously damaged, sometimes exploding.

thermal shock — A stress induced into a system or component due to a rapid temperature change.

thermal stress cracking — In composites, the crazing and cracking of some thermoplastic resins from overexposure to elevated temperatures.

thermal switch — A switch activated by heat.

thermal turbulence — Low-level turbulence (LLT) that is produced dry convection (thermals) in the boundary layer.

thermal wave — Waves, often but not always marked by cloud streets, that are excited by convection disturbing an overlying stable layer. Also referred to as convection waves.

thermals — A rising bubble of warm air. An element of convection.

thermionic — The electron emission caused by an incandescent material.

thermionic current — Current flow in a conductor caused by heat.

thermistor — A semiconductor device with a core material whose electrical resistance changes with a change in temperature.

thermoammeter — An instrument for measuring RF (radio frequency) alternating current in a circuit. The thermoammeter measures the RF alternating current by exposing a thermocouple to the RF energy. As the wire heats, it produces a current proportional to the amount of applied heat. This is displayed as current by the meter instrument.

thermocouple — A temperature measuring system consisting of two dissimilar metal wires joined at a temperature sensing point. A current proportional to the temperature of the junction is produced. In aviation, thermocouples are generally used as a temperature sensing device.

thermocouple exhaust temperature probe — A bimetallic probe, generally of chromel and alumel alloy, located in an exhaust stream. Heat causes a milliampere current to flow to a cockpit indicator that indicates degrees in Celsius.

thermocouple fire detector — A device that produces an electrical current flow by thermal action and which ultimately illuminates a cockpit warning signal.

thermocouple oil temperature bulb — A thermocouple positioned in the oil flow to provide a cockpit indicator with an oil temperature indication. It is generally made up of bimetallic materials such as chromel and constantan.

thermodynamic diagram — A chart presenting isopleths of pressure, temperature, water vapor content as well as dry and saturated adiabats. Various forms exist, the most commonly used in the United States being the Skew-T/Log-P.

thermodynamics — The branch of science that deals with mechanical actions caused by heat.

thermoelectricity — Electrical energy generated by the action of heat on the junction of two dissimilar metals.

thermograph — A thermometer that continuously records temperatures on a chart.

thermometer — An instrument for measuring temperature.

thermonuclear action — The release of heat, light, and electromagnetic radiation from the fusion or fission of nuclear material.

thermopile — A collection of thermocouples connected in a series used to measure minute changes in temperature or changes in the flow of electrical current.

thermoplastic — A plastic material used in advanced composites as a matrix material. Heat is used during the forming operation. It is not a permanent shape, however, if heated again it will soften and flow to form another shape (Plexiglas windshield).

thermoplastic material — A resin-based plastic material that can be softened by heat and cooled many times without losing its tensile strength.

thermoplastic resin — A resin material that softens with the application of heat. Most aircraft windshields and side windows are made of this material.

thermosafety discharge indicator — A red blowout disc located on the outside of an aircraft fuselage or engine nacelle that blows out to vent a cylinder of high-pressure gas in the event of an overpressure condition caused by heat.

thermoset — A plastic material used in advanced composites as a matrix material. Heat is used to form and set the part permanently. Once cured, it cannot be reformed by applying heat. Most composite structural components are made of thermoset plastics.

thermosetting material — A plastic material that remains hard once it is hardened by chemical means or by heat and pressure.

thermosetting resin — A widely used resin that usually sets by chemical means and maintains its hardness even when heat is applied.

thermosphere — The area of the atmosphere above the mesosphere that has little practical influence over the weather. The thermosphere is marked by generally increasing temperature with altitude.

thermostat — A device that functions to establish or maintain a desired temperature produced by a heater or air-conditioning system.

thermostatic bypass valve — A temperature-sensing valve in an engine oil cooler, used to direct the oil either through the core of the cooler or around the inside of the cooler shell to maintain the proper oil temperature.

thermoswitch fire detector — A device that closes an electrical contact in the presence of heat, illuminating a warning light in the cockpit. This is done by thermal expansion of metals.

thickness gauges — A measuring tool consisting of a series of precision-ground steel blades of various thicknesses. It is used to determine the clearance or separation between parts.

thimble — The part of a micrometer caliper that is turned to rotate the spindle.

thinner — A solvent used to reduce the viscosity of dope or paint.

third-class lever — A lever that provides a fractional disadvantage, i.e., one in which a greater force is required than the force of the load lifted. For example, if a muscle pulls with a force of 1,800 lbs. in order to lift a 100 lb. object, a mechanical advantage of $^{100}/_{1,800}$ is obtained. This is a fractional disadvantage, since it is less than 1.

thixotropic — In composites, an agent used to thicken a resin system without adding weight. Makes the resin system less dense. Thixotropic agents include chopped fibers, microballons, and fiber flox. Some agents give more strength than others.

thixotropic agent — A substance added to a resin to increase its resistance to flow.

thread — 1. Projecting helical grooves cut around the outside of a bolt, fitting, or pipe and on the inside of a bolt or fitting. 2. A very fine strand of linen, cotton, or other material used to make yarn or fabric.

thread chaser — A tool used to remove contamination from a threaded device.

thread gauge — A gauge with a series of V-notches cut along one edge. Used to check the number of threads, or the pitch, of a screw or bolt.

thread insert — An internally threaded bushing designed to be molded in or inserted into soft or brittle materials to increase their strength and minimize the wear of the threaded assembly.

thread pitch — The distance from the peak of one thread to the peak of the next thread on a screw, bolt, or other thread fitting. Also referred to as a screw pitch gauge.

thread plug gauge — A go/no-go type gauge to be screwed into internal threads.

thread ring gauge — A ring-type gauge used for checking external threads.

three-axis autopilot — An automatic flight control system that controls an aircraft about all three axes.

three-D (3-D) cam — Part of a linkage system, this multi-labeled cam can rotate and move up and down to allow a cam follower riding on its surface to seek an infinite number of positions. Used to miniaturize linkage systems such as in fuel controls.

three-dimensional object — An object that has three dimensions: length, width, and depth.

three-phase system — An AC electrical system consisting of three conductors, each carrying current 120° out of phase with the others. Three-phase systems are used extensively in modern electrical and electronic actuating systems.

three-point landing — The landing of an airplane in which all three main wheels of the landing system touch the ground at the same time. Three-point landings are not recommended for most tricycle gear aircraft.

three-pole, single-throw switch — An electrical switch with three sets of contactors, or poles, each of which completes only one circuit. The switch is controlled by a single operating toggle.

three-sixty (360) overhead — A series of predetermined maneuvers prescribed for aircraft (often in formation) for entry into the visual flight rules (VFR) traffic pattern and to proceed to a landing. An overhead maneuver is not an instrument flight rules (IFR) approach procedure. An aircraft executing an overhead maneuver is considered VFR and the IFR flight plan is cancelled when the aircraft reaches the "initial point" on the initial approach portion of the maneuver. The pattern usually specifies the following:
a. The radio contact required of the pilot.
b. The speed to be maintained.
c. An initial approach 3 to 5 miles in length.
d. An elliptical pattern consisting of two 180 degree turns.
e. A break point at which the first 180 degree turn is started.
f. The direction of turns.
g. Altitude (at least 500 feet above the conventional pattern).
h. A "Roll-out" on final approach not less than 1/4 mile from the landing threshold and not less than 300 feet above the ground.

three-state buffer — A logic device used to place signals on a data bus where only one signal source is allowed to be active at a time. It can have an output of High or Low or can be Off.

three-view drawing — An orthographic projection drawing that uses three views to portray an object.

three-way light switch — An electrical switch wired to allow the same light to be turned on or off from either of two separate locations.

threshold — The beginning of that portion of the runway usable for landing.

threshold crossing height — The theoretical height above the runway threshold at which the aircraft's glideslope antenna would be if the aircraft maintains the trajectory established by the mean ILS glideslope or MLS glidepath.

threshold lights — See airport lighting.

throat microphone — A special microphone used in areas where there is high background noise. It is applied to a person's throat and operates by picking up vibrations directly from the voicebox.

throatless shear — A heavy-duty cutting implement used to shear large sheets of metal.

throttle — The valve in a carburetor or fuel control unit that determines the amount of fuel-air mixture to be fed to the engine.

throttle body — One of the units of a carburetor system. All air entering the cylinders flows through this body and is measured by volume and weight so that the proper amount of fuel can be added to meet the engine demands under all conditions. The throttle body contains the throttle valves, main venturi, boost venturi, and the impact tubes.

throttle ice — Carburetor ice that forms on the rear side of a throttle valve when the throttle is partially closed. Occurs because of the lower pressure causing a drop in temperature.

through bolts — Long bolts that pass completely through an object to hold it together.

throw — That part of a crankshaft to which the connecting rods are attached.

throwaway part — A part that is not economical to repair if it should fail and must be thrown away and replaced with a new one.

thrust — The forward force produced by a reaction to the exhaust gases escaping the nozzle of a jet engine or by the aerodynamic force of a propeller.

thrust — 1. A forward force that imparts momentum to a mass of air behind it. 2. produced by a reaction to the exhaust gases escaping the nozzle of a jet engine. 3. The forward force produced by the aerodynamic force of a propeller. 4. In helicopters, the force developed by the rotor blades acting parallel to the relative wind and opposing the forces of drag and weight.

thrust bearing — A bearing in a reciprocating aircraft engine that absorbs loads parallel to the length of the crankshaft.

thrust horsepower — The amount of horsepower the engine-propeller combination transforms into thrust.

thrust line — An imaginary line passing through the center of the propeller hub, perpendicular to the plane of the propeller rotation.

thrust loads — Loads imposed upon the engine crankshaft and bearings when the propeller is pulling or pushing the aircraft.

thrust reverser — A device attached to a turbine engine tailpipe to reverse the exhaust gas flow. Reversers assist aircraft brakes and provide aircraft control during landing and during rejected takeoffs.

thrust specific fuel consumption (TSFC) — **1.** An equation, TSFC = $W_f F_n$, where W_f is fuel flow in pounds per hour and F_n is the net thrust in pounds. Used to calculate the fuel consumption as a means of comparison between engines. **2.** The amount of fuel an engine burns in one hour to produce one pound of thrust.

thrust, gross — The thrust developed by an engine, not taking into consideration any pressure of initial air mass momentum. Also referred to as static thrust (Fg).

thrust, net — The effective thrust developed by a jet engine during flight, taking into consideration the initial momentum of the air mass prior to entering the engine.

thruster — A miniature rocket engine fired to change orientation of space vehicles and orbiting satellites.

thumbscrew — A machine screw that has a round flat projection perpendicular to the screw shank. Thumb screws can be turned by hand and are used on access panels where it is necessary to frequently open and close the panel.

thunder — A loud noise heard when lightening occurs between clouds or between clouds and the ground. The lightening generates instantaneous high heat, which causes a violent expansion of the surrounding air. This expanding air causes the shock wave or noise of thunder.

thundercloud — A cumulonimbus cloud.

thunderstorm — Cumulonimbus clouds charged with electricity and producing lightning, thunder, rain, and hail.

thyratron — Gas-filled triode electron tube in which a continuous current is caused to flow by a momentary signal applied to the grid.

thyratron tube — A triode tube into which a gas has been introduced to change its operating characteristics.

thyristor — A semiconductor device that acts as a switch.

tie bolt — Bolts used to assemble wheel halves on split aircraft wheels.

tie rod — Tension rod used for internal and external bracing of various component parts. The ends are threaded for attachment and length adjustment.

tiedown — A special anchoring provision on an airport surface to which airplanes can be secured when parked.

tight-drive fit — An interference fit between parts.

time and opportunity — A perception factor where learning something is dependent on the student having the time to sense and relate current experiences in context with previous events.

time between overhaul (TBO) — A recommendation of the manufacturer of an aircraft engine as to the amount of time that the engine can operate under average conditions before it should be overhauled.

time change item — Any item, component, unit, etc. whose time in service is limited by hours, the number of times the unit is operated, or a calendar basis, and that must be removed and replaced with a new or similar item. Also referred to as a time limited part.

time constant — The time required for the voltage of a capacitor in an RC circuit to increase to 63.2 percent of maximum value or decrease to 36.7 percent of maximum value.

time constant of an electric motor — The time required for a motor to accelerate from off to its rated no-load speed.

time delay relay — A relay that delays the closing or opening of the relay for a specified time after activation.

time group — Four digits representing the hour and minutes from the Coordinated Universal Time (UTC) clock. FAA uses UTC for all operations. The term "ZULU" can be used to denote UTC. The word "local" or the time zone equivalent shall be used to denote local when local time is given during radio and telephone communications. When written, a time zone designator is used to indicate local time; e.g., "0205M" (Mountain). The local time can be based on the 24-hour clock system. The day begins at 0000 and ends at 2359.

time in service — With respect to maintenance time records, means the time from the moment an aircraft leaves the surface of the Earth until it touches it at the next point of landing.

time limited part — Any item, component, unit, etc. whose time in service is limited by hours, the number of times the unit is operated, or a calendar basis, and that must be removed and replaced with a new or similar item. Also referred to as a time change item.

timed turn — The most accurate way to turn to a specific heading without the heading indicator. Use the clock instead of the compass card to determine when to roll out.

Time-Rite Indicator — A piston position indicator, made by Time-Rite, used for locating the position of a piston in the cylinder of a reciprocating engine for the purpose of magneto or valve timing.

timing chain — A metal chain that connects the crankshaft to the camshaft of a reciprocating engine. Causes the valves to open and close in proper coordination with the motion of the pistons.

321

timing disc — A device or tool that can be mounted on an accessory drive or on the propeller to indicate the amount of crankshaft travel for ignition or valve riming.

tin — A soft, silver-white, metallic chemical element with a symbol of Sn and an atomic number of 50. Alloyed with other metals in solders, utensils, and in making tin plate.

tin snips — Hand-operated sheet metal shears.

tinned — Coated by soft solder. Two surfaces that have been tinned can be pressed together and reheated to sweat solder the pieces together.

tinned wire — Electrical wire that has been coated by a thin coating of soft solder.

tinner's rivet — A flat-headed solid rivet that is driven by holding the flat head of the rivet on an anvil and upsetting the rivet by peening the end over with a hammer.

tinplated — Sheet steel coated with a thin layer of tin.

tip cap — A removable tip on the rotor blade tip. The tip cap is often used to hold spanwise balance weights.

tip path plane — The path followed by the tips of a propeller as it rotates.

tip pocket — An area provided at the tip of a helicopter rotor blade to place weight for spanwise balance.

tip speed — The speed of a rotating airfoil. Generally subsonic except for fan blade tips and the tip of centrifugal compression blades.

tip targets — Reflectors placed on helicopter blade tips to determine track with a spotlight or a strobe light.

tip weight — A weight placed in the tip of a helicopter rotor blade for spanwise balance.

tip-path plane — In rotorcraft, the imaginary circular plane outlined by the rotor blade tips as they make a cycle of rotation.

tire — A ring or loop made of rubber compound for toughness and durability. Tires consist of tread, casing plies/cord body, and beads. Tires provide a cushion of air that helps absorb the shocks and roughness of landings and takeoffs. They support the weight of the aircraft while on the ground and provide the necessary traction for braking and stopping aircraft upon landing.

tire bead — Bundles of steel wire embedded in the rubber around the inner circumference of an aircraft tire. Fits onto the wheel to hold the tire in place.

Turbine Inlet Pressure (TIT) — Temperature taken in front of the first stage turbine nozzle vanes. The most critical temperature taken within the engine for use in fuel scheduling.

titanium — A dark-gray, silvery, lustrous, very hard, light, and corrosion-resistant metallic element with a symbol of Ti and an atomic number of 22.

title block — An information block in the lower right-hand corner of an aircraft drawing in which the name of the part, the part number, and other pertinent information is displayed.

toe-in — Aircraft wheels that tend to converge toward the front. Toe-in will cause the tires to try to move closer together.

toe-out — Aircraft wheels that tend to diverge toward the rear of the wheels. Toe-out will cause the tires to try to move apart.

To-From indicator — An indicator on the course deviation indicator (CDI). During VOR operation the vertical needle of the CDI is used as the course indicator. The vertical needle also indicates when the aircraft deviates from the course and the direction the aircraft must be turned to attain the desired course. The "TO-FROM" indicator indicates whether the course set in the indicator will take the aircraft to or from the station.

toggle switch — An electrical switch in which a projecting knob or arm moving through a small arc causes the contacts to open or close rapidly.

tolerance — An allowable variation in the dimensions of a part.

toluene — A colorless, water insoluble, flammable liquid ($C_6H_5CH_3$) used as a solvent, paint remover, and thinner.

toluol — A commercial grade of toluene, which is a liquid aromatic hydrocarbon similar to benzene, but less volatile, flammable, and toxic.

ton of refrigeration — A measure of the cooling capacity of an air conditioning system. It is the same cooling effect as melting one ton of ice in 24 hours.

tone — **1.** A tint or shade of a color. **2.** A variation of a hue. **3.** A sound of a specific frequency.

tool — The mold used in manufacturing a composite component.

tool steel — Hard steel used in the making of tools.

tooling resins — In composites, resins used to make molds.

top dead center (TDC) — The position of the piston within a cylinder when the piston has reached its furthest uppermost position.

top overhaul — The overhaul of the cylinders of an aircraft engine. It consists of grinding the valves, replacing the piston rings, and doing anything else necessary to restore the cylinders to their proper condition. The crankcase of the engine is not opened.

torching — Long plumes of flame extending from the exhaust stack caused by an excessively rich mixture and traces of unburned fuel remaining in the exhaust. This unburned fuel will not ignite until oxygen in the air of the exhaust system mixes with the charge.

tornado — A violently rotating column of air, which appears as a pendant from a cumulonimbus cloud, and nearly always observable as "funnel-shaped." It is the most destructive of all small-scale atmospheric phenomena.

toroidal wound coil — An electrical coil wound around a ring or doughnut-shape core.

torque — 1. A resistance to turning or twisting. 2. Forces that produces a twisting or rotating motion. 3. In rotorcraft, in helicopters with a single, main rotor system, the tendency of the helicopter to turn in the opposite direction of the main rotor rotation.

torque limited — A limitation placed on the drive train of a helicopter in regards to power output of the engine.

torque link — The hinged linkage between the piston and cylinder of an oleo shock strut. The piston moves in and out, but is restrained from rotating. Also referred to as scissors.

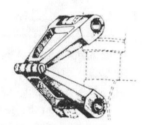

torque nose — A mechanism or apparatus at the nose section of the engine that senses the engine torque and activates a torquemeter.

torque tube — A tubular member of a control system used to transmit torsional movement to the control.

torque wrench — A precision hand tool used to measure the amount of torque applied to a bolt or nut.

torquemeter — An indicator used on some large reciprocating engines or on turboprop engines to indicate the amount of torque the engine is producing.

torquemeter indicator — A turboprop or turboshaft cockpit instrument used to indicate engine power output. The propeller or rotor inputs a twisting force to an electronic or oil operated torquemeter that sends a signal to the indicator.

torsion — An external stress that produces twisting within a body.

torsional force — A twisting force. Torsional forces act on a rotating propeller as the result of aerodynamic twisting force and tensile forces that try to pull it apart.

torsional strength — The strength of a material in a direction that opposes a twisting force.

toss-up — A question where all the alternative answers are equally good.

total drag — The sum of parasite and induced drag.

total estimated elapsed time [ICAO] — For IFR flights, the estimated time required from takeoff to arrive over that designated point, defined by reference to navigation aids, from which it is intended that an instrument approach procedure will be commenced, or, if no navigation aid is associated with the destination aerodrome, to arrive over the destination aerodrome. For VFR flights, the estimated time required from takeoff to arrive over the destination aerodrome.

total pressure — The pressure a moving fluid would have if it were stopped. No losses are considered.

total temperature — A temperature measurement of air in motion. The total of static temperature plus temperature rise due to ram effect.

totalizer — A single fuel quantity gauge that indicates the total of fuel in all of the fuel tanks.

touch and go — An aircraft training technique in which the pilot practices a series of landings and takeoffs without coming to a complete stop. As the airplane lands, the pilot advances the power for another takeoff and go-around.

touch-and-go landing — See touch-and-go.

touchdown — 1. The point at which an aircraft first makes contact with the landing surface. 2. Concerning a precision radar approach (PAR), it is the point where the glide path intercepts the landing surface.

touchdown RVR — The runway visual range (RVR) visibility readout values obtained from RVR equipment serving the runway touchdown zone.

touchdown zone — The first 3,000 feet of the runway beginning at the threshold. The area is used for determination of Touchdown Zone Elevation in the development of straight-in landing minimums for instrument approaches.

touchdown zone [ICAO] — The portion of a runway, beyond the threshold, where it is intended landing aircraft first contact the runway.

touchdown zone elevation — The highest elevation in the first 3,000 feet of the landing surface. TDZE is indicated on the instrument approach procedure chart when straight-in landing minimums are authorized.

touchdown zone lighting — See airport lighting.

toughness — The property of a metal that allows it to be deformed without breaking.

tow — In composites, an untwisted bundle of filaments.

tow hook — A mechanism allowing the attachment and release of a towrope on the glider or tow plane. On

gliders, it is located near the nose or directly ahead of the main wheel. Two types of tow hooks commonly used in gliders are manufactured by Tost and Schweizer.

tower — A terminal facility that uses air/ground communications, visual signaling, and other devices to provide ATC services to aircraft operating in the vicinity of an airport or on the movement area. Authorizes aircraft to land or takeoff at the airport controlled by the tower or to transit the Class D airspace area regardless of flight plan or weather conditions (IFR or VFR). A tower can also provide approach control services (radar or nonradar).

tower enroute control service — The control of IFR enroute traffic within delegated airspace between two or more adjacent approach control facilities. This service is designed to expedite traffic and reduce control and pilot communication requirements.

tower to tower — See tower enroute control service.

tower visibility — Prevailing visibility determined from the control tower.

towering cumulus — A rapidly growing cumulus cloud, it is often typical of the cumulus stage of thunderstorm development. Its top can reach 20,000 feet AGL or more and have a width of three to five miles.

towing eye — A ring or hook on an aircraft structure to which a tow bar can be attached for moving the airplane on the ground.

Townend ring — A cowling used on radial-engine aircraft designed to improve airflow and engine cooling.

toxic — Poisonous.

TPX-42 — A numeric beacon decoder equipment/system. It is designed to be added to terminal radar systems for beacon decoding. It provides rapid target identification, reinforcement of the primary radar target, and altitude information from Mode C.

trace — When precipitation occurs in amounts too small to be measured, less than .01 inches

traceable pressure standard — The facility station pressure instrument, with certification/calibration traceable to the National Institute of Standards and Technology. Traceable pressure standards can be mercurial barometers, commissioned ASOS or dual transducer AWOS, or portable pressure standards or DASI.

tracer — In composites, a fiber, tow, or yarn added to a pre-preg for verifying fiber alignment or for distinguishing warp fibers from fill fibers.

track — 1. The path followed by the tip of a propeller or rotor blade as it rotates. 2. The actual flight path of an aircraft over the ground. Also referred to as ground track.

track [ICAO] — The projection on the Earth's surface of the path of an aircraft, the direction of which path at any point is usually expressed in degrees from North (True, Magnetic, or Grid).

tracking — A process of navigation where a course is placed in the course selector and the course is maintain

tracking flag — A wooden pole that supports a white cotton flag, used to touch the operating rotor blades that have had their tips covered with colored chalk. The marks left on the nag indicate track of the main rotor.

tracking reflectors — Reflectors placed on the blade tips to determine track with a spotlight or a strobe light.

tracking stick — A stick, with a wick on one end, used to touch the operating rotor blades in order to determine track.

tracking targets — Reflectors placed on the blade tips to determine track with a spotlight or strobe light.

TRACON — Acronym for Terminal Radar Approach Control.

tractor propeller — A propeller mounted to the front of an engine that pulls the airplane through the air.

trade winds — Winds that blow toward the equator from the northeast on the north side of the equator and from the southeast on the south side. Trade winds are caused by the friction between the air and the Earth and by the rotation of the Earth.

traffic — 1. A term used by ATC to refer to one or more aircraft. 2. A term used by a controller to transfer radar identification of an aircraft to another controller for the purpose of coordinating separation action. Traffic is normally issued:
a. in response to a handoff or point out,
b. in anticipation of a handoff or point out, or
c. in conjunction with a request for control of an aircraft.

traffic advisories — Advisories issued to alert pilots to other known or observed air traffic that can be in such proximity to the position or intended route of flight of their aircraft to warrant their attention. Such advisories can be based on:
a. Visual observation.
b. Observation of radar identified and nonidentified aircraft targets on an ATC radar display, or
c. Verbal reports from pilots or other facilities.

Note 1: The word "traffic" followed by additional information, if known, is used to provide such advisories; e.g., "Traffic, 2 o'clock, one zero miles, southbound, eight thousand."

Note 2: Traffic advisory service will be provided to the extent possible depending on higher priority duties of the controller or other limitations; e.g., radar limitations, volume of traffic, frequency congestion, or controller workload. Radar/nonradar traffic advisories do not relieve the pilot of his responsibility to see and avoid other aircraft. Pilots are cautioned that there are many times when the controller is not able to give traffic advisories concerning all traffic in the aircraft's proximity; in other words, a pilot should not assume when traffic advisories are requested or received, that all traffic will be issued.

TRAFFIC ALERT (aircraft call sign), TURN (left/right) IMMEDIATELY, (climb/descend) AND MAINTAIN (altitude)

traffic alert and collision avoidance system — An airborne collision avoidance system based on radar beacon signals that operates independent of ground-based equipment. TCAS-I generates traffic advisories only. TCAS-II generates traffic advisories, and resolution (collision avoidance) advisories in the vertical plane.

traffic in sight — Used by pilots to inform a controller that previously issued traffic is in sight.

traffic information — See traffic advisories.

traffic management program alert — A term used in a Notice to Airmen (NOTAM) issued in conjunction with a special traffic management program to alert pilots to the existence of the program and to refer them to either the Notices to Airmen publication or a special traffic management program advisory message for program details. The contraction TMPA is used in NOTAM text.

traffic management unit — The entity in ARTCC's and designated terminals responsible for direct involvement in the active management of facility traffic. Usually under the direct supervision of an assistant manager for traffic management.

traffic no factor — Indicates that the traffic described in a previously issued traffic advisory is no factor.

traffic no longer observed — Indicates that the traffic described in a previously issued traffic advisory is no longer depicted on radar, but can still be a factor.

traffic pattern — The traffic flow that is prescribed for aircraft landing at, taxiing on, or taking off from an airport. The components of a typical traffic pattern are upwind leg, crosswind leg, downwind leg, base leg, and final approach.

a. Upwind Leg — A flight path parallel to the landing runway in the direction of landing.

b. Crosswind Leg — A flight path at right angles to the landing runway off its upwind end.

c. Downwind Leg — A flight path parallel to the landing runway in the direction opposite to landing. The downwind leg normally extends between the crosswind leg and the base leg.

d. Base Leg — A flight path at right angles to the landing runway off its approach end. The base leg normally extends from the downwind leg to the intersection of the extended runway centerline.

e. Final Approach — A flight path in the direction of landing along the extended runway centerline. The final approach normally extends from the base leg to the runway. An aircraft making a straight-in approach VFR is also considered to be on final approach.

traffic situation display (TSD) — TSD is a computer system that receives radar track data from all 20 CONUS ARTCC's, organizes this data into a mosaic display, and presents it on a computer screen. The display allows the traffic management coordinator multiple methods of selection and highlighting of individual aircraft or groups of aircraft. The user has the option of superimposing these aircraft positions over any number of background displays. These background options include ARTCC boundaries, any stratum of en route sector boundaries, fixes, airways, military and other special use airspace, airports, and geopolitical boundaries. By using the TSD, a coordinator can monitor any number of traffic situations or the entire systemwide traffic flows

trailing edge — The rearmost edge of an airfoil.

trailing edge flap — Sections of the trailing edge of an airfoil that can be bent down or extended in flight to 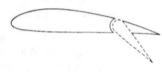 increase the camber of the airfoil, increasing both the lift and drag.

trailing edge mean aerodynamic chord (TEMAC) — The trailing edge of the mean aerodynamic chord that is often used as a location or reference for many aerodynamic measurements in aircraft operations and designs.

trailing finger — An electrode in the distributor of an ignition system that uses a booster magneto for starting. The high voltage from the booster magneto enters through a slip ring and is distributed through a finger that trails the normal ignition finger. In this way, the cylinder would receive a late, hot spark after the normal firing point in order to facilitate starting and to prevent kick-back.

training course outline (TCO) — Within a curriculum, describes the content of a particular course by statement of objectives, descriptions of teaching aids, definition of evaluation criteria, and indication of desired outcome.

training manual — A technical publication used to explain the operation of a system or a component. It is general in nature and is not considered FAA-approved data.

training media — Any physical means that communicates an instructional message to students.

training syllabus — A step-by-step, building block progression of learning with provisions for regular review and evaluations at prescribed stages of learning. The syllabus defines the unit of training, states by objective what the student is expected to accomplish during the unit of training, shows an organized plan for instruction, and dictates the evaluation process for either the unit or stages of learning.

trajectory — A URET CCLD (User Request Evaluation Tool Core Capability Limited Deployment) representation of the path an aircraft is predicted to fly based upon a Current Plan or Trial Plan.

trajectory modeling — The automated process of calculating a trajectory.

trammel points — Sharp points, usually mounted on a long bar. Used to transfer dimensions from one location to another.

tramming — A means of checking the alignment of an aircraft structure by making comparative measurements.

transceiver — A piece of electronic communications equipment in which the transmitter and receiver are housed in the same unit.

transconductance — In bipolar transistors or field-effect transistors, the ratio of a small change in plate current to a small change in grid voltage. The larger the transconductance, the larger the gain.

transcribed weather broadcast (TWEB) — A continuously broadcast weather information service on selected low and medium frequency nondirectional beacons, and on VHF omni-directional ranges, (VOR). The TWEB includes a synopsis and route forecast and is based on a route of flight format specifically prepared by the NWS.

transducer — An electrical device that either takes electrical energy and changes it into mechanical movement or takes mechanical movement and changes it into electrical energy.

transfer — To carry, remove, or send from one place or position to another.

transfer gearbox — In helicopters, a gearbox driven from the main rotor shaft that in turn drives the main (accessory) gearbox.

transfer of control — **1.** That action whereby the responsibility for the separation of an aircraft is transferred from one controller to another. **2.** The act of turning over control of the aircraft from one pilot to another.

transfer punch — A punch having an outside diameter the same as the rivet hole and a sharp point at its exact center. A transfer punch is used to provide a punch mark for starting a drill in the exact center of the hole being transferred.

transferring controller — A controller/ facility transferring control of an aircraft to another controller/facility.

transferring facility — See transferring controller.

transferring unit/controller [ICAO] — Air traffic control unit/air traffic controller in the process of transferring the responsibility for providing air traffic control service to an aircraft to the next air traffic control unit/ air traffic controller along the route of flight.

transformer — An electrical device in which an AC voltage is generated in one winding by mutual induction from AC voltage in another winding. There does not need to be a physical connection between the two windings. Used to change the level of voltage in a circuit. Works only in an AC circuit.

transient conditions — Conditions that can occur briefly while accelerating or decelerating, or while passing through a specific range of engine operations.

transient current — A momentary change in normal current load.

transient voltage — A momentary change in normal voltage.

transistor — A semiconductor device having three electrodes: a base, a collector, and an emitter. Current flow between the collector and emitter is dependant on the voltage applied to the base. The transistor serves as a switch or an amplifier.

transistor voltage regulator — A voltage regulator for DC alternators or generators that uses a transistor to control the field current. A zener diode is used to set the voltage level.

transistorized voltage regulator — A voltage regulator for DC generators or alternators that uses a transistor to control the flow of field current, but uses vibrating points to sense the voltage and control the transistor.

transition — **1.** The general term that describes the change from one phase of flight or flight condition to another; e.g., transition from enroute flight to the approach or transition from instrument flight to visual flight. **2.** A published procedure (DP Transition) used to connect the basic DP to one of several enroute airways/ jet routes, or a published procedure (STAR Transition) used to connect one of several enroute airways/ jet routes to the basic STAR.

transition altitude (QNH) — The altitude in the vicinity of an airport at or below which the vertical position of an aircraft is controlled by reference to altitudes (MSL).

transition height (QFE) — The height in the vicinity of an airport at or below which the vertical position of an aircraft is expressed in height above the airport reference datum.

transition layer — The airspace between the transition altitude and the transition level. Aircraft descending through the transition layer will use altimeters set to local station pressure, while departing aircraft climbing through the layer will be using standard altimeter setting (QNE) of 29.92 inches of Mercury, 1013.2 millibars, or 1013.2 hectopascals.

transition level (QNE) — The lowest flight level available for use above the transition altitude.

transition point — A point at an adapted number of miles from the vertex at which an arrival aircraft would normally commence descent from its enroute altitude. This is the first fix adapted on the arrival speed segments.

transition training — An instructional program designed to familiarize and qualify a pilot to fly types of aircraft not previously flown such as tailwheel aircraft, high-performance aircraft, and aircraft capable of flying at high altitudes.

transitional airspace — That portion of controlled airspace wherein aircraft change from one phase of flight or flight condition to another.

translating cowl — The portion of a turbine engine cowling that moves back to form an exhaust nozzle for thrust reverse air.

translating tendency — The tendency of the single-rotor helicopter to move laterally during hovering flight. Also referred to as tail rotor drift.

translational lift — In helicopters, the additional lift obtained when entering forward flight, due to the increased efficiency of the rotor system.

translucent — The condition of a material to be transparent, letting light pass through it but diffusing it so that objects beyond cannot be seen clearly.

transmission line — A conductor, usually coaxial, used to join a receiver or a transmitter to the antenna.

transmissivity — Ratio of the amount of power transmitted through a radome to the amount of power that would be transmitted with the radome removed.

transmissometer — An apparatus used to determine visibility by measuring the transmission of light through the atmosphere. It is the measurement source for determining runway visual range (RVR) and runway visibility value (RVV).

transmit — In communications, to send information out from a transmitter.

transmitter — Electric device whose function is to collect information from one point and send it electrically to a remote indicator.

transmitting in the blind — A transmission from one station to other stations in circumstances where two-way communication cannot be established, but where it is believed that the called stations can be able to receive the transmission.

transom — In aviation, the vertical bulkhead at the rear end of a seaplane float.

transonic flight — Aircraft in flight approximating 0.8 to 1.2 times the speed of sound in air.

transonic speed — The speed of a body relative to the surrounding fluid that is in some places subsonic and in other places supersonic; usually from Mach 0.8 to 1.2.

transonic speed range — Generally stated as Mach 0.8 to 1.2 speed range, where some portions of the airfoil have subsonic flow and others supersonic flow.

transparent — The property of being easily detected or seen through.

transpiration cooling — Refers to internal cooling air that exits through porous walls of turbine blades and vanes.

transponder — Radar beacon transponder. A radar transmitter-receiver that transmits a coded signal every time it is interrogated by a ground radar facility.

transponder codes — The number assigned to a particular multiple pulse reply signal transmitted by a transponder.

transport category aircraft — An aircraft that is certificated under FAR Part 25.

transverse pitch — The perpendicular distance between two rows of rivets. Also referred to as gauge.

transverse wave — A wave that moves in a perpendicular direction to the direction the wave is moving.

transverse-flow effect — In rotorcraft, a condition of increased drag and decreased lift in the aft portion of the rotor disc caused by the air having a greater induced velocity and angle in the aft portion of the disc.

trapezoid — A plane, four-sided geometric figure having only two sides parallel.

trapped fuel — Undrainable fuel. The amount of fuel remaining in the system after draining. Trapped fuel is considered a part of the empty weight of the aircraft.

traverse — Lying across a body.

triac — A semiconductor device similar to a silicon-controlled rectifier that can be triggered by either a positive or a negative pulse applied to its gate.

trial plan — A proposed amendment that utilizes automation to analyze and display potential conflicts along the predicted trajectory of the selected aircraft.

triangle — A three-sided, enclosed plane figure.

triangulation — In navigation, a method of determining location by drawing lines from two or more known locations in order to determine the position of the unknown point. The lines from the known locations, drawn either by sight or electronically, will cross at the location of the unknown point.

triboelectric series — A list of materials capable of producing static electricity by contact, friction, or induction. Some materials that build up static electricity easily are flannel, silk, rayon, amber, hard rubber, and glass. In the following list, materials at the top of the list become positive to materials lower on the list:

POSITIVE
    Glass
    Mica
    Nylon
    Fur
    Silk
    Paper
    Cotton
    Wood
    Acrylic
    Polystyrene
    Rubber
NEGATIVE

triboelectricity — The production of static electricity by contact or friction between different materials.

trick questions — A question deliberately written in such a way as to cause the student to select the wrong answer even though the student knows the material.

trickle charger — A battery charger that applies a very small constant charge to a battery in order to maintain a full charge on the battery.

trickle charging — A constant current charging method that keeps cells on standby service at full charge by passing a small current through them until they are removed from the charger and returned to service.

tri-cresyl-phosphate — **1.** A plasticizer used as a rejuvenator to restore resilience to dope film. **2.** A toxic substance used in lubricants.

tricycle gear — This type of fixed landing gear consists of two main wheels located on either side of the fuselage and a third wheel, nosewheel, positioned on the nose of the airplane.

tricycle landing gear — The landing gear of an aircraft positioned so the main wheels are behind the center of gravity and the nose is supported with a nose wheel.

trigger pulse — An electric pulse applied to certain electronic circuit elements to start or trigger an operation.

triggering transformer — A high-voltage transformer connected in series with an igniter in a high-energy ignition system for a turbine engine. The transformer places a high voltage across the igniter, ionizing the gap and producing the triggering spark.

trigonometry — The branch of mathematics dealing with the ratios between the sides of a right triangle and the application of these facts in finding the unknown side of any triangle.

trijet — An aircraft propelled by three jet engines.

trim — The adjustment of an aircraft's controls to get a balanced or stable condition of flight.

trim devices — Any device designed to reduce or eliminate pressure on the control stick or yoke. When properly trimmed, an aircraft should fly at the desired airspeed with no control pressure from the pilot (i.e., "hands off"). Trim mechanisms are either external tabs on the control surface or a simple spring-tension system connected to the control stick or yoke.

trim tab — A small auxiliary hinged portion of a movable control surface that can be adjusted during flight to a position resulting in a balance of control forces.

trimetrogon — A system of aerial photography that uses one vertical and two oblique shots of terrain taken simultaneously in order to properly map the contours of the landscape.

trimmer — A potentiometer, generally with a screwdriver-adjusted slider used for fine tuning a circuit.

triode — A vacuum tube having three active electrodes.

trip-free circuit breaker — A circuit protection device that opens a circuit when a current overload exists regardless of the position of the control handle.

triphibian — An aircraft landing gear configuration with the ability to operate from the ground, on snow and ice, and from water.

triplane — An airplane having three main supporting wing surfaces, usually located one above the other.

triple thread — A system using three threads cut into a fastener. The second and third threads begin 120° after the previous thread. This type fastener allows tightly spaced threads without the small lead (distance fastener moves each revolution) that would normally result with a fine threaded, single-thread fastener.

triple-slotted flap — A high lift device used to reduce the takeoff or landing speed by changing the lift characteristics of a wing during the landing or takeoff phases. Triple-slotted flaps extend downward and rearward away from the wing in three sections separated by slots. The slots allow a flow of air over the upper surface of the flap. The effect is to streamline the airflow and improve the efficiency of the flap.

triple-spool engine — Usually a turbofan engine design where the fan is the N_1 compressor, followed by the N_2 intermediate compressor, and the N_3 high pressure compressor, all of which rotate on separate shafts at different speeds.

tri-square — A layout tool that consists of a metal rule set at 90° to a thicker base, usually made of wood or plastic. Used to draw lines perpendicular to an edge of a piece of material.

trituator — Solid waste grinder on airliners.

tropical airmass — An airmass with characteristics developed over low latitudes. Maritime tropical air (mT), the principal type, is produced over the tropical and subtropical seas; very warm and humid. Continental tropical (cT) is produced over subtropical arid regions and is hot and very dry. Compare polar airmass.

tropical cyclone — A general term for a cyclone that originates over tropical oceans. There are three classifications of tropical cyclones according to their intensity:
1. tropical depression: winds up to 34 knots;
2. tropical storm: winds of 35 to 64 knots; and
3. hurricane or typhoon: winds of 65 knots or higher.

tropical storm — See tropical cyclone.

tropopause — The boundary layer between the troposphere and the stratosphere.

tropopause — An area at an average altitude of 36,000 feet that acts as a lid to confine most of the water vapor, and the associated weather, to the troposphere.

tropopause — The boundary between the troposphere and stratosphere, usually characterized by an abrupt change of lapse rate.

troposphere — That portion of the atmosphere from the Earth's surface to approximately 36,000 feet MSL. The troposphere is characterized by decreasing temperature with height, and by appreciable water vapor.

troubleshooting — The systematic analysis of a malfunction in a system or component done in order to determine the cause of malfunction.

trough — In meteorology, an elongated area of relatively low atmospheric pressure and maximum cyclonic curvature of the wind flow (isobars or contours). Also referred to as a trough line.

trough line — See trough.

true air temperature — See outside air temperature (OAT).

true airspeed (TAS) — Calibrated airspeed corrected for non-standard temperature and altitude. The speed at which an aircraft is moving relative to the surrounding air.

true airspeed indicator — An airspeed indicator that takes into consideration dynamic pressure, static pressure, and free air temperature in order to provide a display of the true airspeed.

true altitude — The actual height of an object above mean sea level.

true bearing — The direction measured in degrees clockwise from true north.

true course (TC) — A navigational direction or course of an aircraft in flight as measured in relation to the geographic north pole.

true heading (TH) — The direction the longitudinal axis of the airplane points with respect to true north. True heading is equal to true course plus or minus any wind correction angle.

true north — The direction on the Earth's surface that points toward the geographic north pole.

true power — The power that actually exists in an AC circuit. It is the product of the voltage, current, and power factor; or is the product of the voltage and the current that is in phase with that voltage. True power is expressed in watts.

true wind direction — The direction, with respect to true north, from which the wind is blowing.

true-false — Test item that consists of a statement that the student must determine is true or false.

true-false test items — Consist of a statement followed by an opportunity for the student to determine whether the statement is true or false.

truncated — Cut off as having the angles of an object cut off.

truss — A frame arranged together in such a manner that all members of the truss can carry both tension and compression loads with cross-bracing achieved by using solid rods or tubes.

truss fuselage — Fuselage usually constructed of welded steel tubing to carry the tensile and compressive loads. A superstructure or auxiliary framework is often attached to the truss to give the structure a desirable aerodynamic shape.

truss head — A low, rounded top surface with a flat bearing surface.

TSO (Technical Standard Order) — A standard established by the FAA for quality control in avionics, instruments, and other airborne equipment.

T-square — An instrument used for making aircraft drawings. It consists of a head and a perpendicular blade and is shaped like the letter T.

T-tail — An aircraft with the horizontal stabilizer mounted on the top of the vertical stabilizer, forming a T.

tubing — A rigid, hollow, elongated cylindrical piece of material through which fluids or wiring is passed.

tubing cutter — A tool consisting of a sharp wheel and a set of rollers. The tube to be cut is clamped between the cutter cutting wheel, and the rollers and the tool is rotated around the tube. The cutting wheel is fed into the groove as the tube is cut.

tumble limit — The number of degrees of pitch or roll a gyro can tolerate before it reaches its gimbal stops. Beyond this point, the gyro will tumble.

tumbling — 1. The process of cleaning or abrading parts in a rotating container, either with or without cleaning or abrasive materials. 2. The action of a gyro when it has reached its gimbal stops. The gyro will assume an erroneous indication that does not return to normal until corrected.

tuned circuit — An electrical circuit in which the inductive reactance is equal to the capacitive reactance. They will be equal at a particular frequency called the resonant frequency. The circuit can be made resonant by either varying the frequency until resonance occurs, or holding the frequency constant and varying inductance or capacitance to achieve resonance. Also referred to as a resonant circuit.

tuned radio frequency receiver — A radio receiver in which tuning and amplification are accomplished in the RF section before the signal reaches the detector. After the detector, one or more stages of AF amplification are employed to increase the output sufficiently to operate a loudspeaker.

tungsten — A gray-white, heavy, high melting point, hard metallic element with symbol of W and an atomic number of 74.

tungsten steel — Steel with which tungsten has been alloyed. It is used in the manufacture of cutting tools because of its hardness.

tuning — The process of adjusting circuits to resonance at a particular frequency.

tunnel diode — A semiconductor diode that exhibits a negative resistance characteristic. Under certain conditions, an increase in voltage across the tunnel diode results in a decrease in current through it.

turbine — A rotary wheel device fitted with vane-like airfoils and actuated by impulse or reaction of a fluid flowing through the vanes, or blades. Multiple turbine wheels are arranged along a central shaft.

turbine bucket — The blades on a turbine wheel.

turbine disc — The metal disc to which turbine blades are attached.

turbine discharge pressure (Pt7) — The total pressure at the discharge of the low-pressure turbine in a dual turbine axial flow engine.

turbine efficiency — A ratio of actual work performed by the turbine wheel in ft lbs/Btus and the laboratory standard of 778 ft lbs of work in 1 Btu. Expressed as a percentage.

turbine engine — An aircraft engine that consists of an air compressor, a combustion section, and a turbine. Thrust is produced by increasing the velocity of the air flowing through the engine.

turbine inlet temperature (TIT) — Temperature taken in front of the first stage turbine nozzle vanes. TIT is the most critical temperature taken within the engine and essential to the operation of the fuel control.

turbine nozzle — The orifice assembly through which exhaust gases are directed prior to passing into the turbine blades.

turbine nozzle vanes — The stationary airfoils that precede each turbine blade set. They function to increase gas velocity and direct the gases into the turbine blade at the optimum angle.

turbine stage — In gas turbine engines, a stage consists of a turbine stator vane set followed by a turbine rotor blade set.

turbine wheel — A rotating device actuated by reaction, impulse, or a combination of both. Used to transform some of the kinetic energy of the exhaust gases into shaft horsepower to drive the compressors and accessories.

turbocharger — An air compressor driven by exhaust gases, which increases the pressure of the air going into the engine through the carburetor or fuel injection system.

turbocompound engine — A reciprocating engine that supplements the internal supercharger by an external power recovery turbine (PRT) driven by a portion of the exhaust gas from the aircraft engine. The PRT is connected through a fluid coupling and gear arrangement to help drive the crankshaft.

turbofan — An engine featuring a multi-bladed ducted propeller driven by a gas turbine engine.

turbojet aircraft — An aircraft having a jet engine in which the energy of the jet operates a turbine that in turn operates the air compressor.

turbojet engine — A turbine engine that produces its thrust entirely by accelerating the air through the engine.

turboprop aircraft — An aircraft having a jet engine in which the energy of the jet operates a turbine that drives the propeller.

turboprop engine — A turbine engine that drives a propeller through a reduction gearing arrangement. Most of the energy in the exhaust gases is converted into torque, rather than its acceleration being used to propel the aircraft.

turboshaft — A gas turbine engine geared to an output shaft. Usually for rotorcraft installation, but also for many marine and industrial uses.

turboshaft engine — A gas turbine engine that delivers power through a shaft to operate something other than a propeller.

turbulence — 1. In general, any irregular or disturbed flow in the atmosphere. 2. In aviation, bumpiness in flight.

turbulence in and near thunderstorms (TNT) — That turbulence that occurs within, below, above, and around developing convective clouds and thunderstorms.

turbulent flow — A flow of fluid in an unsteady state.

turbulent gusts — The atmospheric wind and vertical motion fluctuations caused by turbulent eddies.

turbulent wake — The turbulent eddies created near the ground when high surface winds are disrupted by obstacles, such as hangers and other large buildings located near an approach path. Also, the turbulent region downwind of a thunderstorm.

turn — To machine on a lathe.

turn and bank indicator — A flight instrument consisting of a rate gyro to indicate the rate of yaw and a curved glass inclinometer to indicate the relationship between gravity and centrifugal force. The turn and bank indicator indicates the relationship between angle of bank and rate of yaw.

turn and slip indicator — See turn and bank indicator.

turn anticipation — 1. Turning maneuver initiated prior to reaching the actual airspace fix or turn point that is intended to keep the aircraft within established airway or route boundaries. 2. The capability to manually or automatically determine the point along a route of flight, prior to a fly-by fix, where a turn should be initiated to provide a smooth path to the succeeding course. 3. The capability of RNAV systems to determine the point along a course, prior to a turn WP, where a turn should be initiated to provide a smooth path to intercept the succeeding course, and to enunciate the information to the pilot.

turn coordinator — A rate gyro that senses both roll and yaw.

turn WP [turning point] — A WP that identifies a change from one course to another.

turnbuckle — A device used in a control system to adjust cable tension. A turnbuckle consists of a brass barrel with both left- and right-hand threaded terminals.

turning error — One of the errors inherent in a magnetic compass caused by the dip compensating weight. It shows up only on turns to or from northerly headings in the northern hemisphere and southerly headings in the southern hemisphere. Turning error causes the compass to lead turns to the north or south and lag turns away from the north or south.

turning points — Track changes often occur along initial or intermediate approach segments of non-RNAV procedures, some defined by DME and some defined by radials or bearings.

turpentine — A thinner and quick drying agent used in varnishes, enamels, and other oil-based paints.

turret — A tool mount on a lathe. The turret holds several tools that can be selected as needed.

turret lathe — A metal-turning lathe in which the cutting tools are mounted in a turret for purpose of quick change.

tweak test — A test of the output of a wheel speed sensor made by tweaking or flipping the sensor blade with the fingers to rotate it enough for it to generate a voltage.

twelve point head — A standard head form for externally wrenched screws designed for use in counter-bored holes.

twelve-point socket — A socket wrench that has a 12-point double hex opening that makes it easy to position on a bolt or nut.

twenty-minute rating — The ampere hour rating of a battery indicating the amount of current that can be drawn from a battery in order to discharge it in twenty minutes.

twilight — Often called dusk, it is the period of decreasing light from sunset until dark.

twin-row radial engine — A radial engine having two rows of cylinders, one behind the other.

twist drill — A metal cutting tool that has a straight shank and deep spiral flutes in its side and provides a passage for the chips to be removed from the hole it is cutting.

twist grip — In helicopters, the power control on the end of the collective control.

twisted pair — Two insulated wires twisted together. This configuration causes the magnetic fields from each of the conductors to cancel out the magnetic field of the other. These wires are used in circuits such as the lighting circuit for the standby magnetic compass, where magnetic fields would cause compass errors.

two view drawing — An orthographic projection drawing that uses two views to portray an object.

two-axis automatic pilot — An automatic flight control system that operates only in the roll and yaw axes.

two-axis autopilot — An automatic flight control system that controls the airplane about the roll and yaw axes.

two-cycle engine — A reciprocating engine in which a power impulse occurs on each stroke of the piston. As the piston moves outward, fuel-air mixture is drawn into the crankcase below the piston; while above the piston, the mixture is compressed. Near the top of the stroke, ignition occurs. As the piston moves downward, power is produced by the crankshaft. Near the bottom of the stroke, exhaust action takes place on one side of the cylinder and intake action occurs on the opposite side.

two-part adhesive — An adhesive that consists of two parts. One part is the base and the other part is the accelerator. When the two are mixed together in the correct amounts, the adhesive cures in a short time period.

two-state device — An electronic component that can be switched to a high or resistance state by a control signal.

two-stroke-cycle engine — See two-cycle engine.

two-way communications — The ability of both stations involved in communications to transmit and receive signals or information.

two-way radio communications failure — Loss of the ability to communicate by radio. Aircraft are sometimes referred to as NORDO (No Radio). Standard pilot procedures are specified in FAR 91. Radar controllers issue procedures for pilots to follow in the event of lost communications during a radar approach when weather reports indicate that an aircraft will likely encounter IFR weather conditions during the approach.

TYP — A mechanical drawing term that means "typical." It is used to show that the part symbolized by the TYP designator is typical for more than one area or part of the drawing.

type — **1.** As used with respect to the certification, ratings, privileges, and limitations of airmen, means a specific make and basic model of aircraft, including modifications thereto that do not change its handling or flight characteristics. Examples include: DC-7, 1049, and F-27. **2.** As used with respect to the certification of aircraft, means those aircraft similar in design. Examples include: DC-7 and DC-7C; 1049G and 1049H; and F-27 and F-27F. **3.** As used with respect to the certification of aircraft engines means those engines similar in design. For example, JT8D and JT8D-7 are engines of the same type, and JT9D-3A and JT9D-7 are engines of the same type.

Type Certificate Data Sheets — The official specifications of an airplane, engine, or propeller. Type certificate data sheets are issued by the FAA. In order for the device to be airworthy, it must conform to these specifications.

Ty-Rap — A patented nylon cable wrap used to hold wire bundles together.

U

U.S. standards for instrument procedures (TERPS) — An official FAA publication that prescribes standardized methods for use in designing instrument approach procedures.

U-bolt — A U-shaped rod threaded on both ends. U-bolts are used to fasten cables around a thimble.

ultimate load — The amount of load applied to a part beyond which the part will fail.

ultimate strength — The tensile strength of a material that, when exceeded, leads to the material breaking.

ultrahigh frequency (UHF) — The frequency band between 300 and 3,000 MHz. The bank of radio frequencies used for military air/ground voice communications. In some instances this can go as low as 225 MHz and still be referred to as UHF

ultralight vehicle — An aeronautical vehicle operated for sport or recreational purposes that does not require FAA registration, an airworthiness certificate, nor pilot certification. They are primarily single occupant vehicles, although some two-place vehicles are authorized for training purposes. Operation of an ultralight vehicle in certain airspace requires authorization from ATC.

ultrasonic cleaner — A cleaning apparatus that utilizes cleaning fluid agitated by sound waves transmitted through the fluid. This method is widely used for filter and bearing cleaning.

ultrasonic frequencies — Frequencies above 20,000 Hz that cannot be heard by the human ear.

ultrasonic inspection — A nondestructive inspection in which the condition of a material is determined by its ability to conduct ultrasonic vibrations.

ultrasonic soldering — A method of soldering in which the tip of a soldering iron is vibrated at an ultrasonic frequency. The vibration separates the oxide film on the surface of the metal. A flux is melted over the surface of the hot metal to prevent more oxide from forming.

ultrasonics — A branch of science dealing with high frequency sound waves. Normally considered to be above the audio frequency range of 20 to 20,000 hertz.

ultraviolet (UV) — Frequencies higher than blue, having relatively short wavelengths.

ultraviolet lamp — A lamp that produces a light wavelength slightly shorter than the wavelength of visible light.

ultraviolet radiation — Rays of light that are shorter than the wavelength of visible light. Produced by the sun or by special lamps, exposure to these rays can damage organic materials.

ultraviolet rays — See ultraviolet radiation.

umbilical cord — 1. A cable used to carry power or life support to an astronaut for operating outside a space vehicle. 2. A cable used to deliver power and communication to a rocket or spacecraft on the launching pad prior to liftoff.

unable — Indicates inability to comply with a specific instruction, request, or clearance.

unbalanced cell — A condition in a nickel-cadmium battery in which one cell has discharged more than the other cells. This is the initial step in a thermal runaway.

unbalanced transmission line — A cable that has one conductor grounded or exposed to the outside elements in greater magnitudes than the other conductor.

uncontrolled airport — A nontower airport where control of VFR traffic is not exercised.

uncontrolled airspace — Class G airspace.

uncontrolled spin — A spin in an airplane in which the controls are of little or no use in effecting a recovery.

under the hood — Indicates that the pilot is using a hood to restrict visibility outside the cockpit while simulating instrument flight. An appropriately rated pilot is required in the other control seat while this operation is being conducted.

undercarriage — An airplane's entire landing gear.

undercompounded generator — A compound generator (both series and shunt-wound) that is more influenced by the shunt winding than the series winding. Voltage output drops as load increases.

undercurrent relay — An electrical circuit protection device that opens the circuit when the current drops below a predetermined value.

underpowered — An undesirable condition in which an engine does not have enough power to achieve the desired result.

undershoot — A condition of flight in which a landing aircraft touches the ground short of the runway or the landing strip.

underslinging — In helicopters, placing the main rotor hub around and below the top of the mast as is done on semi-rigid rotor systems.

underslung — In rotorcraft, a rotor hub that rotates below the top of the mast, as on semirigid rotor systems.

underspeed condition — A condition resulting from the movement of the blades of a propeller to a higher angle than that required for constant-speed operation. When the speed drops below the RPM for which the governor is set, the resulting decrease in centrifugal force exerted by the governor flyweights permits the speeder spring to lower the pilot valve, thereby opening the propeller-governor metering port. Oil is then directed to the prop in the proper direction to flatten the pitch of the propeller and correct the underspeed condition.

understanding — Level of learning. Ability to comprehend something learned.

understanding — A basic level of learning where a student comprehends or grasps the nature or meaning of something.

undervoltage relay — An electrical circuit protection device that opens the circuit when voltage drops below a predetermined value.

underwing fueling — A method of fueling an aircraft from a single-point pressure fueling port located under the wing.

Underwriter's Laboratories (UL) — An independent, not-for-profit product-safety testing and certification organization.

undrainable fuel — The amount of fuel that remains in the system after draining. Undrainable fuel is considered a part of the empty weight of the aircraft.

undrainable oil — The oil remaining after draining the oil from an engine. Undrainable oil is considered a part of the empty weight.

unfeather — The action of turning the propeller blades to an angle not parallel with the airplane's line of flight.

unfeathered — Low blade angle condition of a propeller.

UNICOM — A nongovernment communication facility that can provide airport information at certain airports. Locations and frequencies of UNICOMs are shown on aeronautical charts and publications.

unidirectional — In composites, a fabric, tape, or laminate with all the major fibers running in one direction, giving strength in that direction.

unidirectional current — A flow of electrons in one direction throughout a circuit. See also direct current.

uniform acceleration — Increasing the speed of an object at an unvaried rate.

uniform surface corrosion — Unvaried corrosion across a surface. No pits or localized damage have formed.

unijunction transistor — A transistor that allows voltage to flow between its two bases when an appropriate voltage is applied to its emitter.

union — Connectors or fittings that attach one item, such as tubing, to another.

unit — A single piece of equipment.

United States — Geographically, the States, the District of Columbia, Puerto Rico, and the possessions, including the territorial waters, and the airspace of those areas.

United States air carrier — A citizen of the United States who undertakes directly by lease, or other arrangement, to engage in air transportation.

universal chuck — The three clamping jaws of a drill used to hold drill bits and other drill accessories.

universal joint — A joint coupling that allows one shaft to drive another shaft at angles to each other.

universal motor — A series-wound motor that operates on either alternating or direct current.

universal propeller protractor — A precision measuring device used to measure the amount of blade angle.

universal time (UT) — In the most common civil usage, UT refers to a time scale called "Coordinated Universal Time" (abbreviated UTC), which is the basis for the worldwide system of civil time. This time scale is kept by time laboratories around the world and is determined using highly precise atomic clocks. The International Bureau of Weights and Measures makes use of data from the timing laboratories to provide the international standard UTC, which is accurate to approximately a nanosecond (billionth of a second) per day. The length of a UTC second is defined in terms of an atomic transition of the element cesium under specific conditions, and is not directly related to any astronomical phenomena. The times of various events, particularly astronomical and weather phenomena, are often given in "Universal Time" (abbreviated UT), which is sometimes referred to, now colloquially, as "Greenwich Mean Time" (GMT). The two terms are often used loosely to refer to time kept on the Greenwich meridian, five hours ahead of Eastern Standard Time. Times given in UT are almost always given in terms of a 24-hour clock. Thus, 14:42 (often written simply 1442) is 2:42 p.m., and 21:17 (2117) is 9:17 p.m. Sometimes a Z is appended to a time to indicate UT, as in 0935Z.

unleaded gasoline — Engine fuel that has not been treated with lead or lead components.

unlimited ceiling — A clear sky or a sky cover that does not constitute a ceiling.

unloaded rotor — In helicopters, the state of a rotor when rotor force has been removed, or when the rotor is operating under a low or negative G condition.

unloading valve — A pressure control valve used in aircraft hydraulic systems to prevent excessive pressures from bursting lines and blowing out seals. When the pressure reaches a predetermined value, the valve reroutes the high output pressure back to the pump inlet.

unmetered fuel — Fuel that enters the fuel control from the fuel pump.

unpublished route — A route for which no minimum altitude is published or charted for pilot use. It can include a direct route between NAVAIDS, a radial, a radar vector, or a final approach course beyond the segments of an instrument approach procedure.

unscheduled maintenance — Maintenance performed as a result of discrepancies found by flight and ground personnel.

unstable — **1.** The characteristic of an aircraft that causes it, when disturbed from a condition of level flight, to depart further from this condition. **2.** In meteorology, a general term to indicate various states of the atmosphere in which spontaneous convection will occur when prescribed criteria are met; indicative of turbulence. See also absolute instability, conditionally unstable air.

unstable air — Air with a temperature lapse rate different than the surrounding air.

unstable oscillation — An oscillation whose amplitude continues to diverge from the initial amplitude with no tendency to return toward the original amplitude.

unusable fuel — The small amount of fuel in the tanks that cannot be safely used in flight or drained on the ground. Unusable fuel is considered part of the empty weight of the aircraft.

unusable oil — Oil that cannot be drained from the engine.

updraft — A localized upward current of air.

updraft carburetor — A carburetor mounted on the bottom of a reciprocating engine. All of the air entering the engine flows upward through the venturi.

upholstery — Materials (fabric, padding, and springs) used to make a soft covering for a seating surface.

upper air temperature — The temperature that is referenced to the height or pressure level where they are measured.

upper deck pressure — The pressure of the air between the compressor and the throttle plate on a supercharged, reciprocating engine. This air is also used for pressurization of the cabin and the fuel injection system.

upper front — A front aloft not extending to the Earth's surface.

upset head — The end of the rivet opposite the manufactured head formed during the riveting process.

upsetting — The process of increasing the cross-sectional area of the end of a rivet or bolt by bucking or hammering it.

upslope fog — Fog formed when air flows upward over rising terrain and is, consequently, adiabatically cooled to or below its initial dewpoint.

upslope wind — The deflection of the air by hills or mountains, producing upward motions along the slopes of a mountain or hill.

upwind — In the direction from which the wind is blowing.

upwind leg — See traffic pattern.

urgency — A condition of being concerned about safety, and of requiring timely but not immediate assistance; a potential distress condition.

urgency [ICAO] — A condition concerning the safety of an aircraft or other vehicle, or of person on board or in sight, but which does not require immediate assistance.

usability — Characteristic of a test being easy to read and clear in the use of directions, figures, and illustrations.

usable fuel — The portion of the total fuel load available to an aircraft in flight.

USAFIB — Acronym for: U. S. Army Aviation Flight Information Bulletin

useful load — **1.** The weight of the occupants, baggage, and usable fuel (Aircraft certified prior to FAR Part 23 include oil). **2.** The difference between the empty weight of the aircraft and the maximum weight of the aircraft. Also referred to as the payload.

useful load — The difference between the empty weight of the airplane and the maximum weight allowed by the manufacturer's specification.

user request evaluation tool core capability limited deployment (URET CCLD) — User Request Evaluation Tool Core Capability Limited Deployment is an automated tool provided at each Radar Associate position in selected En Route facilities. This tool utilizes flight and radar data to determine present and future trajectories for all active and proposal aircraft and provides enhanced, automated flight data management.

user-defined waypoint — User-defined waypoints typically are created by pilots for use in their own random RNAV direct navigation. They are newly established, unpublished airspace fixes that are designated geographic locations/positions that help provide positive course guidance for navigation and a means of checking progress on a flight. They may or may not be actually plotted by the pilot on enroute charts, but would normally be communicated to ATC in terms of bearing and distance or latitude/longitude. An example of user-defined waypoints typically includes those derived from database-driven area navigation (RNAV) systems whereby latitude/longitude coordinate-based waypoints are generated by various means including keyboard input, and even electronic map mode functions used to establish waypoints with a cursor on the display. Another example is an offset phantom waypoint, which is a point in space formed by a bearing and distance from navaids such as VORs, VORTACs, and TACANs, using a variety of navigation systems.

utility category airplane — An airplane certificated for flight that includes limited acrobatics such as spins, lazy eights, chandelles, and steep turns in which the bank angle exceeds 60°.

utility finish — A finish used on aircraft that provides the fabric with the necessary fill and tautness, but lacks the glossy appearance of a showroom finish.

U-tube manometer — A "U"-shaped glass tube filled with a liquid. Pressure is measured by connecting a gas under pressure to one vertical of the manometer. The pressure causes a differential between the levels in the two verticals. This gives a differential pressure between the gas pressure and atmospheric pressure. If the liquid is water, the measurement is in water column inches (WCI).

V

V speeds — Specific airspeeds given for individual aircraft. An example of a V speed would be V_{NE}, or never exceed speed. V speeds are aircraft specific.

vacuum — A negative pressure or pressure below atmospheric pressure, usually expressed in inches of mercury (in. Hg).

vacuum bagging — In composites, a means of applying atmospheric pressure to a part while curing by sealing the part in a plastic bag and removing most of the air.

vacuum bottle — A container for keeping liquids hot or cold. Has inner and outer walls with most of the air taken from the space between the container walls in order to insulate the container.

vacuum distillation — A distillation process that boils liquids at very low pressures. This allows the liquids to be distilled at lower temperatures than would otherwise be possible.

vacuum forming — A thermoplastic forming that uses vacuum and a heated die to form the plastic.

vacuum pump — A device mounted on an aircraft engine that creates negative pressure used to drive some of the gyroscopic flight instruments.

vacuum tube — An evacuated glass or metal envelope containing a cathode, heater, plate, and often one or more grids. It serves as an electron control valve.

vacuum-tube voltmeter — An electronic voltage-measuring instrument used for electronic circuit testing. Its high input impedance prevents it from drawing appreciable power from the circuit being tested.

valence electron — 1. An outer electron that can be given up or gained in the process of ionic compound formation. 2. An outer electron shared with another atom in the process of covalent compound formation.

valence shell — In physics, the electrons in the outer ring of an atom.

validity — Characteristic of a measuring instrument that actually measures what it is supposed to measure and nothing else.

valley breeze — An upslope wind flow caused by the heating of the mountain slope that heats the adjacent air.

valve — A device that regulates the flow of a liquid or gas.

valve clearance — The clearance between a valve head and rocker arm when the valve is seated.

valve core — A spring-loaded, resilient check valve inside a valve stem that, when depressed, allows air to flow into or out of a tire or chamber.

valve duration — The length of time, measured in degrees of crankshaft rotation, a valve in an aircraft engine remains open.

valve face — The inner, tapered surface of an intake or exhaust valve that forms a seal against the valve seat in the cylinder head when the valve is closed.

valve float — In reciprocating engines, a condition in which, at high RPM, the valve lifters lose contact with the cam lobes because the valve springs are not strong enough to overcome the momentum of the various valve train components. This allows the valve to open further and stay open longer than design specifications call for. Extended periods of valve float will damage the valve train.

valve grinding — A process of removing part of a valve face with a precision grinding machine. Grinding improves the seal between the valves and their seats.

valve guide — The component in an aircraft reciprocating engine cylinder head that guides the valve and holds the valve head concentric with the valve seat.

valve lag — The number of degrees of crankshaft rotation after top or bottom center at which the intake or exhaust valves open or close. For example, if the intake valve closes at 60° of crankshaft rotation after the piston passes over top center and starts down on its intake stroke, the exhaust valve lag is 60°.

valve lapping — A process in which a valve is matched to a valve seat by using a fine abrasive compound. Valve lapping removes any rough material to ensure an airtight seal.

valve lead — The number of degrees of crankshaft rotation before top or bottom center at which the intake or exhaust valves open or close. For example, if an intake valve opens 15° before the piston reaches top center on the exhaust stroke, it is said to have a 15° valve lead.

valve lift — The distance a valve is lifted off its seat when it is opened by the cam.

valve overlap — The degrees of crankshaft rotation when both the intake and exhaust valves are open. In a four-stroke-cycle reciprocating engine, valve overlap improves the efficiency of engine operation by allowing the low pressure caused by the exhaust gases leaving the cylinder to help the fresh charge of fuel-air in the induction system to start moving into the cylinder.

valve ports — The intake and exhaust holes in the cylinder of an aircraft reciprocating engine.

valve radius gauge — A gauge to determine if a valve has the proper radius between the stem and the head. An improper radius is an indication of a stretched valve.

valve seat — A hardened ring of steel or bronze, embedded in the cast aluminum cylinder head to provide proper seating for the poppet valve.

valve spring tester — A machine used in engine overhaul to test the condition of valve springs by measuring the force required to compress them to a specified height.

valve springs — Helical-wound steel wire springs used to close the poppet valves in an aircraft engine cylinder.

valve stem — The portion of a poppet valve that rides in the valve guide and maintains concentricity between the head of the valve and the valve seat.

valve stretch — Elongation of the valve stem caused by overheating.

valve stretch gauge — A tool that measures the radius between the valve head and the valve stem.

valve timing — The relationship between the crankshaft rotation and the opening and closing of the intake and exhaust valves.

valve train — Moving parts of the valve operating system. Includes lifters, springs, valves and associated installation parts.

valve-timing clearance — The clearance to which poppet valves using solid valve lifters are adjusted to set the cam for valve timing. After the timing is set, the valve clearance is adjusted to the cold, or running, clearance.

vanadium — A malleable, ductile, silver-white metallic chemical element with a symbol of V and an atomic number of 23. Alloyed with steel, it toughens and adds tensile strength to the steel.

vanadium steel — A steel alloyed with 0.10% to 0.15% vanadium to provide additional hardness and strength. Vanadium steel is used in the manufacture of technicians' tools.

vane — Stationary airfoils within an engine.

vane-type pump — A constant displacement, fluid-moving pump in which a rotor containing sliding vanes turns in an eccentric cavity to force the fluid through the pump.

vanishing point — Points in a drawing that converge to give the appearance of having depth.

vapor — The gaseous state of a material.

vapor degreasing — A method of degreasing a part by treating it with the hot vapors of a solvent such as trichlorethylene.

vapor lock — A condition in a fuel system in which liquid fuel has turned into a vapor in the fuel lines. This vapor prevents the flow of liquid fuel to the carburetor or fuel injectors.

vapor pressure — **1.** The pressure, at a specific temperature, exerted by the vapor above a liquid. Vapor pressure can prevent the release of additional vapor. **2.** In meteorology, the partial pressure of water vapor in the atmosphere.

vapor separator — A device in a pressure-type carburetor regulator unit that prevents air in the fuel from upsetting the metering of the carburetor. The vapor separator consists of a small float and needle valve positioned in the vapor separator chamber. When there are no vapors in the chamber, the fuel level raises the float and holds the needle valve closed. As vapors gather, the fuel level in the chamber drops, lowering the float until the needle valve opens, releasing the vapors back into the fuel tank.

vapor trail — A cloudlike streamer frequently observed behind aircraft flying in clear, cold, humid air. The pressure reduction above the wing surfaces and water vapor in the engine exhaust gases combine to create a visible trail of condensed water vapor.

vapor-cycle air-conditioning system — A system for cooling the air in an aircraft cabin in which the cabin heat is absorbed into a liquid refrigerant, turning it into a vapor. This vaporized refrigerant is carried into a condenser outside the airplane where the heat is released to the outside air, causing the refrigerant to revert back to a liquid to begin the cycle over again.

vaporize — To change a liquid into a vapor.

vaporizing tube — A fuel nozzle that injects a fuel-air mixture into the combustor. This nozzle is in a system that operates at lower pressure than the atomizing type fuel nozzle system.

variable absolute pressure controller (VAPC) — The intelligence of the turbo system. It monitors compressor discharge pressure and limits the maximum pressure. Also maintains the discharge pressure slightly higher than manifold pressure. The VAPC controls the discharge pressure by regulating the oil flow through the waste gate actuator. The variable portion of the pressure controller is mechanically connected to the throttle. Opening the throttle to maintain a higher manifold pressure will also adjust the VAPC to maintain a higher compressor discharge pressure. The system's design prevents the turbocharger from working at maximum output when the excess pressure is not needed.

variable capacitor — A capacitor whose capacity can be changed by varying the distance between the plates. Variable capacitors are commonly used in radio or radar tuning devices.

variable displacement pump — A pump whose output can be varied. For high pressure applications, this is usually done by varying the stroke of a piston-type pump.

variable geometry air inlet duct — The inlet duct on a supersonic turbojet aircraft whose area or shape can be varied in flight. Provides the proper inlet pressure to the first stage of the compressor as the airspeed of the airplane changes.

variable geometry aircraft — An aircraft with the ability to alter the configuration of its wings. The wings are swept forward for takeoff and low speed flight and then swept back for flight at supersonic speeds.

variable inductor — In electricity, an inductor (coil) whose inductance is changed by varying the number of coils or the composition of the core.

variable pitch propeller — A propeller whose blade pitch can be changed while the engine is operating to obtain the most efficient operation. Also referred to as a controllable pitch or constant-speed propeller.

variable resistor — An electronic resistance component that can be varied by rotating an adjusting screw or shaft or adjusting the position of a sliding contact. Rheostats and potentiometers are variable resistors.

variable restrictor — A unit that can be adjusted to control the amount of fluid flow and thereby control the operating speed of a unit.

variable-angle stator vanes — Inlet guide vanes and compressor stator vanes that can change their angle to the accommodate the oncoming airstream. Variable-angle stator vanes are connected to the power lever and control engine stall tendencies during acceleration and deceleration.

variable-frequency oscillator (VFO) — An oscillator in a radio receiver for generating a heterodyne signal that is beat against the incoming signal to create an intermediate frequency. If the intermediate frequency is at an audio rate, it allows a signal to be presented to a speaker or headphones. If the intermediate frequency is a radio frequency, it can be amplified and demodulated.

Variac — The registered trade name for a brand of autotransformer. In general usage it describes a variety of autotransformers. An autotransformer is a single winding transformer having a carbon brush that can tap off any number of turns for the secondary. It produces variable voltage AC output.

variation — The angular difference between the true, or geographic, poles and the magnetic poles at a given point. The compass magnet is aligned with the magnetic poles, while aeronautical charts are oriented to the geographic poles. This variation must be taken into consideration when determining an aircraft's actual geographic location. Indicated on charts by isogonic lines, it is not affected by the airplane's heading.

variometer — Sensitive rate of climb or descent indicator.

vari-ramp — A movable ramp in a turbine engine inlet that controls supersonic diffusions and airflow velocity into the compressor.

varnish — **1.** Surface finish: A preparation of resin dissolved in oil or alcohol. Provides a glossy, protective surface for wood, metal, etc. **2.** Engine deposit: A baked oil deposit formed on reciprocating engine components that have been operated at excessively high temperatures.

varsol — A petroleum product similar to naptha. Used as a solvent.

V-belt — A drive belt that has a cross section in the shape of a "V." V-belts are used to drive generators, pumps, and other accessories from a source of motorized power.

V-block — A block with a V-notch cut across it. Used for shrinking or stretching sheet metal angles or flanges.

vector — A quantity with both magnitude and direction. The length of the line indicates magnitude; the arrowhead indicates direction of motion or force.

vector — **1.** A variable that has magnitude and direction. For example, wind or pressure gradient. The line's length indicates magnitude, and the arrow point represents direction of action. **2.** A heading issued to an aircraft to provide navigational guidance by radar.

vector sum — In mathematics, the resultant of adding two or more vectors.

veering — In weather, the process of wind direction shifting clockwise, as from north to northeast.

velocity — The actual change of distance with respect to time. The average velocity is equal to total distance divided by total time. Usually expressed in MPH or FPS.

vendor — An individual or collective that sells goods or services.

veneer — **1.** Any of the thin layers of wood bonded together to form plywood. **2.** A layer of fine wood glued to the surface of an inferior wood to provide a superior surface or to create a more attractive appearance.

V-engine — An internal combustion engine with cylinders arranged in two rows in the form of a "V," with an angle between the rows, i.e., 45° or 60°.

vent — An opening for the escape of a gas or liquid or for the relief of pressure.

ventilate — To provide fresh air to the inside of a compartment.

ventral fin — A vertical, stabilizing fin on the lower rear portion of an airplane fuselage. Its purpose is to increase the directional stability of the airplane by increasing the area behind the vertical axis.

venturi — A specially shaped restrictor in a fluid flow passage used to increase the velocity of the fluid and decrease its pressure.

venturi tube — A source of vacuum for the operation of gyroscopic instruments. It is a short tube with large openings at both the front and rear with a specially designed restrictor between them. The velocity of the air flowing through this tube is increased and its pressure decreased as it passes the restriction.

verify — Request confirmation of information; e.g., "verify assigned altitude."

verify specific direction of takeoff (or turns after takeoff) — Used by ATC to ascertain an aircraft's direction of takeoff and/or direction of turn after takeoff. It is normally used for IFR departures from an airport not having a control tower. When direct communication with the pilot is not possible, the request and information can be relayed through an FSS, dispatcher, or by other means.

vernier — A means of making extremely small divisions or measurements. It consists of a short scale made to slide along the divisions of a graduated instrument for indicating parts of divisions.

vernier caliper — A precision measuring tool used to measure the inside or outside dimension of an object. An auxiliary or vernier scale is used to accurately divide the increments of the regular scale.

vernier micrometer caliper — A micrometer caliper with a vernier scale that allows each one thousandth-inch increment to be subdivided into ten equal parts of one ten-thousandth of an inch.

vernier scale — See vernier.

vertex — **1.** In geometry, the point of a cone. **2.** The last fix adapted on the arrival speed segments. Normally, it will be the outer marker of the runway in use. However, it can be the actual threshold or other suitable common point on the approach path for the particular runway configuration.

vertex time of arrival — A calculated time of aircraft arrival over the adapted vertex for the runway configuration in use. The time is calculated via the optimum flight path using adapted speed segments.

vertical axis — The axis of an airplane extending vertically through the center of gravity. Also referred to as the yaw axis.

vertical motion — Movement of air parcels in an upward or downward direction.

vertical navigation (VNAV) — A function of area navigation (RNAV) equipment that calculates, displays, and provides vertical guidance to a profile or path.

vertical path angle (VPA) (USA) — The descent angle shown on some non-precision approaches describing the geometric descent path from the Final approach fix (FAF), or on occasion from an intervening stepdown fix, to the Threshold Crossing Height (TCH). This angle can or can not coincide with the angle projected by a Visual Glide Slope Indicator (VASI, PAPI, PLASI, etc.)

vertical S — The basic vertical S begins with a climb at a constant airspeed and rate. Once a particular cardinal altitude is reached, the climb is reversed and a constant rate, constant airspeed descent is begun. The amount of altitude between reversals can be varied. Once students become proficient with the straight-ahead vertical S, turns are added to the problem. The vertical S is a good exercise to use in teaching students to transition from one set of conditions to another. These maneuvers are designed to improve students' cross-check and aircraft control.

vertical separation — Separation established by assignment of different altitudes or flight levels.

vertical speed indicator — A sensitive differential pressure gauge that indicates the rate at which an aircraft is climbing or descending. The vertical speed indicator is connected to the static system and senses the rate of change of ambient pressure measured in feet per minute.

vertical stabilizer — The fixed vertical surface of an aircraft empennage to which the rudder is attached. Also referred to as vertical fin.

vertical takeoff and landing (VTOL) — An aircraft that can takeoff and land without forward motion. Capable of vertical climbs and/or descents and of using very short runways or small areas for takeoff and landings. These aircraft include, but are not limited to, helicopters.

vertical vibration — In helicopters, a vibration in which the movement is in a vertical direction. One cause might be a main rotor that is out of track.

vertical visibility — The distance one can see upward into a surface based obscuration; or the maximum height from which a pilot in flight can recognize the ground through a surface based obscuration.

vertical wind shear — The change in wind speed and/or direction over a vertical distance. See also wind shear.

vertigo — Spatial disorientation caused by the physical senses sending signals to the brain that create a feeling of whirling and dizziness.

very high frequency — The frequency band between 30 and 300 MHz. Portions of this band, 108 to 118 MHz, are used for certain NAVAIDs; 118 to 136 MHz are used for civil air/ground voice communications. Other frequencies in this band are used for purposes not related to air traffic control.

very high frequency omni-directional range station — The omni-directional, or all-directional, range station provides the pilot with a course from any point within its service range It produces 360 usable radials or courses, any one of which is a radio path connected to the station. The radials can be considered as lines that extend from the transmitter antenna like spokes of a wheel. Operation is in the VHF portion of the radio spectrum (frequency range of 109.0-117.95 MHz).

very high frequency omnirange (VOR) navigation equipment — The omni-directional, or all-directional, range station provides the pilot with a course from any point within its service range It produces 360 usable radials or courses, any one of which is a radio path connected to the station. The radials can be considered as lines that extend from the transmitter antenna like spokes of a wheel. Operation is in the VHF portion of the radio spectrum (frequency range of 109.0-117.95 MHz).

very low frequency (VLF) — The radio frequency range between 3 and 30 kHz.

VFR aircraft — An aircraft conducting flight in accordance with visual flight rules.

VFR conditions — Weather conditions equal to or better than the minimum for flight under visual flight rules. The term can be used as an ATC clearance/instruction only when:
a. An IFR aircraft requests a climb/descent in VFR conditions.
b. The clearance will result in noise abatement benefits where part of the IFR departure route does not conform to an FAA approved noise abatement route or altitude.
c. A pilot has requested a practice instrument approach and is not on an IFR flight plan.
All pilots receiving this authorization must comply with the VFR visibility and distance from cloud criteria in FAR 91. Use of the term does not relieve controllers of their responsibility to separate aircraft in Class B and Class C airspace or TRSA's as required by FAA Order 7110.65. When used as an ATC clearance/instruction, the term can be abbreviated "VFR"; e.g., "MAINTAIN VFR," "CLIMB/ DESCEND VFR," etc.

VFR cruising altitude — When flying above 3,000 feet AGL on a magnetic heading of 0° to 179° you must fly at odd thousand-foot altitudes plus 500 feet and on a heading of 180° to 359° you are required to fly on even thousands plus 500 feet up to the flight levels.

VFR flight — Flight in accordance with visual flight rules.

VFR military training routes — Routes used by the Department of Defense and associated Reserve and Air Guard units for the purpose of conducting low-altitude navigation and tactical training under VFR below 10,000 feet MSL at airspeeds in excess of 250 knots IAS. [Jeppesen does not chart these routes]

VFR not recommended — An advisory provided by a flight service station to a pilot during a pre-flight or inflight weather briefing that flight under visual flight rules is not recommended. To be given when the current and/or forecast weather conditions are at or below VFR minimums. It does not abrogate the pilot's authority to make his own decision.

VFR over-the-top — With respect to the operation of aircraft, means the operation of an aircraft over-the-top under VFR when it is not being operated on an IFR flight plan.

VFR terminal area charts — Depict Class B airspace that provides for the control or segregation of all the aircraft within the Class B airspace. The charts depict topographic information and aeronautical information that includes visual and radio aids to navigation, airports, controlled airspace, restricted areas, obstructions, and related data.

VFR waypoint — A predetermined geographical position used for route/instrument approach definition, progress reports, published VFR routes, visual reporting points or points for transitioning, and/or circumnavigating controlled and/or special use airspace. Defined relative to a VORTAC station or in terms of latitude/longitude coordinates.

VFR-on-top — ATC authorization for an IFR aircraft to operate in VFR conditions at any appropriate VFR altitude (as specified in FAR and as restricted by ATC). A pilot receiving this authorization must comply with the VFR visibility, distance from cloud criteria, and the minimum IFR altitudes specified in FAR 91. The use of this term does not relieve controllers of their responsibility to separate aircraft in Class B and Class C airspace or TRSA's as required by FAA Order 7110.65.

VHF omni-directional range /tactical air navigation (VORTAC) — A navigation aid providing VOR azimuth, TACAN azimuth, and TACAN distance measuring equipment (DME) at one site)

vibrating-reed frequency meter — An electronic meter that uses a series of reeds to indicate the tuned frequency. The display has several reeds that are each tuned to vibrate at a specified and marked frequency. When a particular frequency is in tune, the associated reed vibrates and appears as a wider band, thus indicating the frequency.

vibrating-type voltage regulator — A voltage regulator for direct current generators or alternators. Uses vibrating points to sense the voltage and provide a varying resistance for the generator field current.

vibration — A quivering or trembling motion. The motion of the particles of an elastic body or medium is in alternately opposite directions from the position of equilibrium when that equilibrium has been disturbed.

vibration insulator — A resilient support that helps isolate a system from steady state vibrations.

vibration isolator — A flexible, shock-mount support installed between a component and the structure. Vibration isolators reduce damage to electronic units by keeping heavy vibrations in the structure from being transmitted into the unit.

vibration meter — In turbine engines, a meter that senses vibration, measured in MILS (thousandths of an inch) or in./sec. (inches per second) occurring at the engine outer casings. Typical limit: 3 to 5 MILS.

vibration pickup — A small electrical generator that transmits a turbine engine's vibration signal to a vibration meter either in the aircraft or in a test cell.

vibrator — A relay that makes and breaks the flow of direct current to produce pulsating DC, which can be passed through an induction coil to change its voltage.

vibrator ignition system — An older turbine engine ignition system that uses no storage capacitors.

vibratory torque control — The special patented coupling between the crankshaft and propeller shaft of a Continental Tiara engine. It incorporates a quill shaft to absorb torsional vibrations and a centrifugally actuated mechanism to lock out the quill shaft for operation when a solid shaft would be more advantageous.

Victor Airway — An airway system based on the use of VOR facilities. The north-south airways have odd numbers (Victor 11), and the east-west airways have even numbers (Victor 14).

video — The electronic circuit components controlling or producing the visual signals displayed on a CRT.

video amplifier — An electronic component that can amplify signals over a bandwidth sufficient to cover the requirements of color video recording and broadcasting. This bandwidth must range from 5 Hz to in excess of 5 MHz.

video map — An electronically displayed map on the radar display that can depict data such as airports, heliports, runway centerline extensions, hospital emergency landing areas, NAVAIDs and fixes, reporting points, airway/route centerlines, boundaries, handoff points, special use tracks, obstructions, prominent geographic features, map alignment indicators, range accuracy marks, minimum vectoring altitudes.

view limiting device — A device used to limit the viewing field of a pilot. This is typically used during instrument training in order to allow the pilot to view only the instruments directly in front of him or her, entirely blocking references outside the aircraft. Common view limiting devices include plastic hoods and frosted goggles.

virga — Water or ice particles falling from a cloud, usually in wisps or streaks, and evaporating before reaching the ground.

virtual reality (VR) — A computer-based technology that creates a sensory experience that allows a participant to believe and barely distinguish a virtual experience from a real one. VR uses graphics with animation systems, sounds, and images to reproduce electronic versions of real life experience.

viscosimeter — An instrument used to measure the viscosity of a fluid.

viscosity — The resistance to flow. In composites, resins have a viscosity rating that corresponds to how thick they are.

viscosity index — The measure of a fluid's change in viscosity due to a change in temperature.

viscous — Having a thick, glutinous consistency, and a relatively high resistance to flow.

viscous damping — Diminishing the amount of vibration by use of a fluid that turns some of the vibration energy into heat by pushing fluid through an orifice. A fluid viscous damper also dissipates energy by redirecting the force, producing a damping pressure that creates a force that is 90 degrees out of phase with the vibration.

vise — An adjustable tool that holds work and has two jaws that open and close by means of a screw, lever, or cam..

Vise-grip pliers — A brand name of locking pliers. Locking is accomplished by an over-center locking device. The grip width of the pliers is variable through an adjusting screw in the handle.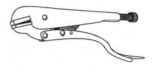

visibility — The ability, as determined by atmospheric conditions and expressed in units of distance, to see and identify prominent unlighted objects by day and prominent lighted objects by night. Visibility is reported as statute miles, hundreds of feet or meters.
a. Flight Visibility — The average forward horizontal distance, from the cockpit of an aircraft in flight, at which prominent unlighted objects can be seen and identified by day and prominent lighted objects can be seen and identified by night.
b. Ground Visibility — Prevailing horizontal visibility near the Earth's surface as reported by the United States National Weather Service or an accredited observer.
c. Prevailing Visibility — The greatest horizontal visibility equaled or exceeded throughout at least half the horizon circle that need not necessarily be continuous.
d. Runway Visibility Value (RVV) — The visibility determined for a particular runway by a transmissometer. A meter provides a continuous indication of the visibility (reported in miles or fractions of miles) for the runway. RVV is used in lieu of prevailing visibility in determining minimums for a particular runway.
e. Runway Visual Range (RVR) — An instrumentally derived value, based on standard calibrations, that represents the horizontal distance a pilot will see down the runway from the approach end. It is based on the sighting of either high intensity runway lights or on the visual contrast of other targets whichever yields the greater visual range. RVR, in contrast to prevailing or runway visibility, is based on what a pilot in a moving aircraft should see looking down the runway. RVR is horizontal visual range, not slant visual range. It is based on the measurement of a transmissometer made near the touchdown point of the instrument runway and is reported in hundreds of feet. RVR is used in lieu of RVV and/or prevailing visibility in determining minimums for a particular runway.
1. Touchdown RVR — The RVR visibility readout values obtained from RVR equipment serving the runway touchdown zone.
2. Mid RVR — The RVR readout values obtained from RVR equipment located midfield of the runway.
3. Rollout RVR — The RVR readout values obtained from RVR equipment located nearest the rollout end of the runway.

visibility (ICAO) — The ability, as determined by atmospheric conditions and expressed in units of distance, to see and identify prominent unlighted objects by day and prominent lighted objects by night. Flight Visibility is the visibility forward from the cockpit of an aircraft in flight. Ground Visibility is the visibility at an aerodrome as reported by an accredited observer. Runway Visual Range (RVR) is the range over which the pilot of an aircraft on the centre line of a runway can see the runway surface markings or the lights delineating the runway or identifying its centre line.

visible light — Light wavelength between 4,000 and 7,700 angstroms (.4 - .77 μM) that can be seen by the eye.

visible line — A line on an engineering drawing that represents a portion of an object that can be seen.

visual approach — An approach conducted on an instrument flight rules (IFR) flight plan that authorizes the pilot to proceed visually and clear of clouds to the airport. The pilot must, at all times, have either the airport or the preceding aircraft in sight. This approach must be authorized and under the control of the appropriate air traffic control facility. Reported weather at the airport must have a ceiling at or above 1,000 feet and visibility of three miles or greater.

visual approach (ICAO) — An approach by an IFR flight when either part or all of an instrument approach procedure is not completed and the approach is executed in visual reference to terrain.

visual approach slope indicator (VASI) — An airport lighting facility providing vertical visual approach slope guidance to aircraft during approach to landing by radiating a directional pattern of high intensity red and white focused light beams that indicate to the pilot that he is "on path" if he sees red/white, "above path" if white/white, and "below path" if red/red. Some airports serving large aircraft have three-bar VASIs that provide two visual glide paths to the same runway.

visual descent point (VDP) — A defined point on the final approach course of a non-precision straight-in approach procedure from which normal descent from the MDA to the runway touchdown point can be commenced, provided the approach threshold of that runway, or approach lights, or other markings identifiable with the approach end of that runway are clearly visible to the pilot.

visual flight rules (VFR) — 1. The procedures for conducting flight under visual conditions according to Federal Aviation Regulations (FARs). The FARs specify minimum cloud clearance and visibility requirements. 2. VFR is also used to describe weather conditions and is often used interchangeably with the term VMC (visual meteorological conditions).

visual holding — The holding of aircraft at selected, prominent geographical fixes that can be easily recognized from the air.

visual inspection — The inspection of a part or component by visual means.

visual learners — Students who learn best with their sense of sight. They prefer to absorb the big picture first, then break the information down into individual parts.

visual meteorological conditions (VMC) — Meteorological conditions expressed in terms of visibility, distance from cloud, and ceiling equal to or better than specified minima.

visual purple — The chemical created by the rods that provides night vision acuity. The proper name is rhodopsin.

visual range — See runway visual range.

visual separation — A means employed by ATC to separate aircraft in terminal areas and enroute airspace in the NAS. There are two ways to effect this separation:
a. The tower controller sees the aircraft involved and issues instructions, as necessary, to ensure that the aircraft avoid each other.
b. A pilot sees the other aircraft involved and upon instructions from the controller provides his own separation by maneuvering his aircraft as necessary to avoid it. This can involve following another aircraft or keeping it in sight until it is no longer a factor.

vitrify — To convert a material into glass or a glass-like substance by heat and fusion.

vivid color — Colors of very high intensity and/or chroma. Sometimes used on aircraft for maximum visibility.

V_{MC} — Minimum controllable airspeed. V_{MC} changes with altitude and is considered the minimum airspeed at which control of a multi-engine airplane can be maintained with one engine inoperative.

V_{MC} demonstration — This demonstration is required during a multi-engine practical test to show the control pressures necessary to maintain directional control with one engine inoperative.

voice switching and control system (VSCS) — The VSCS is a computer controlled switching system that provides air traffic controllers with all voice circuits (air to ground and ground to ground) necessary for air traffic control.

void — 1. Unsatisfactory gaps in a weld. 2. An empty area in the composite laminate. The term void can be used in place of delamination. 3. Internal fissure in ferrous materials. Also referred to as fish eye, chrome check, shatter crack, and snowflake.

volatile — A fluid easily vaporizable at relatively low temperatures.

volatile content — In composites, the percent of volatiles that are driven off as a vapor from a plastic or an impregnated reinforcement.

volatile liquid — A fast-evaporating fluid.

volatile memory — In computers, a memory or storage device that loses its storage capability when power is removed.

volatile mineral spirit — A fast-evaporating petroleum product used as a paint or varnish thinner and for preparing surfaces prior to painting.

volatiles — In composites, materials, such as water and alcohol, in a resin formulation, that are capable of being vaporized at room temperature.

volatility — The ease with which a fluid changes from a liquid to a vapor.

volcanic ash — In general, particulates and gases from a volcanic eruption.

VOLMET broadcast — Routine broadcast of meteorological information for aircraft in flight.

volt — A unit of electromotive force (EMF) or potential difference that is needed to force one ampere of electrical flow through a resistance of one ohm.

voltage — The electric potential or potential difference expressed in volts.

voltage amplifier — A circuit designed to maximize voltage gain at the expense of current gain or power gain.

voltage avalanche — The reverse voltage required to cause a zener diode to break down and begin conducting.

voltage divider — A series of resistors placed across the poles of a source to provide a number of different voltages.

voltage doubler — A circuit that produces an output voltage twice that of the input.

voltage drop — 1. The decrease in voltage in an electrical circuit due to an increase in load. 2. The loss in potential energy in a circuit when a current is made to flow through a load and some of the energy is converted from electrical energy into another form such as heat.

voltage dropping resistor — A resistor placed in series with some other component in order to reduce the terminal voltage across or limit the current through that component.

voltage quadrupler — In electricity, an amplifier whose output current is four times its input current.

voltage regulation — -Maintaining a constant voltage level despite fluctuating load current requirements.

voltage regulator — A device that maintains a constant-level voltage supply despite changes in input voltage or load.

voltage rise — An increase in voltage caused by a decrease in load or by the addition of a source, such as a chemical cell, connected in series aiding with the general current flow in the circuit.

voltage spike — A quick burst of high voltage.

voltage standing-wave ratio — The ratio of the maximum voltage to the minimum voltage along a circuit.

voltaic cell — A device (battery) containing electrodes and an electrolyte for generating electricity by chemical action. Also referred to as a "galvanic" cell.

voltammeter — A d'Arsonval meter movement that can be used either as an ammeter or a voltmeter. Though this type of meter is a current measuring instrument, it indicates voltage by measuring the current flow through a resistance of known value. Measurements are in volts.

volt-amperes — The product of the voltage and current in a circuit.

voltmeter — An electrical measuring instrument used to measure electrical pressure or voltage.

voltmeter sensitivity — A method used to determine the accuracy of a meter. The sensitivity of a voltmeter is given in ohms per volt and is determined by dividing the resistance of the meter plus the series resistance by the full scale reading in volts.

volt-ohm-milliammeter — A multi-range electrical measuring instrument. The instrument can measure volts, amps, or resistance by selecting one of the instrument ranges, which in turn, changes the internal connections to measure a wide range of values.

volume — The space an object occupies. It is measured in cubic units, found by multiplying the area of the base of the container by its height.

volume control — The circuit in a receiver or amplifier that varies loudness of output.

volumetric efficiency — In reciprocating engines, the ratio of the volume of the charge taken into a cylinder, reduced to standard conditions, to the actual volume of the cylinder.

VOR — A ground-based electronic navigation aid transmitting very high frequency navigation signals, 360 degrees in azimuth, oriented from magnetic north. Used as the basis for navigation in the National Airspace System. The VOR periodically identifies itself by Morse Code and can have an additional voice identification feature. Voice features can be used by ATC or FSS for transmitting instructions/ information to pilots.

VOR test signal (VOT) — A ground facility that emits a test signal to check VOR receiver accuracy. Some VOTs are available to the user while airborne, and others are limited to ground use only.

VORTAC — A navigation aid providing VOR azimuth, TACAN azimuth, and TACAN distance measuring equipment (DME) at one site)

vortex — 1. A whirling, circulatory fluid motion. 2. In meteorology, any rotary flow in the atmosphere.

vortex compressor blade tip — Profile-tip. A turbine engine blade that provides a smooth airflow at its tip end.

vortex generators — Vortex generators are small airfoil-like surfaces on the wing, which project vertically into the airstream. Vortices are formed at the tip of these generators just as they are on ordinary wingtips. These vortices add energy to the boundary layer (the layer of air next to the surface of the wing) to prevent airflow separation. This reduces stall speed and can increase takeoff and landing performance.

vortex ring — A microscale circulation cell superimposed on the overall rising motion of a thermal, similar to a smoke ring. It has a relatively narrow core of upward motions surrounded by a broad region of weaker sinking motions.

vortex ring state — In helicopters, a transient condition of downward flight (descending through air after just previously being accelerated downward by the rotor) during which an appreciable portion of the main rotor system is being forced to operate at angles of attack above maximum. Blade stall starts near the hub and progresses outward as the rate of descent increases.

vortices — Circular patterns of air created by the movement of an airfoil through the air when generating lift. As an airfoil moves through the atmosphere in sustained flight, an area of low pressure is created above it. The air flowing from the high pressure area to the low pressure area around and about the tips of the airfoil tends to roll up into two rapidly rotating vortices, cylindrical in shape. These vortices are the most predominant parts of aircraft wake turbulence and their rotational force is dependent upon the wing loading, gross weight, and speed of the generating aircraft. The vortices from medium to heavy aircraft can be of extremely high velocity and hazardous to smaller aircraft.

VOT — A ground facility that emits a test signal to check VOR receiver accuracy. Some VOTs are available to the user while airborne, and others are limited to ground use only.

V-speed — Velocities relating to aircraft operation:

V_1 — Decision speed, up to which it should be possible to abort a takeoff and stop safely within the remaining runway length. After reaching V_1 the takeoff must be continued.

V_A — Design maneuvering speed. The speed below which abrupt and extreme control movements are possible (though not advised) without exceeding the airframe's limiting load factors.

V_{FE} — Maximum flap extension speed (top of white arc on ASI).

V_{FTO} — Final takeoff speed.

V_{MCA} — Minimum control speed (air). The minimum speed at which control of a twin-engine aircraft can be maintained after failure of one engine.

V_{MCG} — Minimum control speed (ground). The minimum speed at which control of a multi-engine aircraft can be maintained after failure of a wing mounted engine on the ground.

V_{NE} — Never-exceed speed, 'redline speed' denoted by a red radial on an ASI.

V_{MO} — Maximum operating speed. Also M_{MO}, Mach limit maximum operating speed.

V_{NO} — Normal operating speed. The maximum structural cruising speed allowable for normal operating conditions (top of green arc on ASI).

V_R — Rotation speed, at which to raise the nose for take-off.

V_{REF} — Reference landing speed.

V_{SO} — Stalling speed at max takeoff weight, in landing configuration with flaps and landing gear down, at sea level, ISA conditions (bottom of white arc on ASI).

V_X — Best angle of climb speed on all engines.

V_{XSE} — Best engine-out angle of climb speed.

V_Y — Best rate of climb speed on all engines.

V_{YSE} — Best engine-out rate of climb speed, 'blue line speed' (blue radial on airspeed indicators of light twins)

V_{SSE} — In multi engine aircraft, intentional one engine inoperative airspeed. V_{SSE} is not an airspeed defined by the FAA, but rather an airspeed developed by the manufacturer. It is considered the minimum speed for intentionally rendering one engine inoperative in flight for pilot training.

V-tail — A glider with two tail surfaces mounted to form a V. V-Tails combine elevator and rudder movements.

V-tail surface — An empennage consisting of two fixed and two movable surfaces arranged in a V shape. These two surfaces have the same aerodynamic function as the more conventional three surfaces.

VTOL aircraft — An aircraft that can takeoff and land without forward motion. Capable of vertical climbs and/or descents and of using very short runways or small areas for takeoff and landings. These aircraft include, but are not limited to, helicopters.

vulcanize — A process of treating crude rubber with sulfur and subjecting it to heat. Vulcanizing increases the strength and elasticity of the rubber.

V_{XSE} — In multi-engine aircraft, best angle of climb, single engine airspeed. V_{XSE} is used for obstruction clearance with engine inoperative. It provides the greatest altitude gain over a specified distance with one engine inoperative.

V_{YSE} — In multi-engine aircraft, best rate of climb, single engine airspeed. V_{YSE} is the speed that produces the greatest gain in altitude in a given amount of time with one engine inoperative.

W

wafer-type selector switch — A multiple-contact switch with contacts arranged around the edge of a wafer with a knob in the center.

waffle piston — A reciprocating engine piston that has fins cast on the bottom of the inside of the piston head that look like the surface of a waffle. This type of forging provides added strength and additional surface area for carrying heat away from the piston.

wake turbulence — The rapidly rotating air that spills over an airplane's wings during flight. The intensity of the turbulence depends on the airplane's weight, speed, and configuration. Also referred to as wing tip vortices.

walk-around bottle — A pressurized container of breathing oxygen that is small enough for a passenger or crewmember to carry around the airplane. It consists of a strap for carrying, a mask, a regulator, and a volume indicator.

wall cloud — The well-defined bank of vertically developed clouds having a wall-like appearance, which form the outer boundary of the eye of a well-developed tropical cyclone. Also the portions of the rain free base of a supercell thunderstorm that is lower in the vicinity of the main updraft. Tornadoes often develop here.

Wankel engine — A brand name of a rotary piston engines where the crankshaft and pistons are replaced by a rotary piston coupled to a rotating shaft. The pistons are replaced with a rotor that rotates inside a chamber causing the fuel-air charge to be compressed. This rotor performs a uniform or variable rotary movement without being affected by alternating inertial forces caused by the constantly reversing direction of a piston/rod/crankshaft assembly. Since the motion produced is rotary, it can be utilized directly without having to be transformed.

warm — A condition of having or emitting heat to a moderate or adequate degree.

warm airmass — An airmass characterized by temperatures warmer than the ground over which it is moving.

warm downslope wind — A warm wind that descends a slope on the lee side of a mountain, often called a chinook or Foehn, is produced by a warm, stable, updraft airmass moving across a range of mountains at high levels.

warm front — The boundary area formed when a warm air mass contacts and flows over a colder air mass. Warm fronts cause low ceilings and rain.

warm front occlusion — Characterized by the warm front remaining on the ground and the cold front moving aloft.

warm sector — The area covered by warm air at the surface and bounded by the warm front and cold front of a wave cyclone.

warm-up time — The time needed for a component and all of its parts to reach operating temperature.

warning area — Airspace of defined dimensions, extending from 3 nautical miles outward from the coast of the United States, that contains activity that can be hazardous to nonparticipating aircraft. The purpose of such warning areas is to warn nonparticipating pilots of the potential danger. A warning area can be located over domestic or international waters or both.

warning lights — Annunciation lights in an aircraft cockpit that warn the flight crew of a dangerous situation or the failure of a system or component.

warp direction — The threads running the length of the fabric as it comes off the bolt. Parallel to the selvage edge.

warp face — The side of the fabric where the greatest number of yarns are parallel to the selvage edge.

warp threads — The threads parallel to the length of a fabric.

warpage — Dimensional distortion in a plastic object.

Warren truss — A truss structure used for aircraft fuselages in which the diagonal members carry both tensile and compressive loads

wash — The disturbance in the air produced by the passage of an airfoil. Also referred to as the "wake."

wash primer — A self-etching primer used on aluminum or magnesium. It is often used to prepare the surface for zinc chromate primer.

washout — Slight twist built in on the outboard portion of the wing, designed to improve the stall characteristics of the wing.

water absorption — The ratio of the weight of water absorbed by a material to the weight of the dry material.

water ballast — System in some gliders for increasing in wing loading, thus enabling increased average cross-country speeds, using releasable water in the wings (via integral tanks or water bags). Some gliders also have a small water ballast tank in the tail for optimizing flying CG.

water break test — Spraying water on a part to be bonded to assure there is no oil or grease contamination on the surface. If oil or grease is present, the water will bead.

water equivalent — The depth of water that would result from the melting of snow or ice.

water jet — Used primarily in the manufacturing process as a cutting tool. A very high-pressure stream of water is used to cut through the component.

water vapor — The gaseous form of H_2O.

water-injection system — A system in which water is injected along with the fuel to avoid damage to the engine. As compression ratios are raised, or supercharging/turbocharging is added to engines, temperatures rise and an increased chance of detonation arises. Adding water injection provides a water/air/fuel mixture that not only burns more efficiently and avoids spontaneous detonation but also provides additional inlet air cooling and, hence, denser air.

waterspout — A tornado that occurs over water.

watt — The basic unit of electrical power. One ampere flowing under a pressure of one volt is equal to one watt. One watt equals $1/746$ HP.

wattage rating — The maximum amount of power an electrical component needs to operate an appliance or device without damaging the device.

watt-hour — A unit of electrical energy equal to one watt acting for one hour.

wattmeter — An instrument designed to measure electric power in watts.

watt-second — A unit of electrical energy equal to one watt acting for one second.

wave carrier — An electromagnetic wave of high-frequency alternating current whose modulations are used to transmit speech, images, and other signals. The more common term is carrier wave.

wave cyclone — A cyclone that forms and moves along a front. The circulation about the cyclone center tends to produce a wavelike deformation of the front.

wave soldering — A method of soldering printed circuit boards where the board is placed slightly above a container of molten solder. A wave is induced in the solder and the resultant rise in level coats the exposed leads, simultaneously soldering all component leads on the board.

wave window — Special areas arranged by Letter of Agreement with the controlling ATC wherein gliders can be allowed to fly under VFR in Class A Airspace at certain times and to certain specified altitudes.

waveform — The shape of an electrical signal as seen on an oscilloscope. Examples of wave forms are sine waves and square waves.

waveguide — A hollow metal tube designed to guide electromagnetic energy.

wavelength — The distance between the crests of a wave of energy. Wavelength is inversely proportional to the frequency of the wave.

way point — A predetermined geographical position used for route/instrument approach definition, progress reports, published VFR routes, visual reporting points or points for transitioning, and/or circumnavigating controlled and/or special use airspace. Defined relative to a VORTAC station or in terms of latitude/longitude coordinates.

waypoint — A specified geographic location used to define an area navigation route or flight path of an aircraft employing area navigation. Waypoints are identified as either fly-by or fly-over.

wear pads — In aircraft brakes, steel pads riveted to the surfaces of the stationary disks, the pressure plate, and the back plate to provide a wearing surface against the sintered material on the rotating disks. It is more economical to replace the wear pads than the disks and the plates themselves.

weather — The instantaneous state of the atmosphere.

weather advisory — In aviation weather forecast practice, an expression of hazardous weather conditions not predicted in the area forecast, as they affect the operation of air traffic and as prepared by the NWS (National Weather Service).

weather vane — A wind vane.

weave — In composites, the particular manner in which a fabric is formed by interlacing yarns. Usually assigned a style number used in ordering for the repair of a component.

web — The portion of any beam or channel that lies between the flanges of a spar, rib, or channel section. Furnishes the strength necessary for longitudinal shear loads.

web browser — A software program that provides access to sites on the World Wide Web (WWW).

web of a beam — The portion of a beam that lies between the flanges of a spar, rib, or channel section. Furnishes the strength necessary for longitudinal shear loads.

weber — A basic metric unit of magnetic flux equal to that flux produced in a single turn of wire when an EMF of one volt is reduced to zero at a uniform rate of one ampere per second.

wedge — A tapered hard piece of wood or metal that can be used for splitting, tightening, securing, or levering.

weft direction — Fibers that are perpendicular to the warp fibers. Sometimes referred to as the woof or fill.

weighing points — The locations on an aircraft where the scales are placed for weighing the aircraft. The manufacturer designates the weighing points.

weight — A measure of the heaviness of an object. The force by which a body is attracted toward the center of the Earth (or another celestial body) by gravity. Weight is equal to the mass of the body times the local value of gravitational acceleration. One of the four main forces acting on an aircraft. Equivalent to the actual weight of the aircraft. It acts downward toward the center of the Earth.

weight and balance records — The aircraft records that provide the required information on the weight of the empty aircraft and the location of its center of gravity (CG).

weight arm — Another term for moment. Weight multiplied by arm equals moment.

weld bead — The metal deposited in a welded joint for reinforcement.

weld fusion zone — The junction area of a weld that has been melted together by heat. The area of base metal melted as seen in the cross section of a weld.

weld procedures — The steps necessary to prepare a weld. These steps can include making certain the necessary equipment is available, the welding connections are properly connected, the equipment is in good working order, and the material to be welded is properly prepared.

welded patch — A patch of thin sheet steel welded over a dent in a steel tubular structure to reinforce the structure at the point of the damage.

welding — A method of joining materials in which a portion of each piece is melted and combined in its molten state. Filler material is usually added for extra mass at the joint.

welding flux — A material used in welding that melts and flows over the weld material to exclude oxygen from the surface of the molten metal and prevent oxides from forming in the weld.

weldment — An assembly that is welded together.

Weston meter — An electronic instrument that utilizes a moving coil.

wet bulb — Contraction of either wet-bulb temperature or wet-bulb thermometer.

wet cell — An electrical power supply with electrodes and a liquid electrolyte for the conversion of chemical into electrical energy.

wet grinder — A precision grinding machine that uses a flow of liquid coolant over the grinding stone to remove the heat caused by grinding, preventing heat damage to the material being ground.

wet head — A helicopter rotor head that uses oil as its lubricant.

wet lay-up — In composites, a method of making a reinforced product by applying the resin system as a liquid when the reinforcement is put in place.

wet sump engine — An engine in which all of the oil supply is carried within the engine itself. A dry sump engine has the majority of the oil contained in a separate tank.

wet takeoff — Takeoff by an aircraft that is equipped with engines that use water injection during takeoff.

wet wing — An integral fuel tank in an aircraft wing made by sealing part of the structure to use as a fuel tank.

wet-bulb temperature — The temperature of the air modified by the evaporation of water from a wick surrounding the thermometer bulb. Wet bulb temperature is used in conjunction with dry bulb temperature to calculate the dew point.

wet-bulb thermometer — A thermometer with a muslin-covered bulb used to measure wet-bulb temperature.

wet-out — In composites, the saturation of an impregnated fabric in which all areas of the fibers are filled with resin.

wet-sump system — An oil system in which the oil is carried in a sump that is an integral part of the engine.

Wheatstone bridge — An electrical measuring circuit in which the current through the indicator is determined by the ratio of the resistances of the four resistors that form the legs of the bridge.

wheel well — The part of the aircraft that receives or encloses the landing gear as it retracts.

when able — When used in conjunction with ATC instructions, gives the pilot the latitude to delay compliance until a condition or event has been reconciled. Unlike "pilot discretion," when instructions are prefaced "when able," the pilot is expected to seek the first opportunity to comply. Once a maneuver has been initiated, the pilot is expected to continue until the specifications of the instructions have been met. "When able," should not be used when expeditious compliance is required.

whetstone — An abrasive stone used for sharpening cutting tools.

whiffletree — A steering bell crank. It allows forces to combine to produce an output.

whip antenna — A quarter-wave antenna usually in the high- or very-high frequency range. It is normally vertically polarized.

whirlwind — A small, rotating column of air; can be visible as a dust devil.

white dew — Frozen dew.

whiteout — A situation where all depth perception is poor. Caused by a low sun angle, and overcast skies over a snow covered surface.

Whitworth thread — A screw thread used principally in Great Britain. Also referred to as the British Standard Whitworth (B.S.W.).

wicking — Occurs when solder flows to the insulation of stranded electrical wire during the soldering process.

wide-area augmentation system (WAAS) — The WAAS is a satellite navigation system consisting of the equipment and software that augments the GPS Standard Positioning Service (SPS). The WAAS provides enhanced integrity, accuracy, availability, and continuity over and above GPS SPS. The differential correction function provides improved accuracy required for precision approach.

Wiggins coupling — A connector that allows fluid lines to be connected and disconnected quickly.

WILCO — I have received your message, understand it, and will comply with it.

winch — A machine that converts rotary motion to a tensile force used for hauling or hoisting. A hand crank or motor is connected to a drum around which a rope or cable winds as the load is lifted.

wind — Air in motion relative to the surface of the Earth; generally used to denote horizontal movement.

wind chill factor — The effect of wind on temperature that causes it to feel colder than the temperature would be without wind.. It blows away the thin layer of warm air that normally surrounds the body, and it draws away body heat by quickly evaporating any moisture that forms on the skin.

wind correction angle (WCA) — The angular difference between the heading of the airplane and the course.

wind direction — The direction from which wind is blowing.

wind grid display — A display that presents the latest forecasted wind data overlaid on a map of the ARTCC area. Wind data is automatically entered and updated periodically by transmissions from the National Weather Service. Winds at specific altitudes, along with temperatures and air pressure can be viewed.

wind shear — A sudden, drastic shift in wind speed and/or direction that occurs over a short distance. Often associated with weather fronts.

wind sock — A truncated cloth cone open at both ends and mounted on a freewheeling pivot to indicate the direction the wind is blowing.

wind triangle — Navigational calculation allowing determination of true heading with a correction for crosswinds on course.

wind tunnel testing — A test that uses a tunnel-like passage through which air is forced at controllable speeds in order to study the effects of wind pressure on and around airfoils, scale models, or other objects.

wind vane — An instrument to indicate wind direction.

wind velocity — A vector that includes wind direction and wind speed.

windmilling — The rotation of an aircraft propeller created by air flowing around it with the engine not operating.

window de-mister — A system of keeping the windows of an aircraft free of condensed moisture by blowing warm air over or between the layers of transparent material.

windshield — A transparent screen device made of plastic or glass located in front of the occupants of a vehicle to protect them from the elements of wind, rain, and cold.

wing — An airfoil whose main function is to provide lift.

wing area — The total wing area measured in square feet by multiplying the wing span by the average wing chord.

wing chord — An imaginary straight line connecting the leading edge and the trailing edge of a wing airfoil section.

wing fillet — A streamlined fairing between a wing and the fuselage. Used to smooth out the airflow and minimize the interference drag caused by the junction.

wing flaps — The movable control surfaces on the trailing edge of a wing inboard of the ailerons. Wing flaps alter the camber and sometimes the area of the wing, increasing both the lift and the drag.

wing heavy — A condition of flight in which one wing has a tendency to fly lower than the other wing about an aircraft's longitudinal axis. This condition is corrected by properly adjusting the flight control rigging system.

wing loading — The ratio of the weight of a fully loaded aircraft to the total wing area.

wing nut — A nut with two wing-like projections that can be gripped with a thumb and forefinger and, thus, turned by hand.

wing panel — A removable access panel or wing section attached with screws, bolts, or rivets.

wing profile — The outline of the wing section.

wing rib — A structural member that gives a wing its desired aerodynamic shape.

wing span — The distance from one wingtip to the other.

wing stations — Points measured from the centerline of an aircraft (buttock line zero) toward the wing tip. Indicates the distance in inches from the centerline.

wing strut — A diagonal brace between the fuselage and the wing of a semi-cantilever wing.

wing tip vortices — The rapidly rotating air that spills over an airplane's wings during flight. The intensity of the turbulence depends on the airplane's weight, speed, and configuration. Also referred to as wake turbulence.

wing tip vortices — Circular patterns of air created by the movement of an airfoil through the air when generating lift. As an airfoil moves through the atmosphere in sustained flight, an area of low pressure is created above it. The air flowing from the high pressure area to the low pressure area around and about the tips of the airfoil tends to roll up into two rapidly rotating vortices, cylindrical in shape. These vortices are the most predominant parts of aircraft wake turbulence and their rotational force is dependent upon the wing loading, gross weight, and speed of the generating aircraft. The vortices from medium to heavy aircraft can be of extremely high velocity and hazardous to smaller aircraft.

wing twist — A change in the aerodynamic shape of a wing that effectively causes the outboard section of the wing to have less angle of attack than the inboard sections. Wing twist allows the outboard section of the wing to avoid stalling at higher angles of attack, allowing ailerons to effectively control the roll of the aircraft.

winglet — A short, almost vertical stabilizing fin projecting from the tip of an aircraft wing. The winglet alters the downwash that normally washes across the wing's surface. This reduces induced drag and actually provides a small amount of forward thrust. These two effects more than cancel the parasitic drag of the winglet and reduce the overall drag of the aircraft.

winglet — A design that nearly blocks or diffuses wing tip vortices. Winglets are nearly vertical extensions on the wingtips, which are actually carefully designed, proportioned, and positioned airfoils with their camber toward the fuselage, and with span, taper, and aspect ratio optimized to provide maximum benefit at a specific speed and angle of attack. Also referred to as tip fins.

wingtip vortices — Circular patterns of air created by an airfoil when generating lift. Vortices from medium to heavy aircraft can be extremely hazardous to small aircraft.

wink Zyglo — A non-destructive inspection method in which the part to be inspected is sprayed with a fluorescent penetrant liquid. The penetrant seeps into any surface cracks in the part. The liquid is then washed from its surface and the part is placed in a vibrating fixture and observed under an ultraviolet light. If the vibration causes a crack that contains the penetrant to open and close, the black light will illuminate the penetrant and give the appearance of winking each time the crack opens up to expose the penetrant.

wiper — A movable electrical contact used in an electrical component.

wire braid — A woven, flexible metal that covers an aircraft's electrical wiring and is used to intercept and ground any radiated electrical energy from the wire to prevent radio frequency interference.

wire bundle — A group of electrical wires tied together and secured to the structure.

wire cloth — A mesh woven of fine wire used for filtering.

wire edge — A sharp burr on the edge of sheet metal that has been cut on a shear.

wire gauge — A gauge used to measure wire diameter.

wire group — Two or more wires going to the same location and tied together to retain their identity.

wire mesh — In composites, a fine wire screen is used to dissipate an electrical charge from lightning or static buildup. It is used as lightning protection usually directly under the top layer of fabric.

wire stripper — A tool designed to remove the insulation from electrical wires.

wireless — British term for radio.

wire-wound resistor — An electrical resistor made up of a winding of high-resistance wire covered with baked-on ceramic material.

wobble pump — A hand-operated fluid pressure pump. The name wobble comes from the movement of the pump handle back and forth as it pulls fluid into one side of the pump and forces it out the other side.

woodruff key — A hardened piece of metal shaped in a half circle on one side and flat on the other side. The key fits into a semi-circular groove to secure a wheel, disk, or gear to a shaft.

Wood's metal — An alloy of lead, tin, bismuth, and cadmium that melts at a temperature of 158°F.

words twice — **1.** As a request: "Communication is difficult. Please say every phrase twice." **2.** As information: "Since communications are difficult, every phrase in this message will be spoken twice."

work — The product of force and the distance through which the force acts.

workability — The ease with which wood, metal, or plastic can be formed or shaped.

work-hardening — Increasing the strength and hardness of a metal by work-hardening or cold-working. The strain-hardening is normally done after a piece of material has been heat treated. If an aluminum alloy is not heat-treatable, strain hardening is the only way it can be hardened.

working life — In composites, the period of time during which a liquid resin or adhesive remains usable.

working memory — The portion of the brain that receives information from the sensory register. This portion of the brain can store information in memory for only a short period of time. If the information is determined by an individual to be important enough to remember, it must be coded in some way for transmittal to long-term memory. Also referred to as short-term memory.

working voltage — The maximum amount of electrical voltage that can safely be applied to an appliance without damaging it.

workload management — Ensures that essential operations are accomplished by planning, prioritizing, and sequencing tasks to avoid work overload.

world aeronautical chart (WAC) — Similar to a sectional chart, but with a scale of 1:1,000,000. Provide a standard series of aeronautical charts covering land areas of the world at a size and scale convenient for navigation by moderate speed aircraft. Topographic information includes cities and towns, principal roads, railroads, distinctive landmarks, drainage, and relief. Aeronautical information includes visual and radio aids to navigation, airports, airways, restricted areas, obstructions and other pertinent data.

world wide web (WWW) — A part of the Internet that provides access through the network by means of graphics and hypertext links to different Web sites. Also referred to as the Web.

worm gear — A gear that consists of a threaded shaft and a toothed wheel that are meshed together.

worm screw — A worm-shaped gear that meshes with a cogged gear set. When turned, it imparts a radial motion on the cogged set, creating a rotary motion 90° from the input.

woven fabric — A material constructed by interlacing yarns, fibers, or filaments to form fabric patterns.

WOXOF — Pronounced "walks off", a slang weather term meaning visibility zero in fog. Also seen as WXOF and WOXOFF.

wrench — A tool with fixed or adjustable jaws for gripping a nut, bolt, or pipe, and a long handle for leverage in turning.

wrinkle — In composites, a surface imperfection in laminated plastics that has the appearance of a crease or fold in one or more outer sheets of the paper, fabric, or other base. Also occurs in vacuum bag molding when the bag is improperly placed, causing a crease.

wrinkle finish — A paint finish that wrinkles as it dries to give a rough appearance.

wrist pin — The hardened steel pin that attaches the small end of a connecting rod to a piston.

written tests — Often used as an evaluation device. They include computerized tests as well as paper-and-pencil tests, and are often referred to as knowledge tests. A test is a set of questions, problems, or exercises used to determine whether your students have a particular knowledge or skill.

wrought iron — An easily welded, forged, or shaped iron that contains very little carbon.

wrought metal — A metal that has been worked by rolling, drawing, or forging, and which has a different grain structure from that of cast metal.

wye connection — An electrical circuit connection that looks like the letter Y.

X

x-axis — The longitudinal axis about which an aircraft rolls. The ailerons are used to control this movement. The vertical axis is the Z-axis and the lateral axis is the Y-axis.

X-band radar — Radar that operates in a frequency band of between 5.2 and 10.9 gigahertz.

xenon — A heavy, colorless, inert chemical gas element with a symbol of Xi and an atomic number of 54.

X-ray — An electromagnetic radiation with an extremely short wavelength. Capable of penetrating solid objects and exposing photographic film.

X-ray inspection — A nondestructive inspection in which high-frequency, high-energy electromagnetic waves pass through the material and expose a photographic film. Defects or discontinuities within the material show up as variations in the density of the image on the film.

xylene — A toxic, flammable, aromatic hydrocarbon, similar to benzene. It is used as a solvent. Also referred to as xylol.

xylol — See xylene.

Y

Yagi antenna — A directional transmitting or receiving antenna that uses one active dipole element and one or more directing dipoles and one or more reflecting dipoles. The passive dipoles, which are not connected electrically, are aligned for physical position. They interact with each other to form a directional field pattern. Most home-type TV antennae are of the Yagi type.

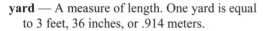

yard — A measure of length. One yard is equal to 3 feet, 36 inches, or .914 meters.

yardstick — A graduated measuring device that is 3 feet long and marked in inches and feet.

yarn — Twisted filaments, fibers, or strands, forming a continuous length that is suitable for use in weaving into materials.

yaw — The movement of an aircraft about its vertical axis.

yaw damper — An automatic control device used to keep an aircraft from yawing. Swept wing aircraft are particularly susceptible to Dutch Roll, an undesirable yawing and rolling motion. Yaw dampers overcome the unwanted yawing condition.

yaw string — A demonstration used to show the existence of slip. To accomplish a yaw string demonstration, tape a piece of yarn to the windscreen or to the top of the nose of the aircraft. It should be placed in an area of undisturbed flow. During flight, you will be able to see the yarn move depending on whether or not the aircraft is in a zero sideslip configuration.

Y-axis — The lateral, pitch axis. Pitch movements are controlled by the elevators.

Y-connected circuit — A three-phase, or polyphase, alternating current circuit that has three single-phase windings spaced so that the voltage induced in each winding is 120° out of phase with the voltages in the other two windings.

yellow arc — A yellow marking on an instrument that indicates a region of caution. On an airspeed indicator, for instance, the yellow arc indicates a range of speeds that are tolerable to fly in calm air, but not in turbulence.

yield point — The load on a material, expressed in lbs./square inch, that causes the initial indication of permanent distortion. Also referred to as yield strength.

yield strength — The load on a material, expressed in lbs./square inch, that causes the initial indication of permanent distortion.

yoke — 1. The control column in an airplane cockpit that connects to and controls the movement of the elevators and ailerons. 2. A cross member in a control system that links or joins something together.

Y-valve — The oil drain valve for a dry sump engine. It derives its name from its shape. One arm of the "Y" goes to the pressure pump inlet, one arm to the oil tank, and the lower arm is fitted with a valve. From this valve, the oil can be drained from the tank. Fuel for oil dilution is also introduced in the Y-valve.

Y-winding — A method of connecting the phase windings of a three-phase AC machine in which one end of each of the three phase windings is connected together to form a common point or a neutral terminal.

Z

Zahn cup — A cup of definite size and shape, with a hole in its bottom used to measure the viscosity of a material by the number of seconds required for the cup to empty.

Z-axis — The vertical axis of an object. Turning the nose of the aircraft, for example, causes the aircraft to rotate about its vertical axis. Rotation of the aircraft about the vertical axis is called yawing. This motion is controlled with the rudder.

zener diode — A diode rectifier designed to prevent the flow of current in one direction until the voltage in the reverse direction reaches a predetermined value. At this time, the diode permits a reverse current to flow.

zenith — The highest point directly overhead.

zephyr — A west wind or a gentle breeze.

Zeppelin — A rigid cylindrical airship supported by internal gas cells. Invented by Count Ferdinand von Zeppelin.

Zerk fitting — A grease fitting with a check valve that allows grease to be pumped through the fitting into a bearing surface. Removal of the grease gun allows the check valve to reseat preventing grease from leaking out and dirt from entering the fitting.

zero — **1.** Numerical: 0. Having no value. Used as a reference point. **2.** Temperature: A reference point on a Fahrenheit thermometer that is 32° below the freezing point of water or the point on a Celsius thermometer where water freezes.

zero adjustment — The adjustment on an instrument to a zero point, or to an arbitrary point from which all negative and positive measurements are to be adjusted.

zero bleed — In composites, a laminate fabrication procedure that does not allow loss of resin during cure.

zero fuel weight — The weight of an aircraft that includes the entire useful load, minus the fuel.

zero fuel weight — The weight of the aircraft to include all useful load except fuel. Limits the ratio of loads between the fuselage and wings. The maximum load that an airplane can carry also depends on the way the load is distributed. The weight of an airplane in flight is supported largely by the wings; therefore, as the load carried in the fuselage is increased, the bending moment on the wings is increased.

zero gravity — The effect of gravity when it has been nullified by parabolic flight.

zero lash — A condition in a valve train in a reciprocating engine in which all of the clearance is kept out of the valve train by the use of hydraulic valve lifters.

zero sideslip — A control technique used following an engine failure in a multi-engine aircraft where the pilot maintains an attitude that minimizes drag, alleviating the sideslip of the airplane. As bank angle exceeds the zero sideslip value, there is a sharp loss of climb performance. Zero sideslip angle varies with the airplane type. Flying at zero sideslip allows adequate directional control with the best climb performance possible.

zero-lash valve lifter — A hydraulic device in a reciprocating engine that reduces the slack between a valve and the valve lifter due to changes in the engine operating temperatures.

zero-lift line — A line through an airfoil, along which a flow of relative wind will produce no lift.

zero-time — An engine overhauled by the factory. Only a factory overhaul can be called a zero-time engine.

zinc — A bluish-white, crystalline metal with a symbol of Zn and an atomic weight of 30. Zinc is ductile in its pure state but quite brittle in its commercial form. It is used in the production of electrical batteries and for coating steel parts to protect them by means of sacrificial corrosion.

zinc chloride cell — In batteries, a chemical cell using powdered manganese dioxide and zinc as its pole pieces and a solution of zinc chloride as its electrolyte.

zinc chromate primer — An alkyd resin, corrosion-inhibiting primer that can be used on almost all metal surfaces. Moisture releases chromate ions to inhibit the formation of corrosion, and the alkyd resin forms a good bond for aircraft finishes.

Z-marker — A radio beacon that radiates in a vertical cone-shaped pattern. Z-markers are placed along airways or approach courses to denote specific locations.

zone numbers — The location marks on an aircraft drawing, both vertical and horizontal. They are used to locate detail parts on the drawing.

Zulu time — The proper radio phraseology when making reference to coordinated universal time.

Zyglo Inspection — A penetrant inspection system in which a fluorescent dye is drawn into surface defects in a material. The defects are made visible by a powder-type developer.

ABBREVIATIONS/ACRONYMS

A — Acceleration

A&P — airframe and powerplant mechanic

A/C — aircraft

A/FD — Airport/Facility Directory

A/G — air to ground

A/H — altitude/height

AAC — Mike Monroney Aeronautical Center

AAF — Army Air Field

AAI — arrival aircraft interval

AAM — air-to-air missile

AAP — advanced automation program

AAP — autoflight annunciator panel

AAR — airport acceptance rate

AAS — Airport Advisory Service

ABC — after bottom center

ABDIS — Automated Data Interchange System Service B

ABS — anti-skid brake system

AC — Advisory Circular

AC — Air Corps

AC — alternating current

AC — convective outlook (weather)

ACAIS — Air Carrier Activity Information System

ACARS — aircraft communications addressing and reporting system

ACAS — aircraft collision avoidance system

ACC — area control center

ACCT — accounting records

ACD — automatic call distributor

ACDO — Air Carrier District Office

ACF — Area Control Facility

ACFO — Aircraft Certification Field Office

ACFT — aircraft

ACID — aircraft identification

ACLS — automatic carrier landing system

ACLT — actual landing time calculated

ACM — air cycle machine

ACO — Aircraft Certification Office

ACW — air crew warning

AD — Airworthiness Directive

AD — ashless dispersant

ADA — Air Defense Area

ADAP — Airport Development Aid Program

ADAS — AWOS Data Acquisition System

ADC — air data computer

ADCCP — Advanced Data Communications Control Procedure

ADCUS — advise customs

ADDA — administrative data

ADF — automatic direction finder

ADI — automatic de ice and inhibitor

ADI — attitude direction indicator

ADIZ — Air Defense Identification Zone

ADL — aeronautical data link -

ADLY — arrival delay

ADM — aeronautical decision making

ADO — Airline Dispatch Office

ADP — automated data processing

ADS — automatic dependent surveillance

ADSIM — airfield delay simulation model

ADSY — administrative equipment systems

ADTN — administrative data transmission network

ADTN2000 — Administrative Data Transmission Network 2000

ADU — attitude direction unit

ADVO — administrative voice

AEG — aircraft evaluation group

AERA — automated enroute air traffic control

AEX — automated execution

AF — airway facilities

af — audio frequency

AFB — Air Force Base

AFC — automatic frequency control

AFCS — automatic flight control system

AFDS — Autopilot Flight Director System

AFIS — automated flight inspection system

AFM — aircraft flight manual

AFP — area flight plan

AFRES — Air Force Reserve Station

AFS — Airways Facilities Sector

AFSFO — AFS Field Office

AFSFU — AFS Field Unit

AFSOU — AFS Field Office Unit (Standard is AFSFOU)

AFSS — Automated Flight Service Station

AFTN — Automated Fixed Telecommunications Network

AGL — above ground level

AHC — auto heading comparator

AHRS — attitude heading reference system
AID — airport information desk
AIG — Airbus Industries Group
AIM — Aeronautical Information Manual
AIP — Aeronautical Information Publication
AIP — Airport Improvement Plan
AIRMET — Airman's Meteorological Information
AIRNET — Airport Network Simulation Model
AIS — Aeronautical Information Service
AIT — automated information transfer
ALD — available landing distance
ALNOT — alert notice
ALP — airport layout plan
ALS — approach lighting system
ALSF1 — ALS with sequenced flashers I
ALSF2 — ALS with sequenced flashers II
ALSIP — approach lighting system improvement plan
ALTRV — altitude reservation
AM — amplitude modulation
AMASS — Airport Movement Area Safety System
AMC — automatic mixture control
AMCC — ACF/ARTCC Maintenance Control Center
AMOS — Automated Meteorological Observation Station
AMP — Airport Master Plan
AMP — ARINC Message Processor
AMS — Aeronautical Material Specification
AMSL — above mean sea level
AMT — aviation maintenance technician

AMVER — Automated Mutual Assistance Vessel Rescue System
AN — Air Force-Navy Standard
ANC — alternate network connectivity
AND — Air Force-Navy Design
ANG — Air National Guard
ANGB — Air National Guard Base
ANMS — automated network monitoring system
ANP — actual navigation performance
ANSI — American National Standards Institute
AOA — angle of attack
AOCC — airline operations control center
AP — acquisition plan
AP — autopilot system
APC — absolute pressure controller
API — American Petroleum Institute
APP — approach
APP — auto-pilot panel
APS — airport planning standard
APU — auxiliary power unit
APV — approach with vertical guidance
AQAFO — Aeronautical Quality Assurance Field Office
ARAC — Army Radar Approach Control
ARAC — Aviation Rulemaking Advisory Committee
ARCTR — FAA Aeronautical Center or Academy
ARF — airport reservation function
ARFF IC — Aircraft Rescue And Fire Fighting Incident Commander
ARINC — Aeronautical Radio Incorporated
ARLNO — airline office

ARO — airport reservation office
ARP — air data reference panel
ARP — airport reference point
ARSA — airport radar service area
ARSR — air route surveillance radar
ARTCC — Air Route Traffic Control Center
ARTS — Automated Radar Terminal System
AS — Aeronautical Standard
ASA — American Standards Association
ASAS — Aviation Safety Analysis System
ASC — AUTODIN switching center
ASCP — aviation system capacity plan
ASD — aircraft situation display
ASDA — accelerate-stop distance available
ASDAR — acft to satellite data relay
ASDE — airport surface detection equipment
ASLAR — aircraft surge launch and recovery
ASM — air-to-surface missile
ASM — available seat mile
ASOS — automated surface observation system
ASP — arrival sequencing program
ASQP — airline service quality performance
ASR — airport surveillance radar
ASTA — airport surface traffic automation
ASTM — American Society of Testing Materials
ASV — airline schedule vendor
AT — air traffic
AT&T — American Telephone and Telegraph

358

AT&T ASDC — AT&T Agency Service Delivery Center

AT&T CSA — AT&T Customer Support Associate

ATA — actual time of arrival

ATA — Air Transport Association

ATAS — Airspace and Traffic Advisory Service

ATC — after top center

ATC — air traffic control

ATCAA — air traffic control assigned airspace

ATCBI — air traffic control beacon indicator

ATCCC — Air Traffic Control Command Center

ATCO — air taxi commercial operator

ATCRB — air traffic control radar beacon

ATCRBS — Air Traffic Control Radar Beacon System

ATCSCC — Air Traffic Control System Command Center

ATCT — airport traffic control tower

ATD — actual time of departure

ATD — along track (straight line) distance

ATE — actual time enroute

ATIS — Automatic Terminal Information Service

ATISR — ATIS recorder

ATM — air traffic management

ATM — air turbine motor

ATM — asynchronous transfer mode

ATMS — Advanced Traffic Management System

ATN — Aeronautical Telecommunications Network

ATODN — AUTODIN terminal (functional unit or system)

ATOMS — Air Traffic Operations Management System

ATOVN — AUOTVON (facility)

ATP — Airline Transport Pilot

ATR — Airline Transport Rating

ATS — air traffic service

ATS — Air Traffic Services (FAA)

ATSCCP — ATS Contingency Command Post

ATT — attitude retention system

ATTIS — AT&T Information Systems

AUL — Approved Unserviceable List.

AUTODIN — DoD Automatic Digital Network

AUTOVON — DoD Automatic Voice Network

AVAIDS — navigational aids

AVC — automatic volume control

AVGAS — Aviation gasoline.

AVN — Aviation Standards National Field Office

AVON — AUTOVON Service

AVTUR — Aviation turbine fuel.

AWC — aviation weather center

AWG — American Wire Gauge

AWIS — airport weather information

AWOS — Automated Weather Observing System

AWP — aviation weather processor

AWPG — aviation weather products generator

AWS — air weather station

BANS — BRITE alphanumeric system

BART — Billing Analysis Reporting Tool (GSA software tool)

BASIC — basic contract observing station

BASOP — military base operations

BBC — before bottom center

BBS — bulletin board system

BC — back course

BCA — benefit/cost analysis

BCD — binary coded decimal

BCR — benefit/cost ratio

BDAT — digitized beacon data

BDC — bottom dead center

BFO — beat frequency oscillator mode of ADF

BFO — beat-frequency oscillator

BHP — brake horsepower

BIM — blade inspection method

BIS — blade inspection system

BL — bend tangent line

BMEP — brake mean-effective pressure

BMP — best management practices

BOC — Bell Operating Company

BPCU — bus power control unit

bps — bits per second

BRI — basic rate interface

BRITE — Bright Radar Indicator Terminal Equipment

BRL — building restriction line

BSFC — brake specific fuel consumption

BTC — before top center

BUEC — back-up emergency communications -

BUECE — back-up emergency communications equipment -

C — Celsius (Centigrade) temperature.

C/A — coarse acquisition

CAA — Civil Aviation Authority

CAB — Civil Aeronautics Board

CADC — central air data comp

CAP — Civil Air Patrol

CARF — Central Altitude Reservation Facility

CAS — calibrated airspeed

CASFO — Civil Aviation Security Office

CAT — category

CAT — clear air turbulence

CAT II — Category II approach

CAU — crypto ancillary unit

CAVU — clear, and visibility unlimited

CAWS — central aural warning system

CBI — computer based instruction

CC&O — customer cost and obligation

CCAS — cockpit crew alerting system

CCC — communications command center

CCCC — staff communications

CCCH — central computer complex host

CCL — convective condensation level

CCLD — Core Capability Limited Deployment

CCS-7NI — Communication Channel Signal-7 Network Interconnect -

CCSD — command communications service designator

CCU — central control unit

CD — common digitizer

CD — controller display

CD — Convergent-Divergent duct (venturi).

CDI — course deviation indicator

CDP — compressor discharge pressure

CDR — cost detail report

CDT — controlled departure time

CDTI — cockpit display of traffic information

CDU — central data unit

CENTX — central telephone exchange

CEQ — Council on Environmental Quality

CERAP — central radar approach

CERAP — combined Center/Rapcon

CFA — controlled firing area

CFC — central flow control

CFCF — central flow control facility

CFCS — central flow control service

CFI — certified flight instructor

CFR — Code of Federal Regulations

CFR — Code of Federal Regulations

CFR — Cooperative Fuel Research

CFWP — central flow weather processor

CFWU — central flow weather unit

CG — center of gravity

CGAS — Coast Guard Air Station

CGCC — center of gravity control computer

CIG — ceiling

CIT — compressor inlet temperature

C_L — coefficient of lift

CLC — course line computer

CLIN — contract line item

CLT — calculated landing time

CM — commercial service airport

CM — circular mils

CMNPS — Canadian Minimum Navigation Performance Specification Airspace

CNS — consolidated NOTAM system

CNSP — consolidated NOTAM system processor

CO — carbon monoxide

CO — central office

CO_2 — carbon dioxide.

COE — U.S. Army Corps of Engineers

COMCO — command communications outlet

Consol — kind of low or medium frequency long range navigational aid

Consolan — a kind of low or medium frequency long range navigational aid

CONUS — Continental United States

CORP — Private Corporation other than ARINC or MITRE

CP — center of pressure

CPDLC — Controller Pilot Data Link Communications

CPE — customer premise equipment

CPMIS — consolidated personnel management information system

CPU — central processing unit

CPU — control protection unit

CRA — conflict resolution advisory

CRDA — converging runway display aid

CRM — crew resource management

CRT — cathode ray tube-

CSA — communications service authorization

CSD — constant-speed drive

CSEU — control system electronic unit

CSG — slide graphic computer

CSIS — centralized storm information system

CSO — customer service office

CSR — communications service request

CSS — central site system

CTA — control area

CTA — controlled time of arrival

CTA — current transformer assembly

CTA/FIR — control area/flight information region

CTAF — Common Traffic Advisory Frequency

CTAS — center TRACON automation system -

CTMA — Center Traffic Management Advisor

CTOT — constant torque on takeoff

CTC — cabin turbo compressor

CUD — course unit display

CUPS — consolidated uniform payroll system

CVFP — charted visual flight procedure

CVFR — controlled visual flight rules

CVR — cockpit voice recorder

CVRS — computerized voice reservation system

CVTS — compressed video transmission service

CW — carrier wave

CW — continuous wave

CW — continuous wave NDB signals

CWA — center weather advisory

CWP — cockpit warning panel

CWS — central warning system

CWSU — center weather service unit

CWY — clearway

DA — decision altitude

DA — decision altitude/decision height

DA — density altitude

DA — descent advisor

DA — direct access

DABBS — DITCO Automated Bulletin Board System

DADC — digital air data computer

DAIR — direct altitude and identity readout

DAR — designated agency representative

DARC — direct access radar channel

dBA — decibels A, weighted

DBCRC — Defense Base Closure and Realignment Commission

DBMS — data base management system

DBRITE — digital bright radar indicator tower equipment

DBU — database unit

DC — direct current

DC — Divergent-Convergent duct.

DCA — Defense Communications Agency

DCA — Ronald Reagan Washington National Airport

DCAA — dual call, automatic answer device

DCAS — digital core avionic system

DCCU — data communications control unit

DCE — data communications equipment

DCP — data collection package

DDA — dedicated digital access

DDD — direct distance dialing

DDM — difference in depth of modulation

DDS — digital data service

DEA — Drug Enforcement Agency

DECM — defensive electronic counter measures

DECU — digital elect control unit

DEDS — data entry and display system

DEIS — draft environmental impact statement

DEP — departure

DEWIZ — Distant Early Warning Identification Zone

DF — direction finder

DFAX — digital facsimile

DFDR — digital flight data recorder

DFGC — direct flight guidance computer

DFGS — direct flight guidance system

DFI — direction finding indicator

DG — directional gyro

DGPS — differential global positioning satellite (system)

DH — Decision Height

DID — direct inward dial

DIN — AUTODIN Service

DIP — drop and insert point

DIRF — direction finding

DITCO — Defense Information Technology Contracting Office Agency

DME — distance

DME — distance measuring equipment

DME/N — standard DME

DME/P — precision distance measuring equipment

DME/P — precision DME

DMN — data multiplexing network

DNL — day light equivalent sound level (also called LDN) -

DoD — Department of Defense

DOD — Department Of Defense

DOD — direct outward dial

DOI — Department of Interior

DOS — Department of State

DOT — Department of Transportation

DOTCC — Department of Transportation Computer Center

DOTS — dynamic ocean tracking system

DP — instrument departure procedure

DPDT — double-pole, double-throw

DPST — double-pole, single-throw

DPU — data processor unit

DR — dead reckoning

DSCS — digital satellite compression service

DSUA — dynamic special use airspace

DTS — dedicated transmission service

DUAT — Direct User Access Terminal

DUATS — Direct User Access Terminal System

DVA — diverse vector area

DVFR — day visual flight rules

DVFR — defense visual flight rules

DVOR — doppler very high frequency omni directional range

DVRCR — differential voltage reverse current relay

DYSIM — dynamic simulator

EADI — electronic attitude direction indicator

EARTS — Enroute Automated Radar Tracking System

EAS — equivalent airspeed

EC — European Community.

ECCM — electronic counter-countermeasures

ECM — electronic countermeasures

ECOM — enroute communications

ECU — engine control unit

ECU — electronic control unit

ECVFP — expanded charted visual flight procedures

EDC — engine driven compressor

EDCT — expect departure clearance time

EDP — engine driven pump

EDP — expedite departure path

EEC — electronic engine computer

EEL — emergency exit lights

EFAS — enroute flight advisory service

EFC — expect further clearance

EFIS — electronic flight information systems

EFIS — electronic flight instrument (indication) system

EGT — exhaust gas temperature

EIAF — expanded inward access features

EICAS — engine indication and crew alerting system

EICAS — engine indication central alert system

EIS — environmental impact statement

ELT — emergency locator transmitter

ELWRT — electrowriter

EMF — electromotive force (voltage).

EMPS — enroute maintenance processor system

EMSAW — Enroute Minimum Safe Altitude Warning

ENAV — enroute navigational aids

EOF — emergency operating facility

EPA — Environmental Protection Agency

EPC — external power container

EPE — estimate of position error

EPE — estimated position error

EPR — engine pressure ratio

EPROM — erasable programmable read only memory

EPS — engineered performance standards

EPSS — enhanced packet switched service

ERAD — enroute broadband radar

ESEC — enroute broadband secondary radar

ESF — extended superframe format

ESFC — equivalent specific fuel consumption

ESP — enroute spacing program

ESV — expanded service volume

ESYS — enroute equipment systems

ETA — estimated time of arrival

ETD — estimated time of departure

ETE — estimated time enroute

ETG — enhanced target generator

ETMS — enhanced traffic management system

ETN — electronic telecommunications network

EU — European Union

EVAS — enhanced vortex advisory system

EVBC — engine variable bleed control

EVC — engine vane control

EVCS — emergency voice communications system

EWCG — empty weight center of gravity

EWR — Newark International Airport

F&E — facility and equipment

F/F — Fuel Flow.

FA — area forecast

FA — aviation area forecast

FAA — Federal Aviation Administration

FAAAC — FAA Aeronautical Center

FAACIS — FAA Communications Information System

FAA-PMA — Federal Aviation Administration Parts Manufacturing Approval

FAATC — FAA Technical Center

FAATSAT — FAA Telecommunications Satellite

FAC — facility

FAD — Fuel Advisory Departure

FAF — final approach fix

FAP — final approach point

FAPM — FTS2000 Associate Program Manager

FAR — Federal Aviation Regulation

FAST — final approach spacing tool

FAWP — final approach waypoint

FAX — facsimile equipment

FB — fly by

FBO — fixed base operator

FBS — fall back switch

FCC — Federal Communications Commission

FCLT — freeze calculated landing time

FCOM — FSS Radio Voice Communications

FCPU — facility central processing unit

FCU — Fuel Control Unit.

FD — flight director system

FD — winds and temperatures aloft forecast

FDAT — flight data entry and printout (FDEP) and flight data service

FDAU — flight data acquisition unit

FDC — Flight Data Center

FDE — flight data entry

FDEP — flight data entry and printout

FDI — integrated flight indicator

FDIO — flight data input/output

FDIOC — flight data input/output center

FDIOR — flight data input/output remote

FDM — frequency division multiplexing

FDP — flight data processing

FED — federal

FEIS — final environmental impact statement

FEP — front end processor

FET — field effect transistor

FFAC — from facility

FFG — fuel flow governor

FFR — fuel flow regulator

FGC — flight guidance computer

FIFO — flight inspection field office

FIG — flight inspection group

FINO — Flight Inspection National Field Office

FIPS — Federal Information Publication Standard

FIR — Flight Information Region

FIRE — fire station

FIRMR — Federal Information Resource Management Regulation

FISDL — flight information services data link

FL — flight level

FLIP — flight information publication

FLOWSIM — traffic flow planning simulation

FM — fan marker

FM — frequency modulation

FMA — final monitor aid

FMC — flight management computer

FMCS — flight management computer system

FMEP — friction mean effective pressure

FMF — facility master file

FMIS — FTS2000 Management Information System

FMS — flight management system

FMSP — flight management system procedure

FNMS — FTS2000 Network Management System

FO — fly over

FOB — fuel on board.

FOD — foreign object damage

FOIA — Freedom of Information Act

FP — flight plan

fpm — feet per minute

FPNM — feet per nautical mile

fps — feet per second

FRC — request full route clearance

FREQ — frequency

FSAS — flight service automation system

FSDO — Flight Standards District Office

FSDPS — flight service data processing system

FSEP — facility/service/equipment profile

FSEU — flap/slat electronic unit

FSN — Fuel Spray Nozzle.

FSP — flight strip printer

FSPD — freeze speed parameter

FSS — Flight Service Station

FSSA — Flight Service Station Automated Service

FSTS — federal secure telephone service

FSYS — flight service station equipment systems

ft/sec — Feet per second.

FTS — Federal Telecommunications System

FTS2000 — Federal Telecommunications System 2000

FUS — functional units or systems

FWCS — flight watch control station

GA — general aviation

GAA — general aviation activity

GAAA — general aviation activity and avionics

363

GADO — General Aviation District Office

gal — Gallon.

GBAS — ground based augmentation system

GCA — ground control approach

GCU — generator control unit

GEO — geostationary satellite

GFP — Ground Fine Pitch.

GLS — GNSS landing system

GNAS — General National Airspace System

GNSS — Global Navigation Satellite System

GNSS — global navigation satellite system

GNSSP — global navigation satellite system panel

GOES — Geostationary Operational Environmental Satellite

GOESF — GOES Feed Point

GOEST — GOES terminal equipment

GPA — gas path analysis

gph — gallons per hour

GPS — global positioning satellite

GPS — global positioning system

GPU — ground power unit

GPWS — ground proximity warning system

GRADE — graphical airspace design environment

GRI — group repetition interval

GS — glide slope

GS — groundspeed

GSA — General Services Administration

GSD — geographical situation display

GSE — ground support equipment

GSI — glide slope indicator

GTE — Gas Turbine Engine.

GUS — ground uplink station

H — Non Directional Radio Homing Beacon (NDB)

HAA — height above airport

HAL — height above landing

HARS — high altitude route system

HASOV — hot air shut-off valve

HAT — height above touchdown

HAT — height above touchdown zone

HAZMAT — hazardous materials

HCAP — high capacity carriers

HDME — NDB with distance measuring equipment

HDQ — headquarters

HDTA — high density traffic airports

HE — High Energy.

HELI — heliport

HF — high frequency

HF — high frequency

Hg — Mercury.

HGS — head-up guidance system

HH — NDB, 2kw or more

HI EFAS — high altitude EFAS (enroute flight advisory service)

HIG — hermetically sealed integrating gyro

HIRL — high intensity runway lights

HIWAS — hazardous inflight weather advisory service

HLDC — high level data link control

HMU — hydromechanical unit

HOC — high oil consumption-jet engine

HOV — high occupancy vehicle

HP — High Pressure.

HP — Horse Power.

hr — hour.

HSI — horizontal situation indicator

HUD — head-up display

HSTCU — horizontal stabilizer trim control unit

HUD — Housing and Urban Development

HVOR — high altitude VOR

HWAS — hazardous in flight weather advisory

HYD ISO — hydraulic isolation

Hz — hertz

I/AFSS — International AFSS

IA — indirect access

IA — inspection authorization

IAF — initial approach fix

IAP — instrument approach procedure

IAPA — instrument approach procedures automation

IAS — indicated airspeed

IAWP — initial approach waypoint

IBM — International Business Machines

IBP — international boundary point

IBR — intermediate bit rate

IC — integrated circuit

ICAO — International Civil Aviation Organization

ICSS — International Communications Switching Systems

IDAT — interfacility data

IDG — integrated drive generator

IEC — integral electronic control

IEPR — integrated engine pressure ratio

IF — intermediate fix

if — intermediate frequency

IFCP — interfacility communications processor

IFDS — interfacility data system

IFEA — in flight emergency assistance

IFF — Identification, Friend or Foe

IFIM — International Flight Information Manual

IFO — International Field Office

IFR — instrument flight rules

IFS — integrated flight system

IFSS — international flight service station

IGFET — insulated gate field effect transistor

IGV — inlet guide vane

IHP — indicated horsepower

ILS — instrument landing system

IM — inner marker

IMC — instrument meteorological conditions

IMEP — indicated mean effective pressure

in. Hg — inches of mercury

INM — integrated noise model

INS — inertial navigation system

Int — intersection

INU — inertial navigation unit

IOAT — indicated outside air temperature

IOC — initial operational capability

IPB — illustrated parts breakdown

IPC — illustrated parts catalog

IPL — illustrated parts list

IPM — illustrated parts manual

IR — IFR military training route

IR — infrared

IRAN — Inspect and Repair As Necessary

IRBM — intermediate-range ballistic missile

IRMP — information resources management plan

IRS — inertial reference system

IRU — inertial reference unit

ISA — International Standard Atmosphere

ISDN — integrated services digital network

ISMLS — interim standard microwave landing system

ISOV — isolation shutoff valve

ISU — inertial sensing unit

ITI — interactive terminal interface

ITT — inlet turbine temperature

ITT — intermediate turbine temperature

ITWS — integrated terminal weather system

IVRS — interim voice response system

IVSI — instantaneous rate of climb indicator

IVSI — instant vertical speed indicator

IW — inside wiring

j — joule (energy unit).

JATO — jet assisted takeoff

JFC — jet fuel control

JFET — junction field effect transistor

JFK — John F. Kennedy international airport

JPT — jet pipe temperature

Kbps — kilobits per second

KE — Kinetic Energy.

Khz — kilohertz

kHz — kilohertz

KIAS — knots indicated airspeed

kt — Knot (1 nm/hr).

KTAS — knots true airspeed

KVAR — kilovolt amperes reactive

KVDT — keyboard video display terminal

L/MF — low/medium frequency

LA — power lever angle

LAA — local airport advisory

LAAS — local area augmentation system

LAAS — Low Altitude Alert System

LABS — leased A B service

LABSC — LABS GS 200 Computer

LABSR — LABS remote equipment

LABSW — LABS switch system

LAHSO — land and hold short operations

LAN — local area network

LASCR — light-activated silicon control rectifier

LAT — latitude

LATA — local access and transport area

LAWRS — limited aviation weather reporting station

lb — pound.

LCD — liquid crystal display

LCF — local control facility

LCN — local communications network

LDA — landing directional aid

LDIN — lead in lights -

LEC — local exchange carrier

LED — light emitting diode

LEMAC — leading edge of the mean aerodynamic chord

LF — low frequency

LFR — low- frequency radio range

LF/MF — low/medium frequency

LGA — LaGuardia Airport

LH — left-hand (threads)

LINCS — Leased Interfacility NAS Communications System

LIRL — low intensity runway lights

LIS — Logistics and Inventory System

LLWAS — Low Level Wind Shear Alert System

LMM — compass locator at middle marker

LMM — locator middle marker

LMS — LORAN monitor site

LNAV — lateral navigation

LOC — ILS localizer

LOCID — location identifier

LOI — letter of intent

LOM — compass locator at outer marker

LONG — longitude

LOP — line of position

LOP — low oil pressure

LORAN — long range navigation system

LOX — liquid oxygen

LP — Low Pressure

LRCO — limited remote communications outlet

LRNAV — long range navigation

LRR — long range radar

LRRA — low range radio altimeter

LSB — least significant bit

LSI — large scale integration

LVOR — low altitude VOR

M — mach

MAA — maximum authorized altitude

MAA — maximum authorized IFR altitude

MAC — mean aerodynamic chord

MAHWP — missed approach holding waypoint

MALS — medium intensity approach lighting system

MALSF — MALS with sequenced flashers

MALSR — MALS with runway alignment indicator lights

MAP — maintenance automation program

MAP — military airport program

MAP — missed approach point

MAP — modified access pricing

MAP — monitor alert parameter

MAWP — missed approach waypoint

MB — magnetic bearing

MB — marker beacon

mb — millibars

Mbps — megabits per second

MCA — minimum crossing altitude

MCAS — Marine Corps Air Station

MCC — Maintenance Control Center

MCD — metal chip detector

MCDU — multipurpose control display unit)

MCL — middle compass locater

MCS — Maintenance and Control System

MDA — minimum descent altitude

MDT — maintenance data terminal

MEA — minimum enroute altitude

MEA — minimum enroute IFR altitude

MEC — main engine control

MEF — maximum elevation figure

MEK — methyl-ethyl-ketone

MEL — minimum equipment list.

METAR — aviation routine weather report

METI — meteorological information

METO — maximum except takeoff power

MF — middle frequency

MFDU — multi function display unit

MFJ — modified final judgment

MFT — meter fix crossing time/slot time

MFQ — main fuel quantity

MH — magnetic heading

MHA — minimum holding altitude

MHz — megahertz

MIA — minimum IFR altitudes

MIDO — Manufacturing Inspection District Office

MIG — metal inert-gas

MILSPEC — military specifications

MIRL — medium intensity runway lights

MIS — meteorological impact statement

MISC — miscellaneous

MISO — Manufacturing Inspection Satellite Office

MIT — miles in trail

MITP — metal in tail pipe

MITRE — Mitre Corporation

MLS — Microwave Landing System

MM — ILS Middle Marker

MMC — maintenance monitoring console

mmf — magnetomotive force

MMS — maintenance monitoring system

MNPS — minimum navigation performance specification

MNPSA — minimum navigation performance specifications airspace

MOA — memorandum of agreement

MOA — military operations area

MOCA — Minimum Obstruction Clearance Altitude

MOCA — minimum obstruction clearance altitude

MODE C — altitude encoded beacon reply

MODE S — mode select beacon system

MOS — metal oxide semiconductor

MOSFET — metal oxide semiconductor field effect transistor

MOU — memorandum of understanding

mph — miles per hour.

MPO — Metropolitan Planning Organization

MPS — maintenance processor subsystem

MPS — master plan supplement

MRA — minimum reception altitude

MRC — monthly recurring charge

MS — military standard

MSA — minimum safe altitude

MSAW — minimum safe altitude warning

MSB — most significant bit

MSD — most significant digit

MSI — medium-scale integration

MSL — mean sea level

MSN — message switching network

MTCS — modular terminal communications system

MTI — moving target indicator

MTR — military training route

MULTICOM — frequency used at airports without a tower, FSS, or UNICOM

MUX — multiplexer

MVA — minimum vectoring altitude

MVFR — marginal VFR

MVFR — marginal visual flight rules

N — Rotational speed, RPM.

n.m. — nautical miles

N$_1$ — Low pressure spool.

N$_2$ — High pressure spool.

N$_2$O — Nitrous oxide.

NAAQS — National Ambient Air Quality Standards

NACO — National Aeronautical Charting Office

NADA — NADIN concentrator

NADIN — National Airspace Data Interchange Network

NADSW — NADIN switches

NAF — Naval Aircraft Factory

NAILS — National Airspace Integrated Logistics Support

NAPRS — National Airspace Performance Reporting System

NAS — National Aircraft Standard

NAS — National Airspace System

NAS — National Aerospace Standard

NAS — Naval Air Station

NASA — National Aeronautics and Space Administration

NASDC — National Aviation Safety Data

NASP — National Airspace System Plan

NASPAC — National Airspace System Performance Analysis Capability

NATCO — National Communications Switching Center

NATO — North Atlantic Treaty Organization.

NAVAID — navigation aid

NAVMN — navigation monitor and control

NAWAU — National Aviation Weather Advisory Unit

NAWPF — National Aviation Weather Processing Facility

NBCAP — National Beacon Code Allocation Plan

NCAR — National Center for Atmospheric Research

NCF — National Control Facility

NCIU — NEXRAD communications interface unit

NCS — National Communications System

NDB — nondirectional radio beacon

NDB(ADF) — nondirectional beacon (automatic direction finder)

NEPA — National Environmental Policy Act

NEXRAD — next generation weather radar

NFAX — National Facsimile Service

NFDC — National Flight Data Center

NFDD — National Flight Data Digest

NFIS — NAS Facilities Information System

NI — network interface

NICS — National Interfacility Communications System

NIDS — National Institute for Discovery Sciences

NIMA — National Imagery and Mapping Agency

NM — nautical mile

nm — nautical mile.

NMAC — near mid air collision

NMC — National Meteorological Center

NMCE — network monitoring and control equipment

NMCS — network monitoring and control system

NOAA — National Oceanic and Atmospheric Administration

NOC — notice of completion

NOPAC — North Pacific

NoPT — no procedure turn required

NOS — National Ocean Service

NOTAM — notice to airmen

NPA — nonprecision approach

NPDES — National Pollutant Discharge Elimination System

NPIAS — National Plan of Integrated Airport Systems

NPN — negative, positive, negative transistor-

NPRM — notice of proposed rule making

NRC — non recurring charge

NRCS — National Radio Communications Systems

NSA — national security area

NSAP — National Service Assurance Plan

NSSFC — National Severe Storms Forecast Center

NSSL — National Severe Storms Laboratory; Norman, OK

NSW — no significant weather

NSWRH — NWS Regional Headquarters

NTAP — Notices to Airmen Publication

NTP — National Transportation Policy

NTS — negative torque sensor

NTSB — National Transportation Safety Board

NTZ — no transgression zone

NWS — National Weather Service

NWS — National Weather Service

NWSR — NWS Weather excluding NXRD

NXRD — advanced weather radar system

O — oxygen.

OAG — Official Airline Guide

OALT — operational acceptable level of traffic

OASIS — Operation and Supportability Implementation System

OAT — outside air temperature

OAW — off airway weather station

OBS — omni bearing selector

ODAL — omnidirectional approach lighting system

ODAPS — Oceanic Display and Processing Station

OEI — one engine inoperative

OFA — object free area

OFDPS — Offshore Flight Data Processing System

OFT — outer fix time

OFZ — obstacle free zone

OM — ILS outer marker

OMB — Office of Management and Budget

ONER — oceanic navigational error report

OPLT — operational acceptable level of traffic

OPSW — operational switch

OPX — off premises exchange

ORD — Chicago O'Hare International Airport

ORD — operational readiness demonstration

OROCA — off-route obstruction clearance altitude

ORTAC — VOR combined with UHF tactical air navigation

OTR — oceanic transition route

OTS — organized track system

P — pressure.

p.s.i. — pounds per square inch

P/CG — pilot/controller glossary

P_1 — pressure at Station 1 (engine nose).

P_7 — tailpipe pressure.

PA — precision approach

PA — pressure altitude

PABX — private automated branch exchange

PAD — packet assembler/disassembler

PAM — peripheral adapter module

PAPI — precision approach path indicator

PAR — precision approach radar

PAR — preferential arrival route

PAR — preferred arrival route

PATWAS — pilots automatic telephone weather answering service

PBCT — proposed boundary crossing time

PBRF — pilot briefing

PBX — private branch exchange

PC — personal computer

PCA — positive control airspace

PCA — positive control area

PCB — printed circuit board

PCL — pilot controlled lighting-

PCM — pulse code modulation

PCU — prop control unit

pd — potential difference

PDAR — preferential arrival and departure route

PDC — performance data computer

PDC — pre departure clearance

PDC — program designator code

PDN — public data network

PDR — preferential departure route

P-factor — an element of asymmetrical thrust

PFC — passenger facility charge

PFD — personal flotation device

PGB — propeller gear box

PIBAL — pilot balloon observation

PIC — pilot in command

PIC — principal interexchange carrier

PIDP — programmable indicator data processor

PinS — point in space

PIREP — pilot weather report

PK — Parker-Kalon screw

PLASI — pulsating approach slope indicator

PM — phase modulation

PMC — power management control

PMS — program management system

PNP — positive, negative, positive transistor-

POB — persons on board

POH — pilot's operating handbook

POI — principal operations inspector

POLIC — police station

POP — point of presence

POT — point of termination

PPI — plan-position indicator

PPIMS — Personal Property Information Management System

PPS — precise positioning service

PR — primary commercial service airport

PRBCV — pressure ratio bleed control valve

PRI — primary rate interface

PRM — precision runway monitor

PROM — programmable read-only memory

PRT — power recovery turbine

PRV — Pressure Relief Valve.

P/S — power section

PSDN — public switched data network

PSEU — proximity sense elect unit

psi — pounds per square inch.

PSN — packet switched network

PSS — packet switched service

PSTN — public switched telephone network

PSU — passenger service unit

PT — procedure turn

PTS — practical test standards

PTU — power transfer unit

PUB — publication

PUP — principal user processor

PV — Pressure-Volume (diagram).

PVC — permanent virtual circuit

PVC — polyvinyl chloride

PVD — plan view display

QC — quality control

QEC — quick engine change

QECA — quick engine change assembly

r.p.m. — revolutions per minute

R-12 — refrigerant 12

RA — resolution advisory

RAF — Royal Air Force.

RAIL — runway alignment indicator lights

RAIM — receiver autonomous integrity monitoring

RAPCO — Radar Approach Control (USAF)

RAPCON — Radar Approach Control (FAA)

RAREP — radar weather report

RAT — ram air temp

RAT — ram air turbine

RATCC — Radar Air Traffic Control Center

RATCF — Radar Air Traffic Control Facility (USN)

RATO — rocket-assisted takeoff

RB — relative bearing

RBC — rotating beam ceilometer

RBDPE — radar beacon data processing equipment

RBDT — ribbon display terminals

RBN — radio beacon

RBSS — Radar Bomb Scoring Squadron

RCAG — remote center air/ground

RCAG — remote communications air/ground

RCC — rescue coordination center

RCCB — remote control circuit breaker

RCCC — Regional Communications Control Centers

RCF — remote communication facility

RCIU — remote control interface unit

RCL — radio communications link

RCLM — runway centerline marking

RCLR — RCL repeater

RCLS — runway centerline lighting system

RCLT — RCL terminal

RCO — remote communications outlet

RCU — remote control unit

RDAT — digitized radar data

RDP — radar data processing

RDSIM — runway delay simulation model

REIL — runway end identifier lights

rf — radio frequency

RF — radio frequency

RFI — radio-frequency interference

RHI — range-height indicator (scope)

RII — required inspection item

RL — general aviation reliever airport

RLIM — runway light intensity monitor

RMCC — remote monitor control center

RMCF — remote monitor control facility

RMI — radio magnetic indicator

RMK — remark

RML — radio microwave link

RMLR — RML repeater

RMLT — RML terminal

RMM — remote maintenance monitoring

RMMS — remote maintenance monitoring system

RMS — remote monitoring subsystem

RMS — root mean square

RMSC — remote monitoring subsystem concentrator

RNAV — area navigation

RNP — required navigation performance

ROD — record of decision

ROSA — report of service activity

ROT — runway occupancy time

RP — restoration priority

RPC — restoration priority code

RPG — radar processing group

RPM — Revolutions per minute.

RPZ — runway protection zone

RR — low or medium frequency radio range station

RRH — remote reading hygrothermometer

RRHS — remote reading hydrometer

RRWDS — remote radar weather display

RRWSS — remote radar weather display (RWDS) sensor site

RSS — remote speaking system

RT — remote transmitter

RT & BTL — radar tracking and beacon tracking level

RTAD — remote tower alphanumeric display

RTCA — Radio Technical Commission for Aeronautics

RTM — resin transfer molding

RTOP — reserve takeoff power

RTR — remote transmitter/receiver

RTRD — remote tower radar display

RVDT — rotary variable display transmitter

RVR — runway visual range as measured in the touchdown zone area

RVV — runway visibility value

RW — runway

RWDS — remote radar weather display

RWP — real time weather processor

rwy — runway

S/G — starter-generator

S/S — sector suite

SAA — special activity airspace

SAC — Strategic Air Command

SAE — Society of Automotive Engineers

SAFI — semi automatic flight inspection

SAI — standby attitude indicator

SALS — short approach light system

SAR — search and rescue

SAS — stability augmentation system

SAT — static air temperature

SATCOM — satellite communications

SAWRS — Supplementary Aviation Weather Reporting System

SBC — surge bleed control

SCAT 1 DGPS — special category 1 differential GPS

SCATANA — Security Control of Air Traffic and Air Navigation Aids

SCC — System Command Center

SCVTS — Switched Compressed Video Telecommunications Service

SD — radar weather report

SDF — simplified direction finding

SDF — simplified directional facility

SDF — simplified directional facility

SDF — software defined network

SDIS — switched digital integrated service

SDP — service delivery point

SDS — switched data service

sec — second (time).

SEL — single event level

SELF — simplified short approach lighting system with sequenced flashing lights

SFAR 38 — Special Federal Aviation Regulation 38

SFC — Specific Fuel Consumption.

sfc — surface

SFL — sequenced flashing lights

SFR — special flight rules

SG — Specific Gravity

SHF — super high frequency

SHP — Shaft Horse Power.

SHPO — State Historic Preservation Officer

SI — international system of units

SIAP — standard instrument approach procedure

SIC — service initiation charge

SID — standard instrument departure

SID — station identifier

SIF — stall indicator failure

SIGMET — significant meteorological information

SIMMOD — airport and airspace simulation model

SIP — state implementation plan

SM — statute mile

SMGC — surface movement guidance and control

SMGCS — surface movement guidance control system

SMPS — sector maintenance processor subsystem

SMS — simulation modeling system

SNR — signal to-noise ratio

SNR or S/N — signal to noise ratio

SOC — service oversight center

SODA — statement of demonstrated ability

SOIR — simultaneous operations on intersecting runways

370

SOIWR — simultaneous operations on intersecting wet runways

SPECI — aviation selected special weather report

SPR — single point refueling

SPS — standard positioning service

SRAP — sensor receiver and processor

SRL — single red line (computer)

SSALF — SSALS with sequenced flashers

SSALR — simplified short approach lighting system

SSALS — simplified short approach light system

SSALSR — simplified short approach light system with runway alignment indicator lights

SSB — single side band

SSB — split system breaker

SSI — small-scale integration

SSU — Saybolt Seconds Universal

STAR — standard terminal arrival

STAR — standard terminal arrival route

STC — Supplemental Type Certificate

STD — standard

STMP — special traffic management program

STMUX — statistical data multiplexer

STOL — short takeoff and landing

STOVL — short takeoff and vertical landing

SURPIC — surface picture

SVFR — special VFR

SVFR — special visual flight rules

SWSL — supplemental weather service

T — Temperature (Absolute).

T — Thrust.

T_1MUX — T_1 multiplexer

T/R — thrust reverser system

TA — traffic advisory

TAA — terminal arrival area

TAAS — terminal advance automation system

TAC — terminal area chart

TACAN — tactical air navigation

TACAN — Tactical Air Navigation System

TACR — TACAN at VOR, TACAN only

TAF — aviation terminal forecast

TAI — tail anti-ice

TARS — terminal automated radar service

TAS — true airspeed

TAT — total air temperature

TATCA — terminal air traffic control automation

TAVT — terminal airspace visualization tool

TBO — time between overhaul

TC — true course

TCA — terminal control area

TCA — Traffic Control Airport or Tower Control Airport

TCACCIS — Transportation Coordinator Automated Command and Control Information System

TCAS — Traffic Alert and Collision Avoidance System

TCAS II — traffic collision avoidance system

TCC — DOT Transportation Computer Center

TCCC — tower control computer complex

TCE — tone control equipment

TCH — threshold crossing height

TCH — threshold crossing height

TCLT — tentative calculated landing time

TCO — Telecommunications Certification Officer

TCOM — terminal communications

TCS — tower communications system

TD — time difference

TDC — top dead center

TDLS — tower data link services

TDMUX — time division data multiplexer

TDWR — terminal Doppler weather radar

TDZL — touchdown zone lights

TEC — tower enroute control

TEHP — Thrust Equivalent Horse Power.

TELCO — telephone company

TELMS — Telecommunications Management System

TEL-TWEB — telephone access to TWEB

TEMAC — trailing edge of the mean aerodynamic chord

TERPS — terminal instrument procedures

TET — Turbine Entry Temperature.

TFAC — to facility

TGT — turbine gas temperature

TH — threshold

THP — Thrust Horse Power.

TIBS — telephone information briefing service

TIG — tungsten inert gas

TIMS — telecommunications information management system

TIPS — terminal information processing system

TIR — total indicator reading

TIT — turbine inlet temperature

TL — taxilane

TLS — transponder landing system

TM&O — Telecommunications Management and Operations

TMA — Traffic Management Advisor

TMC — Traffic Management Coordinator

TMC/MC — Traffic Management Coordinator/Military Coordinator

TMCC — Terminal Information Processing System

TMCC — traffic management computer complex

TMF — traffic management facility

TML — television microwave link

TMLI — television microwave link indicator

TMLR — television microwave link repeater

TMLT — television microwave link terminal

TMP — traffic management processor

TMS — thrust management system

TMS — traffic management system

TMSPS — traffic management specialists

TMU — traffic management unit

TNAV — terminal navigational aids

TODA — takeoff distance available

TOF — time of flight

TOFMS — time of flight mass spectrometer

TOPS — Telecommunications Ordering and Pricing System (GSA software tool)

TORA — take off run available

TPP — terminal procedures publications

TPU — transit pressure unit

TR — telecommunications request

TRACAB — terminal radar approach control in tower cab

TRACON — terminal radar approach control

TRACON — terminal radar approach control facilities

TRAD — terminal radar service

TRF — tuned radio frequency (receiver)

TRI — thrust rating indicator

TRNG — training

TRSA — terminal radar service area

TRU — transformer rectifier unit

TSA — taxiway safety area

TSCU — torque sensing control unit

TSCU — torque signaling condition unit

TSEC — terminal secondary radar service

TSFC — thrust specific fuel consumption

TSO — technical standard order

TSP — telecommunications service priority

TSR — telecommunications service request

TSYS — terminal equipment systems

TTL — total torque limiter

TTMA — TRACON traffic management advisor

TTY — teletype

TVOR — terminal VOR

TW — taxiway

TWEB — transcribed weather broadcast

TWR — tower (non controlled)

TY — Type (FAA Communications Information System)

TYP — typical (mechanical drawing)

U.S. — United States

UAS — uniform accounting system

UBR — utility bus relay

UER — unscheduled engine removal

UFO — unidentified flying object

UHF — ultra high frequency

UNC — United National, coarse

UNF — United National, fine

UNICOM — aeronautical advisory station

URA — Uniform Relocation Assistance and Real Property Acquisition Policies Act of 1970

URET — User Request Evaluation Tool

URET CCLD — User Request Evaluation Tool Core Capability Limited Deployment

USAF — United States Air Force

USAFIB — U. S. Army Aviation Flight Information Bulletin

USCG — United States Coast Guard

USOC — Uniform Service Order Code

USS — United States, standard

UTC — coordinated universal time (Zulu time)

UV — Ultra Violet

UWS — urgent weather SIGMET

V — volume

V_1 — maximum speed in the takeoff at which the pilot must take the first action (e.g., apply brakes, reduce thrust, deploy speed brakes) to stop the airplane within the accelerate-stop distance. V_1 also means the minimum speed in the takeoff, following a failure of the critical engine at V_{EF}, at which the pilot can continue the takeoff and achieve the required height above the takeoff surface within the takeoff distance.

V1 — Pulse jet powered flying bomb.

V2 — Liquid fuel rocket.

V_2 — takeoff safety speed

$V_{2\ MIN}$ — minimum takeoff safety speed

V_A — design maneuvering speed

VAC — volts of alternating current

VAFTAD — volcanic ash transport and dispersion chart

VAPC — variable absolute pressure controller

VAR — volcanic activity reporting

VAR — volt-ampere reactive

VASI — visual approach slope indicator

V_B — design speed for maximum gust intensity

VBV — variable blade vane

V_C — design cruising speed

V_D — design diving speed

VDA — vertical descent angle

VDC — volts of direct current

V_{DF}/M_{DF} — demonstrated flight diving speed

VDME — VOR with distance measuring equipment

VDP — visual descent point

V_{EF} — speed at which the critical engine is assumed to fail during takeoff

V_F — design flap speed

VF — voice frequency

V_{FC}/M_{FC} — maximum speed for stability characteristics

V_{FE} — maximum flap extended speed

VFO — variable-frequency oscillator

VFR — visual flight rules

V_{FTO} — final takeoff speed

VGSI — visual glide slope indicator

V_H — maximum speed in level flight with maximum continuous power

VHF — very high frequency

VHF/DF — VHF direction finder

VIP — very important person (dignitary)

VIP — video integrator and processor (thunderstorm intensity standard)

V_{LE} — maximum landing gear extended speed

VLF — very low frequency

V_{LO} — maximum landing gear operating speed

V_{LOF} — lift-off speed

VLV — valve

V_{MC} — minimum control speed with the critical engine inoperative

VMC — visual meteorological conditions

VMC — visual meteorological conditions

VMC — visual meteorological conditions

Vmini — minimum speed —

V_{MO}/M_{MO} — maximum operating limit speed

V_{MU} — minimum unstick speed

VNAV — vertical navigation

VNAV — visual navigational aids

V_{NE} — never-exceed speed

V_{NO} — maximum structural cruising speed

VNTSC — Volpe National Transportation System Center

VON — virtual on net

VOR — very high frequency omni directional range

VOR/DME — collocated VOR and DME navaids

VORTAC — collocated VOR and TACAN

VOT — VOR test facility

V_R — rotation speed

VR — VFR military training route

V_{REF} — reference landing speed

VRS — voice recording system

V_S — stalling speed or the minimum steady flight speed at which the airplane is controllable

V_{S0} — stalling speed or the minimum steady flight speed in the landing configuration

V_{S1} — stalling speed or the minimum steady flight speed obtained in a specific configuration

VSCS — voice switching and control system

VSI — vertical speed indicator

V_{SR} — reference stall speed

V_{SR1} — reference stall speed in a specific configuration

V_{SR0} — reference stall speed in the landing configuration

VSTOL — vertical or short takeoff and landing

VSV — variable stator vane

V_{SW} — speed at which onset of natural or artificial stall warning occurs

VSWR — voltage standing-wave ratio

VTA — vertex time of arrival

VTAC — VOR collocated with TACAN

VTOL — vertical takeoff and landing

V_{TOSS} — takeoff safety speed for Category A rotorcraft

VTS — voice telecommunications system

VTVM — vacuum-tube voltmeter

VV — vertical visibility

VVI — vertical velocity indicator

V_X — speed for best angle of climb

V_Y — speed for best rate of climb

w/m — Water/methanol coolant mix.

WA — AIRMET

WAAS — wide area augmentation system

WAC — world aeronautical chart

373

WAN — wide area network

WARP — weather and radar processor

WC — work center

WCA — wind correction angle

WCP — weather communications processor

WECO — Western Electric Company

WESCOM — Western Electric Satellite Communications

WEU — warning electronic unit

WFO — weather forecast office

WGS84 — World Geodetic System of 1984

WH — hurricane advisory

whr — watthour

WMO — World Meteorological Organization

WMS — wide area master station

WMSC — weather message switching center

WMSCR — weather message switching center replacement

WOW — weight on wheels (squat switch)

WOXOF — weather, zero visibility in fog

WP — waypoint

WRS — wide area ground reference station

WS — SIGMET

WSCMO — weather service contract meteorological observatory

WSFO — Weather Service Forecast Office

WSMO — Weather Service Meteorological Observatory

WSO — Weather Service Office

WSP — weather system processor

WST — convective SIGMET

WSW — wind shear warning

WTHR — weather

WW — Severe Weather Watch Bulletin

WX — weather

WXR — weather radar

WST — convective significant meteorological information

WTHR — weather

WW — severe weather watch bulletin

WX — weather

WXOF — weather, zero visibility in fog

GREEK LETTERS

A	α	alpha		N	ν	nu
B	α	beta		Ξ	ξ	xi
Γ	γ	gamma		O	o	omicron
Δ	δ	delta		Π	π	pi
E	ε	epsilon		P	ρ	rho
Z	ζ	zeta		Σ	σ	sigma
H	η	eta		T	τ	tau
Θ	θ	theta		Y	υ	upsilon
I	ι	iota		Φ	φ	phi
K	κ	kappa		X	χ	chi
Λ	λ	lamda		Ψ	ψ	psi
M	μ	mu		Ω	ω	omega

STANDARD ABBREVIATIONS/SYMBOLS

α — temperature coefficient of resistance
CO — carbon monoxide
CO_2 — carbon dioxide
F_g — static thrust or gross thrust
F_n — net thrust
G_m — transconductance
H_2. — hydrogen
HCl — hydrochloric acid
k — dielectric constant
N_f — RPM of a free turbine
N_1 — RPM of a low-pressure compressor
N_2 — RPM of a high-pressure compressor
P — power
Pb — lead
P_b — bumer pressure
Pt_2 — inlet pressure
Pt_4 — compressor discharge pressure
PT_7 — turbine discharge pressure
R — resistance
Tt_2 — inlet temperature
V_1 — takeoff decision speed
V_2 — minimum takeoff safety speed
X_c — capacitive reactance
Z — impedance
+ — positive
- — negative
Ω — ohm
° — degree
" — inch(es)
' — foot (feet)
amp — ampere
Btu — British thermal unit
C — Celsius
cal — calorie(s)

Cal — large calorie(s)
cm — centimeter
cos — cosine
cu cm — cubic centimeter(s)
cu in — cubic inch(es)
cu ft — cubic foot (feet)
cu m — cubic meter(s)
dB — decibel
deg — degree
ESHP — equivalent shaft horsepower
f — farad
F — Fahrenheit
ft — foot (feet)
ft lb — foot-pound
ft lbs — foot-pounds
g — gram
gal — gallon
HP — horsepower
hr — hour
Hz — hertz
in — inch(es)
in hg — inch(es) of mercury
IPS — inches per second
k — kilo
K — Kelvin
kg — kilogram
kHz — kilohertz
km — kilometer
kM — kilomega
kw — kilowatt
kw-hr — kilowatt hour
l — liter
lb — pound
lbs — pounds
m — meter
mb — milibar
mf or μf — microfarad
mHz — millihertz

MHz — megahertz
mi — mile(s)
mm — millimeter
MPH — miles per hour
mv — millivolt
neg — negative
oz — ounce(s)
pf or μμf — picofarad
pos — positive
PPH — pounds per hour
PPM — parts per million
psi — pounds per square inch
PSIA — pounds per square inch absolute pressure
PSID — pounds per square inch differential pressure
PSIG — pounds per square inch gauge
pt — pint
qt — quart
R — Rankine
rev — revolution(s)
RPM — revolutions per minute
sec — second
SHP — shaft horsepower
sin — sine
sq cm — square centimeter(s)
sq in — square inch(es)
sq ft — square foot (feet)
sq m — square meter
sq mi — square mile(s)
sq mil — square mil
tan — tangent
TEHP — thrust equivalent horsepower
THP — thrust horsepower
U.S. — United States
V — volt
yd — yard

CHEMICAL ELEMENTS

actinium — Ac
aluminum — Al
americum — Am
antimony — Sb
argon — Ar
arsenic — As
astatine — At
barium — Ba
berkelium — Bk
beryllium — Be
bismuth — Bi
bohrium — Bh
boron — B
bromine — Br
cadmium — Cd
calcium — Ca
californium — Cf
carbon — C
cerium — Ce
cesium — Cs
chlorine — Cl
chromium — Cr
cobalt — Co
columbium — Cb
copper — Cu
curium — Cm
dubnium — Db
dysprosium — Dy
einsteinium — Es
emanation — Em
erbium — Er
europium — Eu
fermium — Fm
fluorine — F
francium — Fr
gadolinium — Gd
gallium — Ga
germanium — Ge
glucinium — Gl

gold — Au
hafnium — Hf
hahnium — Ha
hassium — Hs
helium — He
holmium — Ho
hydrogen — H
illnium — Il
indium — In
iodine — I
iridium — Ir
iron — Fe
joliotium — Jl
krypton — Kr
lanthanum — La
lawrencium — Lr
lead — Pb
lithium — Li
lutetium — Lu
magnesium — Mg
manganese — Mn
meitnerium — Mt
mendelevium — Md
mercury — Hg
molybdenum — Mo
neodymium — Nd
neon — Ne
neptunium — Np
nickel — Ni
niobium — Nb
nitrogen — N
nobelium — No
osmium — Os
oxygen — O
palladium — Pd
phosphorus — P
platinum — Pt
plutonium — Pu
polonium — Po

potassium — K
praseodymium — Pr
promethium — Pm
protactinium — Pa
radium — Ra
radon — Rn
rhenium — Re
rhodium — Rh
rubidium — Rb
ruthenium — Ru
rutherfordium — Rf
samarium — Sm
scandium — Sc
seaborgium — Sg
selenium — Se
silicon — Si
silver — U
sodium — Na
strontium — Sr
sulfur — S
tantalum — Ta
technetium — Tc
telllurium — Te
terbium — Tb
thallium — Tl
thorium — Th
thulium — Tm
tin — Sn
titanium — Ti
tungsten — W
uranium — U
vanadium — V
wolfram — W
xenon — Xe
ytterbium — Yb
yttrium — Y
zinc — Zn
zirconium — Zr

ATA/JASC CODES

(AIR TRANSPORT ASSOCIATION [ATA] SPECIFICATION 100 or JOINT AIRCRAFT SYSTEM COMPONENT [JASC] CODES)

Note: General Aviation is gradually beginning to use ATA/JASC Codes, but presently might also be using either GAMA (General Aviation Maintenance Association) Codes which are similar to ATA/JASC Codes, or manufacturer specific codes.

05 MAINTENANCE CHECKS
(OPERATOR DERIVED)

10 PARKING AND MOORING
(OPERATOR DERIVED)

11 PLACARDS AND MARKINGS

1100 — PLACARDS AND MARKINGS

12 — SERVICING

1210 — FUEL SERVICING

1220 — OIL SERVICING

1230 — HYDRAULIC FLUID SERVICING

1240 — COOLANT SERVICING

18 — HELICOPTER VIBRATION

1800 — HELICOPTER VIB/NOISE ANALYSIS

1810 — HELICOPTER VIBRATION ANALYSIS

1820 — HELICOPTER NOISE ANALYSIS

21 — AIR CONDITIONING

2100 — AIR CONDITIONING SYSTEM

2110 — CABIN COMPRESSOR SYSTEM

2120 — AIR DISTRIBUTION SYSTEM

2121 — AIR DISTRIBUTION FAN

2130 — CABIN PRESSURE CONTROL SYSTEM

2131 — CABIN PRESSURE CONTROLLER

2132 — CABIN PRESSURE INDICATOR

2133 — PRESSURE REGUL/OUTFLOW VALVE

2134 — CABIN PRESSURE SENSOR

2140 — HEATING SYSTEM

2150 — CABIN COOLING SYSTEM

2160 — CABIN TEMPERATURE CONTROL SYSTEM

2161 — CABIN TEMPERATURE CONTROLLER

2162 — CABIN TEMPERATURE INDICATOR

2163 — CABIN TEMPERATURE SENSOR

2170 — HUMIDITY CONTROL SYSTEM

22 — AUTO FLIGHT

2200 — AUTO FLIGHT SYSTEM

2210 — AUTOPILOT SYSTEM

2211 — AUTOPILOT COMPUTER

2212 — ALTITUDE CONTROLLER

2213 — FLIGHT CONTROLLER

2214 — AUTOPILOT TRIM INDICATOR

2215 — AUTOPILOT MAIN SERVO

2216 — AUTOPILOT TRIM SERVO

2220 — SPEED-ATTITUDE CORRECT. SYSTEM

2230 — AUTO THROTTLE SYSTEM

2250 — AERODYNAMIC LOAD ALLEVIATING

23 — COMMUNICATIONS

2300 — COMMUNICATIONS SYSTEM

2310 — HF COMMUNICATION SYSTEM

2311 — UHF COMMUNICATION SYSTEM

2312 — VHF COMMUNICATION SYSTEM

2320 — DATA TRANSMISSION AUTO CALL

2330 — ENTERTAINMENT SYSTEM

2340 — INTERPHONE & PA SYSTEM

2350 — AUDIO INTEGRATING SYSTEM

2360 — STATIC DISCHARGE SYSTEM

2370 — AUDIO/VIDEO MONITORING

24 — ELECTRICAL POWER

2400 — ELECTRICAL POWER SYSTEM

2410 — ALTERNATOR-GENERATOR DRIVE

2420 — AC GENERATION SYSTEM

2421 — AC GENERATOR-ALTERNATOR

2422 — AC INVERTER

2423 — PHASE ADAPTER

2424 — AC REGULATOR

2425 — AC INDICATING SYSTEM

2430 — DC GENERATING SYSTEM

2431 — BATTERY OVERHEAT WARN. SYSTEM

2432 — BATTERY/CHARGER SYSTEM

2433 — DC RECTIFIER-CONVERTER

2434 — DC GENERATOR-ALTERNATOR

2435 — STARTER-GENERATOR

2436 — DC REGULATOR

2437 — DC INDICATING SYSTEM

2440 — EXTERNAL POWER SYSTEM

2450 — AC POWER DISTRIBUTION SYSTEM

2460 — DC POWER/DISTRIBUTION SYSTEM

25 — EQUIPMENT/FURNISHINGS

2500 — CABIN EQUIPMENT/FURNISHINGS

2510 — FLIGHT COMPARTMENT EQUIPMENT

2520 — PASSENGER COMPARTMENT EQUIPMENT

2530 — BUFFET/GALLEYS

2540 — LAVATORIES

2550 — CARGO COMPARTMENTS

2551 — AGRICULTURAL SPRAY SYSTEM

2560 — EMERGENCY EQUIPMENT

2561 — LIFE JACKET

2562 — EMERGENCY LOCATOR BEACON

2563 — PARACHUTE

2564 — LIFE RAFT

2565 — ESCAPE SLIDE

2570 — ACCESSORY COMPARTMENT

2571 — BATTERY BOX STRUCTURE

2572 — ELECTRONIC SHELF SECTION

26 — FIRE PROTECTION

2600 — FIRE PROTECTION SYSTEM

2610 — DETECTION SYSTEM

2611 — SMOKE DETECTION

2612 — FIRE DETECTION

2613 — OVERHEAT DETECTION

2620 — EXTINGUISHING SYSTEM

2621 — FIRE BOTTLE, FIXED

2622 — FIRE BOTTLE, PORTABLE

27 — FLIGHT CONTROLS

2700 — FLIGHT CONTROL SYSTEM

2701 — CONTROL COLUMN SECTION

2710 — AILERON CONTROL SYSTEM

2711 — AILERON TAB CONTROL SYSTEM

2720 — RUDDER CONTROL SYSTEM

2721 — RUDDER TAB CONTROL SYSTEM

2722 — RUDDER ACTUATOR

2730 — ELEVATOR CONTROL SYSTEM

2731 — ELEVATOR TAB CONTROL SYSTEM

2740 — STABILIZER CONTROL SYSTEM

2741 — STABILIZER POSITION INDICATING

2742 — STABILIZER ACTUATOR

2750 — TE FLAP CONTROL SYSTEM

2751 — TE FLAP POSITION IND. SYSTEM

2752 — TE FLAP ACTUATOR

2760 — DRAG CONTROL SYSTEM

2761 — DRAG CONTROL ACTUATOR

2770 — GUST LOCK/DAMPER SYSTEM

2780 — LE FLAP CONTROL SYSTEM

2781 — LE FLAP POSITION IND. SYSTEM

2782 — LE FLAP ACTUATOR

28 — FUEL

2800 — AIRCRAFT FUEL SYSTEM

2810 — FUEL STORAGE

2820 — ACFT FUEL DISTRIB. SYSTEM

2821 — ACFT FUEL FILTER/STRAINER

2822 — FUEL BOOST PUMP

2823 — FUEL SELECTOR/SHUTOFF VALVE

2824 — FUEL TRANSFER VALVE

2830 — FUEL DUMP SYSTEM

2840 — ACFT FUEL INDICATING

2841 — FUEL QUANTITY INDICATOR

2842 — FUEL QUANTITY SENSOR

2843 — FUEL TEMPERATURE INDICATING

2844 — FUEL PRESSURE INDICATOR

29 — HYDRAULIC POWER

2900 — HYDRAULIC POWER SYSTEM

2910 — HYDRAULIC, MAIN SYSTEM

2911 — HYDRAULIC POWER-ACCUMULATOR-MAIN

2912 — HYDRAULIC FILTER-MAIN SYSTEM

2913 — HYDRAULIC PUMP. ELECT-ENG.-MAIN

2914 — HYDRAULIC HANDPUMP-MAIN

2915 — HYDRAULIC PRESSURE RELIEF VLV-MAIN

2916 — HYDRAULIC RESERVOIR-MAIN

2917 — HYDRAULIC PRESSURE REGULATOR-MAIN

2920 — HYDRAULIC, AUXILIARY SYSTEM

2921 — HYDRAULIC ACCUMULATOR-AUXILIARY

2922 — HYDRAULIC FILTER-AUXILIARY

2923 — HYDRAULIC PUMP-AUXILIARY

2925 — HYDRAULIC PRESSURE RELIEF-AUXILIARY

2926 — HYDRAULIC RESERVOIR-AUXILIARY

2927 — HYDRAULIC PRESSURE REGULATOR-AUX.

2930 — HYDRAULIC SYSTEM INDICATING

2931 — HYDRAULIC PRESSURE INDICATOR

2932 — HYDRAULIC PRESSURE SENSOR

2933 — HYDRAULIC QUANTITY INDICATOR

2934 — HYDRAULIC QUANTITY SENSOR

30 — ICE AND RAIN PROTECTION

3000 — ICE/RAIN PROTECTION SYSTEM

3010 — AIRFOIL ANTI/DE-ICE SYSTEM

3020 — AIR INTAKE ANTI/DE-ICE SYSTEM

3030 — PITOT/STATIC ANTI-ICE SYSTEM

3040 — WINDSHIELD/DOOR RAIN/ICE REMOVAL

3050 — ANTENNA/RADOME ANTI-ICE/DE-ICE SYSTEM

3060 — PROP/ROTOR ANTI-ICE/DE-ICE SYSTEM

3070 — WATER LINE ANTI-ICE SYSTEM

3080 — ICE DETECTION

31 — INSTRUMENTS

3100 — INDICATING/RECORDING SYSTEM

3110 — INSTRUMENT PANEL

3120 — INDEPENDENT INSTRUMENTS (CLOCK, ETC.)

3130 — DATA RECORDERS (FLT/MAINT)

3140 — CENTRAL COMPUTERS (EICAS)

3150 — CENTRAL WARNING

3160 — CENTRAL DISPLAY

3170 — AUTOMATIC DATA

32 — LANDING GEAR

3200 — LANDING GEAR SYSTEM

3201 — LANDING GEAR/WHEEL FAIRING

3210 — MAIN LANDING GEAR

3211 — MAIN LANDING GEAR ATTACH SECTION

3212 — EMERGENCY FLOTATION SECTION

3213 — MAIN LANDING GEAR STRUT/AXLE/TRUCK

3220 — NOSE/TAIL LANDING GEAR

3221 — NOSE/TAIL LANDING GEAR ATTACH SECTION

3222 — NOSE/TAIL LANDING GEAR STRUT/AXLE

3230 — LANDING GEAR RETRACT/EXT. SYSTEM

3231 — LANDING GEAR DOOR RETRACT SECTION

3232 — LANDING GEAR DOOR ACTUATOR

3233 — LANDING GEAR ACTUATOR

3234 — LANDING GEAR SELECTOR

3240 — LANDING GEAR BRAKE SYSTEM

3241 — BRAKE ANTI-SKID SECTION

3242 — BRAKE

3243 — MASTER CYL/BRAKE VALVE

3244 — TIRE

3245 — TIRE TUBE

3246 — WHEEL/SKI/FLOAT

3250 — LANDING GEAR STEERING SYSTEM

3251 — STEERING UNIT

3252 — SHIMMY DAMPER

3260 — LANDING GEAR POSITION & WARNING

3270 — AUXILIARY GEAR (TAIL SKID)

33 — LIGHTS

3300 — LIGHTING SYSTEM

3310 — FLIGHT COMPARTMENT LIGHTING

3320 — PASSENGER COMPARTMENT LIGHTING

3330 — CARGO COMPARTMENT LIGHTING

3340 — EXTERIOR LIGHTING

3350 — EMERGENCY LIGHTING

34 — NAVIGATION

3400 — NAVIGATION SYSTEM

3410 — FLIGHT ENVIRONMENT DATA

3411 — PITOT/STATIC SYSTEM

3412 — OUTSIDE AIR TEMP. IND./SENSOR

3413 — RATE OF CLIMB INDICATOR

3414 — AIRSPEED/MACH INDICATING

3415 — HIGH SPEED WARNING

3416 — ALTIMETER, BAROMETRIC/ENCODER

3417 — AIR DATA COMPUTER

3418 — STALL WARNING SYSTEM

3420 — ATTITUDE AND DIRECTION DATA SYSTEM

3421 — ATTITUDE GYRO & IND. SYSTEM

3422 — DIRECTIONAL GYRO & IND. SYSTEM

3423 — MAGNETIC COMPASS

3424 — TURN & BANK/RATE OF TURN INDICATOR

3425 — INTEGRATED FLT. DIRECTOR SYSTEM

3430 — LANDING & TAXI AIDS

3431 — LOCALIZER/VOR SYSTEM

3432 — GLIDE SLOPE SYSTEM

3433 — MICROWAVE LANDING SYSTEM

3434 — MARKER BEACON SYSTEM

3435 — HEADS UP DISPLAY SYSTEM

3436 — WIND SHEAR DETECTION SYSTEM

3440 — INDEPENDENT POS. DETERMINING SYSTEM

3441 — INERTIAL GUIDANCE SYSTEM

3442 — WEATHER RADAR SYSTEM

3443 — DOPPLER SYSTEM

3444 — GROUND PROXIMITY SYSTEM

3445 — AIR COLLISION AVOIDANCE SYSTEM (TCAS)

3446 — NON RADAR WEATHER SYSTEM

3450 — DEPENDENT POSITION DETERMINING SYSTEM

3451 — DME/TACAN SYSTEM

3452 — ATC TRANSPONDER SYSTEM

3453 — LORAN SYSTEM

3454 — VOR SYSTEM

3455 — ADF SYSTEM

3456 — OMEGA NAVIGATION SYSTEM

3457 — GLOBAL POSITIONING SYSTEM

3460 — FLIGHT MANAGE. COMPUTING SYSTEM

35 — OXYGEN

3500 — OXYGEN SYSTEM

3510 — CREW OXYGEN SYSTEM

3520 — PASSENGER OXYGEN SYSTEM

3530 — PORTABLE OXYGEN SYSTEM

36 — PNEUMATIC

3600 — PNEUMATIC SYSTEM

3610 — PNEUMATIC DISTRIBUTION SYSTEM

3620 — PNEUMATIC INDICATING SYSTEM

37 — VACUUM

3700 — VACUUM SYSTEM

3710 — VACUUM DISTRIBUTION SYSTEM

3720 — VACUUM INDICATING SYSTEM

38 — WATER/WASTE

3800 — WATER & WASTE SYSTEM

3810 — POTABLE WATER SYSTEM

3820 — WASH WATER SYSTEM

3830 — WASTE DISPOSAL SYSTEM

3840 — AIR SUPPLY (WATER PRESS. SYSTEM)

45 — CENTRAL MAINT. SYSTEM

4500 — CENTRAL MAINT. COMPUTER

49 — AIRBORNE AUXILIARY POWER

4900 — AIRBORNE APU SYSTEM

4910 — APU COWLING/CONTAINMENT

4920 — APU CORE ENGINE

4930 — APU ENGINE FUEL & CONTROL

4940 — APU START/IGNITION SYSTEM

4950 — APU BLEED AIR SYSTEM

4960 — APU CONTROLS

4970 — APU INDICATING SYSTEM

4980 — APU EXHAUST SYSTEM

4990 — APU OIL SYSTEM

51 — STANDARD PRACTICES/STRUCTURES

5100 — STANDARD PRACTICES/STRUCTURES

5101 — AIRCRAFT STRUCTURES

5102 — BALLOON REPORTS

52 — DOORS

5200 — DOORS

5210 — PASSENGER/CREW DOORS

5220 — EMERGENCY EXIT

5230 — CARGO/BAGGAGE DOORS

5240 — SERVICE DOORS

5241 — GALLEY DOORS

5242 — E/E COMPARTMENT DOORS

5243 — HYDRAULIC COMPARTMENT DOORS

5244 — ACCESSORY COMPARTMENT DOORS

5245 — AIR CONDITIONING COMPART. DOORS

5246 — FLUID SERVICE DOORS

5247 — APU DOORS

5248 — TAIL CONE DOORS

5250 — FIXED INNER DOORS

5260 — ENTRANCE STAIRS

5270 — DOOR WARNING SYSTEM

5280 — LANDING GEAR DOORS

53 — FUSELAGE

5300 — FUSELAGE STRUCTURE (GENERAL)

5301 — AERIAL TOW EQUIPMENT

5302 — ROTORCRAFT TAIL BOOM

5310 — FUSELAGE MAIN STRUCTURE

5311 — FUSELAGE MAIN FRAME

5312 — FUSELAGE MAIN BULKHEAD

5313 — FUSELAGE MAIN LONGERON/STRINGER

5314 — FUSELAGE MAIN KEEL

5315 — FUSELAGE MAIN FLOOR BEAM

5320 — FUSELAGE MISCELLANEOUS STRUCTURE

5321 — FUSELAGE FLOOR PANEL

5322 — FUSELAGE INTERNAL MOUNT STRUCTURE

5323 — FUSELAGE INTERNAL STAIRS

5324 — FUSELAGE FIXED PARTITIONS

5330 — FUSELAGE MAIN PLATE/SKIN

5340 — FUSELAGE MAIN ATTACH FITTINGS

5341 — WING ATTACH FITTINGS (ON FUSELAGE)

5342 — STABILIZER ATTACH FITTINGS

5343 — LANDING GEAR ATTACH FITTINGS

5344 — FUSELAGE DOOR HINGES

5345 — FUSELAGE EQUIPMENT ATTACH FITTINGS

5346 — POWERPLANT ATTACH FITTINGS

5347 — SEAT/CARGO ATTACH FITTINGS

5350 — FUSELAGE AERODYNAMIC FAIRINGS

54 — NACELLES/PYLONS

5400 — NACELLE/PYLON STRUCTURE

5410 — MAIN FRAME (ON NACELLE/PYLON)

5411 — FRAME/SPAR/RIB(NACELLE/PYLON)

5412 — BULKHEAD/FIREWALL (NAC/PYLON)

5413 — LONGERON/STRINGER (NAC/PYLON)

5414 — PLATE SKIN (NAC/PYLONS)

5415 — ATTACH FITTINGS (NAC/PYLON)

55 — STABILIZERS

5500 — EMPENNAGE STRUCTURE

5510 — HORIZONTAL STABILIZER STRUCTURE

5511 — HORIZONTAL STABILIZER SPAR/RIB

5512 — HORIZONTAL STABILIZER PLATE/SKIN

5513 — HORIZONTAL STABILIZER TAB STRUCTURE

5520 — ELEVATOR STRUCTURE

5521 — ELEVATOR SPAR/RIB STRUCTURE

5522 — ELEVATOR PLATES/SKIN STRUCTURE

5523 — ELEVATOR TAB STRUCTURE

5530 — VERTICAL STABILIZER STRUCTURE

5531 — VERTICAL STABILIZER SPAR/RIB STRUCTURE

5532 — VERTICAL STABILIZER PLATES/SKIN

5533 — VENTRAL STRUCTURE (ON VERT. STAB)

5540 — RUDDER STRUCTURE

5541 — RUDDER SPAR/RIB STRUCTURE

5542 — RUDDER PLATE/SKIN STRUCTURE

5543 — RUDDER TAB STRUCTURE

5550 — EMPENNAGE FLT. CONT. ATTACH FITTING

5551 — HORIZONTAL STABILIZER ATTACH FITTING

5552 — ELEVATOR/TAB ATTACH FITTINGS

5553 — VERT. STAB. ATTACH FITTINGS

5554 — RUDDER/TAB ATTACH FITTINGS

56 — WINDOWS

5600 — WINDOW/WINDSHIELD SYSTEM

5610 — FLIGHT COMPARTMENT WINDOWS

5620 — PASSENGER COMPARTMENT WINDOWS

5630 — DOOR WINDOWS

5640 — INSPECTION WINDOWS

57 — WINGS

5700 — WING STRUCTURE

5710 — WING MAIN FRAME STRUCTURE

5711 — WING SPAR STRUCTURE

5712 — WING RIB STRUCTURE

5713 — WING LONGERON/STRINGER

5714 — WING CENTER BOX

5720 — WING MISCELLANEOUS STRUCTURE

5730 — WING PLATES/SKINS

5740 — WING ATTACH FITTINGS

5741 — WING, FUSELAGE ATTACH FITTINGS

5742 — WING, NAC/PYLON ATTACH FITTINGS

5743 — WING, LANDING GEAR ATTACH FITTINGS

5744 — CONTROL SURFACE ATTACH FITTINGS

5750 — WING CONTROL SURFACE STRUCTURE

5751 — AILERON STRUCTURE

5752 — AILERON TAB STRUCTURE

5753 — TE FLAP STRUCTURE

5754 — LEADING EDGE DEVICE STRUCTURE

5755 — SPOILER STRUCTURE

61 — PROPELLERS/PROPULSORS

6100 — PROPELLER SYSTEM

6110 — PROPELLER ASSEMBLY

6111 — PROPELLER BLADE SECTION

6112 — PROPELLER DE-ICE BOOT SECTION

6113 — PROPELLER SPINNER SECTION

6114 — PROPELLER HUB SECTION

6120 — PROPELLER CONTROL SYSTEM

6121 — PROPELLER SYNCHRONIZER SECTION

6122 — PROPELLER GOVERNOR

6123 — PROPELLER FEATHERING/REVERSING

6130 — PROPELLER BRAKING

6140 — PROPELLER INDICATING SYSTEM

62 — MAIN ROTOR

6200 — MAIN ROTOR SYSTEM

6210 — MAIN ROTOR BLADES

6220 — MAIN ROTOR HEAD

6230 — MAIN ROTOR MAST/SWASHPLATE

6240 — MAIN ROTOR INDICATING SYSTEM

63 — MAIN ROTOR DRIVE

6300 — MAIN ROTOR DRIVE SYSTEM

6310 — ENGINE/TRANSMISSION COUPLING

6320 — MAIN ROTOR GEARBOX

6321 — MAIN ROTOR BRAKE

6322 — ROTORCRAFT COOLING FAN SYSTEM

6330 — MAIN ROTOR TRANSMISSION MOUNT

6340 — ROTOR DRIVE INDICATING SYSTEM

64 — TAIL ROTOR

6400 — TAIL ROTOR SYSTEM

6410 — TAIL ROTOR BLADE

6420 — TAIL ROTOR HEAD

6440 — TAIL ROTOR INDICATING SYSTEM

65 — TAIL ROTOR DRIVE

6500 — TAIL ROTOR DRIVE SYSTEM

6510 — TAIL ROTOR DRIVE SHAFT

6520 — TAIL ROTOR GEARBOX

6540 — TAIL ROTOR DRIVE INDICATING SYSTEM

67 — ROTORS FLIGHT CONTROL

6700 — ROTORCRAFT FLIGHT CONTROL

6710 — MAIN ROTOR CONTROL

6711 — TILT ROTOR FLIGHT CONTROL

6720 — TAIL ROTOR CONTROL SYSTEM

6730 — ROTORCRAFT SERVO SYSTEM

71 — POWERPLANT

7100 — POWERPLANT SYSTEM

7110 — ENGINE COWLING SYSTEM

7111 — COWL FLAP SYSTEM

7112 — ENGINE AIR BAFFLE SECTION

7120 — ENGINE MOUNT SECTION

7130 — ENGINE FIRESEALS

7160 — ENGINE AIR INTAKE SYSTEM

7170 — ENGINE DRAINS

72 — TURBINE/TURBOPROP ENGINE

7200 — ENGINE (TURBINE/TURBOPROP)

7210 — TURBINE ENGINE REDUCTION GEAR

7220 — TURBINE ENGINE AIR INLET SECTION

7230 — TURBINE ENGINE COMPRESSOR SECTION

7240 — TURBINE ENGINE COMBUSTION SECTION

7250 — TURBINE SECTION

7260 — TURBINE ENGINE ACCESSORY DRIVE

7261 — TURBINE ENGINE OIL SYSTEM

7270 — TURBINE ENGINE BYPASS SECTION

73 — ENGINE FUEL & CONTROL

7300 — ENGINE FUEL & CONTROL

7310 — ENGINE FUEL DISTRIBUTION

7311 — ENGINE FUEL-OIL COOLER

7312 — FUEL HEATER

7313 — FUEL INJECTOR NOZZLE

7314 — ENGINE FUEL PUMP

7320 — FUEL CONTROLLING SYSTEM

7321 — FUEL CONTROL/ELECTRONIC

7322 — FUEL CONTROL/CARBURETOR

7323 — TURBINE GOVERNOR

7324 — FUEL DIVIDER

7330 — ENGINE FUEL INDICATING SYSTEM

7331 — FUEL FLOW INDICATING

7332 — FUEL PRESSURE INDICATING

7333 — FUEL FLOW SENSOR

7334 — FUEL PRESSURE SENSOR

74 — IGNITION

7400 — IGNITION SYSTEM

7410 — IGNITION POWER SUPPLY

7411 — LOW TENSION COIL

7412 — EXCITER

7413 — INDUCTION VIBRATOR

7414 — MAGNETO/DISTRIBUTOR

7420 — IGNITION HARNESS (DISTRIBUTION)

7421 — SPARK PLUG/IGNITER

7430 — IGNITION SWITCHING

75 — AIR

7500 — ENGINE BLEED AIR SYSTEM

7510 — ENGINE ANTI-ICING SYSTEM

7520 — ENGINE COOLING SYSTEM

7530 — COMPRESSOR BLEED CONTROL

7531 — COMPRESSOR BLEED GOVERNOR

7532 — COMPRESSOR BLEED VALVE

7540 — BLEED AIR INDICATING SYSTEM

76 — ENGINE CONTROLS

7600 — ENGINE CONTROLS

7601 — ENGINE SYNCHRONIZING

7602 — MIXTURE CONTROL

7603 — POWER LEVER

7620 — ENGINE EMERGENCY SHUTDOWN SYSTEM

77 — ENGINE INDICATING

7700 — ENGINE INDICATING SYSTEM

7710 — POWER INDICATING SYSTEM

7711 — ENGINE PRESSURE RATIO (EPR)

7712 — ENGINE BMEP/TORQUE INDICATING

7713 — MANIFOLD PRESSURE (MP) INDICATING

7714 — ENGINE RPM INDICATING SYSTEM

7720 — ENGINE TEMP. INDICATING SYSTEM

7721 — CYLINDER HEAD TEMP (CHT) INDICATING

7722 — ENG. EGT/TIT INDICATING SYSTEM

7730 — ENGINE IGNITION ANALYZER SYSTEM

7731 — ENGINE IGNITION ANALYZER

7732 — ENGINE VIBRATION ANALYZER

7740 — ENGINE INTEGRATED INSTRUMENT SYSTEM

78 — ENGINE EXHAUST

7800 — ENGINE EXHAUST SYSTEM

7810 — ENGINE COLLECTOR/TAILPIPE/NOZZLE

7820 — ENGINE NOISE SUPPRESSOR

7830 — THRUST REVERSER

79 — ENGINE OIL

7900 — ENGINE OIL SYSTEM (AIRFRAME)

7910 — ENGINE OIL STORAGE (AIRFRAME)

7920 — ENGINE OIL DISTRIBUTION (AIRFRAME)

7921 — ENGINE OIL COOLER

7922 — ENGINE OIL TEMP. REGULATOR

7923 — OIL SHUTOFF VALVE

7930 — ENGINE OIL INDICATING SYSTEM

7931 — ENGINE OIL PRESSURE

7932 — ENGINE OIL QUANTITY

7933 — ENGINE OIL TEMPERATURE

80 — STARTING

8000 — ENGINE STARTING SYSTEM

8010 — ENGINE CRANKING

8011 — ENGINE STARTER

8012 — ENGINE START VALVES/CONTROLS

81 — TURBOCHARGING

8100 — EXHAUST TURBINE SYSTEM (RECIP)

8110 — POWER RECOVERY TURBINE (RECIP)

8120 — EXHAUST TURBOCHARGER

82 — WATER INJECTION

8200 — WATER INJECTION SYSTEM

83 — ACCESSORY GEARBOXES

8300 — ACCESSORY GEARBOXES

85 — RECIPROCATING ENGINE

8500 — ENGINE (RECIPROCATING)

8510 — RECIPROCATING ENGINE FRONT SECTION

8520 — RECIPROCATING ENGINE POWER SECTION

8530 — RECIPROCATING ENGINE CYLINDER SECTION

8540 — RECIPROCATING ENGINE REAR SECTION

8550 — RECIPROCATING ENGINE OIL SYSTEM —